PREFACE

From June 6 to 9, 1976, about 140 participants (physiologists, chemists, ecologists, animal behaviorists, and psychologists) gathered in the Gideon Putnam Hotel at Saratoga Springs, New York for a symposium entitled "Chemical Signals in Vertebrates". The focus of this symposium, sponsored by the United States National Science Foundation, was on chemical communication in higher animals, most notably mammals. This included the chemical nature, production, and reception of chemical signals, and their modulating effects on behavior.

Almost all the world's laboratories working in this area were represented. It was the first meeting of its kind, and although the physiological aspects of taste and smell on the one hand and insect pheromones on the other have previously been treated in several fine symposia, they have not before been treated as a backdrop to chemical communication in vertebrates. The field of insect pheromones is well developed, with hundreds of active compounds identified. By contrast, in vertebrates only six mammalian phero- mones in as many species had been identified chemically by 1976.

The new aspects of chemical signalling which this symposium addressed included: its evolution in higher vertebrates, including man, its adaptive significance to the animal, the social contexts in which the chemical signals are operative, and their ecological determinants. An important objective of the symposium was to bring together the physical scientists and the biological scientists in- vestigating vertebrate chemical communication so that these workers may develop a greater understanding of each other's theoretical con- cepts and practical problems. Advancement in this comparatively new field appears to require the expertise of several different types of scientists, who, to be most effective, must be aware of each other's findings. The Organizing Committee of the symposium and the Editors of the resultant volume feel that this conference did achieve con- siderable progress in these areas. It gave a needed impetus to an evolving interaction between the physical and biological scientists, who are beginning to approach chemocommunication problems as inte- grated teams rather than as practitioners of separate disciplines

v

working in parallel. The symposium provided a forum for several
newly emerging questions that need clarification: 1) the advisa-
bility of recasting the definition of pheromones (or of even
abandoning the term) because of several apparent differences between
insect and vertebrate chemical communication systems, 2) the effect
of learning upon chemical communication, 3) the role of diet in the
origin of pheromones, 4) the effect upon chemocommunication of the
multisensory context in which it must operate, 5) the elucidation
of multicomponent pheromones.

Although the young research field of vertebrate chemical
communication poses its own unique problems, it must still be
aware of the concepts developed by investigators of other animal
communication channels, such as vision and audition, where research
has reached more advanced stages. On the other hand, investigators
of these other channels are becoming aware of chemical signals. An
integrated approach is vital for a comprehensive view of animal
communication and its evolution.

M.M.M. edited Parts 7 and 8; D.M.S. the rest.

We thank Christine Müller-Schwarze and Ronald G. Butler for
preparing the indices, and Cathy Northway, Mary Lent, and Mrs. Ruth
Jackson for typing the indices.

Syracuse, March 1977   D. Müller-Schwarze
                                             M.M. Mozell

CONTENTS

Part Three: Behavior:Reviews

Part Four: Behavior:Laboratory Studies

Part Five: Ecology

Part Six: Bioassay

Part Seven: Reception of Chemical Signals

Part Eight: Central Processes

STRUCTURE AND FUNCTION OF SKIN GLANDS

W. B. Quay

Department of Zoology, and Endocrinology-Reproductive

Physiology Program, University of Wisconsin, Madison,

Wisconsin 53706

INTRODUCTION

It is fitting that in this symposium on "Chemical Signals in Vertebrates" we concern ourselves first with the tissue and glandular sources of chemical signals. The particular chemical signals selected for consideration here are the *pheromones*. These have been defined as chemical signals secreted to the outside of the body by one individual, and received by a second individual of the same species, in which, as a result, a specific reaction occurs. This reaction can be manifested as a definite and characteristic behavior or behavior pattern, or the modification of a particular physiological or developmental process (Karlson and Lüscher, 1959; Butler, 1970). Although the definition of pheromones and appreciation of their biological significance derived primarily from chemical signals in insects, work of both earlier and recent date demonstrates the importance of pheromones for vertebrate animals of different groups from fish to "higher" mammals.

On the basis of presently available evidence the designation of *pheromones* as the chemical signals with which we are concerned remains tenable. However, there are signs that through the adaptive radiation and inventiveness of vertebrate evolution, external chemical signals have originated, been transmitted and received in ways sometimes that do not fit precisely the definition *sensu stricto* of pheromones. These are areas in which disagreements could probably be elicited, but to no constructive purpose. Our concern should most effectively be directed to mechanisms of chemical communication, without limitations as to what these are called

and whether they fall within previous and necessarily arbitrary
definitions. Elements of divergence from a narrow definition of
pheromones might be suggested for such situations as the use of environmentally derived odors for chemical communication and the
functionally significant reception being by other taxa or species
than that of the sender (Hawes, 1976). Examples of these and other
differences surely will be provided during this symposium.

The title of the present paper is not meant to imply that the
cell and tissue sources of vertebrate pheromones are all to be found
within "skin glands." Especially within the mammals have other
sources of vertebrate pheromones been demonstrated or suggested,
such as vagina (Michael et al., 1971, 1974; Singer et al., 1976),
urine, possibly from or via the kidneys (Vandenburgh, 1975; Vandenburgh et al., 1975; and Hoppe, 1975), male accessory glands of
the reproductive system (Jones and Nowell, 1973), contents of the
caecum (Leon, 1975), and others. Nevertheless the "skin glands"
are worthy of emphasis as both actual and potential sources of
pheromones for at least two categories of reasons: (1) Skin glands
are widespread in distribution through the integuments of most vertebrates and at the same time show marked local variation in structure and composition in relation to skin region as well as species
or other taxonomic groupings. (2) Vertebrate skin glands are most
advantageously located for the role of selective release and placement of chemical signals onto environmental substrates or media.
This pertains not only to the animals' being able to "mark" the
substrate most quickly and selectively with skin scent, but also
to the utility of other integumentary structures and derivatives
in such processes. This includes neural and muscular components
of the glandular effector mechanism, and the scales, hair and other
products for transfer and sometimes further processing of the secretions.

Although much remains to be learned about the mechanisms and
evolutionary origins of pheromonal and other functions of skin
glands in vertebrates, several generalizations appear to be worthy
of suggestion at this time. These are offered as general hypotheses to aid in the focus and plan of observational and experimental
studies:

I. It can be suggested that in essentially all cases known
so far, the scenting and pheromonal functions of various vertebrate
skin glands are adaptively and evolutionarily secondary or derived.
The particular glands are likely in all cases to have, or to have
had, other more generalized functions before their involvement in
pheromonal functions. These more general functions would include
different kinds of protective activities related to maintaining the
physical integrity of the skin and to controlling the passage of
materials through it. In some groups of animals it seems likely

that other modes of communication than the olfactory one preceded
the pheromonal function of a particular integumentary organ or glandular area. This might be suggested for the flank "glands" of microtine rodents (Quay, 1968), and some of the differently colored
glandular regions of ungulates, among others, wherein flashing a
visually distinctive patch of hair preceded the pheromonal function
of the glands in the same region.

      II. It is apparent in the case of some of the scents associ-
ated with particular skin glands, that the acquisition of odor or
scent is subsequent to the process of secretion by the skin gland.
Thus the skin gland secretion, as perhaps that of some ungulate
tarsal glands, is not the source of the pheromone, but the urine
directed onto the glandular region is. In other instances, as for
example some apocrine gland secretions in some mammals, the gland
secretion is not scented or effective as a pheromone until acted
upon externally by bacterial or other agents.

      III. Evolutionary specialization of particular skin gland
regions or localized gland units as scent- or pheromone-emitting
structures typically involves other aspects of integumentary structure, physiology and hormonal relations. These secondary specializations often may be progressive in the evolution of a pheromone-
producing skin gland organ or region. Within a series of closely
related species differences can often be observed in the degree or
magnitude of these secondary specializations as well as in radial
adaptational differences. In both of these kinds of interspecific
differences qualitative as well as quantitative functional changes
may occur. The qualitative changes may relate not only to the chemistry of the secretion itself, but to its dependency on hormonal
stimulation, mode and timing of secretion, and influence from dietary factors, among others. Three kinds or areas of secondary
specialization in pheromone-producing integumentary organs can be
noted: (A) Accessory structures of the skin may be enlarged and/
or modified in distinctive ways related to the local shift in functional emphasis to pheromone production and secretion. Modifications, and especially increases in size, are seen most commonly in
size and structure of hair follicles, arrector pili muscles, and
thickness and activities of the overlying epidermis. (B) Modifications in other integumentary constituents in and around a glandular
region may be more suggestive of modifications of levels and timing
of metabolic and transport activities. This includes most promi-
nently the differences observed in comparison with neighboring un-
specialized skin, of the dermal and subdermal adipose tissue, the
richness of the blood supply, the richness and proportional composition of the innervation, the relative numbers of certain connective tissue cell types, often including mast cells, histiocytes,
and pigment cells. (C) Specializations in responses to hormonal
controls are sometimes observed, especially in relation to so-called
sex steroids. These specializations must involve changes in the

receptor characteristics of the gland cells, but these have not been studied to the detail lavished upon uterine and certain other steroid receptor cells of the reproductive tract.

The remainder of this review will have two major topics and interrelated purposes. First a general survey will be made of the structure and function of skin glands in the major groups of vertebrate animals. This can be only a general outline, with an emphasis on pheromonal relations, since several volumes could be written on merely the describing of the diverse skin glands and their modifications within vertebrates. The purpose of this survey or outline will be to provide a visual and mechanistic basis for conceptualizing major sources of vertebrate pheromones. The last portion of this review will suggest some investigative approaches to defining the functional significance of vertebrate skin glands as producers of pheromones, and to experimentally reveal the cellular and molecular mechanisms by which they are controlled.

OUTLINE OF VERTEBRATE SKIN GLANDS

Primarily Aquatic Groups

Lower Chordates. Chordate animals lacking vertebral structures are small and thin-skinned. As in their larger aquatic vertebrate distant relatives their skin glands are mostly of the unicellular type--scattered or continuous, mostly mucus-secreting cells. In amphioxus a single layer of small mucus-secreting epidermal cells with microvilli is observed (Olsson, 1961). More specialized developments are sometimes seen, such as the luminous slime secreted by some hemichordates (Baxter and Pickens, 1964).

Cyclostomes. Secretory cells in the skins of lampreys and hagfishes usually appear to represent four types: (1) small and superficial mucous cells; (2) large and more specialized mucous cells; (3) granular cells (Körnerzellen); and (4) thread cells (club cells, Kolbenzellen) (Fig. 1). Mechanical and protective functions have been suggested for the secretions of cyclostome skin glands, but more specialized functions have received little attention as possibilities. Since this applies also to possible pheromonal significance we will refer the reader to an earlier review for more details and for references (Quay, 1972).

Cartilaginous Fishes. Four kinds of integumentary glands have been described in the better known Chondrichthyes: (1) unicellular mucous cells scattered over the body surface and within the stratified epidermis; (2) multicellular venom glands associated with dermal spines; (3) multicellular glands modified as "light organs" or

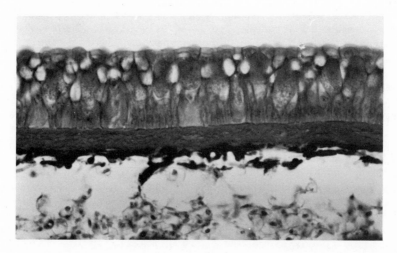

Fig. 1. Vertical section through the lateral abdominal skin of
a brook lamprey (*Lampetra lamottei*) showing unicellular gland cells
of three types: thread cells (3, at center and left and basal within the epidermis), granular cells (about 8, scattered at mostly intermediate levels), and superficial and numerous mucous cells (pale
cytoplasm) with microvilli.

photophores; and (4) multicellular mucous or goblet cell glands at
the base of the claspers of males (Daniel, 1935; Quay, 1972). The
general impression is that the glandular secretions involved in
these are mostly mucoid, but very little has been published concerning their chemistry. Insofar as known, protective functions appear
to be most frequently pertinent, but the luminescent organs and
the sexually dimorphic clasper gland are clearly more specialized.
Experimental studies on the functions of skin glands in this group
concern most usually the venom glands, with little evidence available as yet concerning possible pheromonal secretions.

   Bony Fishes. Among the very numerous and diverse groups of
bony fishes we see a wide variety of integumentary glandular specializations, including some that are clearly pheromonal in nature.
The complexity is increased with the realization that here, as well
as in many amphibian groups, embryonic, larval and adult life
stages may have different epidermal glandular specializations and
functions. The embryonic and larval stages of some teleosts have
transitory secretory cells, usually in the cephalic epidermis
(Brinley and Eulberg, 1953; Peters, 1964). Such glands in embryos
have been suggested to secrete an enzyme that dissolves the egg
shell or membrane prior to hatching (Bourdin, 1926), or to increase the viscosity of the perivitelline fluid (Brinley and Eulberg, 1953). More convincing is the function ascribed to such

secretory cells after hatching, when they have been designated as
adhesive, cement or attachment organs (Ilg, 1952; Peters, 1964).
The life span of such a gland in one species, *Esox lucius*, is but
21 days (Georges, 1964).

External, and apparently mostly mucous, secretions in fishes
have evolved with specialized functions in diverse taxa. A few of
the many possible examples are: (1) the nocturnally secreted and
protective mucous envelope of some parrot fishes (Scaridae) and
wrasses (Labridae) (Winn, 1955); (2) the seasonally secreted mucous
cocoon believed significant in preventing dehydration of the lungfish (*Protopterus*) (Smith and Coates, 1936; Kitzan and Sweeny,
1968); (3) the skin mucus secreted by parental cichlids of certain
species for feeding their young (Hildemann, 1959; Ward and Barlow,
1967); (4) the epidermal mucus produced by the lining of the male
seahorse's (*Hippocampus*) brood pouch (Linton and Soloff, 1964); (5)
the increased mucous secretory cells and activity in spawning female salmonids (Stoklosowa, 1966); and (6) the reduced drag or resistance during swimming provided by the mucous secretions or slimes
of some species (Rosen and Cornford, 1971).

Pheromones in bony fish have been reviewed recently by Bardach
and Todd (1970), who noted six general categories of pheromones on
the basis primarily of behavioral responses and relationship: (1)
sexual pheromones, important in pair attraction and mating (Leiner,
1930; Eggert, 1931; Nelson, 1964); (2) pheromones emitted by young
that influence parental behavior and aid in identification of brood;
(3) non-sexual social pheromones, as involved in individual recognition (Todd et al., 1967), and the setting up of hierarchies and
territories; (4) "aggregating" pheromones, notable in schooling behavior; (5) "alarm" pheromones discovered by von Frisch (1938, 1942)
and recently most extensively investigated by Pfeiffer and coworkers (1960, 1963, 1967, 1971, 1973); and (6) interspecific chemical
communication.

Amphibians. In addition to the "hatching," "frontal," "adhesive" or "cementing" glands in some anuran embryos and posthatch-
ing larvae (Bergeot and Wintrebert, 1926) there are present in
some anuran larvae an alarm substance, extractable from their dorsal skin and responsible for a flight reaction (Hrbaček, 1950; Kulzer, 1954). Adult amphibians of both anuran and urodele types have
multicellular saccular (alveolar) glands, and these are most often
separable microscopically as being composed of either mucous or
granular (serous) secretory cells. A third kind of alveolar gland,
one having a lipoid secretion, has been described recently in phyllomedusine frogs and appears to be involved in a wiping behavior
reducing evaporative water loss through the skin (Blaylock et al.,
1976). A wide variety of glandular specializations occurs in the
skins of adult amphibians and the chemical diversity and complexity

of their secretions largely defies, as yet, functional explanation.
Further details and references may be found in an earlier review
(Quay, 1972).

Reptiles. Living reptiles possess a remarkable diversity of
skin glands, much in contrast with the impression fostered by text-
books and casual observation. These glands, however, are usually
small, hidden and not always easily distinguished from horny out-
growths, pits and pores of diverse sorts associated with regions of
modified epidermal keratinization. Excluding possibly the deeper
proctodeal glands, nearly all reptilian skin glands are holocrine
and have close histogenetic relations with keratinizing epidermal
cells. Two categories of glands have been distinguished in lizards
by Maderson and Chiu (1970; Maderson, 1970, 1972): (1) "generation
glands," whose secretory activity is linked to adjacent keratinizing epidermis and periodic sloughing; and (2) "preanal organs,"
morphologically more gland-like and more differentiated from epidermal keratinization and related processes. Early as well as recent investigations on glandular structures in the skins of other
reptile groups demonstrate diverse, mostly holocrine, secretory units
with chemical and probably functional individuality (Reese, 1921;
Müller, 1961; Rose et al., 1969; Rose, 1970; Kroll and Reno, 1971;
Winokur and Legler, 1975; Oldak, 1976; and many others). Additional
references, especially on lacertilian glands, can be found in Cole
(1966), Maderson (1972) and Quay (1972). A review of chemical perception in reptiles (Burghardt, 1970) suggests that while territorial marking behavior had not yet been demonstrated in any reptile,
the femoral glands of lizards appear to be likely prospects for
such a functional relation. Behavioral evidence is stronger for
some of the holocrine skin glands, for example some of those in
chelonians, functioning in courtship or related activities.

Birds. Uropygial or preen glands are usually assumed to be
the only notable or significant skin glands in birds. However,
this is surely an oversimplification. The uropygial glands, lying
medially over the base of the tail, structurally and chemically resemble holocrine lipid-secreting glands of reptiles and mammals
(Weitzel and Lennert, 1951). Uropygial gland secretions are demon-
strably important in the maintenance of the structural integrity
and water-repellant capacity of the plumage (Elder, 1954), and relative gland size is greater in aquatic than in terrestrial birds
(Kennedy, 1971). Nevertheless, it should be noted that epidermal
lipid secretion can be found within epidermal cells of the general
skin surface in some regions, and other holocrine sebaceous-type
glands occur in the external ear canal and certain other regions
in some species (Lucas and Stettenheim, 1972; Quay, unpublished
observations).

More difficult to interpret as yet, both evolutionarily and
functionally, are the cloacal and anal glands of birds. The former

is a useful indicator of testicular development in some species
(Siopes and Wilson, 1975) and the latter, showing variability in
distribution and composition, is poorly understood functionally
(Quay, 1967).

Since the olfactory centers and sensory structures are poorly developed in most birds, it can be argued that there is little
likelihood of pheromonal communication being significant in this
vertebrate group. However, such a blanket exclusion of all birds
from pheromone research may be premature. At least for some of the
pelagic petrels and shearwaters there is experimental evidence for
their ability to use olfactory cues during foraging behavior at
night as well as in daylight (Grubb, 1972).

Mammals. Mammalian skin glands present a wide assortment of
structural types and chemical characteristics, but they are essentially all members or mixtures of two basic types, the sebaceous
(alveolar, holocrine) and sudoriferous (tubular, merocrine and apocrine) (Schaffer, 1940). Mammary glands appear to be histogenic
relatives of primitive "apocrine" sudoriferous glands. Sebaceous
and primitive apocrine sudoriferous glands have structural and histogenetic relations with hair follicles and may have been derived
from lipoidal secretory alveoli of some ancestral reptilian line.
Many sebaceous glands and some apocrine sudoriferous glands share

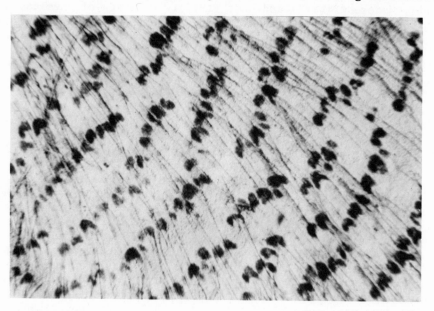

Fig. 2. Whole mount of haired skin of the hind foot of a red-backed
vole (*Clethrionomys gapperi*) stained with Oil Blue N to show the
sebaceous glands.

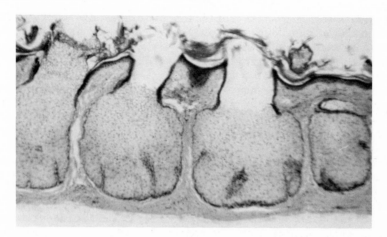

Fig. 3. Vertical section through the dorsal holocrine, sebaceous glandular region of the kangaroo rat (*Dipodomys panamintinus*).

with the lipoidal skin glands of reptiles frequent sexual dimorphism, seasonal changes related to breeding or reproductive activity, responses to sex steroid hormones, and involvement in olfactory or pheromonal communication.

The most ubiquitous of mammalian skin glands are the sebaceous

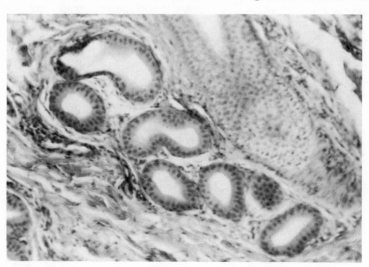

Fig. 4. Section through the secretory portion of a coiled apocrine sudoriferous gland of a bovine (*Bos taurus*) rectoanal junction. An oblique section of the associated hair follicle appears in the upper right.

glands, secreting into hair follicles over the body surface (Fig.
2) and into hairless ducts in some mucocutaneous junctions and
other specialized gland regions (Fig. 3). It is questionable
whether the generally distributed sebaceous glands are significant
in olfactory communication. It is clear, however, that some of the
specialized sebaceous glands, such as the preputial glands of some
rodents, are important sources of pheromones.

The tubular or sudoriferous glands of mammalian skin are of
two general types. The more primitive of these is often designated
as apocrine (Fig. 4), in relation to an early view, probably erroneous and based on artifactual changes, that the apical tips of the
secretory cells were pinched off to form the secretion. More significant characteristics of these apocrine sudoriferous glands are
their more saccular structure (especially in primitive mammals),
closer association with hair follicles, thicker and chemically
more complex and variable secretion, and wider distribution in
skins of lower mammalian groups.

The other type of sudoriferous gland is the true sweat gland
or eccrine sweat gland (Fig. 5). This is relatively widespread in
human and some primate skins, but it, or a gland type very much
like it, in lower animals is largely restricted to the plantar and
palmar pads. Here, they apparently function to reduce slippage between foot and substrate during running and climbing, especially
during flight or escape (Adelman, et al., 1975). Although the so-
called "apocrine" sudoriferous glands are probably the ones of primary significance in olfactory communication in most mammals, the
eccrine glands are not without chemical individuality on a regional
basis and from one individual to another. This has been suggested

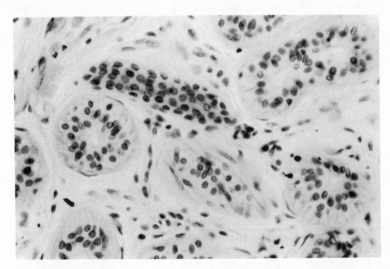

Fig. 5. Section through an eccrine gland in human scalp skin.

from chromatographic studies of the amino acid composition of human
sweat (Liappis and Jäkel, 1975).

The structure and function of skin glands in mammals is too
large and diverse a topic for more than rudimentary coverage here.
Any attempt at a sequel to Schaffer's (1940) still unique and useful compendium would require several volumes alone. However, at
this time different and more selective treatments of the topics
seem more appropriate. Examples of the roles played by skin glands
in mammalian communication have been provided and reviewed from
time to time (Mykytowycz, 1970) and will be further developed during this symposium. It is hoped that such information will be
viewed, at least in part, in relation to the development and testing of such principles and generalizations as have been noted in
the introductory section of this review.

APPROACHES AND PROBLEMS FOR INVESTIGATION

An attempt to look ahead, to outline worthy approaches and
problems for future investigators, can emphasize either one of two
different orientations, one concerned chiefly with application of
new techniques in skin gland research, such as diverse ultrastructural and immunochemical techniques, and the other directed more
toward basic problems in biological science. The latter orientation
is selected for concern here. Within this problem orientation one
can see a partial dichotomy between studies devoted primarily to
attaining greater understanding of skin glands and other structures
as they relate to mechanisms of chemical communication, and those
that may take advantage of the attractive possibilities of utiliz-
ing skin glands for more effective attack on broader or other kinds
of basic biological problems. Studies aiming to advance knowledge
concerning chemical communication by means of skin gland secretions
seem on one hand to be potentially aided by intensive comparative
studies of example species within selected genera or larger taxa,
and to be aided on the other hand by combining the approaches,
techniques and results of different specialties, especially those
of ethology, morphology, physiology and chemistry.

Four general and basic problem areas in biology appear to this
reviewer to be open to advances through appropriate selection and
study of mechanisms in specialized scent glands of the skin: (1)
Evolutionary mechanisms--particularly in speciation, may be revealed
by intensive and coordinated study of sensory, secretory, and behavioral mechanisms involved in the derivation and function of a
pheromonal system. (2) Mechanisms of differentiation and induction
at cellular and molecular levels could be profitably and efficiently
studied with the externally accessible and diversely modified skin
gland cells. This includes the possible study of mechanisms of

hormone action on such cells and the receptor and related components of mediation of hormone action. (3) <u>Mechanisms of secretory
cellular change during periodic function</u> (secretory cycles, reproductive cycles, and others) <u>and aging</u>. The last of these seems
particularly evident as a possibility with the apocrine sudoriferous glands, wherein regional and species aging differences have
been observed in animal models (Quay, 1955, 1959; Quay and Müller-
Schwarze, 1971). (4) <u>Mechanisms of metabolic interrelations of
different integumentary cell and tissue types</u>. It is often noted
that in specialized skin gland regions the specializations and
differences in levels and timing of cellular activity extend to
other adjacent tissues, such as hair follicles, adipose tissue and
others. There are also the basic developmental interactions of
epidermal and dermal cells or tissue layers which can be both observed and experimentally manipulated in skin gland regions. In
summary, the diversity of cellular specializations and activities
in skin gland regions, and their easy accessibility for biopsy
and experimental manipulation, makes them a neglected but promising workshop for biologists interested in any one of a number of
fundamentally important problems.

REFERENCES

ADELMAN, S., TAYLOR, C. R., and HEGLUND, N. C. 1975. Sweating on
    paws and palms: what is its function? *Am. J. Physiol.* 229:
    1400-1402.
BARDACH, J. E., and TODD, J. H. 1970. Chemical communication in
    fish. *Adv. in Chemoreception* 1:205-240.
BAXTER, C. H., and PICKENS, P. E. 1964. Control of luminescence
    in hemichordates and some properties of a nerve net system.
    J. Exp. Biol. 41:1-14.
BERGEOT, P., and WINTREBERT, P. 1926. Le déterminisme de l'éclo-
    sion chez l'Alyte (*Alytes obstetricans,* Laur.). *CR Soc. Biol.*
    95:1326-30.
BLAYLOCK, L. A., RUIBAL, R., and PLATT-ALOIA, K. 1976. Skin struc-
    ture and wiping behavior of phyllomedusine frogs. *Copeia* 2:
    283-295.
BOURDIN, J. 1926. Le mecanisme de l'éclosion chez les teleosteans.
    III. Morphologie et repartition des glandes sereuses du
    tegument. *Compt. rend. Soc. Biol.* 95:1239-1240.
BRINLEY, F. J., and EULBERG, L. Embryological head glands of the
    cichlid fish *Aequidens portalegrensis*. *Copeia* 1953:24-26.
BURGHARDT, G. M. 1970. Chemical perception in reptiles. *Adv. in
    Chemoreception* 1:241-308.
BUTLER, C. G. 1970. Chemical communication in insects: behavior-
    al and ecologic aspects. *Adv. in Chemoreception* 1:35-78.
COLE, C. J. 1966. Femoral glands of the lizard, *Crotaphytus col-
    laris*. *J. Morph.* 118:119-136.

DANIEL, J. F. 1934. The Elasmobranch Fishes, pp. 26-30. University of California Press, Berkeley.

EGGERT, B. 1931. Die Geschlechtsorgane der Gobiiformes und Blenniformes. *Z. Wiss. Zool.* 193:249-558.

ELDER, W. H. 1954. The oil gland of birds. *Wilson Bull.* 66:6-31.

FRISCH, K. V. 1938. Zur Psychologie des Fisch-Schwarmes. *Naturwissenschaften* 26:601-606.

FRISCH, K. V. 1942. Über einen Schreckstoff der Fischhaut und seine biologische Bedeutung. *Zeit. vergl. Physiol.* 29:46-145.

GEORGES, D. 1964. Evolution morphologique et histologique des organes adhésifs du brochet (*Esox lucius* L.). *Trav. Lab. Hydrobiol. Grenoble* 56:7-16.

GRUBB, T. C., JR. 1972. Smell and foraging in shearwaters and petrels. *Nature* 237:404-405.

HAWES, M. L. 1976. Odor as a possible isolating mechanism in sympatric species of shrews (*Sorex vagrans* and *Sorex obscurus*). *J. Mammal.* 57:404-406.

HILDEMANN, W. H. 1959. A cichlid fish, *Symphosodon discús*, with unique nurture habits. *Am. Naturalist* 93:27-34.

HOPPE, P. C. 1975. Genetic and endocrine studies of the pregnancy-blocking pheromone of mice. *J. Reprod. Fert.* 45:109-115.

HRBAČEK, J. 1950. On the flight reaction of tadpoles of the common toad caused by chemical substances. *Experientia* 6:100-102.

ILG, L. 1952. Über larvale Haftorgane bei Teleostiern. *Zool. Jahrb. (Anat.)* 72: 577-600.

JONES, R. B., and NOWELL, N. W. 1973. The coagulating glands as a source of aversive and agression-inhibiting pheromone(s) in the male albino mouse. *Physiol. and Behavior* 11:455-462.

KARLSON, P., and LÜSCHER, M. 1959. "Pheromones": a new term for a class of biologically active substances. *Nature* 183:55-56.

KENNEDY, R. J. 1971. Preen gland weights. *Ibis* 113:369-372.

KITZAN, S. M., and SWEENY, P. R. 1968. A light and electron microscope study of the structure of *Protopterus annectens* epidermis. I. Mucus production. *Can. J. Zool.* 46:767-772.

KROLL, J. C., and RENO, H. W. 1971. A re-examination of the cloacal sacs and gland of the blind snake, *Leptotyphlops dulcis*, (Reptilia: Leptotyphlopidae). *J. Morph.* 133:273-280.

KULZER, E. 1954. Untersuchungen über die Schreckreaktion der Erdkrotenkaulquappen (*Bufo bufo* L.). *Zeit. vergl. Physiol.* 36: 443-463.

LEINER, M. 1930. Fortsetzung der ökologischen Studien an *Gasterosteus aculeatus*. *Z. Morphol. Okol. Tiere* 16:499-522.

LEON, M. 1975. Dietary control of maternal pheromone in the lactating rat. *Physiol. and Behavior* 14:311-319.

LIAPPIS, N., and JÄKEL, A. 1975. Über die Ausscheidung der freien Aminosäuren im menschlichen ekkrinen Schweiss. *Arch. Dermatological Res.* 254:185-203.

LINTON, J. R., and SOLOFF, B. L. 1964. The physiology of the brood pouch of the male sea horse *Hippocampus*. *Bull. Marine Sci. of the Gulf and Caribbean* 14:45-61.

LUCAS, A. M., and STETTENHEIM, P. R. 1972. Avian Anatomy. In-
    tegument. Part II. U. S. Government Printing Office, Wash-
    ington, D. C.
MADERSON, P. F. A. 1970. Lizard glands and lizard hands: models
    for evolutionary study. *Forma et Functio* 3:179–204.
MADERSON, P. F. A. 1972. The structure and evolution of holocrine
    epidermal glands in sphaerodactyline and eublepharine gekkonid
    lizards. *Copeia* 1972:559–571.
MADERSON, P. F. A., and CHIU, K. W. 1970. Epidermal glands in
    gekkonid lizards: evolution and phylogeny. *Herpetologica* 26:
    233–238.
MICHAEL, R. P., BONSALL, R. W., and WARNER, P. 1974. Human vaginal
    secretions: volatile fatty acid content. *Science* 186:1217–
    1219.
MICHAEL, R. P., KEVERNE, E. B., and BONSALL, R. W. 1971. Phero-
    mones: isolation of male sex attractants from a female pri-
    mate. *Science* 172:964–166.
MÜLLER, G. 1961. Die Moschusdrüsen von *Clemmys caspica* Gmelin
    1774. *Zool. Anz.* 167:151–158.
MYKYTOWYCZ, R. 1970. The role of skin glands in mammalian commu-
    nication. *Adv. in Chemoreception* 1:327–360.
NELSON, K. 1964. Behavior and morphology in the Glandulocaudine
    fishes (Ostariophysi, Characidae). *Univ. Calif. Publ. Zool.*
    75:59–152.
OLDAK, P. D. 1976. Comparison of the scent gland secretion lipids
    of twenty-five snakes: implications for biochemical system-
    atics. *Copeia* 1976:320–326.
OLSSON, R. 1961. The skin of Amphioxus. *Z. Zellforsch.* 54:90–104.
PETERS, H. M. 1964. Über larvale Haftorgane bei *Tilapia* (Cich-
    lidae, Teleostei) und ihre Rückbildung in der Evolution.
    Zool. Jahrb. 71:287–300.
PFEIFFER, W. 1960. Über die Schreckreaktion bei Fischen und die
    Herkunft des Schreckstoffes. *Z. vergl. Physiol.* 43:578–614.
PFEIFFER, W. 1963. Vergleichende Untersuchungen über die Schreck-
    reaktion und den Schreckstoff der Ostariophysen. *Zeit. f.
    vergl. Physiol.* 47:111–147.
PFEIFFER, W. 1967. Schreckreaktion und Schreckstoffzellen bei
    Ostariophysi und Gonorhynchiformes. *Z. vergl. Physiol.* 56:
    380–396.
PFEIFFER, W., and LEMKE, J. 1973. Untersuchungen zur Isolierung
    und Identifizierung des Schreckstoffes aus der Haut der El-
    ritze, *Phoxinus phoxinus* (L.) (Cyprinidae, Ostariophysi, Pis-
    ces). *J. Comp. Physiol.* 82:407–410.
PFEIFFER, W., SASSE, D., and ARNOLD, M. 1971. Die Schreckstoff-
    zellen von *Phoxinus phoxinus* und *Morulius chrysophakedion*
    (Cyprinidae, Ostariophysi, Pisces). *Z. Zellforsch.* 118:203–
    213.
QUAY, W. B. 1959. Microscopic structure and variation in the
    cutaneous glands of the deer, *Odocoileus virginianus*. *J.
    Mammal.* 40:114–128.

QUAY, W. B. 1955. Histology and cytochemistry of skin gland areas in the caribou, *Rangifer*. *J. Mammal*. 36:187-201.

QUAY, W. B. 1967. Comparative survey of the anal glands of birds. *Auk* 84:379-389.

QUAY, W. B. 1968. The specialized posterolateral sebaceous glandular regions in microtine rodents. *J. Mammal*. 49:427-445.

QUAY, W. B. 1972. Integument and the environment: glandular composition, function and evolution. *Am. Zool*. 12:95-108.

QUAY, W. B., and MÜLLER-SCHWARZE, D. 1971. Relations of age and sex to integumentary glandular regions in Rocky Mountain mule deer (*Odocoileus hemionus hemionus*). *J. Mammal*. 52:670-685.

REESE, A. M. 1921. The structure and development of the integumental glands of the Crocodilia. *J. Morph*. 35:581-611.

ROSE, F. L. 1970. Tortoise chin gland fatty acid composition: behavioral significance. *Comp. Biochem. and Physiol*. 32:577-580.

ROSE, F. L., DROTMAN, R., and WEAVER, W. G. 1969. Electrophoresis of chin gland extracts of *Gopherus* (tortoises). *Comp. Biochem. and Physiol*. 29:847-851.

ROSEN, M. W., and CORNFORD, N. E. 1971. Fluid friction of fish slimes. *Nature* 234:49-51.

SCHAFFER, J. 1940. Die Hautdrüsenorgane der Säugetiere mit besonderer Berücksichtigung ihres histologischen Aufbaues und Bemerkungen über die Proktodäaldrüsen. Urban and Schwarzenberg, Berlin and Wien.

SINGER, A. G., AGOSTA, W. C., O'CONNELL, R. J., PFAFFMANN, C., BOWEN, D. V., and F. H. FIELD. 1976. Dimethyl disulfide: an attractant pheromone in hamster vaginal secretion. *Science* 191:948-950.

SIOPES, T. D., and WILSON, W. O. 1975. The cloacal gland--an external indicator of testicular development in coturnix. *Poultry Sci*. 54:1225-1229.

SMITH, G. M., and COATES, C. W. 1936. On the histology of the skin of the lungfish *Protopterus annectens* after experimentally induced aestivation. *Quart. J. Microscop. Sci*. 79:487-491.

STOKLOSOWA, S. 1966. Sexual dimorphism in the skin of sea-trout, *Salmo trutta*. *Copeia* 1966:613-614.

TODD, J. H., ATEMA, J., and BARDACH, J. E. 1967. Chemical communication in the social behavior of a fish, the yellow bullhead, *Ictalurus natalis*. *Science* 158:672-673.

VANDENBERGH, J. G. 1975. Pheromonal stimulation of puberty in female mice. *J. Endocrinology* 64:38p.

VANDENBERGH, J. G., WHITSETT, J. M., and LOMBARDI, J. R. 1975. Partial isolation of a pheromone accelerating puberty in female mice. *J. Reprod. Fert*. 43:515-523.

WARD, J. A., and BARLOW, G. W. 1967. The maturation and regulation of glancing off the parents by young orange chromides (*Etrophus maculatus*: Pisces-Cichlidae). *Behavior* 29:1-56.

WEITZEL, G., and LENNERT, K.   1951.   Untersuchungen über die Bür-
        zeldrüse der Vögel.   I. Mitteilung.   Die Fettstoffe der Bür-
        zeldrüsen von Enten.   II. Mitteilung.   Morphologie und Histo-
        chemie der Bürzeldrüsen von Enten.   *Hoppe-Seyl. Zeit.* 288:4-6;
        251-272.
WINN, H. E.   1955.   Formation of a mucous envelope by parrot fish-
        es.   *Zoologica* 40:145-148.
WINOKUR, R. M., and LEGLER, J. M.   1975.   Chelonian mental glands.
        J. Morph. 147:275-292.

HORMONAL CONTROL OF MAMMALIAN SKIN GLANDS

F. John Ebling

Professor of Zoology

Zoology Department, The University, Sheffield, U.K.

INTRODUCTION

Apart from the footpad glands, and the eccrine glands of certain primates, both of which develop independently of hair follicles and neither of which are primarily sources of odour, mammalian glands are of two kinds. Each arises in development as an outpushing from the hair follicle. In the holocrine glands, by definition, a lipid secretion is formed by complete disintegration of the cells which are replenished by cell division at the alveolar periphery; in the tubular apocrine glands the secretory process allegedly involves destruction of the cell apices at the luminal border (Schiefferdecker, 1917, 1922). Individual holocrine or sebaceous glands are dispersed throughout the hairy skin in all mammals except whales, and dispersed apocrine glands occur in some mammalian families. Specialized scent glands are aggregations of units of either or both types. Both dispersed or aggregated glands are influenced by hormones. The endocrine control of several species has been studied; previous reviews include those of Ebling (1972) and Strauss and Ebling (1970).

Among holocrine glands investigated are the sebaceous glands of the rat, rabbit and man, the preputial glands of the rat, the flank organ of the hamster, the supracaudal gland of the guinea pig, the abdominal glands of the gerbil and the mid-ventral gland of Peromyscus. Purely apocrine models include the chin, inguinal and anal glands of the rabbit and the post-auricular glands of the shrew. The side glands of the shrew contain both types of secretory unit.

HOLOCRINE GLANDS

The Sebaceous Glands of the Rat

The responses of the sebaceous glands of the rat to hormones have been assessed by changes in gland size or mitotic activity (Ebling, 1963) or by sebum production, measured either in terms of the amount of lipid removable by total immersion of the animal in solvents (Shuster and Thody, 1974) or as the rate of build up of fat extractable by ether from hair clipped at intervals after shampooing the whole rat (Ebling and Skinner, 1967, 1975; Ebling, 1974).

There is general agreement that testosterone increases the size of the sebaceous glands and stimulates sebum production. The effect involves an increase in mitoses in the glands (Figure 1), which can be prevented by anti-androgenic steroids such as 17α-methyl-B-nortestosterone (Ebling, 1967) or cyproterone acetate (Ebling, 1973) in a dosage about ten times that of administered testosterone.

Estradiol, in one thousandth of that dose, has an even greater effect in depressing sebaceous secretion, apparently without significantly lowering mitotic activity. It must thus be presumed to act by inhibiting intracellular lipid synthesis.

The claim that progesterone has a stimulating effect equal to that of testosterone (Haskin et al., 1953) has not subsequently been substantiated (Ebling, 1961). But the possibility that it could have some effect in very large doses cannot be ruled out (Ebling, 1963).

The effect of testosterone on the sebaceous glands is greatly diminished by hypophysectomy (Ebling, 1957; Ebling et al., 1969a). A large series of rats treated with testosterone were compared with untreated litter-mate controls, and the responses in hypophysecto-mized-castrated rats were only about 20 per cent of those of cast-rated rats (Ebling, 1974). Such a small effect may not prove statistically significant in small groups of rats. The effect has been completely restored with prolactin or porcine growth hormone (Ebling et al., 1969a), pure bovine growth hormone (Figure 2) or synthetic α-MSH (Ebling et al., 1975c; Thody and Shuster, 1975). The bovine growth hormone and α-MSH were each also shown to have some independent effect (see below). 5α-dihydrotestosterone, 5α-androstane-3β,17β-diol, and androstenedione each proved more effec-tive than testosterone in stimulating sebum production in hypophy-sectomized-castrated rats (Ebling, 1974; Ebling et al., 1971, 1973).

The thyroid gland has also been shown to affect the sebaceous glands in the rat. Thyroidectomy diminishes sebaceous secretion and thyroxine (Shuster and Thody, 1974) or thyrotropic hormone (Ebling, 1974; Ebling et al., 1970a) increases it.

The Human Sebaceous Glands

As studied in the face, the human sebaceous glands are

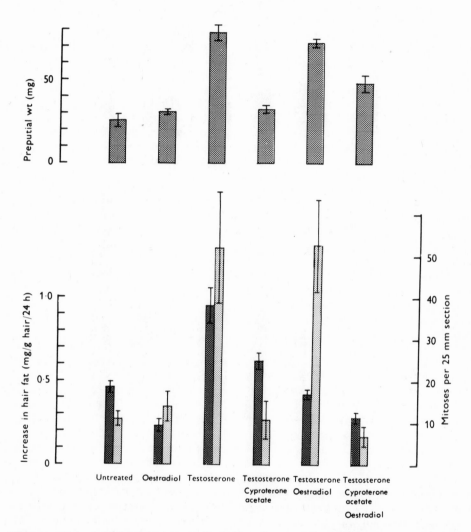

Figure 1.   Independent and combined effects of estradiol (2-4µg/ 24 hr), testosterone (0.2 mg/24 hr) and cyproterone acetate (2.0 mg/24 hr) on absolute weight of preputial glands (upper diagram), sebum secretion (lower diagram, left-hand columns) and mitosis in the sebaceous glands (right-hand columns) in castrated rats.   Means ±S.E. for groups of 6 rats.   Data from Ebling, 1973.

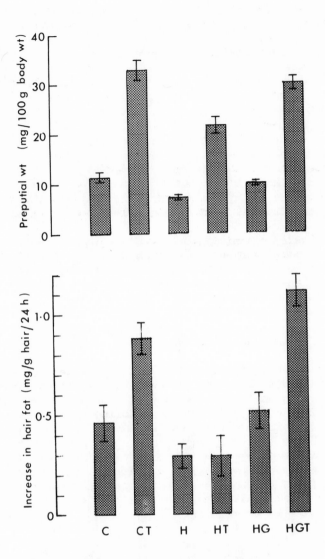

Figure 2. Relative preputial glad weights and sebum production
as measured by increase in hair fat. Means ± S.E. for groups of
11 rats. C = castrated; CT = castrated and treated with testosterone; H = hypophysectomized and castrated; HT = hypophysectomized-
castrated and treated with testosterone; HG = hypophysectomized-
castrated and treated with bovine growth hormone; HGT = hypophy-
sectomized-castrated and treated with testosterone and bovine
growth hormone. Data from Ebling et al., 1975a,b.

moderately large at birth, but then become small and undifferen-
tiated (Strauss and Pochi, 1963). They enlarge again in both sexes
at puberty and secrete increased quantities of sebum. In men secretion is well maintained until the age of 70 or more, but in women,
in whom average sebum production is somewhat lower, there is a
decline after the menopause (Pochi and Strauss, 1974; Strauss and
Pochi, 1963). Administration of testosterone has been shown to
increase sebum secretion, as measured by absorption from the forehead, in a prepuberal boy and in eunuchs, but not in normal men
in whom the sebaceous glands are presumed to be maximally stimulated
by endogenous androgens (Strauss and Pochi, 1963). Androstenedione
and dehydroepiandrosterone, under appropriate conditions, are also
effective stimulants (Pochi and Strauss, 1969). It seems unlikely
that progesterone in physiological doses has any effect on human
sebaceous secretion (Jarrett, 1959; Strauss and Kligman, 1961).

Systemically administered estrogens depress sebaceous secretion in both males and females (Jarrett, 1955; Strauss and Pochi,
1961; Strauss et al., 1962). Topically administered estrogens
have proved less effective and it has not been possible to demonstrate a local as distinct from a general action.

The Inguinal Sebaceous Glands of the Rabbit

That the sebaceous glands of the ear of the rabbit respond to
testosterone was first shown by Montagna and Kenyon (1949). In
the inguinal region a discrete mass of sebaceous glands can be
dissected out and weighed (Wales and Ebling, 1971). In rabbits
aged 24 weeks which had been castrated at 13 weeks, the inguinal
sebaceous glands were only about half the size of those in intact
males of similar body weight (Figure 3). In rabbits which had
been implanted with testosterone immediately after castration the
sebaceous glands were of normal weight. Estradiol reduced the
weight by about one half, both in intact males and in castrated
males treated with testosterone.

The Preputial Glands of the Rat

Preputial glands occur in many species of rodents (for review
see Clevedon Brown and Williams, 1972), including the musk rat
(Ondatra zibethicus), in which they are responsible for the
musky odour, the rat and the mouse. In males they consist of a
pair of discrete sebaceous organs, lying between the prepuce and
the penis, or between the pubic skin and the body wall. Each
opens by a relatively long duct alongside the tip of the glans
penis. Similar glands are present in the female, where they are
sometimes known as clitoral glands.

As measured by changes in weight, the preputial glands of castrated rats are stimulated by testosterone (Figure 1). However, in contrast to the sebaceous glands, estradiol has no effect, whether or not the animals are also treated with testosterone. On the other hand, anti-androgenic steroids, for example cyproterone acetate, markedly inhibit the effect of testosterone.

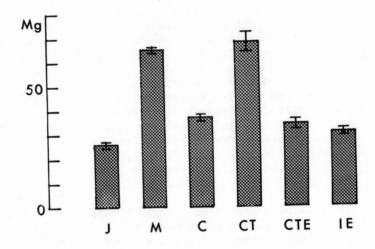

Figure 3.   Weights of inguinal sebaceous glands (mg) from 13 week old juvenile (J) and 21 week old mature (M) male rabbits. C = castrated; CT = castrated and treated with testosterone; CTE = castrated and treated with testosterone and estradiol; IE = intact treated with estradiol. Means + S.E. for groups of 6-8 animals. Data from Wales and Ebling, 1971.

The response to testosterone is significantly reduced by hypophysectomy, though not to the same extent as that of the sebaceous glands. Growth hormone acts synergistically with testosterone (Huggins et al., 1955). In addition to a very significant synergistic effect on relative preputial gland weight, bovine growth hormone has a smaller independent effect (Figure 2). A similar synergistic action has been demonstrated for synthetic α-MSH (Krähenbuhl and Desaulles, 1969) which also has a small independent action (Ebling et al., 1975c; Thody and Shuster, 1975).

In contrast to the cutaneous sebaceous glands, the preputial glands can be stimulated by progesterone. This response is reduced by hypophysectomy. The existence of a separate pituitary hormone which acts synergistically with progesterone has been claimed (Lorincz and Lancaster, 1957; Woodbury et al., 1965). However, a porcine growth hormone which restored the response of the sebaceous glands to testosterone showed no synergistic action with progesterone as measured by the response of the preputial glands (Ebling et al., 1969b).

Flank Organ of the Golden Hamster

The flank organs of the golden hamster consist of two dorsolateral spots containing dermal melanocytes, hair follicles and sebaceous glands (Hamilton and Montagna, 1950). Not only the sebaceous glands, but also growth of the hair and pigmentation are androgen dependent, so that the effects of hormonal stimulation can readily be observed and measured in the living animal. Moreover, since it is possible to apply hormone unilaterally and to maintain the contralateral gland as a control, the gland provides a model for testing the topical as distinct from systemic effect of a hormone (Burdick and Hill, 1970; Frost et al., 1973).

Supra-caudal Gland of Guinea Pig

The supra-caudal gland is a pigmented oval area, covered with sparse hair, situated mid-dorsally above the caudal vertebrae. It consists of an aggregation of complex sebaceous glands emptying into hair follicles which are full of sebum (Martan, 1962).

In male guinea pigs, castrated at nine weeks, the glands diminished rapidly in area and lost their cushioned appearance (Martan and Price, 1967). The acini of the gland became reduced in number. Spayed females also showed a less marked reduction in gland size. Odour gradually disappeared in both sexes.

Restoration of glandular size and activity in either sex was achieved by injection for three weeks of testosterone propionate, or - though the effect was less - of progesterone.

Abdominal Gland of the Gerbil (Meriones ungulatus)

The abdominal gland of the gerbil is a discrete pad of sebaceous glands in the midline of the abdomen, about 25 mm long and 5 mm wide in the adult male, and somewhat smaller in the female. It has a porous appearance and from each pore emerges a

cluster of sharp pointed hairs, which have flattened curved
surfaces in cross section and resemble miniature troughs. These
hairs appear to conduct the secretion to the surface of the pad.
Each pad contains 200 - 300 or more holocrine alveoli.

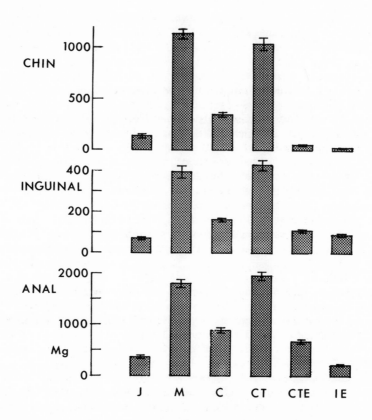

Figure 4.   Weights (mg) of submandibular (chin), inguinal apocrine
and anal glands from 13 week old juvenile (J) and 21 week old
mature (M) male rabbits. C = castrated; CT = castrated and treated
with testosterone; CTE = castrated and treated with testosterone
and estradiol; IE = intact treated with estradiol. Means $\pm$ S.E.
for groups of 6-8 animals. Data from Wales and Ebling, 1971.

The sebaceous gland pad of the immature gerbil has been
shown to respond to a range of hormones (Glenn and Gray, 1965).
The most effective was testosterone which increased gland weight,
as well as cholesterol and fatty acid content in both males and
females. However, the glands of the intact, though not the

ovariectomized, female responded also to gonadotrophic hormone (PMS) in parallel with its effect on the uterus. Either estradiol-17-β or progesterone similarly produced a good response. ACTH did not, even though dehydroepiandrosterone was reported to stimulate the pads of mature male gerbils.

The hair follicles of the gland pad are similarly under hormonal control. According to Mitchell and Butcher (1966) each gland complex contains as many as three hair follicles in various phases of growth. They remain in almost continuous anagen since, although clubs are formed, there is virtually no resting phase before the follicle enters the next anagen. Castration of male gerbils caused almost immediate cessation of follicular growth as well as involution of the sebaceous elements. All follicles went into catagen or telogen. Administration of testosterone restored activity within a few days.

Midventral Sebaceous Glands of Peromyscus

Mature males of Peromyscus polionatus and P. maniculatus exhibit a well marked glandular area in the midventral region (Doty and Kart, 1972) The gland was found to diminish in size with age and to be present, though it was less marked, in females of these species. The gland was absent from both sexes in a number of other species and sub-species. The gland is entirely holocrine in nature and, according to the authors, did not contain any hair follicles or arrector pili muscles in any of the specimens examined.

The glandular area disappeared after castration, but treatment with testosterone propionate restored it to normal in a few days. Treatment of mature ovariectomized females with estradiol benzoate and progesterone appeared to diminish sebaceous development.

APOCRINE GLANDS

The Apocrine Glands of the Rabbit

The structure and function of the anal, inguinal and submandibular (chin) glands of the rabbit have been described in detail by Mykytowycz (1966a,b,c) who showed that growth of the glands was inhibited by castration during early life.

The weight of the chin gland of the rabbit (New Zealand White) increases nearly ten fold between 13 and 24 weeks of age, even though body weight does no more than double (Figure 4).

Castration prevents this allometric increase, whereas treatment
with testosterone restores it in castrated rabbits. Estradiol
has a dramatic inhibitory effect both in intact males and in
castrated males implanted with testosterone (Wales and Ebling, 1971).

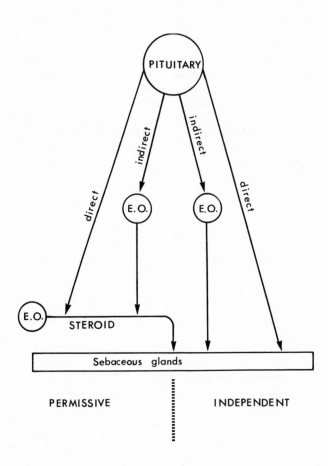

Figure 5.   Possible ways in which pituitary hormones may affect
the sebaceous glands. The effect may be direct or indirect, and
in addition may be either independent or permissive, in the sense
that the pituitary hormone facilitates the response to another
hormone, for example a steroid. E.O., Endocrine organ. From
Ebling et al., 1975a, by permission of the British Journal of
Dermatology.

The testosterone-stimulated gland has tubules of great diameter
and a taller secretory epithelium than the glands in immature or
castrated males. Estradiol reduces both tubule diameter and epithelium height.

The inguinal and anal glands are similarly stimulated by
testosterone and inhibited, though to a lesser extent, by
estradiol.

The Sideglands of the Shrew (Suncus murinus)

The side glands of Suncus are a pair of padlike structures
consisting of batteries of sebaceous elements with underlying
apocrine glands. In males, both elements atrophy after castration and can be restored by testosterone or progesterone (Dryden
and Conaway, 1967). Estradiol appears to have no effect. Shrews
also have post-auricular fields of apocrine units which show
similar hormonal responses.

THE ROLE OF THE PITUITARY

Hormones of the pituitary may affect the skin glands in
several possible ways. First the effect may be indirect by way of
another endocrine organ or direct (see Figure 5). If the skin
glands are affected by hormones of the testes, ovaries, adrenal
cortex, or thyroid, then, provided the requisite organs are
present, such effect will similarly be produced by the appropriate
tropic hormones. Thus the sebaceous glands of the rat are
stimulated by gonadotropin (Shuster and Thody, 1974), ACTH (Ebling
et al.,1970) or thyrotropin (Ebling, 1974; Thody and Shuster, 1972)
and the abdominal glands of the intact female gerbil, which are
estrogen sensitive, by gonadotropin (Glenn and Gray, 1965).

On the other hand, the pituitary peptides may have a direct
action at the skin target; for example, both bovine growth hormone
and α-MSH have been shown to stimulate sebum production in hypophysectomized-castrated rats (Ebling et al., 1975a,c). It is
possible that other glands are sensitive to pituitary influence,
for example both the Harderian and lachrymal glands of the rat are
considerably diminished by hypophysectomy (Ebling et al., 1975b,c).

Apart from these independent effects,whether indirect or
direct, pituitary peptides may also act permissively or synergistically with steroid hormones. For example, the response of
the sebaceous glands of the hypophysectomized-castrated rat to
testosterone can be enhanced by prolactin, growth hormone or α-
MSH, and similar effects have been demonstrated in the preputial
glands (see above).

METABOLISM OF ANDROGEN AT TARGET SITE

Some, at least, of the effects of testosterone involve its transformation to active metabolites. The importance of 5α-dihydrotestosterone as an active androgen was first established by studies on the prostate (Bruchovsky and Wilson, 1968). However, both male and female human skin (Wilson and Walker, 1969) and rat skin (Flamigni et al., 1970) are similarly capable of producing 5α-dihydrotestosterone from testosterone.

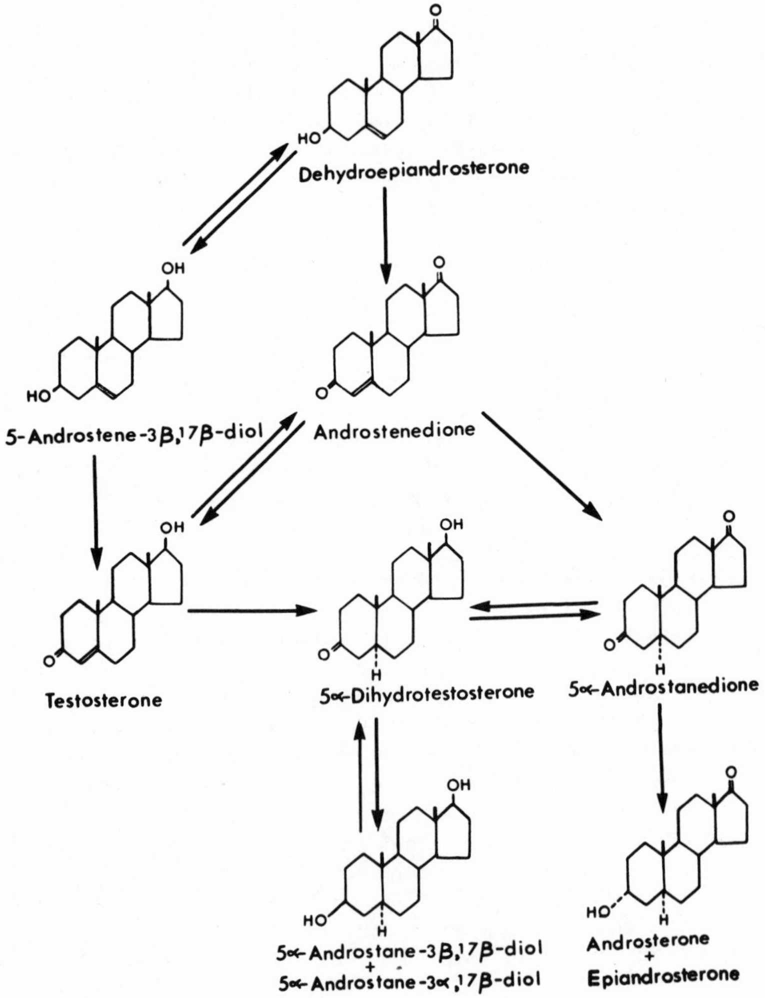

Figure 6.   Possible pathways of androgen metabolism in skin.

This is not the only metabolite; androstenedione and 5α-andro
stanediols are also formed; the possible pathways of androgen
metabolism in skin have been reviewed by several authors (see, for
example Ebling, 1970; King and Mainwaring, 1975; Milne, 1969;
Sansone-Bazzano and Reisner, 1974; Strauss and Pochi, 1969) and
are summarized in Figure 6. Of particular interest is the demonstration that hamster flank organ in vitro has about four times
the capacity of adjacent skin for 5α-reduction of testosterone.
The metabolites androstenedione, androsterone, androstanedione,
dihydrotestosterone and 3α-androstanediol were all identified
(Gomez et al., 1974).

The evidence (see above) that, while hypophysectomy reduces
the response of the rat sebaceous glands to testosterone, the
glands remain somwehat more responsive to 5α-dihydrotestosterone,
5α-androstane-3β,17β-diol and androstenedione suggests that
pituitary peptides may mediate the metabolic transformation of
testosterone, but this hypothesis is far from proven.

SUMMARY AND CONCLUSIONS

Sebaceous glands, whether dispersed or aggregated, are in
general under hormonal control, as are the apocrine glands such as
the chin, inguinal and anal glands of the rabbit, and the post-
auricular glands of shrews.

Testosterone appears to be a universal stimulant and estrogens
generally have an inhibitory action. However, the response of
different glands in the same or different species are not necessarily identical. For example, in the rat, the preputial gland
differs from the cutaneous sebaceous gland in at least three
respects: it is markedly less dependent on the presence of the
pituitary; it responds well to progesterone whereas the cutaneous
gland does not, and it does not appear to be inhibited by estrogens.
Several differences between species are notable. Whereas estrogens
have been shown to inhibit sebaceous glands in man, rat and mouse,
and both sebaceous and apocrine derivatives in rabbit, they may
actually stimulate development of the abdominal gland of the gerbil.
In addition to steroid hormones, thyroxine has, in the rat, been
shown to stimulate sebaceous secretion.

The overriding role of the hypothalamo-hypophyseal system in
the control of skin glands, of relevance to seasonal changes in
activity, may be exercised in three ways. Firstly, indirect control
can be mediated by way of thyroid, adrenals and gonads. Second,
peptide hormones may have a direct and independent effect on the
skin glands. Third, pituitary hormones may act synergistically to
alter the response of the skin glands to steroids.

REFERENCES

Bruchovsky, N. and Wilson, J.D. 1968. Conversion of testosterone
     to 5α-androstan-17β-ol-3-one by rat prostate in vivo and in
     vitro. J. Biol. Chem. 243: 2012-2021.
Burdick, K.H. and Hill, R. 1970. The topical effect of the anti-
     androgen chlormadinone acetate and some of its chemical
     modifications on the hamster costovertebral organ. Br. J.
     Derm. 82 (Suppl. 6): 19-25.
Clevedon Brown, J. and Williams, J.D. 1972. The rodent
     preputial gland. Mamm. Rev. 2: 105-147.
Doty, R.L. and Kart, R. 1972. A comparative and developmental
     analysis of the mid-ventral sebaceous glands in 18 taxa of
     Peromyscus, with an examination of gonadal steroid influ-
     ences in Peromyscus maniculatus Bairdii. J. Mammal. 53:
     83-99.
Dryden, G.L. and Conaway, C.H. 1967. The origin and control of
     scent production in Suncus murinus. J. Mammal. 48:420-
     427.
Ebling, F.J. 1957. The action of testosterone on the sebaceous
     glands and epidermis in castrated and hypophysectomized male
     rats. J. Endocr. 15: 297-306.
Ebling, F.J. 1961. Failure of progesterone to enlarge sebaceous
     glands in the female rat. Br. J. Derm. 73: 65-68.
Ebling, F.J. 1963. Hormonal control of the sebaceous gland in
     experimental animals. Pages 200-219. in W. Montagna, R. A.
     Ellis and A. F. Silver (ed.), Advances in Biology of Skin;
     Volume 4: Sebaceous Glands. Pergamon Press, Oxford.
Ebling, F.J. 1967. The action of an antiandrogenic steroid, 17α-
     methyl-B-nortestosterone, on sebum secretion in rats treated
     with testosterone. J. Endocr. 38: 181-185.
Ebling, F.J. 1970. Steroid hormones and sebaceous secretion.
     in M. H. Briggs (ed.) Advances in Steroids, Volume 2.
     Academic Press, London.
Ebling, F.J. 1972. The response of the cutaneous glands to
     steroids. Gen. Comp. Endocr. Suppl. 3: 228-237.
Ebling, F.J. 1973. The effects of cyproterone acetate and
     estradiol upon testosterone stimulated sebaceous activity
     in the rat. Acta Endocr. 72: 361-365.
Ebling, F.J. 1974. Hormonal control and methods of measuring
     sebaceous gland activity. J. invest. Derm. 62: 161-171.
Ebling, F.J., Ebling,E., McCaffery, V. and Skinner, J. 1971.
     The response of the sebaceous glands of the hypophysectomized-
     castrated male rat to 5α-dihydrotestosterone, androstenedione,
     dehydroepiandrosterone and androsterone. J. Endocr. 51: 181-
     190.
Ebling, F.J., Ebling, E., McCaffery, V. and Skinner, J. 1973.
     The responses of the sebaceous glands of the hypophysecto-
     mized-castrated male rat to 5α-androstanedione and 5α-andro-
     stane-3β,17β-diol. J. invest. Derm. 60: No. 2, 183-187.

Ebling, F.J., Ebling, E., Randall, V. and Skinner, J. 1975a. The
    sebotrophic action of bovine growth hormone (BGH) in the rat.
    Br. J. Derm. 92: 325-332.
Ebling, F.J., Ebling, E., Randall, V. and Skinner, J. 1975b. The
    effects of hypophysectomy and of bovine growth hormone on the
    responses to testosterone of prostate, preputial, Harderian
    and lachrymal glands and of brown adipose tissue in the rat.
    J. Endocr. 66: 401-406.
Ebling, F.J., Ebling, E., Randall, V. and Skinner, J. 1975c. The
    synergistic action of α-MSH and testosterone on the sebaceous,
    prostate, preputial, Harderian and lachrymal glands and brown
    adipose tissue in hypophysectomized rats. J. Endocr. 66:
    407-412.
Ebling, F.J., Ebling, E. and Skinner, J. 1969a. The influence of
    pituitary hormones on the response of the sebaceous glands of
    the rat to testosterone. J. Endocr. 45: 245-256.
Ebling, F.J., Ebling,E. and Skinner, J. 1969b. The influence of
    the pituitary on the response of the sebaceous and preputial
    glands of the rat to progesterone. J. Endocr. 45: 257-263.
Ebling, F.J., Ebling, E. and Skinner, J. 1970a. The effects of
    thyrotrophic hormone and of thyroxine on the response of the
    sebaceous glands of the rat to testosterone. J. Endocr. 48:
    83-90.
Ebling, F.J. and Skinner, J. 1967. The measurement of sebum
    production in rats treated with testosterone and oestradiol.
    Br. J. Derm. 79: 386-393.
Ebling, F.J. and Skinner, J. 1975. The removal and restitution
    of hair fat in the rat. Br. J. Derm. 92: 321-324.
Ebling, F.J., Ebling, E., Skinner, J., and White, A. 1970b. The
    response of the sebaceous glands of hypophysectomized-
    castrated male rats to adrenocorticotrophic hormone and to
    testosterone. J. Endocr. 48: 73-81.
Flamigni, C.A., Collins, W.P., Koullapis, E.N. and Sommerville,
    I.F. 1970. Formation and metabolism of testosterone in
    rat skin. Endocrinology 87: 764-770.
Frost, P., Giegel, J.L., Weinstein, G.D. and Gomez, E.C. 1973.
    Biodynamic studies of hamster flank organ growth: hormonal
    influences. J. invest. Derm. 61: 159-167.
Glenn, E.M. and Gray, J. 1965. Effect of various hormones on
    the growth and histology of the gerbil (Meriones ungui-
    culatus) abdominal sebaceous gland pad. Endocrinology,
    76: 1115-1123.
Gomez, E.C., Llewellyn, A. and Frost, P. 1974. Metabolism of
    testosterone-4-^{14}C by hamster skin. J. invest. Derm. 63:
    383-387.
Hamilton, J.B. and Montagna, W. 1950. The sebaceous glands of
    the hamster. I. Morpholgical effects of androgens on
    integumentary structures. Am. J. Anat. 86: 191-234.

Haskin, D., Lasher, N. and Rothman, W. 1953. Some effects of
    ACTH,cortisone, progesterone and testosterone on sebaceous
    glands in the white rat. J. invest. Derm. 20: 207-211.

Huggins, C., Parsons, F.M. and Jensen, E.V. 1955. Promotion of
    growth of preputial glands by steroids and the pituitary
    growth hormone. Endocrinology, 57: 25-32.

Jarrett, A. 1955. The effects of stilboestrol on the surface
    sebum and upon acne vulgaris. Br. J. Derm. 67: 165-179.

Jarrett, A. 1959. The effects of progesterone and testosterone
    on surface sebum and acne vulgaris. Br. J. Derm. 71: 102-
    116.

King, R.J.B. and Mainwaring, W.I.P. 1975. Steroid-cell inter-
    actions. Butterworths, London.

Krähenbühl, C. and Desaulles, P.A. 1969. Interactions between
    α-MSH and sex steroids on the preputial glands of female
    rats. Experientia, 25: 1193-1195.

Lorincz, A.L. and Lancaster, G. 1957. Anterior pituitary
    preparation with tropic activity for sebaceous, preputial
    and Harderian glands. Science 126:209-222.

Martan, J. 1962. Effect of castration and androgen replacement
    on the supracaudal gland of the male guinea pig. J. Morph.
    110: 285-298.

Martan, J. and Price, D. 1967. Comparative responsiveness of
    supracaudal and other sebaceous glands in male and female
    guinea pigs to hormones. J. Morph. 121: 209-222.

Milne, J.A. 1969. The metabolism of androgens by sebaceous
    glands. Br. J. Derm. 81: 23-28.

Mitchell, O.G. and Butcher, E.O. 1966. Growth of hair in the
    ventral glands of castrate gerbils following testosterone
    administration. Anat. Rec. 156: 11-18.

Montagna, W. and Kenyon, P. 1949. Growth potentials and mitotic
    division in the sebaceous glands of the rabbit. Anat. Rec.
    103: 365-380.

Mykytowycz, R. 1966a. Observations on odoriferous and other
    glands in the Australian wild rabbit, Oryctolagus
    cuniculus (L.) and the hare, Lepus europaeus P. I.
    The anal gland. CSIRO Wildl. Res. 11: 11-29.

Mykytowycz, R. 1966b. Observations on the odoriferous and other
    glands in the Australian wild rabbit, Oryctolagus
    cuniculus (L.), and the hare, Lepus europaeus P. II.
    The inguinal glands. CSIRO Wildl. Res. 11: 49-64.

Mykytowycz, R. 1966c. Observations on odoriferous and other
    glands in the Australian wild rabbit, Oryctolagus
    cuniculus (L.), and the hare, Lepus europaeus P. III.
    Harder's, lachrymal and submandibular glands. CSIRO Wildl.
    Res. 11: 65-90.

Pochi, P.E. and Strauss, J.S. 1969. Sebaceous gland response in
    man to the administration of testosterone, Δ^4-androstenedione,
    and dehydroisoandrosterone. J. invest. Derm. 52: 32-36.

Pochi, P.E. and Strauss, J.S. 1974. Endocrinological control of the development and activity of the human sebaceous gland. J. invest. Derm. 62: 191-201.

Sansone-Bazzano, G. and Reisner, R. M. 1974. Steroid pathways in sebaceous glands. J. invest. Derm. 62: 211-216.

Schiefferdecker, P. 1917. Die Hautdrüsen des Menschen und der Säugetiere, ihre biologische und rassenanatomische Bedeutung, sowie die muscularis sexualis. Biol. Zbl. 37: 534-562.

Schiefferdecker, P. 1922. Die Hautdrüsen des Menschen und der Säugetiere, ihre biologische und rassenanatomische Bedeutung, sowie die muscularis sexualis. Zoologica, 27: 1-154.

Shuster, S. and Thody, A.J. 1974. The control and measurement of sebum secretion. J. invest. Derm. 62: 172-190.

Strauss, J.S. and Ebling, F.J. 1970. Control and function of skin glands in mammals. Mem. Soc. Endocr. 18: 341-371.

Strauss, J.S. and Kligman, A.M. 1961. The effect of progesterone and progesterone-like compounds on the human sebaceous gland. J. invest. Derm. 36: 309-318.

Strauss, J.S., Kligman, A.M. and Pochi, P.E. 1962. The effect of androgens and estrogens on human sebaceous glands. J. invest. Derm. 39: 139-155.

Strauss, J.S. and Pochi, P.E. 1961. The quantitiative gravimetric determination of sebum production. J. invest. Derm. 36: 293-298.

Strauss, J.S. and Pochi, P.E. 1963. The human sebaceous gland: its regulation by steroidal hormones and its use as an end organ for assaying androgenicity in vivo. Recent Prog. Horm. Res. 19: 385-444.

Strauss, J.S. and Pochi, P.E. 1969. Recent advances in androgen metabolism and their relation to the skin. Archs. Derm. 100: 621-636.

Thody, A.J. and Shuster, S. 1972. A study of the relationship between the thyroid gland and sebum secretion in the rat. J. Endocr. 54: 239-244.

Thody, A.J. and Shuster, S. 1975. Control of sebaceous gland function in the rat by $\propto$-melanocyte stimulating hormone. J. Endocr. 64: 503-510.

Wales, N.A.M. and Ebling, F.J. 1971. The control of apocrine glands of the rabbit by steroid hormones. J. Endocr. 51: 763-771.

Wilson, J.D. and Walker, J.D. 1969. The conversion of testosterone to $5\propto$-androstan-17β-ol-3-one (dihydrotestosterone) by skin slices of man. J. clin. Invest. 48: 371-379.

Woodbury, L.P., Lorincz, A. L. and Ortega, P. 1965. Studies on pituitary sebotropic activity. II. Further purification of a pituitary preparation with sebotropic activity. J. Invest. Derm. 45: 364-367.

BACTERIA AS A SOURCE OF CHEMICAL SIGNALS IN MAMMALS

Eric S. Albone[1], Pauline E. Gosden[2], and Georges C. Ware[2]

[1]Department of Animal Husbandry, University of Bristol, Landford, Bristol BS18 7DU, UK

[2]Department of Bacteriology, University of Bristol, Bristol BS8 1TD, UK

Although it has long been known that microorganisms are, at least partially, responsible for odor, such as axillary odor (Shelley et al., 1953) and halitosis, or breath odor (Tonzetich et al., 1967, 1971), in man, only recently has it been suggested that microbially-derived odors might assume a chemical communicatory significance in mammals. This presents new opportunities for bacteriologists to contribute to multidisciplinary studies in mammalian chemical communication.

Odor production is most likely to result from incomplete substrate oxidation associated with anaerobic processes, so that anatomical structures, such as cavities and pouches, which restrict the access of oxygen and in which at least local regions of anaerobiosis may be established by facultative organisms first eliminating the residual oxygen, warrant particular attention. Examples of anaerobic odor production are numerous and include the formation of volatile fatty acids, of amines and of mercaptans by Clostridia (Brooks and Moore, 1969; Anema et al., 1973; Doelle, 1975) and other organisms.

Cholesterol is known to yield cholestanols in anaerobic
environments, but evidence also suggests that estrogen may
be obtained from cholesterol microbially under certain conditions (Goddard and Hill, 1974). Much additional information concerning chemical transformations occurring in compounds in anaerobic microbial environments has emerged from
studies on the organic geochemistry of recent sediments
(Eglinton, 1975).

An examination of the anatomical literature (Schaffer,
1940; Ewer, 1973) reveals that mammals possess a variety of
sites appropriate to microbial odor production, although in
few cases have bacteriological studies been undertaken.
Thus, the sheep, Ovis aries, in common with certain other
ungulates, possesses infraorbital pouches (invaginations
some 1.5cm deep rostral to the medial angle of the eye),
inguinal pouches and, on each foot, an interdigital pouch,
a reservoir about 3cm long and 0.7cm diameter (Sisson, 1975).
All accumulate fatty secretions, partially if not entirely,
as the result of the activity of sebaceous and sudoriferous
glands lining these structures. Similar infraorbital and
interdigital glands have been described in the black-tailed
deer, Odocoileus hemionus columbianus (Quay and Müller-
Schwarze, 1970). In the pig, Sus scrofa, a substantial contribution to boar odor arises from the preputial diverticulum,
a large (about 150ml) pouch, glandless except for sebaceous
glands around its opening, in which stagnate decomposing
urine and epithelial cells (Sisson, 1975). Volatile fatty
acids, phenols and aromatic acids such as phenylacetic acid
are among the odorous components identified in boar preputial
fluid (Patterson, 1967, 1968). Similarly, carnivores possess a variety of scent-producing organs comprising sac or
pouch structures. These include the anal sacs of the canids and felids and the perfume pockets of such viverrids as
the civets and genets.

Other sites of bacteriological interest include the
vagina (Hurley et al., 1974) and the prepuce. Much research
has been conducted on the microbiology of human skin
(Marples, 1974; Noble and Somerville, 1974; Woodroffe and
Shaw, 1974). Here, anaerobes such as Propionibacterium
acnes inhabiting the sebaceous follicular canals have been

implicated in the formation of free fatty acids from sebum triglycerides (Puhvel et al., 1975).

Among the relatively few sites which have received bacteriological study in relation to odor production are the anal pocket of the Indian mongoose, Herpestes auropunctatus (Gorman et al., 1974) and the anal sacs of the red fox, Vulpes vulpes, and of the lion, Panthera leo (Albone et al., 1974; Gosden and Ware, 1976). In all cases, odorous secretions were found to be rich in volatile fatty acids. In addition, ammonia, putrescine and cadaverine are major contributors to the anal sac secretions of both the lion and the red fox (Albone and Perry, 1976). All these compounds are known to be (or are suspected of being) microbial products, derived in the fox from the apocrine secretions and desquamated cells which constitute the inputs to the sacs from the sac walls. In the lion, a substantial sebum input is also expected (Albone and Gronneberg, 1976).*

The red fox anal sac supports a restricted and relatively invariant microflora. Proteus, Streptococci and Clostridia are commonly present in large numbers, while faecal coliforms and Staphylococci, common skin species, are excluded. Although the precise exclusion mechanisms remain to be studied, it is interesting that 5-aminovaleric acid, a major component of fox anal sac secretion for which a microbial origin has also been suggested, is reported to exhibit antistaphylococcal activity (Albone et al., 1976).

Thus, all inputs to the sac from the sac wall are exposed to the activity of a simple, structured microbial eco-

---

*Although no bacteriological studies have been reported on the skunk, it would be most remarkable if microorganisms did not contribute to anal sac odor production in this species also.

system which substantially influences the odor of the secretion finally voided by the fox. Although with such a
scent source species differences can occur, as between the
anal sac secretions of the red fox and the striped skunk,
Mephitis mephitis (Andersen and Bernstein, 1975), microbial
odor production offers considerable possibilities for cross-
species similarities, as between the fox and the lion.

The Indian mongoose can discriminate between different
mixtures of the same volatile fatty acids that occur microbially in its anal pocket secretion (Gorman, 1976). Presumably many, if not all, mammals can detect and distinguish
between differences in such odors. The functions that such
scents, which are the direct product of commensal microorganisms rather than of the mammal itself, can assume are
necessarily dependent on the properties of the microbial
systems involved. Properties of microbial systems in this
context are largely unresearched.

Functions may be classified into three groups.
a) Functions for which limited variations in fermentation product composition with time are relatively unimportant.
Anaerobic processes frequently generate (subjectively) repulsive odorous products and if such a product were utilized
as a defense secretion, its efficacy would be unlikely to
depend sensitively on minor variations in composition.

b) Functions for which time modulation of fermentation product composition is required, as in signalling
the occurrence of a particular physiological state in a
mammal. The mammal is able to control the fermenting system by varying substrate composition and availability,
and providing that microorganisms are present which have
adapted to utilize the new substrate, the response in
terms of signal production will be rapid.

Thus, glycogen is deposited in the human vaginal epithelium under the influence of increased estrogen levels, so
that microbial substrate availability is controlled in relation to mammalian physiological state. Lactic acid is
formed and vaginal pH is lowered. Volatile fatty acids may

also be formed. It is claimed that volatile fatty acids
microbially formed in the vaginal secretion of the rhesus
monkey, Macaca mulatta, reflecting the hormonal status of
the female, act as a male sex attractant (Michael et al.,
1972). It is implied that such volatile fatty acids have
a similar function in other primates and man. Several
species of Lactobacillus with Staphylococcus albus act
together in generating these acids (Michael et al, 1975).
Preti and Huggins (1975) have critically discussed these
findings in relation to man and have documented the occurrence of other classes of odorous compound in human vaginal secretion.

        Other factors which it might be available to the mam-
mal to vary in order to regulate microbial activity include, for example, the temperature, water input or the pH
of a particular fermenting site.

        c) Functions for which stability of fermentation pro-
duct composition with time is important. To be employed as
a basis for individual recognition, an animal's scent must
be sufficiently constant with time to become representative
of that animal. Gorman (1976) has suggested that the
Indian mongoose employs volatile fatty acids microbially
produced in its anal pocket secretion as a basis for individual recognition, different individuals being distinguished
by different acid profiles. A fermentation hypothesis of
chemical recognition has also been discussed for the red
fox (Albone and Perry, 1976).

        Stability of fermentation product composition with time
requires constance of input substrate composition. It
also requires a stable microflora. An established microflora
will be expected to be stable for it will have adapted for
the most efficient exploitation of the particular micro-
environment and introduced organisms will have little chance
of competing successfully. This is evidenced by the reported difficulties encountered in implanting bacteria in
the gut (Drasar and Hill, 1974) and our own problems in intro-
ducing coliforms into the anal sac of the fox. Finally it
requires a through-put of secretion so that conditions ap-

proximating to open continuous culture are established
(Herbert, 1961). Such systems are characteristically stable
and resistant to change once a true steady state has been
established. In the steady state, the growth rate of the
microbial population adapts to, and is proportional to, the
rate of input of nutrient and all intermediate and final
metabolites are present in constant concentrations. How far
these conditions apply in particular cases are matters for
future observation.

We thank Dr. P.F. Flood for useful discussions and the
Nuffield Foundation (ESA) and the Science Research Council
(PEG/GCW) for financial support.

REFERENCES

ALBONE, E.S., EGLINTON, G., WALKER, J.M., and WARE, C.G.
1974. The anal sac secretion of the red fox, Vulpes
vulpes; its chemistry and microbiology. A comparison
with the anal sac secretion of the lion, Panthera leo.
Life Sci. 14:387-400.

ALBONE, E.S., and GRONNEBERG, T.O. 1976. Lipids of the anal
sac secretions of the red fox, Vulpes vulpes, and of
the lion, Panthera leo. Submitted for publication.

ALBONE, E.S., and PERRY, G.C. 1976. Anal sac secretion of
the red fox, Vulpes vulpes; volatile fatty acids and
diamines. Implications for a fermentation hypothesis
of chemical recognition. J. Chem. Ecol. 2:101-111.

ALBONE, E.S., ROBINS, S.P. and PATEL, D. 1976. 5-Aminovaleric
acid, a major amino acid component of the anal sac secretion of the red fox, Vulpes vulpes. Comp. Biochem.
Physiol., 55B: 483-486.

ANDERSEN, K.K., and BERNSTEIN, D.T. 1975. Some chemical
constituents of the scent of the striped skunk, Mephitis
mephitis. J. Chem. Ecol., 1:493-499.

ANEMA, P.J., KOOIMAN, W.T., and GEERS, J.M. 1973. Volatile acid production by Clostridium sporogenes under controlled conditions. J. Appl. Bact., 36:683-687.

BROOKS, J.B., and MOORE, W.E.C. 1969. Gas chromatographic analysis of amines and other compounds produced by several species of Clostridium. Can. J. Microbiol., 15:1433-1447.

DOELLE, H.W. 1975. Bacterial metabolism. Academic Press, New York and London.

DRASAR, B.S., and HILL, M.J. 1974. Human intestinal flora, Academic Press, London and New York. 22-23.

EGLINTON, G. (Ed.). 1975. Environmental Chemistry, Vol. 1, Specialist Periodical Report, The Chemical Society, London.

EWER, R.F. 1973. The carnivores, Weidenfeld and Nicholson, London. 90-100.

GODDARD, P., and HILL, M.J. 1974. The in vivo metabolism of cholesterol by gut bacteria in the rat and guineapig. J. Steroid Biochem., 5:569-572.

GORMAN, M.L., NEDWELL, D.B., and SMITH, R.M. 1974. An analysis of the anal scent pockets of Herpestes auropunctatus. J. Zool. 172:389-399.

GORMAN, M.L. 1976. A mechanism for individual recognition by odour in Herpestes auropunctatus. Anim. Behav. 24:141-145.

GOSDEN, P.E., and WARE, G.C. 1976. The aerobic bacterial flora of the anal sac of the red fox. J. Appl. Bact. 41:271-275.

HERBERT, D. 1961. A theoretical analysis of continuous culture systems, in Continuous culture of microorganisms, Society of Chemical Industry, London, 21-53.

HURLEY, R., STANLEY, V.C., LEASK, B.G.S. and DE LOUVOIS, J. 1974. Microflora of the vagina during pregnancy, in The normal microbial flora of man (F.A. Skinner and J.G. Carr, eds.) Acedemic Press, New York and London, 155-185.

MARPLES, M.J.   1974.   The normal microbial flora of the skin, in The normal microbial flora of man (F.A. Skinner and J.G. Carr, eds.) Academic Press, New York and London, 7-12.

MICHAEL, R.P., ZUMPE, D., KEVERNE, E.B., and BONSALL, R.W. 1972. Neuroendocrine factors in the control of primate behavior. Recent Progr. Horm. Res., 28:665-706.

MICHAEL, R.P., BONSALL, R.W., and WARNER, P.   1975.   Primate sexual pheromones, in Olfaction and Taste, Vol. 5, (D.A. Denton and J.P. Coghlan, eds.) Academic Press, New York and London, 417-424, 433-434.

NOBLE, W.C., and SOMERVILLE, D.A.   1974.   Microbiology of human skin, W.B. Saunders Co., London, Philadelphia, Toronto.

PATTERSON, R.L.S.   1967.   A possible contribution of phenolic components to boar odour. J. Sci. Fd. Ag., 18:8-10.

PATTERSON, R.L.S.   1968.   Acidic components of boar preputial fluid. J. Sci. Fd. Ag., 19:38-40.

PRETI, G., and HUGGINS, G.R.   1975.   Cyclical changes in volatile acidic metabolites of human vaginal secretions and their relation to ovulation. J. Chem. Ecol., 1:361-376.

PUHVEL, S.M., REISNER, R.M., and SAKAMOTO, M.   1975.   Analysis of lipid composition of isolated human sebaceous gland homogenates after incubation with cutaneous bacteria. Thin layer chromatography. J. Invest. Derm., 64: 406-411.

QUAY, W.B., and MÜLLER-SCHWARZE, D.   1970.   Functional histology of integumentary glandular regions in black-tailed deer, Odocoileus hemionus columbianus. J. Mammal., 51:675-694.

SCHAFFER, J.   1940.   Die Hautdrusenorgane der Saugetiere, Urban and Schwartzenberg, Berlin and Vienna.

SHELLEY, W.B., HURLEY, H.J., and NICHOLS, A.C.   1953.   Axillary odor. Arch. Derm. Syph., 68:430-446.

SISSON, S. 1975. Sisson and Grossman's The Anatomy of the Domestic Animals. (R. Getty, ed.) 5th edn., W.B. Saunders Co., London, Philadelphia and Toronto, p1210, 1301.

TONZETICH, J., EIGEN, E., KING, W.J., and WEISS, S. 1967. Volatility as a factor in the ability of certain amines and indole to increase the odour of saliva. Archs. Oral Biol., 12:1167-1175.

TONZETICH, J., and CARPENTER, P.A.W. 1971. Production of volatile sulphur compounds from cysteine, cystine, and methionine by human dental plaque. Archs. Oral Biol., 16:599-607.

WOODROFFE, R.C.S., and SHAW, D.A. 1974. Natural control and ecology of microbial populations of skin and hair, in The normal microbial flora of man (F.A. Skinner and J.G. Carr, eds.) Academic Press, New York and London 13-34.

CHEMICAL ATTRACTANTS OF THE RAT PREPUTIAL GLAND

Anthony M. Gawienowski

Department of Biochemistry
University of Massachusetts
Amherst, Massachusetts

The interest in olfaction and the relationship to animal behavior has brought attention to the preputial and other glands as
a source of chemical attractants. The molecules, synthesized by
one animal, affect the physiology and behavior of another of the
same species.

Many studies relating pheromones to endocrine changes have
been performed on laboratory mice and usually involved urine as
the attractant source (Whitten, 1969). Some studies gave clear
indication that the preputial gland secretions was responsible for
the behavioral responses (Stanley and Powell, 1941; Bronson and
Caroom, 1971).

Rats have been reported to recognize the odors of stressed
and non-stressed rats (Valenta and Rigby, 1968; Stevens and Koster,
(1972) also to differentiate the sex of the animal. Sexually
experienced male rats prefer the odor of receptive over non-receptive females. Both sexually experienced and virgin females prefer
the odor of intact males over that of castrate males, while non-
receptive females had no preference (Carr et al., 1966; Carr et al.
1970, Orsulak and Gawienowski, 1972). Additional literature on
this subject was covered by Mykytowycz (1974), Brown and Williams
(1972), and Bronson (1974).

The bioassay apparatus used in our studies was similar to
that described by Orsulak and Gawienowski (1972). A square
chamber (61 x 61 x 35.5 cm) was made of galvanized metal, painted
black with removable tunnels on two sides. An earlier model had
four tunnels. Each tunnel was equipped with a photocell sample

holder, air inlet tubes connected to a fine control valve, a two
stage regulator and a compressed air tank. Time and event indicators were wired to an interval clock which started and stopped
them simultaneously.

The galvanized wire grid floor of the test chamber was routinely washed and the sawdust underneath replaced after each
group of animals had been tested. Each rat was tested 5 min for
olfactory preference. Some of the earlier experiments involved
3 to 10 min periods. Cumulative investigating time and frequency
of each tunnel was recorded for the test rat. The data were analyzed by the chi square test, two-way analysis of variance and
Tukey's procedure for comparing several means (Steel and Torrie,
1960).

The animals were individually caged and isolated in a small
well ventilated room with a reverse light-dark cycle (12/12).
Tests were usually made during the beginning of the dark cycle.
Animals of the opposite sex in separate cages were placed into
the bioassay room over the weekend in order to keep conditions
reasonably normal as to odors.

In one of the preliminary experiments the olfactory preference of male and female rats was assayed for preputial glands,
submaxillary-sublingual glands, and foot pads taken from the
opposite sex. The homogenized samples, 50 mg tissue/ml, were
applied to the sample wick in the test tunnels. A control homogenate of muscle, fat or liver was placed in the second tunnel. The
male preputial odor caused a significant increase in investigating
time, Figure 1, of the female rat over the control. The odor of
the female preputial glands brought on a significant increase in
investigating time of the male versus the control, Figure 2
(Orsulak and Gawienowski, 1972).

The endocrine relationship and attraction of the female rat
preputial gland are illustrated in Table 1. When ether extracts
of preputial glands from receptive and nonreceptive females were
presented to the male rats, they preferred the former. In order
to investigate the chemical nature of the attractant, male preputial glands and an equal weight of muscle-fat control were
homogenized separately in water and acidified. The samples were
extracted with ether (extract A). A portion of extract A was
washed with NaOH and designated extract B. The alkaline solution
was acidified and extracted with ether (extract C). The results
which indicate the preference of female rats toward neutral
extract B are presented in Table 2. These findings were followed
by gas chromatography to separate the volatile male preputial

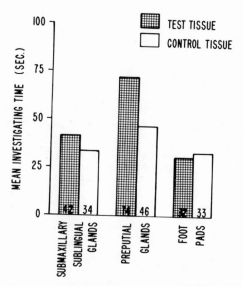

Fig. 1. Investigation time for female rats presented saline homogenates of various odor sources from male rats.

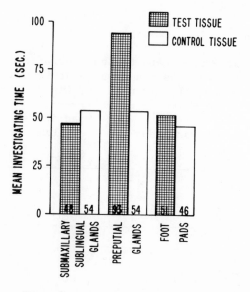

Fig. 2. Investigation time for male rats presented saline homogenates of various odor sources from female rats.

Table 1. Male Rat Response to the Odor of Ether Extract of Preputial Glands From Receptive and Non-Receptive Female Rats

Source of preputial gland	Number of rats preferring odor	Mean investigating time (sec)
Receptive females	11	1.5
Non-receptive females	3	0.8
p	< 0.05*	< 0.01†

NS, Not significant.
* Probability was determined from χ^2 value.
† Probability was determined using using the F value obtained
  from the two-way analysis of variance.

Table 2. Response of Female Rats to the Odor of Ether Extracts
(A) from Male Preputial Glands and Muscle-Fat Control Tissue,
Fractionated into Neutral (B) and Acid (C) Lipids

Source of odor	No. of rats preferring odor			Mean investigating time (sec)		
	Sample	Control	P*	Sample	Control	P†
Extract A	8	2	< 0.10	9.3	5.4	< 0.05
Extract B	9	1	< 0.025	6.4	3.6	< 0.01
Extract C	6	4	NS	3.8	3.0	NS

*, † See Table 1

lipids into three fractions, Table 3. The female rats preferred fraction 3. A gas chromatogram of the fractions is given in Figure 3, which together with Tables 2 and 3 were reported by Gawienowski et al.,(1975).

A significant preference was demonstrated by male rats for the total lipid extract of the female preputial gland and the neutral B fraction, Table 4. The free fatty acids are apparently not attractants since they would be in fraction C.

Volatile and nonvolatile portions of the female rat preputial gland and muscle-fat control were presented to male rats, as noted in Table 5. A significant preference was given to the volatile fraction. Tables 1, 4, 5, 6 and Figure 4 were from an article by Gawienowski et al.,(1976).

Gas liquid chromatography analysis of the volatile preputial gland lipids indicated a number of compounds, Figure 4. In order to separate the components, four fractions were trapped and assayed for attractant activity, Table 6. The male rats preferred the second fraction. Up to this point the evidence indicated that the rat preputial gland attractants probably are ether soluble neutral compounds. Therefore, it was important to review the neutral lipid compounds found in rodent preputial glands.

In 1969, Spener et al. reported that lipid extracts of male mice contain less than 1% alkyl acetates. Size, lipid content and relative amounts of alkyl acetates in female mice preputial glands were increased by testosterone injections. The composition of the alkyl acetates is given in Table 7.

Sansone-Bazzano et al. (1972) indicated that mouse preputial gland lipids varied in time of appearance and amounts of the different lipid classes. These could be correlated with the age and sex of the mice. Injections of testosterone propionate to female mice, immature male mice or to mature castrated male mice caused marked synthesis and accumulation of lipids containing fatty alcohol moieties, Table 8.

Testosterone treatment increased the incorporation of carbon-14 labeled glucose into all lipid fractions of the preputial glands, Figure 5. The glands of castrated mice had virtually no incorporation of the radioisotope into alkyl acetates, alkyl-1-enyl diacyl glycerols, alkyl glyceryl diesters and little into other lipid components. One week after testosterone treatment, the incorporation of carbon-14 was increased approximately 10 fold into wax esters. The radioisotope was detected in the alkyl acetates, alkyl-1-enyl diacyl glycerols and alkyl glyceryl diesters fraction only after testosterone treatment and indicated de novo synthesis.

Table 3. Response of Female Rats to the Odor of the Volatile
G.L.C. Fractions of Male Preputial Gland Extract

Source of odor	No. of rats preferring odor				Mean investigating time (sec)		
	Sample	Control	No pre-ference	P*	Sample	Control	P†
Fraction 1	8	1	1	< 0.10	2 0		< 0.10
Fraction 2	6	2	2	NS	1.4	0.6	< 0.10
Fraction 3	9	0	1	< 0.025	3.3		< 0.01
Total extract	10	0	0	< 0.005	6.4		< 0.01

*, † See Table 1

Table 4. Male Rat Response to the Odor of Ether Extract of Female
Preputial Glands (A), Fractionated into Neutral (B) and Acidic (C)
Lipids

Extract	No. of rats preferring odor			Mean investigating time (sec)		
	Sample	Control	P*	Sample	Control	P†
Total lipids(A)	11	1	< 0.005	7.6	1.4	< 0.05
Neutral (B)	10	2	< 0.025	4.3	2.0	< 0.05
Acidic (C)	6	6	NS	2.4	2.0	NS

*, † See Table 1.

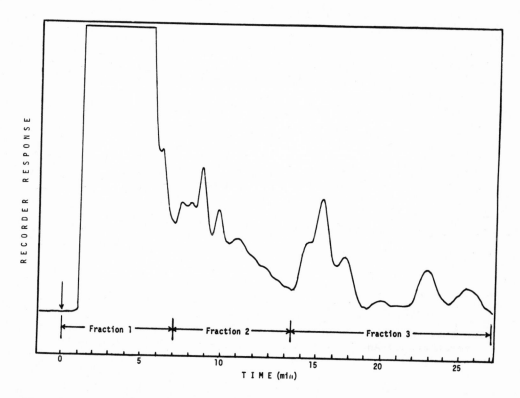

Fig. 3. Gas chromatogram of volatile preputial gland lipids from male rats. Column, 3·66 m x 3·2 mm of 15% Carbowax 1500 on Chromosorb W (HP); N_2 flow rate, 17 ml/min. Column temperature, 100°C; detector temperature, 155°C; injector temperature, 125°C. Three collected fractions are indicated under the chromatogram.

Table 5. Male Rat Response to the Odor of Volatile (A) and Non-
volatile (B) Fractions from Female Rat Preputial Gland and Muscle-
Fat Control

Sample source	No. of rats preferring odor			Mean investigating time (sec)		
	A	B	P*	A	B	P†
Female preputial glands 1	12	4	< 0.05	2.1	1.4	< 0.05
Female preputial glands 2	17	3	< 0.005	5.3	3.4	< 0.001
Female preputial glands 3	13	4	< 0.05	3.7	2.1	< 0.05
Muscle-fat control 1	5	12	< 0.10	5.2	6.3	NS
Muscle-fat control 2	8	9	NS	2.7	2.7	NS

*, † See Table 1.

Table 6. Male Rat Response to the Odor of Fractions from Gas
Chromatography Analysis of Volatile Lipids from Female Preputial
Gland

Fraction number (see Fig. 1)	Number of rats preferring odor	Mean investigating time (sec)
1	4	1.6 a ‡
2	15	2.0 a
3	1	1.0 b
4	6	1.6 a
p	< 0.005*	< 0.01†

*, † See Table 1.

‡ Means within column not followed by the same letter are different at 5% level according to Tukey's procedure.

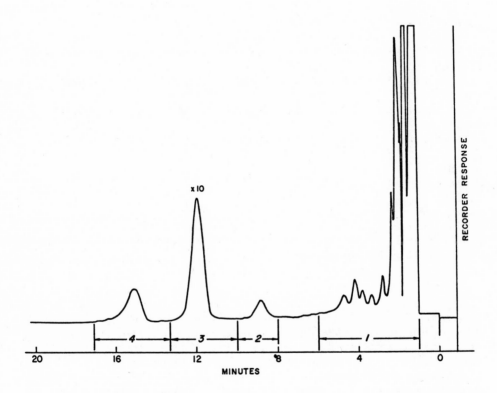

Fig. 4.   Gas chromatogram of volatile preputial gland lipids from female rats.   Column 6.10 m x 3.2 mm, 15% Carbowax 1500.   N$_2$, flow rate, 30 ml/min.   Column temperature 120°C, detector temperature 80°C, injector temperature 220°C.   Four collected fractions are indicated under the chromatogram.

Table 7. Composition of the Alkyl Acetates in the Preputial Glands of Mice

Chain length and number of double bonds of alkyl moieties	Male mice strain CD-1* (weight %)	Female mice strain CD-1** (weight %)	Male mice strain C-57** (weight %)
14:0	0.4	0.7	1.5
14:1	0.4	3.4	4.2
14:2			Trace
15:0	0.3	0.7	0.5
16:0 br	0.7	3.1	1.7
16:0	75.2	63.8	68.0
16:1	9.6	17.6***	15.6***
16:2			0.5
17:0 br	3.4		
17:0	2.8	2.8	1.8
18:0 br	0.5	0.1	0.1
18:0	4.8	5.9	4.4
18:1	1.9	1.6	1.7

*Analysis of pooled preputial glands from 50 mice
**Mean value of analyses of four single glands
***Including 17:0 branched.

(Spener et al., 1969)

Table 8. Effect of Sexual Maturation and Testosterone Administration on Lipid Constituents of Mouse Preputial Glands

Sex	Male	Male	Male	Male	Female	Treated* female
Age (in weeks)	3	4	5	10	8	8
Wax	43	58	153	243	50	240
Alk-1-enyl diacyl glycerols	31	30	58	87	48	120
Alkyl glyceryl diesters	34	35	65	110	53	158
Triglyceride**	100	100	100	100	100	100
Lipid to protein ratio	0.8	0.8	2.4	9.35	0.9	8.7

*Testosterone propionate 1 mg per day subcutaneously for 7 days.
**Triglyceride concentration set equal to 100. All other lipid concentrations expressed relative to triglyceride.

(Sansone-Bazzano et al., 1972)

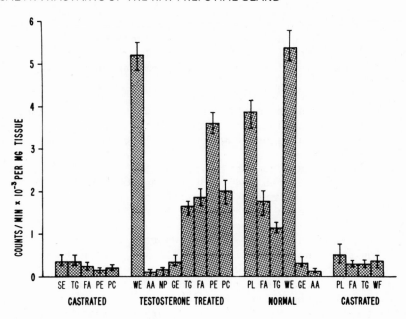

Fig. 5. Incorporation of $[6-^{14}C]$ glucose into lipids of preputial glands from male mice castrated at 2 weeks of age, testosterone-treated castrated mice, normal male mice and mice castrated at 6 weeks. Abbreviations: WE, predominantly wax; SE, predominantly sterol esters; TG, triglycerides; FA, fatty acids; PE, phosphatidyl ethanolamine fraction; PC, phosphatidyl choline fraction; AA, alkyl acetates; GE, alkyl glyceryl diesters; NP, neutral plasmalogens alk-1-enyl diacyl glycerols; PL, phospholipids.
                              (Sansone-Bazzano et al., 1972).

Table 9. Response of Male and Female Rats to the Odor of Alkyl Acetates

Acetates of the Alcohols	Time of Investigation ♂	♀	Frequency ♂	♀
Ethyl				
Propyl		+++		+++
Isopropyl	+			
Pentyl		+++		++
Decyl		+++		

Probabilities were determined using F values obtained from the two-way analysis of variance.
   + p < 0.1   ++ p < 0.05   +++ p < 0.01

In our laboratory vacuum distillation of ether extracts of male rat preputial glands resulted in a fraction containing among other compounds--ethyl, propyl, isopropyl, pentyl and decyl acetate. These were analyzed by a gas chromatography-mass spectrometer system.

A computer search for the analysis of the mass spectra, identified the alkyl acetates and a number of other compounds, Table 9, (Stacewicz-Sapuntzakis and Gawienowski, 1976).

We are presently completing a study of saturated and unsaturated aliphatic acetates and find that approximately one third of the compounds have an attractant effect on the female rat. The male rat was attracted to only a few of the acetates.

Sansone-Bazzano et al. (1972) reported that testosterone treatment stimulates the synthesis of wax, alkyl acetates and alkyl glycerols in the mouse preputial gland. They postulate on the following sequence of events: testosterone stimulation leads to increase pentose phosphate shunt activity which in turn increases the level of NADPH. NADPH is a necessary co-factor for fatty acyl precursors, according to Snyder et al. (1969). They believe this order of events may be the mechanism by which testosterone acts as the hormonal inducer of alkyl glycerols, wax and alkyl acetates in the preputial gland of mice.

According to Snyder and Blank (1969), wax esters, glyceryl ether diesters, triglycerides and sterol esters are the major lipid classes present in normal preputial glands of mice, Table 10. The chain lengths of the alkyl and alk-1-enyl hydrocarbon moieties of glyceryl ether diesters, phosphatidyl choline, and phosphatidyl ethanolamine were similar to one another and to the alcohol moieties of the waxes. They found the location of double bonds in the 16:1 and 14:1 acyl moieties of waxes and the alkyl moieties of glycerol ether diesters was similar.

Snyder and Blank (1969) suggested that in preputial glands of mice the intermediary pathways or enzymatic sites responsible for the biosynthesis of fatty alcohol in the waxes were identical to those pathways or sites responsible for the biosynthesis of alkyl chains of glyceryl ether diesters. They believed because of the similarities of alkyl and alk-1-enyl ether linked hydrocarbon moieties that these two classes of lipids may be interconverted enzymatically.

Singer et al. (1976) found that dimethyl disulfide occurs in female hamster vaginal secretion. The compound triggers sniffing and digging behavior in male hamsters. They suggest that such behavior is related to sexual attraction. We also found dimethyl

Table 10.   Lipid Class Composition of Preputial Glands in Mice[a]

Lipid class	% of Total lipids
Sterol esters	5
Waxes	48
Alk-1-enyl GEDE	8
Alkyl GEDE	6
Triglycerides	17
Free fatty acids	tr
Sterols + diglycerides	3
GEME[b]	1
Alk-1-enyl PE	2
Alkyl PE	tr
Diacyl PE	tr
Alk-1-enyl PC	tr about 1%
Alkyl PC	tr
Diacyl PC	6
Other phospholipids	3

[a] A duplicate analysis of a pooled sample (100 glands).

[b] Consisted of a ratio of 1.7 alk-1-enyl to alkyl glyceryl ethers, entirely 1-isomer.

Abbreviations: GEDE, glyceryl ether diesters; GEME, glyceryl ether monoesters.   Additional abbreviations are noted in Figure 5.

(Snyder and Blank, 1969)

disulfide extremely attractive to the male rat. Our search for sex attractants in the preputial gland extract revealed another sulfur compound, dimethyl sulfite, which was attractive only to male rats (Gawienowski and Stacewicz-Sapuntzakis, 1976).

It has been demonstrated that the lipid fraction of the pre-putial gland contain attractants for the opposite sex. Our research also revealed the attraction of females to alkyl acetates which were identified in the preputial gland. Investigation of the hormonal mechanism for sex attractant production by the pre-putial gland should aid in the future investigations of the neuro-biochemical processes involving olfaction.

Acknowledgements—Supported in part by a U. S. Army Natick Development Center research contract DAAG 17-72-C-0025 and a U. S. Army Research Office research grant DAHC 04-75-C-0073.

REFERENCES

Bronson, F. H., and Caroom, D. 1971. Preputial gland of the male
    mouse: attractant function. J. Reprod. Fertil. 279-282.

Bronson, F. H. 1974. Pheromonal influences on reproductive activi-
    ties, pp. 344-365 in Birch, M. C. (ed.), Pheromones. Elsevier,
    New York.

Brown, J. C., and Williams, J. D. 1972. The rodent preputial
    gland. Mammal Review 2:105-147.

Carr, W. J., Loeb, L. S., and Wylie, N. R. 1966. Responses to
    feminine odors in normal and castrated male rats. J. Comp.
    Physiol. Psychol. 62:336-338.

Carr, W. J., Krames, L., and Costanzo, D. J. 1970. Previous
    sexual experience and olfactory preference for novel versus
    original sex partners in rats. J. Comp. Physiol. Psychol.
    71:216-222.

Gawienowski, A. M., Orsulak, P. J., Stacewicz-Sapuntzakis, M., and
    Joseph, B. M. 1975. Presence of sex pheromone in preputial
    glands of male rats. J. Endocrinol. 67:283-288.

Gawienowski, A. M., Orsulak, P. J., Stacewicz-Sapuntzakis, M., and
    Pratt, J. Jr. 1976. Attractant effect of female preputial gland
    extracts on the male rat. Psychoneuroendocrinol. 1:411-418.

Gawienowski, A. M., and Stacewicz-Sapuntzakis, M. 1976b. Attraction
    of male rats to dimethyl disulfide and dimethyl sulfite, sub-
    mitted for publication.

Mykytowycz, R. 1974. Odor in the spacing behavior of mammals,
    pp. 327-343 in Birch, M. C. (ed.). Pheromones. Elsevier,
    New York.

Orsulak, P. J., and Gawienowski, A. M. 1972. Olfactory preferences
    for the rat preputial gland. Biol. Reproduction 6:219:223.

Sansone-Bazzano, G., Bazzano, G., Reisner, R. M., and Hamilton,
    J. G. 1972. The hormonal induction of alkyl glycerol, wax and
    alkyl acetate synthesis in the preputial gland of the mouse.
    Biochim. Biophys. Acta 260:35-40.

Singer, A. G., Agosta, W. C., O'Connell, R. J., Pfaffman, C.,
    Bowen, D. V., and Field, F. H. 1976. Dimethyl disulfide: an
    attractant pheromone in hamster vaginal secretion. Science
    191:948-950.

Snyder, F., Wykle, R. L., and Malone, B. 1969. A new metabolic
    pathway: biosynthesis of alkyl ether bonds from glyceralde-
    hyde-3-phosphate and fatty alcohols by microsomal enzymes.
    Biochem. Biophys. Res. Comm. 34:315-321.

Snyder, F., and Blank, M. L. 1969. Relationships of chain lengths
    and double bond locations in O-alkyl, O-alk-1-enyl, acyl, and
    fatty alcohol moieties in preputial glands of mice. Arch.
    Biochem. Biophys. 130:101-110.

Spener, F., Mangold, H. K., Sansone, G., and Hamilton, J. G. 1969.
    Long-chain alkyl acetates in the preputial gland of the mouse.
    Biochim. Biophys. Acta 192:516-521.

Stacewicz-Sapuntzakis, M., and Gawienowski, A. M. 1977. Rat ol-
    factory response to aliphatic acetates, J. Chem. Ecol. (in
    press).
Stanley, A. J., and Powell, R. A. 1941. Studies on the preputial
    gland of the white rat. Proc. Louisiana Acad. Sci. 5:28-29.
Steel, R. G. P., and Torrie, H. H. 1960. "Principles and Proce-
    dures of Statistics." McGraw Hill, New York.
Stevens, D. A., and Koster, E. P. 1972. Open-field responses of
    rats to odors from stressed and nonstressed predecessors.
    Behavioral Biol. 7:519-525.
Valenta, J. G., and Rigby, M. K. 1968. Discrimination of the odor
    of stressed rats. Science 161:599-601.
Whitten, W. K. 1966. Pheromones and Mammalian Reproduction.
    Advances in Reproductive Physiology, Vol. 1, pp. 155-177, A.
    McLaren (ed.). Academic Press, New York.

PROPERTIES OF COMPOUNDS USED AS CHEMICAL SIGNALS

James W. Wheeler

Howard University

Washington, D.C. 20059

The most important property of a compound used as a chemical signal is an appropriate volatility. This limits its upper molecular weight to ~300 and the effective lower molecular weight appears to be ~50 (Wilson, 1963; Wilson et al., 1969). Many different functional groups have been observed in pheromones and defensive materials from nonpolar molecules such as alkanes and alkenes to very polar compounds which may be acidic (acids, phenols) or basic (amines) (Weatherston, 1967; Law and Regnier, 1971).

As more mammalian systems are investigated, there develops some overlap between the chemicals from secretions in arthropods and those in mammals. The recent report (Singer, et al., 1976) of dimethyl disulfide (1) from vaginal secretions of hamsters follows a relatively old report of this same compound (and dimethyl trisulfide) in ponerine ants (Casnati et al., 1967). Another sulfur compound which is secreted from the anal scent gland of the striped hyena 2 (Wheeler et al., 1975) finds a parallel in a thioester 3 isolated from human urine after eating asparagus (White, 1975). The skunk, which had been reported to use n-butyl mercaptan and dicrotyl sulfide for defensive purposes, (Stevens, 1945) has been shown to use crotyl mercaptan (4), isopentyl mercaptan (5) and methyl crotyl sulfide (6) (Anderson and Bernstein, 1975). The thiomethyl group appears to be easily synthesized in these insect and mammalian systems.

Acids and phenols are widely distributed in mammalian and arthropod systems. Low molecular weight acids have been identified in the anal gland secretion of the red fox (Albone and Fox, 1971) as well as rhesus monkey (Michael et al., 1971) and human (Michael et al., 1974) vaginal secretions. The subauricular scent of the male

Sulfur Compounds

CH_3SSCH_3 $CH_3SCH_2CH_2COCOCH_3$ $CH_3SCH_2CH_2COSCH_3$

1   2   3

$CH_3CH=CHCH_2SH$ $(CH_3)_2CHCH_2CH_2SH$ $CH_3CH=CHCH_2SCH_3$

4   5   6

pronghorn contains 2-methylbutanoic acid (7), 3-methylbutanoic
acid (8), and two different methyl tetradecanols (9 and 10) in addition to mixed esters of these acids and alcohols (Müller-Schwarze
et al., 1974). The isovaleric acid (8) has been identified as the
active principle and the ester may be a method of storing this
acid until needed. Phenylacetic acid (11) has been identified as
the ventral scent marking pheromone of male Mongolian gerbils
(Thiessen et al., 1974). Another phenol, 2-(p-hydroxyphenyl)
ethanol (12) has been isolated from the chest gland of a bushbaby
(Galago) (Wheeler et al., 1976a). Low molecular weight acids such
as these are common defensive chemicals and attractants in insects
(Schildknecht, 1970; Jacobson et al., 1968) while phenols have
been identified as insect pheromones in ticks and beetles (Berger,
1972; Henzell and Lowe, 1970).

Acids, Phenols and Alcohols

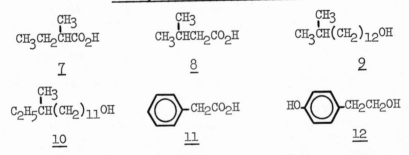

Other oxygen-containing compounds which are important contain
a lactone functional group. One of the components of the tarsal
scent of the black-tailed deer is the γ-lactone 13 (Brownlee et al.,
1969). Similar lactones (14 and 15) have been identified in the
defensive secretions of staphylinid beetles (Wheeler et al., 1972)
and Australian ants (Cavill et al., 1968). A δ-lactone 16 is
used by the Oriental hornet as a pheromone (Ikan et al., 1969),
while other δ-lactones are found in Camponotus ants (17), (Brand et
al., 1973) Xylocopa bees (18) (Wheeler et al., 1976b) and various
species of ants (19) (Cavill, 1960). Macrocyclic lactones have been
isolated from solitary bees (20,21,22) (Andersson et al., 1967;
Bergstrom, 1974) and the anal gland secretion of the striped hyena
(23) (Wheeler et al., 1976d), the latter four being analogous to

ambrettolide (24) and exaltolide (25) which are found in plants
(Kershbaum, 1927). All of these lactones contain an even number of
carbon atoms ranging between C_{10} and C_{20} (with the exception of 18
and 25). Although 19 is terpenoid, most of the others appear to be
derived from straight chain fatty acids.

Lactones

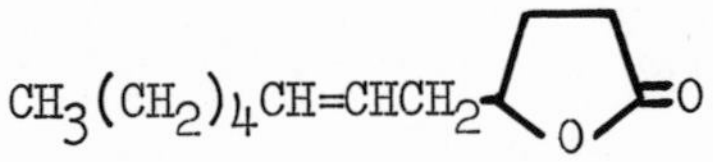

$CH_3(CH_2)_4CH=CHCH_2$ [structure] O=O

13

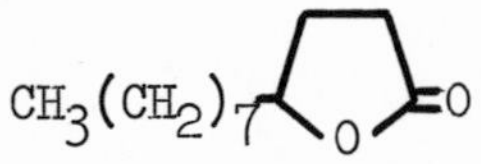

$CH_3(CH_2)_7$ [structure] O=O

14

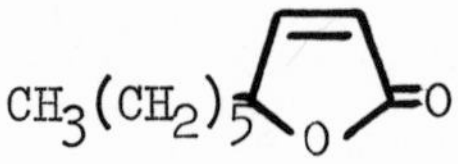

$CH_3(CH_2)_5$ [structure] O=O

15

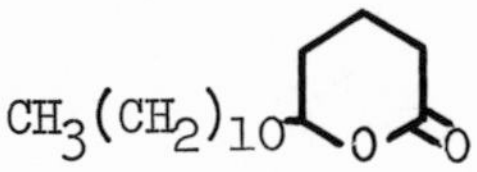

$CH_3(CH_2)_{10}$ [structure] O O

16

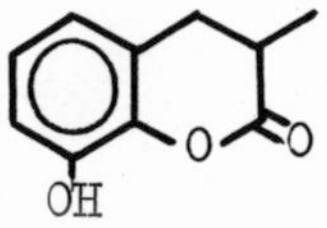

[structure with OH] O=O

17

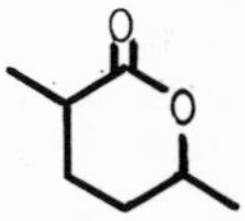

[structure]

18

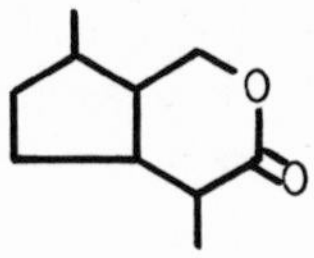

[structure]

19

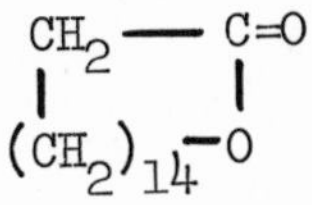

CH_2 — $C=O$
$(CH_2)_{14}$ —O

20

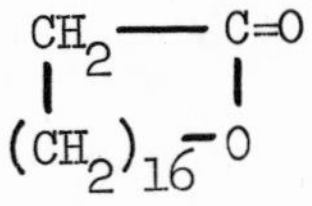

CH_2 — $C=O$
$(CH_2)_{16}$ —O

21

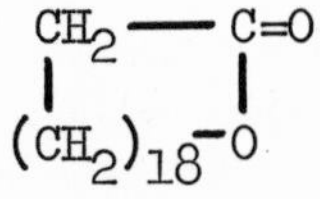

CH_2 — $C=O$
$(CH_2)_{18}$ —O

22

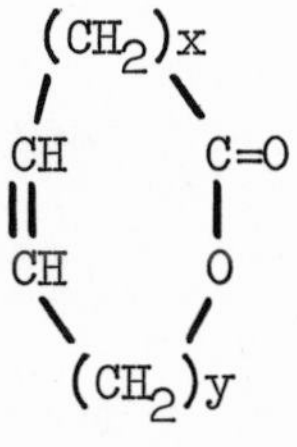

$(CH_2)x$
CH $C=O$
CH O
$(CH_2)y$

23   x+y=11

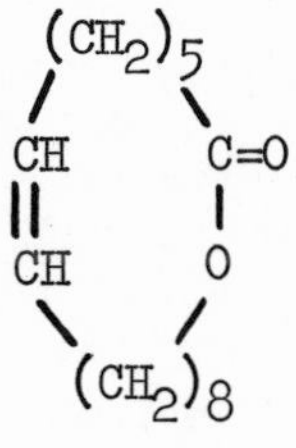

$(CH_2)_5$
CH $C=O$
CH O
$(CH_2)_8$

24

CH_2 — $C=O$
$(CH_2)_{13}$ —O

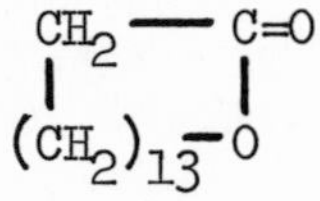

25

Reinvestigation of the scent of the musk rat (Stevens and Erickson, 1942) has provided the first cyclic alkynes, 5-cyclohepta-decynone (26) and 7-cis-cycloheptadecen-5-ynone (27) (Van Dorp et al., 1973), in addition to cycloheptadecanone (28) and 5-cis-(29) and 7-cis-heptadecenone (30) analogous to the original civetone (31) found in civet cats (Ruzicka, 1926a). Two bisnor-homologs cyclo-pentadecanone (32) and 5-cis-cyclopentadecenone (33) were also isolated. Ruzicka had also isolated muscone (34) from musk deer in 1926 (Ruzicka, 1926b).

Cyclic Ketones

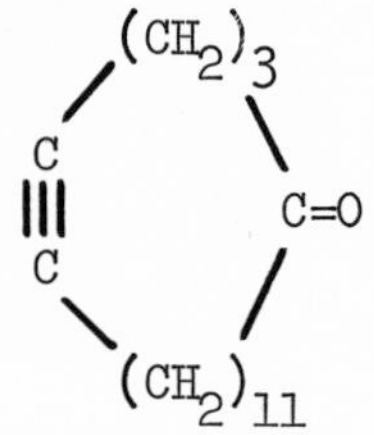

26

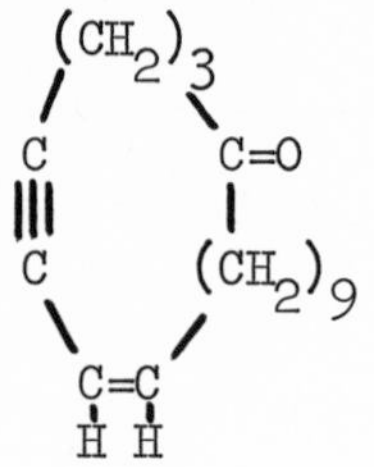

27

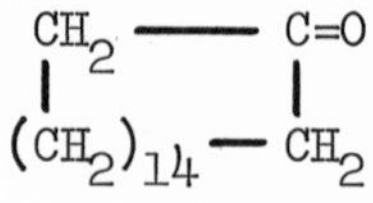

28

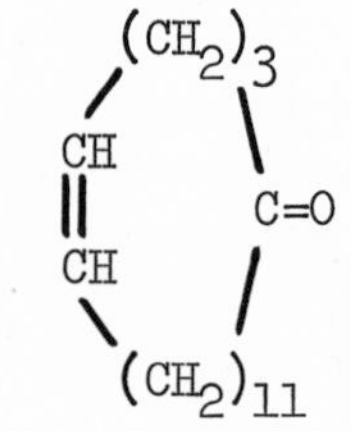

29

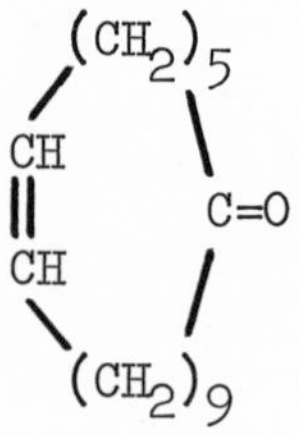

30

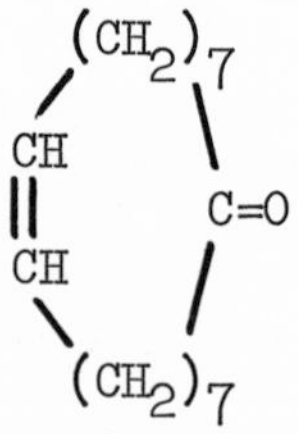

31

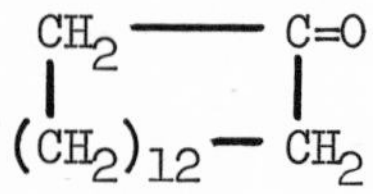

32

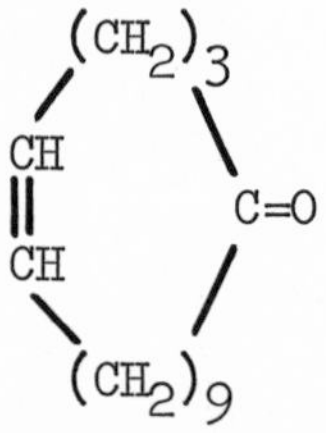

33

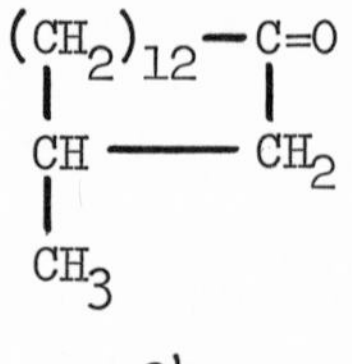

34

Although these compounds, as the macrocyclic lactones mentioned previously, appear to be derived from long chain fatty acids, the muskrat uses ketones containing an odd number of carbon atoms. The double bonds, on the other hand, fall at positions 5,7, and 9 as in fatty acids. Muscone contains an additional methyl group and an even number of carbons (16).

Other molecules which appear to be derived from fatty acids have an important function as sex attractants in moths and butter-flies. All of these contain an even number of carbon atoms from C_{10} to C_{18} , most with twelve or fourteen carbons and contain at least one double bond (Evans and Green, 1973). Practically no alcohols of this type have been isolated outside lepidoptera.

Although their biological function is unknown, two high molec-ular weight amines have been isolated from scent glands of mammals. Castoramine (35) is found in the Canadian beaver (Valenta and Khaleque, 1959), and muscopyridine (36) is found in the musk deer Biemann et al., 1957).

<u>Amines</u>

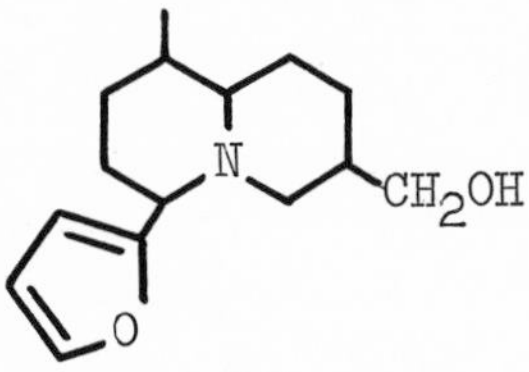

35

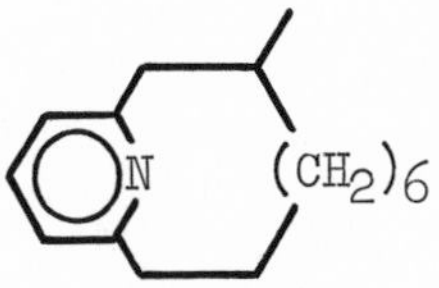

36

Nitrogen-containing compounds are becoming fairly commonplace in insects: piperidines in fire ants (37) (MacConnell et al., 1970), pyrazines in ponerine ants (38) (Wheeler and Blum, 1973), indolizines (39) in Pharaoh's ants (Ritter et al., 1973), a pyrrole in leaf-cutting ants (40) (Tumlinson et al., 1971), and an alkaloid 41 in the common lady beetle (Tursch et al., 1971).

<u>Nitrogen-Containing Compounds from Insects</u>

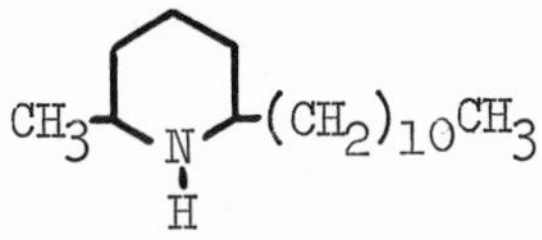

37

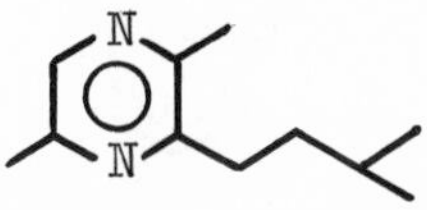

38

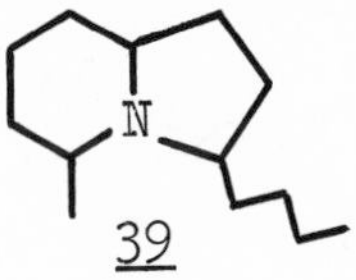

39

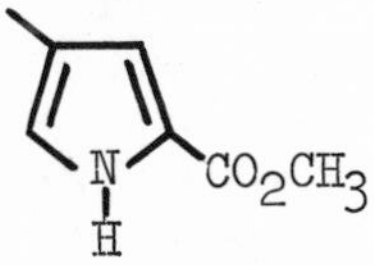

40

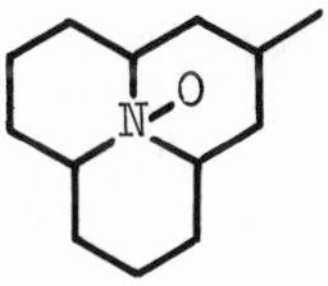

41

In many of the molecules used as chemical signals, at least
one chiral center is present. Although little is known about the
specific nature of the enantiomer in many cases, it is well known
that humans can distinguish between some enantiomeric isomers by
smell (Russell and Hills, 1971; Friedman and Miller, 1971). Very
recent work has shown that both enantiomers of 6-methyl-5-hepten-
2-ol need to be present for maximum response in the ambrosia beetle
(Borden, et al., 1976) while only one enantiomer is preferred in
other cases (Wood, et al., 1976; Riley, et al., 1974).

Terpenes and terpene derived molecules also play an important
part in this scheme in insects (though not in mammals) from the common 6-methyl-5-hepten-2-one (42) in various species of ants (Blum
and Brand, 1972) to some of the more complex cyclopentanoid monoterpenes found in ants (43, 44, 19) (Cavill, 1960) and in walking
sticks (44) (Meinwald et al., 1966). A similar compound, nepeta-
lactone (45) from catnip is an attractant for cats (Waller et al.,
1969). A related alkaloid actinidine (46) has recently been found
in ants (Wheeler et al., 1976c) as well as in staphylinid beetles
(Bellas et al., 1974). Another nitrogen-containing terpene polyzonimine (47) has been isolated from a millipede (Smolanoff et al.,
1975).

Examples of Terpenes and Terpene Derived Molecules

Even a group commonly used by chemists for protecting purposes
is present in some of the signal producers. The bark beetle attractants brevicomin (47) (Silverstein et al., 1968) and frontalin (48)
(Kinzer et al., 1969; Kinzer et al., 1971) are both ketals formed
by the intramolecular reaction of two alcohol groups with a ketone.

Although diverse functional groups are represented in the
molecules known to be signal producers, almost all have fairly
polar functional groups in the molecule. There are exceptions,

however, such as the sex attractant of the common housefly which
has been identified as <u>cis</u>-9-tricosene (Carlson et al., 1971).
Much more diversity will undoubtedly be found as new species of
insects and mammals are studied.

47   48

REFERENCES

ALBONE, E.S. and M.W. FOX. 1971. Anal Gland Secretion of the Red
    Fox. <u>Nature</u> (<u>London</u>) 233:569.
ANDERSON, K.K. and D.T. BERNSTEIN. 1975. Some Chemical Constituents
    of the Scent of the Striped Skunk (<u>Mephitis mephitis</u>). <u>J</u>. <u>Chem</u>.
    Ecol. 1:493.
ANDERSSON, C.-O., G. BERGSTROM, B. KULLENBERG and S. STALLBERG-
    STENHAGEN. 1967. Studies on Natural Odouriferous Compounds I.
    Identification of Macrocyclic Lactones as Odouriferous Compon-
    ents of the Scent of the Solitary Bees <u>Halictus</u> <u>calceatus</u> and
    <u>H. albipes</u> <u>Ark</u>. <u>Kemi</u>. 26:191.
BELLAS, T.E., W.V. BROWN and B.P. MOORE. 1974. The alkaloid Actini-
    dine and Plausible Precursors in Defensive Secretions of Rove
    Beetles. <u>J</u>. <u>Insect</u> <u>Physiol</u>. 20:277.
BERGER, R.S. 1972. 2,6-Dichlorophenol, Sex Pheromone of the Lone
    Star Tick. <u>Science</u> 177:704.
BERGSTROM, G. 1974. Studies on Natural Odouriferous Compounds X.
    Macrocyclic Lactones in the Dufour Gland Secretion of the Sol-
    itary Bees <u>Colletes</u> <u>cunicularius</u> L. and <u>Halictus</u> <u>calceatus</u>
    Scop. <u>Chemica</u> <u>Scripta</u> 5:39.
BIEMANN, K., G. BUCHI and B.H. WALKER. 1957. The Structure and
    Synthesis of Muscopyridine. <u>J</u>. <u>Am</u>. <u>Chem</u>. <u>Soc</u>. 79:5558
BLUM, M.S. and J.M. BRAND. 1972 Social Insect Pheromones: Their
    Chemistry and Function. <u>Am</u>. <u>Zool</u>. 12:553.
BORDEN, J.H., L. CHONG, J.A. MCLEAN, K.N. SLESSOR and K. MORI.
    1976. <u>Gnathotrichus</u> <u>sulcatus</u>: Synergistic Response to Enan-
    tiomers of the Aggregation Pheromone Sulcatol. <u>Science</u> 192:894.
BRAND, J.M., H.M. FALES, E.A. SOKOLOSKI, J.G. MacCONNELL, M.S. BLUM,
    and R.M. DUFFIELD. 1973. Isolation of Mellein in the Mandibular
    Gland Secretions of Carpenter Ants. <u>Life</u> <u>Sci</u>. 13:201.
BROWNLEE, R.G., R.M. SILVERSTEIN, D. MÜLLER-SCHWARZE and A.G.
    SINGER. 1969. Isolation, Identification and Function of the
    Chief Component of the Male Tarsal Scent in Black-Tailed Deer.
    <u>Nature</u> (<u>London</u>) 221:284.

CARLSON, D.A., M.S. MAYER, D.L. SILHACEK, J.D. JAMES, M. BEROZA and
    B.A. BIERL. 1971. Sex Attractant Pheromone of the Housefly:
    Isolation, Identification and Synthesis. Science 174:76.
CASNATI, G., A. RICCA and M. PAVAN. 1967. Sulla Secrezione Difen-
    siva della glandole mandibolari de Paltothyreus tarsatus. Chim.
    Ind. (Milan) 49:57.
CAVILL, G.W.K. 1960. The Cyclopentanoid Monoterpenes, Rev. Pure and
    App. Chem. 10:169.
CAVILL, G.W.K., D.V. CLARK and F.B. WHITFIELD. 1968. Insect Venom
    Attractants and Repellants XI. Massoilactone from two Species
    of Formicine Ants, and Some Observations on Constituents of
    the Bark Oil of Cryptocarya massoia. Aust. J. Chem. 21:2819.
EISNER, T. 1964. Catnip Its Raison d'Etre. Science 146:1318.
EVANS, D.A. and C.L. GREEN. 1973. Insect Attractants of Natural
    Origin. Chemical Society Reviews 2:75.
FRIEDMAN, L. and J.G. MILLER. 1971. Odor Incongruity and Chirality.
    Science 172:1044.
HENZELL, R.F. and M.D. LOWE. 1970. Sex Attractant of the Grass
    Grub Beetle. Science 168:1005.
IKAN, R., R. GOTTLIEB, E.D. BERGMANN and J. ISHAY. 1969. The Pher-
    omone of the Queen of the Oriental Hornet, Vespa orientalis.
    J. Insect Physiol. 15:1709.
JACOBSON, M., C.E. LILLY and C. HARDING. 1968. Sex Attractant of
    the Sugar Beet Wireworm: Identification and Biological Acti-
    vity. Science 159:208.
KERSHBAUM, M. 1927. Über Lactone mit grossen Ringen--die Tragen
    des vegetabilischen Moschus-Duftes. Ber 60:902.
KINZER, G.W., A.F. FENTIMAN, T.F. PAGE, R.L. FOLTZ, J.P. VITÉ and
    G.B. PITMAN. 1969. Bark Beetle Attractants: Identification,
    Synthesis and Field Bioassay of a New Compound Isolated from
    Dendrooctonus. Nature (London) 221:477.
KINZER, G.W., A.F. FENTIMAN, R.L. FOLTZ and J.A. RUDINSKY. 1971.
    Bark Beetle Attractants: 3-Methyl-2-cyclohexen-1-one Isolated
    from Dentrooctonus pseudotsugae. J. Econ. Entomol. 64:970.
LAW, J.H. and F.E. REGNIER. 1971. Pheromones. Annu. Rev. Biochem.
    40:533.
MacCONNELL, J.G., M.S. BLUM and H.M. FALES. 1970. Alkaloid from
    Fire Ant Venom: Identification and Synthesis. Science 168:840.
MEINWALD, J., G.M. HAPP, J. LABOWS and T. EISNER. 1966. Cyclopen-
    tanoid Terpene Biosynthesis in a Phasmid Insect and in Cat-
    mint. Science 151:79.
MICHAEL, R.P., E.B. KEVERNE and R.W. BONSALL. 1971. Pheromones:
    Isolation of Male Sex Attractants from a Female Primate.
    Science 172:964.
MICHAEL, R.P., R.W. BONSALL and P. WARNER. 1974. Human Vaginal Se-
    cretions: Volatile Fatty Acid Content. Science 186:1217.
MÜLLER-SCHWARZE, D., C. MÜLLER-SCHWARZE, A.G. SINGER and R.M. SIL-
    VERSTEIN. 1974. Mammalian Pheromone: Identification of Active
    Component in the Subauricular Scent of the Male Pronghorn.
    Science 183:860.

RILEY, R.G., R.M. SILVERSTEIN and J.C. MOSER. 1974. Biological Responses of Atta texana to its Alarm Pheromone and the Enantiomer of the Pheromone. Science 183:760.

RITTER, F.J., I.E.M. ROTGANS, E. TALMAN, P.E.J. VERWEIL and F. STEIN. 1973. 5-Methyl-3-butyloctahydroindolizine, A Novel Type of Pheromone Attractive to Pharaoh's Ants (Monomorium pharaonis L.). Experientia 29:530.

RUSSELL, G.F. and J.I. HILLS. 1971. Odor Differences Between Enantiomeric Isomers. Science 172:1043.

RUZICKA, L. 1926a. Zur Kenntnis des Kohlenstoffringes I. Über die Konstitution des Zibetons. Helv. Chim. Acta. 9:230.

RUZICKA, L. 1926b. Zur Kenntnis des Kohlenstoffringes VII. Über die Konstitution des Muscons. Helv. Chim. Acta. 9:715; VIII. Weitere Beitrage zur Konstitution des Muscons. ibid. 9:1008.

SCHILDKNECHT, H. 1970. The Defensive Chemistry of Land and Water Beetles. Agnew. Chem. Int. Ed. 9:1.

SILVERSTEIN, R.M., R.G. BROWNLEE, T.E. BELLAS, D.L. WOOD and L.E. BROWNE. 1968. Brevicomin: Principal Sex Attractant in the Frass of the Female Western Pine Beetle. Science. 159:889.

SINGER, A.G., W.C. AGOSTA, R.J. O'CONNELL, C. PFAFFMANN, D.V. BOWEN, and F.H. FIELD. 1976. Dimethyl Disulfide: An Attractant Pheromone in Hamster Vaginal Secretions. Science 191:950.

SMOLANOFF, J., A.F. KLUGE, J. MEINWALD, A. McPHAIL, R.W. MILLER, K. HICKS and T. EISNER. 1975. Polyzonimine: A Novel Insect Repellant Produced by a Millipede. Science 188:734.

STEVENS, P.G. and J.L.E. ERICKSON. 1942. American Musk I. The Chemical Constitution of the Musk of the Louisiana Muskrat. J. Am. Chem. Soc. 64:144.

STEVENS, P.G. 1945. American Musk III. The Scent of the Common Skunk. J. Am. Chem. Soc. 67:407.

THIESSEN, D.C., F.E. REGNIER, M. RICE, M. GOODWIN, N. ISAAKS and N. LAWSON. 1974. Identification of a Ventral Scent Marking Pheromone in the Male Mongolian Gerbil (Meriones unguiculatas). Science 184:83.

TUMLINSON, J.H., R.M. SILVERSTEIN, J.C. MOSER, R.G. BROWNLEE and J.M. RUTH. 1971. Identification of the Trail Pheromone of a Leaf-cutting Ant, Atta texana. Nature (London) 234:348.

TURSCH, B., D. DALOZE, M. DUPONT, J.M. PASTEELS and M. -C. TRICOT. 1971. A Defense Alkaloid in a Carnivorous Beetle. Experientia 27:1380.

VALENTA, Z. and A. KHALEQUE. 1959. The Structure of Castoramine. Tetrahedron Lett. 1.

VAN DORP, D.A., R. KLOK and D.H. NUGTEREN. 1973. New Macrocyclic Compounds from the Secretions of the Civet Cat and the Musk Rat. Recl. Trav. Chim. Pays-bas. 92:915.

WALLER, G.R., G.H. PRICE and E.D. MITCHELL. 1969. Feline Attractant cis, trans-Nepetalactone: Metabolism in the Domestic Cat. Science 164:1281.

WEATHERSTON, J. 1967. The Chemistry of Arthropod Defensive Substances. Quart. Rev. (London) 21:287.

WHEELER, J.W., G.M. HAPP, J. ARAUJO and J.M. PASTEELS. 1972. γ-
    Dodecalactone from Rove Beetles. Tetrahedron Lett. 4635.
WHEELER, J.W. and M.S. BLUM. 1973. Alkylpyrazine Alarm Pheromones
    in Ponerine Ants. Science 182:501.
WHEELER, J.W., D.W. VON ENDT and C. WEMMER. 1975. 5-Thiomethyl-
    pentan-2,3-dione: A Unique Natural Product from the Striped
    Hyena. J. Am. Chem. Soc.,97:441.
WHEELER, J.W., M.S. BLUM and A. CLARK. 1976a. Unpublished results.
WHEELER, J.W., S.L. EVANS and M.S. BLUM. 1976b. Unpublished
    results.
WHEELER, J.W., T. OLAGBEMIRO, A. NASH and M.S. BLUM. 1976c Unpub-
    lished results.
WHEELER, J.W., D.W. VON ENDT and C. WEMMER. 1976d. Unpublished
    results.
WHITE, R.H. 1975. S-Methylthioesters from Human Urine After Eating
    Asparagus. Science 189:810.
WILSON, E.O. 1963. Pheromones. Scientific American 208:100.
WILSON, E.O., W.H. BOSSERT and F.E. REGNIER. 1969. A General Method
    for Estimating Threshold Concentrations of Odorant Molecules.
    J. Insect Physiol. 15:597.
WOOD, D.L., L.E. BROWNE, B. EWING, K. LINDAHL, W.D. BEDARD, P.E.
    TILDEN and K. MORI. 1976. Western Pine Beetle: Specificity
    Among Enantiomers of Male and Female Components of an Attrac-
    tant Pheromone. Science 192:896.

CHEMICAL METHODOLOGY IN THE STUDY OF MAMMALIAN COMMUNICATIONS

Alf Claesson[1] and Robert M. Silverstein

College of Environmental Science and Forestry

Syracuse, New York   13210

Many of the interactions among plants and animals are mediated by chemical signals, and vertebrates are no exception. By far the largest amount of work has been devoted to interactions among insects, economically important pests in particular, and much of the chemical methodology applied to mammalian studies has been carried over from the insect field (Young and Silverstein, 1975).

Nomenclature and classification in this relatively new and rapidly developing field of chemical communication are also in a state of development to cope with the complexities encountered. The most useful general term is __semiochemical__ (Gr. __semion__, a mark or signal), proposed by Law and Regnier (1971) to describe a chemical that delivers any message to any organism.

Semiochemicals are divided into two groups: __pheromones__ (Karlson and Butenandt, 1959), which are used for intraspecific communications, and __allelochemics__ (Whittacker and Feeny, 1971), which are used for interspecific communications. Allelochemics have been further subdivided by Brown __et al__. (1970) into __allomones__, which confer adaptive advantage on the emitter (e.g., spray of a skunk, venom of a wasp or snake, repellant secretions of many arthropods), and __kairomones__, which confer adaptive advantage on the receiver (e.g., signals that enable predators to locate prey). Nordlund and Lewis (1976) have recently proposed two additional categories of allelochemics: __synomones__ (Gr. __syn__, with or jointly) and __apneumones__ (Gr. __a-pneum__, breathless or lifeless). __Synomones__ evoke a response adaptively favorable to both emitter and receiver (e.g., floral

---

[1] Current address: Department of Organic Pharmaceutical Chemistry, University of Uppsala, Box 574, 75123 Uppsala, Sweden

scents and nectars). <u>Apneumones</u> are emitted by a non-living material
that evokes a response adaptively favorable to a receiving organism,
but detrimental to an organism of another species that may be found
in or on the non-living material; for example, several predators are
attracted to a non-living material (e.g., meat, oatmeal) containing
the prey, the attractant being emitted by the meat or oatmeal rather
than by the prey.

Two observations must be stated at the outset: First, a response may be evoked by a single compound, or several compounds presented simultaneously may be required to evoke the response. Numerous
examples are known from insect communications; it appears that
multicomponent pheromones are the more common (Silverstein and Young,
1976). Second, a compound or group of compounds may carry more than
one message; thus the multicomponent aggregation pheromone (or one
or more components thereof) of several bark beetles also attracts
predators and so functions as a kairomone.

It is probably too early to set up a satisfactory classification scheme for mammalian semiochemicals in view of the complexity
of the social interactions, including the complications of recognition
of individual animals and of the highly developed learning abilities.
The latter factor also is involved in some of the highly developed
social insects. In both insects and mammals, studies have been directed largely to defensive secretions and pheromones. Mammalian
pheromones are reviewed by Stoddart (1974), Müller-Schwarze (1974),
Mykytowycz (1974), Bronson (1974), Epple (1974), Comfort (1974),
Eisenberg and Kleiman (1972), Ralls (1971), Johnson (1973), Thiessen
and Rice (1976). A conceptual approach to pheromones, based on
energetics, has been proposed (Thiessen, 1976).

The problems involved in development of insect bioassays have
been briefly reviewed (Young and Silverstein, 1975). Müller-Schwarze,
O'Connell, and Thiessen deal with mammalian bioassays in separate
chapters in this volume.

Although chemical signals can be transmitted by direct contact
as well as through space, by far the bulk of the studies have dealt
with communication through space.

The task of the chemist is to devise methods for collection,
isolation, and identification of biologically active components;
he then synthesizes the molecules to confirm structure and to provide sufficient material for field tests; finally he assists in
preparing appropriate formulations for field tests.

In the collection and isolation steps, he works closely with
his biological collaborators, and again in the formulation step,
the chemist is guided by the biologists' suggestions for appropriate
modes of presenting the material to the animal.

As mentioned above, a certain amount of volatility has been assumed in most studies. Also implicit in the chemist's approach is the assumption that several components will probably be involved in the message. In this connection, it is interesting to note that most investigators of mammalian signals have submitted reports based on a single dominant component — a situation reminiscent of the earlier insect studies.

This review is not meant to be exhaustive in itself. Rather, it is designed to be read in the context of the other chapters in the volume.

Collection

Collection of semiochemicals from insects frequently involves extracting large numbers of whole insects or excising and extracting segments thereof (review by Young and Silverstein, 1975). This approach is hardly feasible in most studies of mammals, for which the following procedures are among those reported:

> Swabbing skin glands with an absorbent material, with or without squeezing (Thiessen et al., 1974; Berüter et al., 1974). Squeezing the gland or pouch and collecting the expressed material in glass capillaries or vials (Goodrich and Mykytowycz, 1972; Albone and Fox, 1971; Gorman, 1976; Albone et al., 1974; Patterson, 1968, and Preti et al., 1976). Collecting vaginal secretion on filter paper (Singer et al., 1976), on tampons (Preti and Huggins, 1975; Michael et al., 1974), and by lavage (Curtis et al., 1971; Goldfoot et al., 1976). Collecting directly from the gland with an inserted cannula or syringe (Albone et al., 1974; Albone and Perry, 1976) or by scraping the surface (Stoddart et al., 1975). Collecting deposited scent marks (Yarger et al., in press; Wheeler et al., 1975). Aerating from swabs and cold-trapping volatiles (Thiessen et al., 1974). Aerating from swabs or gland secretion and trapping on an absorbent (Singer et al., 1976; Claesson et al., unpublished). Collecting feces (Mykytowycz and Gambale, 1969) or urine (Dixon and Mackintosh, 1971; Berüter et al., 1973; Vandenbergh et al., 1975).

Killed game animals may be intercepted at hunter check stations, and glands may be excised (Brownlee et al., 1969; van Dorp et al., 1973) or urine taken from the bladder (Müller-Schwarze, unpublished). Other investigators have sacrificed animals and excised glands (e.g. Goodrich and Mykytowycz, 1972).

The ever present problem of obtaining enough material is usually overcome by repetitive collection and pooling samples. Singer et al, (1976) aerated filter paper swabs and accumulated the volatiles on Tenax (Applied Science Laboratory), which is a polymer of 2,6-diphenyl-p-phenylene oxide. Zlatkis et al. (1973a, 1973b, 1974) and Dowty et al. (1975) aerated urine or blood plasma and collected the volatiles on Tenax, and Politzer et al. (1975) collected volatiles from tissue in the same way. In our studies of the black-tailed deer, we have used this technique to accumulate volatiles from swabs of the metatarsal gland and from the waxy secretion scraped from the antorbital gland. The Tenax, after conditioning (Versino et al., 1974), is thermally stable and the accumulated volatiles may be desorbed by heating in an inert gas stream. If desired, desorption can be accomplished in the carrier gas stream and the volatiles can thus be transferred to the inlet end of a gas chromatography column (see under Isolation). Obviously, there are upper and lower volatility limits for absorption and desorption. Singer et al. (1976) isolated dimethyl disulfide; Russell (1975) and Pellizari et al. (1975a, b) recovered several chlorinated phenols; we have run C_9-C_{11} ketones and C_{12}-C_{14} normal hydrocarbons through the entire absorption-desorption procedure with excellent recoveries. Zlatkis et al. (1973b) recovered p-cresol and some monoterpenoid ketones from urine. In our experience, Tenax can be heated to about 200°C without excessive bleeding; however since it is somewhat soluble in a number of solvents, it is usually not feasible to remove the absorbed compounds by solvent extraction. Leoni et al. (1975) absorbed aqueous pollutants on Tenax and extracted with ether. For compounds that are difficult to heat-desorb or that are thermally unstable, absorption on Porapak-Q (a cross-linked styrene polymer) and desorption by solvent extraction may be a better procedure (Byrne et al., 1975; Peacock et al., 1975; Hill et al., 1975; Silverstein and Young, 1976; Look, in press). Butler and Burke (1976) have compared the characteristics of several absorbents.

None of the aerations described have been applied directly to mammals, but it should be quite feasible to aerate caged small animals, or even a large animal confined in a room if the semiochemicals are sufficiently volatile. Of course, a medley of odors will be collected, but the subsequent fractionation can be monitored with a bioassay and only the fractions of interest selected. This approach has been used in several insect studies and has resulted in the isolation of compounds that could not be isolated by direct extraction of the whole insect (Silverstein and Young, 1976) of frass (Peacock et al., 1975) or of glands (Hill et al., 1975). Evidently some insects only produce and release semiochemicals on demand; for example, during the calling period of a female (Hammack et al., 1976). Bierl et al. (1970) found large amounts of the probable precursor but only small amounts of the pheromone itself in gland extracts of the gypsy moth. Similar complications may be encountered in investigations of mammals. Most mammalian secretions investi-

gated do contain a variety of chemical compounds, and we may speculate that each component serves a purpose and that the proportions
of some of them convey a message or several messages. Not all of
the compounds need be volatile messengers; some of the high-boiling
compounds identified, for example, in the subauricular gland of the
pronghorn antelope (Müller-Schwarze et al., 1974) may serve as
"keepers" to regulate the volatility of the messenger compounds.
Alternatively the compounds identified may serve as precursors to
the pheromone compounds, which are produced on demand.

     Sometimes the best choice of source material is not obvious –
in fact, in one of our experiences with the black-tailed deer, the
apparent source of the "tarsal scent" was not the true source of the
pheromone component. The tarsal gland of the black-tailed deer is
one of central importance in social behavior in the herd. We collected and extracted tarsal hair tufts from a large number of
hunter-killed deer and identified the most active component of this
"social" scent as cis-4-hydroxydodec-6-enoic acid lactone (Brownlee
et al., 1969). We did, however, recognize the urine as a possible
source because of the occurrence of "rub-urination". Recently, we
identified this compound in a urine sample, and we have been unable
to find any material of similar volatility in the oily secretion expressed from the freshly cleaned tarsal gland. Our resourceful
biological collaborators obtained an unsoiled urine sample by pa-
tiently waiting until the deer urinated into a fresh patch of snow.
Other samples were obtained by catheterization of anesthetized does.
Dr. Müller-Schwarze, in his chapter, describes the role of the tarsal
tuft and the tarsal secretion in dissemination of the social odor.

     With animals in the field, variability of diet with season and
location is a complicating factor. Nothing is known of the biogenesis
of mammalian semiochemicals, and ingested material cannot be excluded
as a source of semiochemicals either by sequestration or by transformation. In one of the early insect pheromone studies, a host
tree hydrocarbon was found to be one of the components of an aggregation pheromone (Silverstein et al., 1968), and it was recently
demonstrated that the larva of the pine sawfly sequesters resin,
which it expels defensively (Eisner et al., 1974). Thus, a collection procedure must consider temporal and geographic factors for
field animals. The diet for small cage animals is usually consistent
throughout the year, but large penned animals can supplement their
diet with browse in season. It seems likely that the compounds in
many "secretions" are the result of microbial action, and interactions
among some mammals may actually depend on the population of the
microorganisms harbored (Gorman, 1976; Albone and Perry, 1976).

Isolation

     In many cases, the isolation procedure poses a formidable chall-
enge. The first problem is that the collection procedure may furnish

marginal amounts of the compounds of interest so that even when the
compounds are isolated, their purity is difficult to confirm, and
the identification methods are pushed to their limits or beyond.
The chemist — especially one who has worked with insects — is be-
guiled by the relative size of mammals, but quickly learns that
mammals may also communicate with minute amounts of semiochemicals.
The entomological collaborator can be pressed into delivering several
thousand more insects, but the mammalogist may be reluctant to expand
his penned herd of twenty black-tailed deer by several orders of
magnitude.

The second problem involves a difficult choice of several
possible approaches. The preferred modus operandi is to monitor
each step of the chemical isolation with a bioassay. (Bioassays are
discussed in the chapters by O'Connell, Thiessen, and Müller-Schwarze.
This approach has been highly successful in a large number of insect
studies, even with the complications posed by synergistic effects,
which are handled by testing all combinations of the fractions in a
laboratory bioassay (Young and Silverstein, 1975). This approach
should be feasible with many small mammals, and was used by Singer
et al. (1976) to identify the major component of a hamster attractant
pheromone. It was also used by Curtis et al. (1971) to identify a
mixture of attractant fatty acids from a female primate. Thiessen
(1974) isolated the major component of the marking pheromone of a
gerbil by following the characteristic odor through the fractionation; the assumption that the human nose can substitute for the test
animal was justified in this case, although extrapolations seem
risky.   Electroantennograms also have been used to monitor fractions
of insect pheromones (Roelofs and Comeau, 1971; Roelofs et al., 1971).

But the preferred approach may not always be feasible, especiall
with large animals that require testing under field conditions,
which may be difficult to control and therefore require large amounts
of material for replicated testing. Many responses are only valid
in the proper physical and temporal context. The most important component of the "tarsal" scent (see above) was selected because it was
the compound present in the largest amount from the buck and was
much less conspicuous in the doe (Brownlee et al., 1969). Again, one
might be reluctant to generalize that the compounds present in
largest amounts are necessarily the most important. Significant compounds may be selected because they are present, or predominate, in
one sex, may appear at a specific time such as at estrous, or may
appear, increase or decrease in response to manipulation such as
castration or hormone treatment. The consistent appearance of a
compound through wide dietary variations may be significant (see the
chapter by Stoddart).

Sometimes the only recourse is to identify every compound presen
synthesize them, and then systematically bioassay the synthetic compounds in all combinations. This approach — brute force — was used

to designate isovaleric acid as the most important component of the subauricular scent of the male pronghorn antelope (Müller-Schwarze et al., 1974). The risk here, of course, is that expediency may dictate that the minor components be disregarded. This approach has also been used in an investigation of the reindeer tarsal scent (Andersson et al., 1975), the anal sac of the red fox (Albone et al., 1974; Albone and Perry, 1976), the preputial glands of the muskrat and the civet cat (van Dorp et al., 1973), and the marmoset scent mark (Yarger et al., in press; see also the chapter by Gawienowski).

As techniques improve, it will probably turn out that many mammalian semiochemicals are complex mixtures. This would be expected on the basis of the complex responses evoked and the extraordinarily highly developed receptor organs. It seems worth noting that six components have been implicated in the sex attractant pheromone of several dermestid beetles (Silverstein and Young, 1976); this is more complex than any mammalian pheromone yet reported. Individual recognition is a feature unlikely to be encountered in insect studies, but surely must be reckoned with in mammalian studies. In one case, the Indian mongoose, individual recognition has been ascribed to characteristic ratios of lower fatty acids in the anal pockets (Gorman, 1976).

The isolation procedure is designed for two purposes: to furnish fractions and ultimately pure compounds for bioassay if the monitoring regime is followed, and to furnish enough of each pure compound for identification. It is well to bear in mind that the criteria for purity in the bioassay of a compound may be much more stringent than those required for identification. A fraction of one percent of a highly active impurity may lead to erroneous designation of an inactive material as the active semiochemical component.

Since purity requirements are high and amounts are small, and since most of the semiochemicals so far investigated are at least somewhat volatile, gas chromatography is the most widely used isolation technique. An extract usually requires a preliminary cleanup to eliminate very high boiling materials; a sample collected by aeration (Byrne et al., 1975; Peacock et al., 1975; Hill et al., 1975; Singer et al., 1976) usually does not, and can be subjected directly to gas chromatography. Generally, short-path vacuum distillation (Silverstein et al., 1966) or gel permeation chromatography (Tumlinson and Heath, 1976) are rapid and convenient procedures for excluding high boiling material. These procedures may be followed by other liquid chromatographic separations, which utilize modern high pressure techniques (Tumlinson and Heath, 1976), or by thin layer chromatography. Chemical separation into acidic, basic, and neutral fractions is a useful general procedure (Berüter et al., 1974; Hesterman et al., 1976). Other authors have dialyzed urine samples (Vandenbergh et al., 1975).

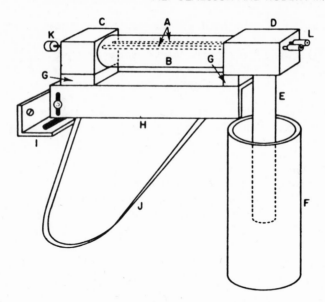

Fig 1.   Thermal gradient collector for gas chromatography.   (A)
         Aluminum tube to accept a 30 cm x 1.6 mm o.d. glass capillary
         and a 30 cm x 3 mm o.d. glass tube.   (B)   Ceramic thermal
         insulator.   (C)   Aluminum heating black.   (D)   Aluminum
         block.   (E)   Aluminum rod.   (F)   Cooling bath.   (I)   Mounting
         bracket.   (J)   Support.   (K)   Teflon ferrule.   (L)   Teflon
         rod to prevent frost accumulation.

        The heart of preparative gas chromatography is an efficient
splitter and collector.   The bugaboo of "fogging" is effectively
exorcised by a thermal-gradient collector (Brownlee and Silverstein
1968) (Fig. 1).   This device permits efficient recovery (∿90%) of
samples in the microgram range in long capillary tubes; these can
be sealed and stored, and subsequently rinsed with solvent.   The
glass capillary can also be inserted into an attachment to the inlet
system of a mass spectrometer and broken in situ.   Very volatile
compounds may be collected on an absorbent column (Singer et al.,
1976).

        Isolation techniques used for insect semiochemicals have been
recently reviewed (Young and Silverstein, 1975); obviously the same
techniques apply to isolation of semiochemicals from mammals.   However, it may be worthwhile to give some detail on several recent
procedures for transferring materials to a gas chromatographic
column.

        Figure 2 shows an apparatus for transferring material from
cotton swabs or gland material to Tenax, thence to a gas chromatog-

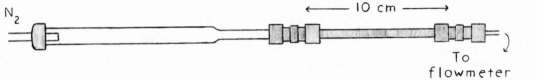

Fig 2.   Apparatus for transferring volatiles from cotton swabs to
         Tenax.

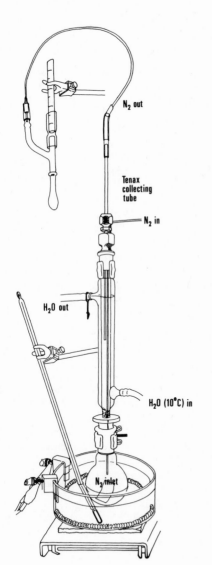

Fig 3.   Apparatus for transfer-
         ring volatiles from wetted
         filter paper swabs to
         Tenax. Singer et al.
         (1976). (Picture courtesy
         of Dr. Singer).

raphic column. It consists of a glass tube, made from a screw
thread joint (Quickfit, SQ 13) on which is fused a 6 mm (o.d.) glass
tube. This is connected via Swagelock fittings to a stainless steel
tube containing the Tenax. The screw cap is fitted with a silicone
ring and a Teflon liner. When the swabs are heated in the glass
tube under a slow stream of nitrogen or helium, the volatiles are
trapped on the Tenax. When sufficient material has accumulated on
the Tenax after a number of transfers, condensed water is blown out
of the Tenax container with dry nitrogen, and the tube is fitted
with an injection needle soldered to the fitting. The needle is
inserted through the septum, and the Tenax container is placed in
the carrier gas line of the gas chromatography inlet. The gas flow
through the Tenax is reversed from the direction of flow during the
collection. When the Tenax is heated (block or heating tape), the
volatiles are transferred to the inlet end of the gas chromatography
column, which may be kept at room temperature (Parsons and Mitzner,
1975) or above (our experiments), or may be cooled for low boiling
compounds (Singer et al., 1976; Zlatkis et al., 1973a, 1973b). When
the transfer is complete, the gas chromatography column is temperature programmed to start the fractionation. The apparatus used by
Singer et al. (1976) for accumulating material on Tenax by aerating
filter paper swabs (wetted to facilitate dispersion) is shown in
Fig 3. Other designs of adsorbent columns are described by Bertsch
et al. (1974), Leoni et al. (1975), Parsons and Mitzner (1975),
Pellizzari et al. (1975b), Politzer et al. (1975), Russell (1975),
Versino et al. (1974), and Zlatkis et al. (1973a).

Bergström (1973) has used a "pre-column tube" in which he has
placed such "odor" sources as bumble-bee labial glands, wing scales
of butterflies, and an absorbent material (silicone grease on
Chromosorb) on which volatile compounds had been absorbed. When
the pre-column tube is heated in the carrier gas stream, volatile
compounds are transferred to the cooled anterior portion of the gas
chromatography column. This apparatus has been coupled to a mass
spectrometer (see below) and components of the gland of a single
bumble-bee gland have been identified. A similar system with a
pre-column outside the oven has been described by Andersson et al.
(1975); see also designs by Karlsen (1972) and Ställberg-Stenhagen
(1972).

For compounds that have low volatility or are thermally unstable,
recently developed techniques of high pressure liquid chromatography
are extremely useful (Walton, 1976). The feasibility of interfacing
a liquid chromatograph with a mass spectrometer has been demonstrated
but units are not yet commercially available (Burlingame et al.,
1976).

Identification

Identification of the small amounts of semiochemicals usually obtained rests mainly on four spectrometric methods (mass, NMR, infrared, and ultraviolet; for the most recent developments see Burlingame et al., 1976; Wasson and Lorenz, 1976; McDonald, 1976; Hummel and Kaufman, 1976; Ayling, 1974) and on several chemical procedures that are applicable at the microgram level; these are hydrogenation, hydrogenolysis, and ozonolysis. Compounds may be tested for certain functional groups by "subtractive loops" used as a fore-column on a gas chromatography column, or derivatives may be prepared and tested for loss or retention of biological activity. References are given in the review by Young and Silverstein (1975). The use of specific gas chromatograph detectors can be useful for compound identification. Nitrogen, sulfur, phosphorus, and halogen detectors are commercially available (Cram and Juvet, 1976).

Major new developments have been in the areas of Fourier transform nuclear magnetic resonance (both ^{1}H and ^{13}C), Fourier transform infrared spectrometry, and mass spectrometry (chemical ionization in particular). Infrared and mass spectrometers can be coupled with gas chromatography to obviate the difficulties of sample handling.

Ordinarily a nuclear magnetic resonance spectrum is obtained by scanning through a range of frequencies. In the Fourier transform mode, all frequencies are used simultaneously and all nuclei respond simultaneously. The resulting complex mixture of frequencies emitted by the relaxing nuclei is transformed into a conventional spectrum by a mathematical operation -- the Fourier transform -- carried out by a computer. The advantage is greatly increased sensitivity; proton spectra can now be obtained on a sample size in the range of several micrograms, and carbon-13 spectra on several hundred micrograms by using a 1.7 mm o.d. capillary tube and a special probe.

The ability to obtain ^{13}C spectra on submilligram samples will have a tremendous impact on natural products chemistry. Two books for the organic chemist that deal with interpretation of ^{13}C spectra are available (Stothers, 1972; Levy and Nelson, 1972). Although there are some parallels with ^{1}H spectrometry, the organic chemist must develop another set of interpretive skills.

Conventional infrared spectrometers with a beam condenser and microcells can give satisfactory spectra on samples in the range of 10-50 µg, whereas with Fourier transform, the sample size can be reduced to about 0.5 µg (King, 1973). A gas chromatograph can be connected directly to the Fourier transform infrared spectrometer.

One of the major difficulties in the interpretation of electron impact mass spectrometry has been the inability to see a molecular ion peak (M) in many classes of compounds. This has been largely overcome by chemical ionization mass spectrometry (Field, 1972), in which ionization is accomplished by interaction with charged

species of a reagent gas such as methane or isobutane. A spectrum
of a 10 ng sample of a terpenoid alcohol has been recorded by
Tumlinson and Heath (1976). The reagent gases can be varied to
give spectra that give the M + 1 and M - 1 peaks characteristic of
those obtained with the hydrocarbon gases and also give the useful
fragmentation information of an electron impact spectrum (see for
example the use of nitric oxide as a reagent gas; Hunt and Ryan,
1972).

     A complete chemical identification must include specification
of the stereochemistry, both geometric and enantiomeric; ideally
the absolute configuration at each chiral center should be described. Many insects not only respond differently to geometric
isomers, but actually use precise mixtures of geometric isomers as
attractant pheromones; closely related sympatric species may maintain species integrity by using a different ratio of the same compounds (Silverstein and Young, 1976; Persoons and Ritter, 1975).
Since geometric isomers are grossly different from one another -
different physical and chemical properties - it is not surprising
that humans readily distinguish them on the basis of odor. Recently,
it was shown that the black-tailed deer can distinguish between the
cis (Z)- and trans (E) isomers of the "tarsal gland" pheromone component which is the "cis lactone"; the E lactone does not seem to
be present as a pheromone component (Müller-Schwarze et al., 1976).

     Enantiomers differ from each other at a more subtle level.
Their chemical and physical properties are identical in the usual
laboratory tests, but they differ in "handedness". One might expect
odor receptor sites to be chiral, but this has been the subject of
a long standing controversy because of the experimental rigor required for an unequivocal answer. In 1971, Friedman and Miller, and
Russell and Hills in a series of careful experiments demonstrated
that the human nose indeed acknowledges chirality. And it was only
recently that two species of ants were shown to distinguish between
the compound used as the alarm pheromone and the enantiomer, which
is an artifact (Riley et al., 1974a and 1974b). Wood et al., 1976,
studied the responses of a species of bark beetle to a three-component
pheromone containing two chiral components; substitution by the unnatural enantiomer diminished the response. Borden et al. (1976)
demonstrated a synergistic response to a mixture of enantiomers that
approximated the natural composition of the attractant pheromone of
a species of ambrosia beetle. The presence of both enantiomers may
serve as a species isolating mechanism. No similar studies of enantiomeric components of mammalian pheromones have been reported.
We made an unsuccessful attempt to demonstrate such an effect with
the enantiomers of the black-tailed deer "tarsal" pheromone, and will
make another attempt during the next rutting season.

Because pheromone components are usually isolated in very small amounts, it may be difficult to obtain an accurate optical rotation (required amount > 1 mg), and an estimate of enantiomeric composition by optical rotation requires a knowledge of the rotation of one of the pure enantiomers. Another approach is to treat the enantiomer or mixture of enantiomers with a chiral derivatizing agent or a chiral shift reagent and examine the ^{1}H, ^{13}C or ^{19}F NMR spectra. Application of these techniques to several insect pheromone components has recently been demonstrated (Plummer et al., 1976; Stewart, et al., in press; Pearce et al., in press). Enantiomeric compositions of alcohols and bicyclic ketals were determined on isolated compounds in the range of 5-500 µg.

Synthesis

Synthesis of a semiochemical is carried out to confirm the proposed structure and to furnish material for field tests. The general principles have been summarized with appropriate references (Young and Silverstein, 1975).

Of particular interest here is the question of enantiomeric composition, since receptor sites may discriminate between enantiomers. There are three major approaches to the synthesis of enantiomers: (1) resolution of a racemate (Whilen, 1971), (2) utilization of a compound with a chiral center (usually a natural product) as the starting material (for an example of a synthesis of both enantiomers of an insect pheromone, see Mori, 1975) or (3) introduction of the chiral center in an asymmetric reaction (Morrison and Mosher, 1971).

Formulations

Ideally, formulated synthetic materials should be presented to the test animal in as natural a context as possible, but compromises are usually necessary (Chapter by Müller-Schwarze).

From the chemical point of view there are several considerations. Many compounds degrade in the presence of air; addition of commercially available antioxidants usually solves this problem for the duration of the test. Rate of evolution of volatile compounds can be controlled by evaporation from various wicks or containers with appropriate surface areas, by use of non-volatile "keepers" such as oils or polymeric materials, or by diffusion through a membrane. Longevity of the dispenser depends on stability and release rate and on the capacity of the container. Most of the technology has been developed in connection with pesticides and insect pheromones and are summarized by Cardarelli (1974, 1976) and Harris (1975).

Acknowledgements

    We wish to thank R. N. Roberts, D. N. Hildebrand and N. J.
Volkman for several literature references. Our investigations were
supported by the National Science Foundation A. C. acknowledges a
grant from Stiftelsen B. Boncompagni-Ludovisi, Stockholm.

References

Albone, E. S., Eglinton, G., Walker, J. M., and Ware, G. C. 1974. The anal sac secretion of the red fox (Vulpes vulpes); its chemistry and microbiology. A comparison with the anal sac secretion of the lion (Panthera leo). Life Sci. 14:387-400.

Albone, E. S. and Fox, M. W. 1971. Anal gland secretion of the red fox. Nature 233:569-570.

Albone, E. S. and Perry, G. C. 1976. Anal sac secretion of the red fox, Vulpes vulpes; volatile fatty acids and diamines: implications for a fermentation hypothesis of chemical recognition. J. Chem. Ecol. 2:101-111.

Andersson, G., Andersson, K., Brundin, A., and Rappe, C. 1975. Volatile compounds from the tarsal scent gland of reindeer (Rangifer tarandus). J. Chem. Ecol. 1:275-281.

Ayling, G. M. 1974. Spectroscopic methods of identification of microquantities of organic materials. in Brame, Jr., E. G. (ed.). Applied Spectroscopy Reviews, Vol. 8, Part A. Marcel Dekker, N. Y.

Bergström, G. 1973. Studies on natural odoriferous compounds. VI. Use of a pre-column tube for the quantitative isolation of natural, volatile compounds for gas chromatography/mass spectrometry. Chemica Scripta. 4:135-138.

Bertsch, W., Chang, R. C., and Zlatkis, A. 1974. The determination of organic volatiles in air pollution studies: characterization of profiles. J. Chromatog. Sci. 12:175-182.

Berüter, J., Beauchamp, G. K., and Muetterties, E. L. 1973. Complexity of chemical communication in mammals: urinary components mediating sex discrimination by male guinea pigs. Biochem. Biophys. Res. Commun. 53:264-271.

Berüter, J., Beauchamp, G. K., Muetterties, E. L. 1974. Mammalian chemical communication: perineal gland secretion of the guinea pig. Physiol. Zool. 47:130-136.

Bierl, B. A., Beroza, M., and Collier, C. W. 1970. Potent sex attractant of the gypsy moth: its isolation, identification, and synthesis. Science 170:87-89.

Borden, J.H., Chong, L., McLean, J.A., Slessor, K.N., and Mori, K. 1976. Gnathotrichus sulcatus: synergistic response to enantiomers of the aggregation pheromone, sulcatol. Science 192:894-896.

Bronson, F.H. 1974. Pheromonal influences on reproductive activities in rodents, in Birch, M. (ed.). Pheromones, North-Holland Publishing Co., Amsterdam.

Brown, W.L., Jr., Eisner, T., and Whittacker, R.H. 1970. Allomones and kairomones, transspecific chemical messengers. Bioscience 20:21-22.

Brownlee, R.G. and Silverstein, R.M., 1968. A micro-preparative gas chromatograph and a modified carbon skeleton determinator. Anal. Chem. 40:2077-2079.

Brownlee, R.G., Silverstein, R.M., Müller-Schwarze, D., and Singer, A.G. 1969. Isolation, identification, and function of the chief component of the male tarsal scent in black-tailed deer. Nature 221:284-285.

Burlingame, A.L., Kimble, B.J., and Derrick, P.J. 1976. Mass spectrometry. Anal. Chem. 48:368R-403R.

Butler, L.D. and Burke, M.F. 1976. Chromatographic characterization of porous polymers for use as adsorbents in sampling columns. J. Chromatog. Sci. 14:117-122.

Byrne, K.J., Gore, W.E., Pearce, G.T., and Silverstein, R.M. 1975. Porapak-Q collection of airborne organic compounds serving as models for insect pheromones. J. Chem. Ecol. 1:1-7.

Cardarelli, N. (ed.). 1974. Controlled release pesticide symposium. Engineering and Science Division, Community and Technical College, University of Akron, Akron, Ohio   44325.

Cardarelli, N. 1976. Controlled release pesticides formulations. CRC Press, Cleveland.

Comfort, A. 1974. The likelihood of human pheromones, in Birch, M. (ed). Pheromones, North-Holland Publishing Co., Amsterdam.

Cram, S.P. and Juvet, R.S. 1976. Gas chromatography. Anal. Chem. 48:411R-442R.

Curtis, R.F., Ballantine, J.A., Keverne, E.B., Bonsall, R.W., and Michael, R.P. 1971. Identification of primate sexual pheromones and the properties of synthetic attractants. Nature 232:396-398.

Dixon, A.K. and Mackintosh, J.H. 1971. Effects of female urine upon the social behavior of adult male mice. Anim. Behav. 19:138–140.

Dowty, B., Gonzales, R., and Laseter, J.L. 1975. Biomed. Mass Spec. 2:142.

Eisenberg, J.F. and Kleiman, D.G. 1972. Olfactory communication in mammals. Ann. Rev. Ecol. and System. 3:1–32.

Eisner, T., Johnessee, J.S., Carrel, J., Hendry, L.B., and Meinwald, J. 1974. Defensive use by an insect of a plant resin. Science 184:996–999.

Epple, G. 1974. Primate pheromones, in Birch, M. (ed.). Pheromones, North-Holland Publishing Co., Amsterdam.

Field, F.H. 1972. Chemical ionisation mass spectrometry, in Maccoll, A. (ed.). Mass spectrometry, Physical Chemistry Series One, 5:133–181.

Friedman, L. and Miller, J.G. 1971. Odor incongruity and chirality. Science 172:1044–1046.

Goldfoot, D.A., Kravetz, M.A., Goy, R.W., and Freeman, S.K. 1976. Lack of effect of vaginal lavages and aliphatic acids on ejaculatory responses in rhesus monkeys: behavioral and chemical analyses. Horm. Behav. 7:1–27.

Goodrich, B.S. and Mykytowycz, R. 1972. Individual and sex differences in the chemical composition of pheromone-like substances from the skin glands of the rabbit, Oryctolagus cuniculus. J. Mammal. 53:540–548.

Gorman, M.L. 1976. A mechanism for individual recognition by odour in Herpestes auropunctatus (Carnivora: Viverridae). Anim. Behav. 24, 141–145.

Hammack, L., Ma, M., and Burkholder, W.E. 1976. Sex pheromone-releasing behaviour in females of the dermestid beetle Trogoderma glabrum. J. Insect. Physiol. 22:555–561.

Harris, F.W. (ed.). 1975. Proceedings, 1975 International Controlled Release Pesticides Symposium. Wright State University College of Science and Engineering. Dayton, Ohio 45431.

Hesterman, E.R., Goodrich, B.S., and Mykytowycz, R. 1976. Behavioral and cardiac responses of the rabbit, Oryctolagus cuniculus, to chemical fractions from anal gland. J. Chem. Ecol. 2:25–37.

Hill, A.S., Cardé, R.T., Kido, H., and Roelofs, W.L. 1975. Sex pheromones of the orange tortrix moth (Argyrotaenia citrana

Lepidoptera:Tortricidae). J. Chem. Ecol. 1:215-224.

Hummel, R. and Kaufman, D. 1976. Ultraviolet spectrometry. Anal. Chem. 48:268R-273R.

Hunt, D.F. and Ryan, J.F. 1972. Chemical ionization mass spectrometry studies. Nitric oxide as a reagent gas. J. Chem. Soc. Chem. Commun. 620-621.

Johnson, R.P. 1973. Scent marking in mammals. Anim. Behav. 21:521-535

Karlsen, J. 1972. Microanalysis of volatile compounds in biological material by means of gas liquid chromatography. J. Chromatog. Sci. 10:642-644.

Karlson, P. and Butenandt, A. 1959. Pheromones (ectohormones) in insects. Ann. Rev. Entomol. 4:39-58.

King, S.S.T. 1973. Application of infrared Fourier transform spectroscopy to analysis of micro samples. J. Agric. Food Chem. 21: 526-530.

Law, J.H. and Regnier, F.E. 1971. Pheromones. Ann. Rev. Biochem. 40:533-548.

Leoni, V., Puccetti, G., Grella, A. 1975. Preliminary results on the use of Tenax for the extraction of pesticides and polynuclear aromatic hydrocarbons from surface and drinking waters for analytical purposes. J. Chromatogr. 106:119-124.

Levy, G.C. and Nelson, G.L., 1972. Carbon-13 magnetic resonance for organic chemists. John Wiley and Sons, Inc., N.Y.

Look, M. in press. Determining release rates of 3-methyl-2-cyclohexen-1-one, anti-aggregation pheromone of Dendroctonus pseudotsugae. J. Chem. Ecol.

McDonald, R.S. 1976. Infrared spectrometry. Anal. Chem. 48:196R-216R.

Michael, R.P., Bonsall, R.W., and Warner, P. 1974. Human vaginal secretions: volatile fatty acid content. Science 186:1217-1219.

Mori, K. 1975. Synthesis of optically active forms of frontalin, the pheromone of Dendroctonus bark beetles. Tetrahedron 31:1381-1384

Morrison, J.D. and Mosher, H.S. 1971. Asymmetric Organic Reactions. Prentice-Hall, Inc., Englewood Cliffs, N. J.

Müller-Schwarze, D. 1974. Olfactory recognition of species, groups, individuals, and physiological states among mammals, in Birch M. (ed.). Pheromones, North-Holland Publishing Co., Amsterdam.

Müller-Schwarze, D., Müller-Schwarze, C., Singer, A.G., and Silverstein, R.M. 1974. Mammalian pheromone: identification of active component in the subauricular scent of the male pronghorn. Science. 183:860-862.

Müller-Schwarze, D., Silverstein, R.M., Müller-Schwarze, C., Singer, A.G., and Volkman, N.J. 1976. Response to a mammalian pheromone and to its geometric isomer. J. Chem. Ecol. 2:389-398.

Mykytowycz, R. 1974. Odor in the spacing behavior of mammals, in Birch, M. (ed.). Pheromones, North-Holland Publishing Co., Amsterdam.

Mykytowycz, R. and Gambale, S. 1969. The distribution of dung-hills and the behavior of free-living rabbits, Oryctolagus cuniculus (L.), on them. Forma et Functio 1:333-349.

Nordlund, C.H. and Lewis, W.J. 1976. Terminology of chemical releasing stimuli in intraspecific and interspecific interactions. J. Chem. Ecol. 2:211-220.

Parsons, J.S. and Mitzner, S. 1975. Gas chromatographic method for concentration and analysis of traces of industrial organic pollutants in environmental air and stacks. Environ. Sci. Technol. 9:1053-1058.

Patterson, R.L.S. 1968. Acidic components of boar preputial fluid. J. Sci. Fd. Agric. 19:38-40.

Peacock, J.W., Cuthbert, R.A., Gore, W.E., Lanier, G.N., Pearce, G.T. and Silverstein, R.M. 1975. Collection on Porapak-Q of the aggregation pheromone of Scolytus multistriatus (Coleoptera:Scolytidae). J. Chem. Ecol. 1:149-160.

Pearce, G.T., Gore, W.E., and Silverstein, R.M. 1976. Synthesis and absolute configuration of multistriatin. J. Org. Chem. 41:2797-2803.

Pellizzari, E.D., Carpenter, B.H., Bunch, J.E. and Sawicki, E. 1975a. Collection and analysis of trace organic vapor pollutants in ambient atmospheres. Thermal desorption of organic vapors from sorbent media. Environ. Sci. Technol. 9:556-560.

Pellizzari, E.D., Bunch, J.E., Carpenter, B.H. and Sawicki, E. 1975b. Collection and analysis of trace organic vapor pollutants in ambient atmospheres. Technique for evaluating concentration of vapors by sorbent media. Environ. Sci. Technol. 9:552-555.

Persoons, C.J., and Ritter, F.J. 1975. Binary sex pheromone mixtures in Tortricidae. Role of positional and geometrical isomers. Z. Angew. Entomol. 77:342-346.

Plummer, E.L., Stewart, T.E., Byrne, K.J., Gore, W.E., Pearce, G.T., Silverstein, R.M. 1976. Determination of the enantiomeric purity of several insect pheromone alcohols. J. Chem. Ecol. 2: in press.

Politzer, I.R., Githens, S., Dowty, B.J., Laseter, J.L. 1975. Gas chromatographic evaluation of the volatile constituents of lung, brain and liver tissues. J. Chromatog. Sci. 13:378-379.

Preti, G. and Huggins, G.R. 1975. Cyclical changes in volatile acidic metabolites of human vaginal secretions and their relation to ovulation. J. Chem. Ecol. 1:361-376.

Preti, G., Muetterties, E.L., Furman, J.M., Kennelly, J.J., and Johns, B.E. 1976. Volatile constituents of dog (Canis familiaris) and coyote (Canis Latraus) anal sacs. J. Chem. Ecol. 2:177-186.

Ralls, K. 1971. Mammalian scent marking. Science 171:443-449.

Riley, R.G., Silverstein, R.M., and Moser, J.C. 1974a. Biological responses of Atta texana to its alarm pheromone and the enantiomer of the pheromone. Science 183:760-762.

Riley, R.G., Silverstein, R.M., and Moser, J.C. 1974b. Isolation, identification, synthesis, and biological activity of volatile compounds from the heads of Atta ants. J. Insect Physiol. 20:1629-1637.

Roelofs, W.L. and Comeau, A. 1971. Sex pheromone perception: electroantennogram responses of the red-banded leaf roller moth. J. Insect Physiol. 17:1969-1982.

Roelofs, W.L., Comeau, A., Hill, A. and Milicevic, G. 1971. Sex attractant of the codling moth: characterization with electroantennogram technique. Science 174:297-299.

Russell, J.W. 1975. Analysis of air pollutants using sampling tubes and gas chromatography. Environ. Sci. Technol. 9:1175-1178.

Russell, G.F. and Hills, J.I. 1971. Odor differences between enantiomeric isomers. Science 172: 1043-1044.

Singer, A.G., Agosta, W.C., O'Connell, R.J., Pfaffmann, C., Bowen, D.V., and Field, F.H. 1976. Dimethyl disulfide: an attractant pheromone in hamster vaginal secretion. Science 191:948-950.

Silverstein, R.M., Brownlee, R.G., Bellas, T.E., Wood, D.L., and

Browne, L.E. 1968. Brevicomin:principal sex attractant in the frass of the female western pine beetle. Science 159:889-891.

Silverstein, R.M., Rodin, J.O., Wood, D.L., and Browne, L.E. 1966. Identification of two new terpene alcohols from frass produced by Ips confusus boring in ponderosa pine. Tetrahedron 22:1929-1936.

Silverstein, R.M. and Young, J.C. 1976. Insects generally use multicomponent pheromones, in Beroza, M. (ed.). Pest management with insect sex attractants and other behavior-controlling chemicals, ACS Symposium Series, No. 23, American Chemical Society, Washington, D.C.

Ställberg-Stenhagen, S. 1972. Splitter-free all glass intake system for glass capillary gas chromatography of volatile compounds from biological material. Chemica Scripta 2:97-100.

Stewart, T.E., Plummer, E.L., McCandless, L.L., West, J.R., and Silverstein, R.M., in press. Determination of enantiomer composition of several bicyclic ketal insect pheromone components. J. Chem. Ecol.

Stoddart, D.M. 1974. The role of odor in the social biology of small mammals, in Birch, M. (ed.). Pheromones. North-Holland Publishing Co., Amsterdam.

Stoddart, D.M., Aplin, R.T., and Wood, M.J. 1975. Evidence for social difference in the flank organ secretion of Arvicola terrestris (Rodentia:Microtinae). J. Zool., Lond. 177:529-540.

Stothers, J.B. 1972. Carbon-13 NMR Spectroscopy. Academic Press, N. Y.

Thiessen, D.D., Regnier, F.E., Rice, M., Goodwin, M., Isaacks, N., and Lawson, N. 1974. Identification of a ventral scent marking pheromone in the male Mongolian gerbil (Meriones unguiculatus). Science 184:83-85.

Thiessen, D.D. and Rice, M. in press. Mammalian scent gland marking and social behavior. Psycholog. Bull.

Thiessen, D.D. 1976. Thermoenergetics and the evolution of phero-mone communication, in Spraguet, J.M. and Epstein, A.N. (eds.). Progress in Psychobiological and Physiological Psychology. Academic Press.

Tumlinson, J.H., and Heath, R.R. 1976. Structure elucidation of insect pheromones by microanalytical methods. J. Chem. Ecol. 2:87-99.

Vandenbergh, J.G., Whitsett, J.M., Lombardi, J.R. 1975. Partial isolation of a pheromone accelerating puberty in female mice. J. Reprod. Fert. 43:515-523.

van Dorp, D.A., Klok, R., and Nugteren, D.H. 1973. New macrocyclic compounds from the secretions of the civet cat and the muskrat. Recl. Trav. Chim. Pays-Bas 92:915-928.

Versino, B., de Groot, M., Geiss, F. 1974. Air pollution - sampling by adsorption columns. Chromatographia 7:302-304.

Walton, H.F. 1976. Ion exchange and liquid column chromatography. Anal. Chem. 48:52R-66R.

Wasson, J.R. and Lorenz, D.R. 1976. Nuclear magnetic resonance spectroscopy. Anal. Chem. 48:246R-261R.

Wheeler, J.W., von Endt, D.W., Wemmer, C. 1975. 5-Thiomethylpentan-2, 3-dione. A unique natural product from the striped hyena. J. Amer. Chem. Soc. 97:441-442.

Whilen, S.H. 1971. Resolving agents and resolution in organic chemistry, in Allinger, N.L. and Eliel, E.L. (eds.). Topics in Stereochemistry, Vol. 6. Wiley-Interscience, N. Y.

Whittacker, R.H. and Feeny, P.P. 1971. Allelochemics: chemical interaction between species. Science 171:757-770.

Wood, D.L., Browne, L.E., Ewing, B., Lindahl, K., Bedard, W.D., Tilden, P.E., Mori, K., Pitman, G.B., and Hughes, P.R. 1976. Western pine beetle: specificity among enantiomers of male and female components of an attractant pheromone. Science 192:896-898.

Yarger, R.G., Smith III, A.B., Preti, G., and Epple, G. in press. The major volatile constituents of the scent mark of a South American primate (Saguinus fusicollis, Callithricidae). J. Chem. Ecol.

Young, J.C. and Silverstein, R.M. 1975. Biological and chemical methodology in the study of insect communication, in Moulton, D.G., Turk, A., and J.W. Johnson, Jr., (eds.). Methods in Olfactory Research. Academic Press, N. Y.

Zlatkis, A., Lichtenstein, H.A., Tishbee, A. 1973a. Concentration and analysis of trace volatile organics in gases and biological fluids with a new solid adsorbent. Chromatographia 6:67-70.

Zlatkis, A., Bertsch, W., Lichtenstein, H.A., Tishbee, A., Shunbo, F., Liebich, H.M., Coscia, A.M., and Fleischer, N. 1973b. Profile of volatile metabolites in urine by gas chromatography - mass spectrometry. Analyt. Chem. 45:763-767.

Zlatkis, A., Bertsch, W., Bafus, D.A., Liebich, H.M. 1974. Analysis of trace volatile metabolites in serum and plasma. J. Chromatogr. 91:379-383.

CHEMICAL AND BEHAVIORAL COMPLEXITY IN MAMMALIAN CHEMICAL COMMUNICATION SYSTEMS: GUINEA PIGS (CAVIA PORCELLUS), MARMOSETS (SAGUINUS FUSCICOLLIS) AND HUMANS (HOMO SAPIENS)

George Preti, Amos B. Smith, III, and Gary K. Beauchamp
Monell Chemical Senses Center* and Departments of
Chemistry and Otorhinolaryngology and Human Communication
University of Pennsylvania
*3500 Market Street
Philadelphia, Pennsylvania 19104

The odorous secretions produced by mammals have long interested man. In some ancient civilizations, the scents produced by sacred or powerful animals were worn by men in an effort to capture some of its power (Kingston, 1965). In more recent times, organic chemists and perfumers have sought to identify and employ the volatile constituents of mammalian secretions and excretions in perfumes and colognes while animal behaviorists and reproductive biologists have explored the role such constituents play in olfactory chemical communication. With the establishment that volatiles in excreted body fluids play a significant role in mammalian communication organic chemists have become interested in the isolation and identification of these compounds. Such studies require an interdisciplinary approach.

The odorous compounds being sought are found within the complex mixture of organic compounds present in mammalian secretions and excretions. Using the appropriate collection, concentration and high resolution chromatographic techniques, it is possible for the organic chemist to separate the compounds in bodily fluids employing their properties of volatility, molecular weight and/or functional group. Once a separation is effected, the volatiles of the mixture may be identified, quantitated and examined for male-female differences. In addition, profiling of the metabolites found in the secretion during the estrous or menstrual cycle may reveal cyclical changes in these compounds which correlate with the changing levels in sex steroids in different physiological states; however, without behavioral evidence the chemist can only speculate that a compound(s) specific to one sex or a constituent(s) which increase during estrous and/or ovulation are good candidates for communicating this information to conspecifics.

Those who study the animal's behavior may show that certain information (i.e., species, sex, individuality, social and reproductive status) is communicated by odor. They may also identify the secretion or excretion which carries the odor of interest. In addition, a behavioral or endocrinological change caused by the olfactory message may be documented and subsequently serve as a bioassay. However, the behaviorist is unlikely to be able to identify the compound(s) which control the behavior of interest.

Consequently, scientists from both disciplines, working independently, may obtain information about a mammal which is of traditional interest to their area of specialty. However, only through a collaborative effort will the information-carrying constituents be established with the bioassay guiding the chemist's analytical efforts. Joint studies have, in the past, been difficult to arrange due in part to the departmental structure within which most academic scientists operate. Recently, this situation has begun to change and several interdisciplinary groups are currently investigating a variety of mammalian species.

Table I lists the studies which report the identification of volatile organic compounds eliciting a specific behavior. The first of these studies by Brownlee, Silverstein, Müller-Schwarze and Singer (1969) identified cis-4-hydroxydodic-6-enoic acid lactone as the major component of the male tarsal scent of the black-tailed deer (Odocoileus hemionus columbianus). Synthetic as well as naturally occuring lactone, when offered to the deer, elicited responses typically shown when a new or strange deer approaches the herd. More recently, Müller-Schwarze, Müller-Schwarze, Singer and Silverstein (1974) have identified the 8 major constituents of the subauricular gland from the male prong-horn (Antilocapra americana). One of these compounds, isovaleric acid, was found to be as effective as the composite mark in eliciting the marking response.

Estrous female domestic pigs assume the lordosis posture when presented with the odor of boar's saliva. Melrose, Reed and Patterson (1971) have identified two androgen metabolites (see Table I) which elicit this behavior. Thiessen and co-workers (1974) have found phenylacetic acid to be the major component of the ventral scent mark of the male Mongolian gerbil (Meriones unguiculatus). The gland is used in scent marking and phenylacetic acid elicited the same amount of exploratory interest from males as did the mark itself. Very recently, Singer et al. (1976) found dimethyl disulfide, isolated from estrous hamster vaginal secretions is an attractant for male hamsters. The disulfide was from 60-80% as effective as natural vaginal secretions in eliciting approaching, digging and sniffing behavior in males.

Table I

Mammalian Chemical Communication: Collaborative Investigations

Investigators	Animal/ Secretion	Active Compound(s)	Reported Function
Brownlee et al., 1969	Male Black-tailed Deer/ Tarsal Gland	Cis-4-hydroxy-dodec-6-enoic acid lactone	Individual and sexual recognition
Melrose et al., 1971	Boar/Salivary Secretions	5α-androst-16-en-3-one; 3α-hydroxy-5α-androst-16-ene	Induces Lordosis
Michael et al., 1971	Rhesus Monkey/ Vaginal Secretions	Acetic, Propionic, Isobutric, Butyric and Isovaleric	Attracts and arouses males
Müller-Schwarze et al., 1974	Male Pronghorn Deer/Subauricular Gland	Isovaleric Acid	Individual recognition; Territorial maintenance
Thiessen et al., 1974	Male Mongolian Gerbil/Ventral Sebaceous Glands	Phenylacetic Acid	Territorial maintenance
Goldfoot et al., 1976	Rhesus Monkey/ Vaginal Secretions	Acetic, Propionic, Isobutyric, Butyric, Isovaleric	No effect
Singer et al., 1976	Hamster / Vaginal Secretions	Dimethyl disulfide	Attracts males

One of the most intriguing investigations in Table I is that of Michael, Keverne and Bonsall (1971). They reported that short chain aliphatic acids (see Table I) in the vaginal secretions of estrous female rhesus monkeys stimulated mating behavior in males. A synthetic mixture of these acids, when applied to the sexual skin of unreceptive females, attracted and stimulated males to attempt copulation. Goldfoot, Goy, Kravitz and Freeman (1976), however, were not able to confirm the earlier findings by Michael and co-workers (1971), thereby adding controversy to the primate chemical communication area.

A recent review (Beauchamp, Doty, Moulton and Mugford, 1976) has questioned the use of the term "pheromone" as applied to mammalian behavior and suggests that the compounds listed in Table I may not be solely responsible for the specific biological activities ascribed to them. Thus, both the controversy and the criticism generated by investigators engaged in this new research area suggest that it is still evolving, and with many questions yet to be answered.

Table II shows studies which investigated organic constituents of mammalian excretions and secretions thought to contain olfactorily active compounds. Several of these reports are preliminary in nature with bioassays of identified compounds in progress or the active constituents proving difficult to isolate, even with bioassays in hand. Studies by Smith, Yarger and Epple (1976) and Beauchamp, Beruter and Muetterties (1974), (Table II), reflect the behavioral and chemical complexities involved in studying olfactory communication. The problems inherent in such investigations are magnified in examining the possibility of chemical signaling in man as discussed below.

Guinea Pig Urine

Male guinea pigs (Cavia porcellus) are attracted to conspecific urine (Beauchamp, 1973) with the male discriminating between male and female urine on the basis of olfaction (Beauchamp, Magnus, Shmunes and Durham, 1976). Urine from castrated males cannot be differentiated from female urine by the male test subjects, however, treatment of the castrates with testosterone propionate restores the discrimination (Beauchamp, unpublished, 1974). Consequently, the molecules responsible for the differential responses to male and female urine appears to be under androgenic control.

A simple chemical approach to this problem's solution would appear to involve gas chromatographic (GC) analysis of volatiles from males and females. Volatiles present in females' but not in males' urine might be likely attractants and/or volatiles present in the male but not female urine would be likely substances reducing the attractiveness of male urine. These naive assumptions, however,

Table II

Investigations of Mammalian Secretions and Excretions

Investigators	Animal	Compounds	Reported Function
Lederer, 1949	Canadian Beaver	Complex Mixture	Used in perfumes Individual recognition?
Ruzicka, 1926	Abyssinian Civet Cat Musk Deer	Civetone Muscone	Musk Odors – used in perfumes
Goodrich and Mykytowycz, 1972; Hesterman et al., 1976	Rabbit – Anal Chin and Inguinal Glands	Hydrocarbons, fatty acids, esters and glycerides	Scent Marking, Individual recognition
Albone and Fox, 1971	Red Fox Anal Sacs	C_2 to C_6 acids, trimethylamine	Marking Feces, Signal Alarm?
Albone et al., 1974, 1976	Red Fox and Lion Anal Sacs	4 Aromatic Acids	
Berüter et al., 1974	Domestic Guinea Pig Perineal Gland	C_2 to C_5 acids	Relates to Social Dominance, Individual Recognition
Berüter et al., 1973	Domestic Guinea Pig Urine	Varied Molecular Types	Discriminates Sex, species
Gorman et al., 1974, 1976	Indian Mongoose Anal Sacs	C_2 to C_5 acids	Individual Recognition
Wheeler et al., 1975	Striped Hyena Anal Sacs	2-Thiomethylpentane-2,3-dione	Individual Recognition
Anderson et al., 1975	Reindeer	C_7-C_{10} aldehydes C_{12}-C_{16} alchols, hydrocarbons	
Vanderberg et al., 1975	Mouse		

Table II - Continued

Anderson and Bernstein, 1975	Striped Skunk	Sulfur Compounds	Defense
Preti et al., 1976	Dog and Coyote Anal Sacs	C_2-C_6 aliphatic acids, 2-piperi-done, aldehydes, trimethylamine	Signal Alarm?
Smith et al., 1976	Marmoset Monkey Scent Mark	SAT'D and UN-SAT'D C_{20}-C_{28} n-butyric acid esters	Sex, Social Status, Repro-ductive Status

are shown not to hold.

Extensive ion-exchange chromatographic separations aimed at isolating the attractive constituents show compounds of varied type and molecular weight to be involved in the discrimination and preferences of males (Berüter, Beauchamp and Muetterties, 1973). Gas chromatography of volatiles from whole urine as well as ion-exchange separated fractions which males prefer reveal little or no differences in the resultant chromatograms from males and females. In addition, exhaustive dialysis of the male and female urine with subsequent bioassay of the nondialysable constituents revealed a preference for the female components by the male. This latter finding suggests that a protein-small molecule(s) complex plays a role in the discrimination (Berüter, Beauchamp and Muetterties, 1973; Smith, Byrne and Beauchamp, unpublished, 1975). A similar situation may be operating in the case of the puberty-accelerating substance in male rat urine (Vandenbergh, Whitsett and Lombardi, 1975).

These results show that the male guinea pig preference for female urine may be mediated by a variety of chemically different compounds. The failure to find consistent differences in the GC patterns of whole urine or fractions thereof, which relate to the male's preference, suggests that the discriminating compounds may be below the detection level of the flame ionization detector. Alternatively, in only looking at the male-female difference via GC, there is the possibility of different compounds from either sex having a similar GC retention time. They, therefore, may be wrongly assumed to be the same compounds. High resolution GC, employing open tubular columns and/or obtaining the mass spectral cracking patterns of each constituent present in male and female fractions, is currently being used to check this possibility. However, the latter is particularly laborious without a gas

chromatograph-mass spectrometer-computer facility with attendent
library generating and searching capacity (Hertz, Hites and
Biemann, 1971).

The evidence suggesting a protein-small molecule(s) complex
retaining substantial activity is a situation which until very
recently (Vandenbergh et al., 1975) has not been previously encountered in other insect or mammalian pheromone investigations.
These molecular complexes represent another potentially confusing
aspect to be resolved aside from the task of isolating compounds
from a mixture as complex as mammalian urine.

Marmoset Scent Mark

Chemical-behavioral studies on chemical cues in sexual and
social communication in non-human primates have focused primarily
on the rhesus monkey (Michael et al., 1971; Curtis et al., 1971;
Goldfoot et al., 1976). As discussed above, the status of the
compounds originally thought to be attractants has now become
clouded. Another interdisciplinary group has recently published
findings from a study of the South American marmoset monkey,
Saguinus fuscicollis (Smith, Yarger and Epple, 1976). This primate
utilizes its scent mark to communicate a variety of information
including the sex, social status and individual identity to conspecifics (Epple, 1974). Of particular interest is the role the
odors of the scent mark may play in the reproduction of these
animals. Social dominance appears to be a prerequisite for reproductive success since only dominant females become pregnant
and bear offspring (Epple, 1975). The presence of a dominant
female inhibits reproduction in all other females of a group and
may delay sexual maturation in her own female offspring. Since
scent marking is used to demonstrate and maintain social dominance
it is possible that the odor of dominant females affects either
follicular maturation, ovulation, conception or the ability to
carry to term.

Scent marking behavior in the marmoset consists of rubbing
circumgenital and sternal body regions against objects in the
environment. This behavioral pattern results in the deposition of
material consisting mainly of secretions from the specialized
sebaceous and apocrine skin glands (Perkins, 1966) in the circumgenital area, mixed with urine and, in the case of the female,
possibly vaginal discharge (Epple, 1975).

The scent mark is collected on frosted glass plates, washed off
with CH_2Cl_2 and CH_3OH, concentrated to an oil and redissolved in
hexane. Marmosets respond to the hexane soluble materials in a
manner similar to intact marks (Smith, Epple and Yarger, unpublished,
1975). High resolution GC analysis followed by micropreparative GC
with subsequent gas chromatography-mass spectrometry (GC-MS), NMR,

IR and chemical transformations of the collected fractions, showed
the major volatiles of the mark to be long-chain butyrate esters.
These esters ranged in size from C_{20}–C_{28} and contained from 0 to 2
cis–double bonds in the positions shown in Figure 1.   Squalene was
also present.

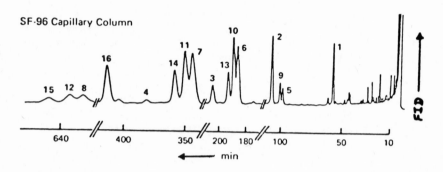

Peak	Molecular Formula	Identification
		SATURATED
1	$C_{20}H_{40}O_2$	
2	$C_{22}H_{44}O_2$	
3	$C_{24}H_{48}O_2$	
4	$C_{26}H_{52}O_2$	
		MONOENES–TYPE A
5	$C_{22}H_{42}O_2$	
6	$C_{24}H_{46}O_2$	
7	$C_{26}H_{50}O_2$	
8	$C_{28}H_{54}O_2$	
		MONOENES–TYPE B
9	$C_{22}H_{42}O_2$	
10	$C_{24}H_{46}O_2$	
11	$C_{26}H_{50}O_2$	
12	$C_{28}H_{52}O_2$	
		DIENES
13	$C_{24}H_{44}O_2$	
14	$C_{26}H_{48}O_2$	
15	$C_{28}H_{52}O_2$	
		SQUALENE
16	$C_{30}H_{50}$	

Figure 1 – Major constituents of marmoset scent mark.   Chromatogram
produced on a 500 ft. x 0.03 in. wall coated capillary column.

The butyrate esters and squalene represent >96% by weight of the marmoset mark (Smith, Yarger and Epple, 1976) and may originate from the circumgenital glands rather than from the urogenital secretions. Currently, the function the esters play in olfactory communication between the marmosets is not known, though several possibilities exist. A characteristic ester profile may serve as an individual's distinguishing odor. With 15 different esters available, the possible combinations which could be made from these compounds is very large. If particular concentrations of the esters serve to identify a marmoset, then the dominant females may have no unique features present in their scent mark (Smith, Epple and Yarger, unpublished, 1976). Their own individual pattern, which animals of the group learn to associate with the dominant female would convey all the information needed. Alternatively, the esters may serve as a fixative or diluent for more volatile compounds present at much lower concentrations. Such constituents may be seen in Figure 1 eluting between the solvent front and the first C_{20} ester.

In the studies described above, volatile substances are being sought which carry two basic pieces of information concerning an animal: its sex and its individual identity. The latter may be based on concentration variations of volatile metabolites common to all members of the species. Different concentrations of C_2-C_5 aliphatic acids found in the perineal gland (Guinea pigs) and anal sacs (Mongoose) of these animals may be characteristic of an individual and thus code individual identity. While this has yet to be demonstrated in any species, Gorman (1976) has shown that this could occur in the mongoose, since these animals can be trained to respond differentially to mixtures containing different ratios of C_2-C_5 acids. One prerequisite for this mechanism to work is that the ratios of acids remain relatively constant within an individual over time. Gorman (1976) does not report this; however, Berüter et al., (1974) found that individual male guinea pigs exhibited large fluctuations in relative concentrations of acids over 30 to 60 day periods. While these variations could be due to an unspecified defect in the collection-fractionation procedure, they suggest caution in assigning to these acids a role in coding of individual olfactory identity, at least for the guinea pig.

Volatiles conveying an animal's gender and reproductive status may be by-products derived from metabolic pathways controlled by sex hormones. Detailed studies of the same secretions or excretions from males and females of the animal have not been numerous (Goodrich and Mykytowycz, 1972). In much of our work (Berüter, Beauchamp and Muetterties, 1973; Smith, Yarger and Epple, 1976; Preti et al., 1976) we have not found consistent qualitative differences in the volatile constituents. Much of the information to be deciphered identifying sex, time of estrous (or ovulation) may thus be coded by a quantitative difference in a variety of constituents - as

discussed above for individual recognition. This would then require subsequent careful quantitative work to accurately define the
amount of each compound present in males and females, individuals,
or stage in the estrous cycle. In addition, minor components, whose
presence are obvious only to the animal's olfactory system may be
critically important. Such complexities reflect the need for the
chemist to rely closely on the behavioral/chemical collaboration and
not solely on his instruments.

Human Secretions

Since chemical communication occurs in such a diverse number
of mammals, many have wondered whether this acts in some subliminal
fashion to effect human reproductive biology. Much of the anecdotal
evidence suggesting the notion of human chemical communication has
been reviewed in this volume (see Chapter by R.L. Doty) and elsewhere (Doty, 1972; Doty, 1976; Comfort, 1974; Schneider, 1971). In
addition, recent studies have appeared which discuss the small
organic constituents in human secretions with respect to their
acting as chemical communicators (Waltman et al., 1973; Michael,
Bonsall and Warner, 1974; Brooksbank, Brown and Gustafsson, 1974;
Preti and Huggins, 1975). Both axillary sweat and vaginal secretions have been suggested as possible carriers of volatiles used in
olfactory communication (Comfort, 1974; Brooksbank, Brown and
Gustafsson, 1974).

The effects of odors in regulating social interactions in
humans, in a manner analagous to that seen in other mammals (see
above), has not yet been adequately demonstrated, experimentally.
Hence, no bioassays exist with which to test volatiles in human
secretions or excretions for "activity". In addition, a recent
psychophysical study using human vaginal secretions demonstrated
that the time of ovulation could not be reliably predicted from
odors of individual menstrual cycles (Doty, Ford, Preti and
Huggins, 1975). Consequently, the chemist investigating human
secretions can only look for "good candidates" based on their
structure, appearance in males or females and/or cyclical appearance in response to changes in circulating sex steroids.

Although no detailed studies have appeared which survey all
volatiles in human axillary sweat, Brooksbank and co-workers (1974)
have identified 5α-androst-16-en-3α -ol in this secretion. They
feel it is a likely candidate for a "human pheromone" but do not
speculate as to what message it may convey.

Human vaginal secretions are thought to consist of the constituents listed in Table III (Moghissi, 1972; Cohen, 1969). Components b, c and e are known to change in type and amount under the
influence of sex steroid levels; consequently, metabolites in these
components may also vary. These changes in addition to Michael and

Table III

Components of Human Vaginal Secretions

a) Vulval secretions from sebaceous, sweat, Bartholin's and Skeen's glands.

b) Mucus secretions from the cervix.

c) Uterine and oviductal fluids.*

d) Transudate through the vaginal walls.**

e) Exfoliated cells of the vaginal mucosa.

* Doyle, Ewers and Sapit, 1960; Hafez and Black, 1969

** Masters and Johnson, 1966

co-workers (1971) findings with rhesus monkey vaginal secretions (see Table I) suggest vaginal secretions as likely candidates for carrying "biologically active" volatiles.

The rhesus monkey results appear to have had directing influence on some workers regarding the compounds investigated in the vaginal milieu. Both Michael and co-workers (1974, 1975) as well as Waltman et al. (1973) have both examined vaginal secretions but only with respect to its C_2-C_5 aliphatic acid content using GC only. Michael et al. (1974, 1975) found larger amounts of these acids in normally cycling subjects when compared to subjects on birth control pills. These investigators also report the acids increase to maximum concentration at midcycle and then decrease; however, no documentation of the day of ovulation or correlating the amounts of acids produced with ovulation was reported.

More detailed studies of the vaginal secretions have shown them to be more complex than simple mixtures of C_2-C_5 aliphatic acids (Keith et al. 1975; Preti and Huggins, 1975). Profiling the secretion's small organic compounds by GC and GC-MS, with documentation of the time of ovulation, reveal different types of compounds, some of which increase in concentration near ovulation (Preti and Huggins, 1975; Huggins and Preti, 1976). These detailed analysis show that the C_2-C_5 acids are not produced by all women. Figures 2 and 3 show representative GC traces from acid producers and non-producers, respectively. Also listed are the constituents which have been identified.

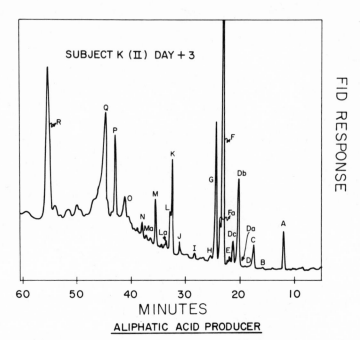

* NOT PRESENT IN NON-ACID PRODUCERS

Figure 2 — Gas chromatogram (10% Carbowax 20M) of volatiles from an "acid producer". Sample obtained 3 days after ovulation.

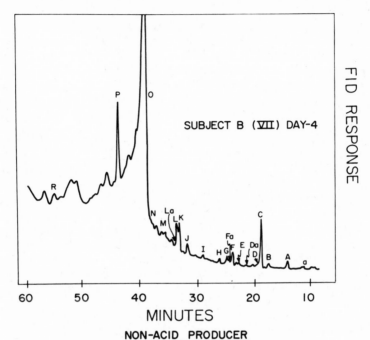

NON–ACID PRODUCER

PEAK	COMPOUND(S)	PEAK	COMPOUND(S)
a	Pyridine	H	N-C17 hydrocarbon
A	3-Hydroxy-2-butanone	I	N-C18 hydrocarbon
B	Maleic Anhydride (T)	J	Dimethylsulfone
	Nonanal (T)		N-C19 hydrocarbon
C	Acetic Acid	K	Internal STD -Heptanoic Acid
D	Furfural	L	N-Dodecanol
Da	Benzaldehyde	La	Phenol
E	Propylene Glycol	M	P-Cresol
F	Ethylene Glycol	N	N-Tetradecanol
Fa	2-Isopentyl Furan (T)	O	Lactic Acid
G	Furfuryl Alcohol	P	N-Hexadecanol
	Phenylacetaldehyde*	R	Unknown

*NOT PRESENT IN ACID PRODUCERS

Figure 3 – Gas chromatogram (10% Carbowax 20M) of volatiles from a "non-acid producer".   Sample obtained 4 days before ovulation.

In complex biological mixtures it is common to find several constituents having very similar or identical GC retention times on the same chromatographic substrate. In such cases, mass spectrometry is needed to elucidate a potentially confusing situation. Such a situation is depicted in Figure 4 which show the portions of two chromatograms where C_2-C_5 acids elute. Peaks F and G in both subjects have relative retention times identical to n-butyric acid and the C_5 methyl substituted acids (see Figure 2). However, analysis by GC-MS demonstrates the compounds in subject A (a non-acid producer) to be composed of ethylene glycol (peak F) and phenylacetaldehyde/furfuryl alcohol (peak G) while in subject E they correspond to the acids noted above. Consequently, speculation on the nature of volatiles present is difficult if it is based solely on GC retention times.

When the concentration of C_2-C_5 acids are plotted with respect to the day of ovulation, maximum amounts are seen to occur during mid-cycle or during the luteal phase. Figure 5 illustrates this with data from 8 representative cycles from 4 acid producers (Huggins and Preti, 1976). These data show a great deal of individual heterogeneity in the distribution of these acids. In addition, the acid producers tend to produce larger amounts of other volatiles not found in the non-producers (see Figure 2). The above results show that these acids are not present in all woman and when present, need not be maximized at midcycle. In addition, those women who produce these also produce large amounts of other compounds whose odor undoubtedly add to and modify the odor of the C_2-C_5 acids. Consequently, no one type of compound can assume to be a sole contributor to vaginal odor. In the absence of appropriate bioassays, any or all volatiles present may contribute to any alleged communicatory function the odors may have.

Figure 6 illustrates some of the compounds having larger molecular weights, and in some cases little or no volatility, also present in the secretions. Lactic acid and urea (chromatographed as their trimethylsilyderivatives) are present in all women's secretions and increase in a characteristic manner around ovulation (Huggins and Preti, 1976; Preti and Huggins, 1975). However, due to the low volatility of these compounds they are not "good candidates" for olfactorily communicating ovulation.

Several effects in both human children and adults have been speculated as being mediated by odors (Comfort, 1974; Anon, 1971). However, due to the complexities of human behavior and the number of modes of signaling we use, considerable work will be needed to prove odors are involved. Until that time, speculation is bound to continue, given the interest in the subject. In addition, detailed characterizations of the structure and abundance of the vaginal secretions small organic compounds shall continue due to

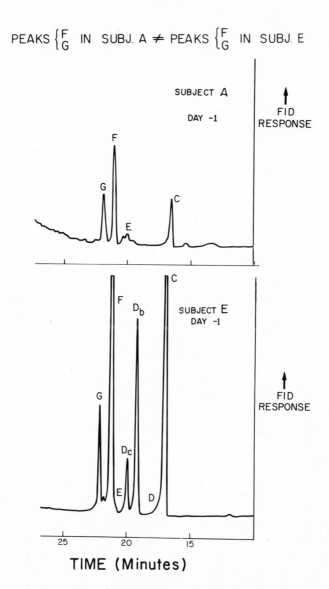

Figure 4 – Portions of gas chromatograms (10% Carbowax 20M) shown are where C_2-C_5 aliphatic acids elute. Compounds F and G in subject A (non-acid producer) have identical retention time to F and G in subject E (acid producer) but are different compounds.

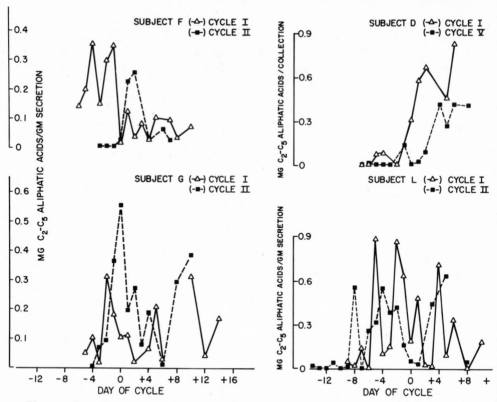

Figure 5 – Concentration of C_2–C_5 aliphatic acids during the menstrual cycle of 4 subjects. Day 0 equals day of ovulation.

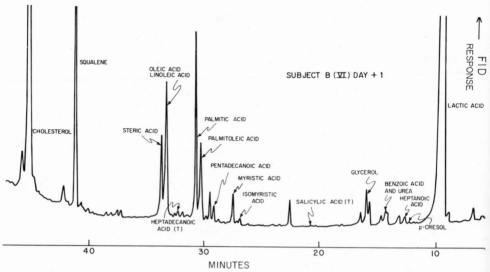

Figure 6 – Compounds chromatographed (3% SE–30) as their trimethyl-silyl derivatives. Subject B is a "non-acid producer".

the diagnostic value for prediction of ovulation (Huggins and Preti, 1976) and possibly reproductive track pathologies.

Acknowledgements – The authors wish to acknowledge support of their research through NSF grants GB-34879; BNS 76-01642; GB 33104x; BMS 75-13164; BNS 75-17119 and by Rockefeller Foundation Grant RF 72018.

References

Albone, E.S. and Fox, M.W., 1971. Anal gland secretion of the red fox. Nature, 233:569-570.

Albone, E.S., Eglinton, G., Walker, J.M. and Ware, G.C., 1974. The anal sac of the red fox (Vulpes vulpes): its chemistry and microbiology. A comparison with the anal sac secretion of the lion (Panthera leo). Life Sciences, 14:387-400.

Albone, E.S. and Perry, G.C., 1976. Anal sac secretion of the red fox, Vulpes vulpes; volatile fatty acids and diamines: implications for a fermentation hypothesis of chemical recognition. J. Chem. Ecology, 2:101-111.

Andersen, K.K. and Bernstein, D.T., 1975. Some chemical constituents of the scent of the striped skunk (Mephitis mephitis). J. Chem. Ecology, 1:493-499.

Andersson, G., Andersson, K., Brudin, A. and Rappe, C., 1975. Volatile compounds from the tarsal scent gland of reindeer (Rangifer tarandus). J. Chem. Ecology, 1:275-281.

Anonymous, 1971. A human pheromone? Lancet 1:279.

Beauchamp, G.K., 1973. Attraction of male guinea pigs to conspecific urine. Physiol. Behav., 10:589-594.

Beauchamp, G.K., Doty, R.L., Moulton, D.G. and Mugford, R.A., 1976. The pheromone concept in mammalian chemical communication: a critique, in R.L. Doty (ed), Mammalian Olfaction, Reproductive Processes, and Behavior. Academic Press, New York, in press.

Beauchamp, G.K., Magnus, J.G., Shmunes, N.T. and Durham, T., 1976. Effects of olfactory bulbectomy on social behavior of male guinea pigs (Cavia porcellus). Physiol. Behav.: in press.

Berüter, J., Beauchamp, G.K. and Muetterties, E.L., 1973. Complexity of chemical communication in mammals: Urinary components mediating sex discrimination by male guinea pigs. Biochem. Biophys. Res. Comm., 53:264-271.

Berüter, J., Beauchamp, G.K. and Muetterties, E.L., 1974. Mammalian chemical communication: perineal gland secretion of the guinea pig. Physiol. Zool., 47:130-136.

Brooksbank, B.W.L., Brown, R. and Gustafsson, 1974. The detection of 5$\propto$-Androst-16-en-3α-ol in human male axillary sweat. Experientia, 30:864-865.

Brownlee, R.G., Silverstein, R.M., Müller-Schwarze, D. and
    Singer, A.G., 1969. Isolation, identification and function
    of the chief component of the male tarsal scent in black-
    tailed deer. Nature, 221:284-285.
Cohen, L., 1969. Influence of pH on vaginal discharges. Brit. J.
    Vener. Dis., 45:241-245.
Comfort, A., 1974. The likelihood of human pheromones, p. 386-394,
    in M.C. Birch (ed), Pheromones. North Holland Research
    Monograph, Frontiers in Biology, Vol. 32, North Holland
    Publishing Company, Amsterdam - London.
Curtis, R.F., Ballantine, J.A., Keverne, E.G., Bonsall, R.W. and
    Michael, R.P., 1971. Identification of primate sexual
    pheromones and the properties of synthetic attractants.
    Nature, 232:396-398.
Doty, R.L., 1972. The role of olfaction in man - sense or non-
    sense? P. 143-159, in S.H. Barthey (ed), Preception in
    Everyday Life. Harper and Row, New York.
Doty, R.L., Ford, M., Preti, G. and Huggins, G.R., 1975. Changes
    in the intensity and pleasantness of human vaginal odors
    during the menstrual cycle. Science, 190:1316-1318.
Doty, R.L., 1976. Reproductive endocrine influences upon human
    nasal chemoreception: a review, in R.L. Doty (ed),
    Mammalian Olfaction, Reproductive Processes, and Behavior.
    Academic Press, New York.
Doyle, J.B., Ewers, F.J. and Sapit, D., 1960. The new fertility
    testing tape: a predictive test of the fertile period.
    J. Am. Med. Assoc., 172:1744-1750.
Epple, G., 1974. Pheromones in primate reproduction and social
    behavior, p. 131-155, in W. Montagna and W.A. Sadler (eds)
    Reproductive Behavior. Plenum Publishing Corp., New York.
Epple, G., 1975. The behavior of marmoset monkeys (Callithricidae)
    p. 195-235, in L.A. Rosenblum (ed) Primate Behavior:
    Developments in Field and Laboratory Research, Vol. 4.
    Academic Press, New York.
Goldfoot, D.A., Kravetz, M.A., Goy, R.W. and Freeman, S.K., 1976.
    Lack of effect of vaginal lavages and aliphatic acids on
    ejaculatory responses in rhesus monkeys: Behavioral and
    chemical analyses. Horm. Behav.,7:1-27.
Goodrich, B.S. and Mykytowycz, R., 1972. Individual and sex differ-
    ences in the chemical composition of the pheromone-like
    substances from the skin glands of the rabbit. J. Mamm.,
    53:540-548.
Gorman, M.L.,Nedwell, D.B. and Smith, R.M., 1974. An analysis of
    the contents of the anal pockets of Herpestes auropunctatus
    (Carnivora: Viverridae). J. Zool., Lond., 172:389-399.
Gorman, M.L., 1976. A mechanism for individual recognition by
    odour in Herpestes auropunctatus (Carnivora: Viverridae)
    Anim. Behav., 24:141-145.

Hafez, E.S.E. and Black, D.L., 1969. The mammalian uterotubal
    junction, p. 114, in E.S.E. Hafez and D.L. Black (eds)
    The Mammalian Oviduct: Comparative Biology and Methodology.
    The University of Chicago Press.

Hertz, H.S., Hites, R.A. and Biemann, K., 1971. Identification of
    mass spectra by computer-searching a file of known spectra.
    Anal. Chem., 43:681-690.

Hesterman, E.R., Goodrich, B.S. and Mykytowycz, R., 1976. Behavioral
    and cardiac responses of the rabbit, Oryctolagus cuniculus,
    to chemical fractions from anal gland. J. Chem. Ecology,
    2:25-37.

Huggins, G.R. and Preti, G., 1976. Volatile constituents of
    human vaginal secretions. Am. J. Obstet. Gynecol., 124:
    385-401.

Keith, L., Stromberg, P., Krotoszynski, B.K., Shah, J. and
    Dravnieks, A., 1975. The odors of the human vagina.
    Arch. Gynäk, 220:1-10.

Kingston, B.H., 1965. The chemistry and olfactory properties of
    musk, civet and castoreum, Internal Proceedings of the
    Second International Congress of Endocrinology, 209-214.

Lederer, E., 1949. Chemistry and biochemistry of some mammalian
    secretions and excretions. J. Chem. Soc., 2115-2125.

Masters, W.H. and Johnson, V.E., 1966. Human Sexual Response.
    Little, Brown and Company, Boston. p. 68-100.

Michael, R.P., Keverne, E.B. and Bonsall, R.W., 1971. Pheromones:
    isolation of male sex attractants from a female primate.
    Science, 172:964-966.

Michael, R.P., Bonsall, R.W. and Warner, P., 1974. Human vaginal
    secretions: volatile fatty acid content. Science, 186:
    1217-1219.

Michael, R.P., Bonsall, R.W. and Kutner, M., 1975. Volatile fatty
    acids, "copulins," in human vaginal secretions. Psychoneuro-
    endo., 1:153-163.

Melrose, D.R., Reed, H.C.B. and Patterson, R.L.S., 1971. Androgen
    steroids associated with boar odour as an aid to the
    detection of oestrous in pig artificial insemination.
    Brit. Vet. J.,127:497-501.

Moghissi, K.S., 1972. The effect of steroidal contraceptives on the
    reproductive system, p. 559, in E.S.E. Hafez and P.H. Evans
    (eds), Human Reproduction: Conception and Contraception.
    Harper and Row, Hagerstown, Maryland.

Müller-Schwarze, D., Müller-Schwarze, C., Singer, A.G. and
    Silverstein, R.M., 1974. Mammalian Pheromone: Identification
    of active component in the subauricular scent of the male
    pronghorn. Science, 183:860-862.

Perkins, E.M., 1966. The skin of the black-collared tamarin
    (Tamarinus nigricollis). Am. J. Phys. Anthrop., 25:41-69.

Preti, G. and Huggins, G.R., 1975. Cyclic changes in volatile
    acidic metabolites of human vaginal secretions and their
    relation to ovulation. J. Chem. Ecology, 1:361-376.

Preti, G., Muetterties, E.L., Furman, J., Kennelly, J.J. and
    Johns, B.E., 1976. Volatile constituents of dog (Canis
    familiaris) and coyote (Canis latrans) anal sacs. J. Chem.
    Ecology, 2:179-188.
Ruzicka, L., 1926. Helv. Chim Acta, 9:230, 715 and 1008.
Schneider, R.A., 1971. The sense of smell and human sexuality.
    Med. Aspects Hum. Sex., 5:157-168.
Singer, A.G., Agosta, W.C., O'Connell, R.J., Pfaffmann, C.,
    Bowen, D.V. and Field, F.H., 1976. Dimethyl disulfide:
    an attractant pheromone in hamster vaginal secretion.
    Science, 191:948-950.
Smith, A.B., Yarger, R.G. and Epple, G., 1976. The major volatile
    constituents of the marmoset (Saguinus fuscicollis) scent
    mark. Tet. Letters, 983-986.
Thiessen, D.D., Regnier, F.E., Rice, M., Goodwin, M., Issacks, N.
    and Lawson, N., 1974. Identification of a vential scent
    marking pheromone in the male Mongolian gerbil (Meriones
    unguidulatus). Science, 184:83-84.
Vandenbergh, J.G., Whitsett, J.M. and Lombardi, J.R., 1975.
    Partial isolation of a pheromone accelerating puberty in
    female mice. J. Reprod. Fertil., 43:515-523.
Waltman, R., Tricomi, V., Wilson, G.E., Jr., Lewin, A.H., Goldberg,
    N.L. and Chang, M.M.Y., 1973. Volatile fatty acids in vaginal
    secretions: human pheromones? Lancet, 2:496.
Wheeler, J.W., von Endt, D.W. and Wemmer, C., 1975. 5-Thiomethyl-
    pentane-2,3-dione. A unique natural product from the
    striped hyena. J. Am. Chem. Soc., 97:441-442.

ON THE CHEMICAL AND ENVIRONMENTAL MODULATION OF PHEROMONE RELEASE

FROM VERTEBRATE SCENT MARKS

Fred E. Regnier and Michael Goodwin

Department of Biochemistry, Purdue University

West Lafayette, IN 47907 U.S.A.

ABSTRACT

An equilibrium model of pheromone release from scent marks is developed employing equations from physical chemistry and chromatography. Experimental evaporation curves of phenylacetic acid, an odorous component of gerbil scent marks are given for a variety of conditions treated in the model. It is shown that the chemical nature of the pheromone, the non-pheromonal component of the scent mark, the type of surface to which the mark is applied and relative humidity have pronounced effects on the temporal pattern of pheromone release.

Key words: Scent mark, pheromones, odor release, phenylacetic acid, Meriones unguiculatus.

INTRODUCTION

As literature rapidly accumulates on the pheromone systems of both vertebrates and invertebrates, it is apparent that in many species multiple compounds are used in semiochemic communication (Silverstein and Young, 1976). In addition to pheromonal material, numerous other compounds of a relatively non-volatile and non-pheromonal character are also found in glandular secretions. Since these compounds apparently exhibit no specific biological activity, little effort has been directed toward defining their chemical nature and even less attention has been given to their effect on the active space of pheromones associated with them.

Vertebrate scent marks are known to originate from a variety
of sources including urine, feces, vaginal secretions, and an array
of cutaneous scent glands. In those few cases that have been examined in chemical detail, it is clear that most vertebrate scent
marks are composed of a set of constituents. Broadcasting a chemical signal to a second organism is regulated by the volatility of
semiochemicals in the scent mark. Some of the chemical and environmental factors that modulate odorant release will be examined below.

Intuitively we know that the chemical and physical properties
of a given pheromone molecule govern its volatility and therefore
its temporal pattern viewed as a chemical signal. It is reasonable
to expect that a very volatile pheromone will evaporate in a few
minutes while a compound of low volatility will last for days. The
importance of temporal patterning is obvious. For example, an alarm
signal must be communicated quickly and fade quickly while a trail
or territorial mark must endure for a long time in order to be effective.

The question of the control of pheromone emission rate from a
scent mark is intriguing. Using experimental data from studies on
scent marking as well as a wealth of data on the physical chemistry
of phase transitions between gas-solid and gas-liquid phases, we
will attempt to build a theoretical model of how the total chemical
composition of a scent mark modulates the release of specific pheromones.

The first systematic mathematical analysis of chemical communication in animals was that of Bossert and Wilson (1963). Their
treatment was limited primarily to four general cases of pheromone
diffusion from an odor source. Those cases in which:
1) the pheromone is released in still air from a static source
    as a puff;
2) the pheromone is released continuously in still air;
3) the pheromone is released continuously from a moving source
    as an odor trail; and
4) the pheromone is released continuously in a wind.

The analysis of Bossert and Wilson asserts that the diffusion
of a pheromone from a point source in still air results in an ex-
panding sphere of pheromone. When dealing with a continuously emitting pheromone source after a long period of time, the pheromone
concentration at any point in space is given by the equation

$$U(r) = \frac{Q}{2D\pi r} \tag{1}$$

where U(r) is the pheromone concentration at the radius (r), D is
the diffusion coefficient of the pheromone, and Q is the pheromone
emission rate.

The Bossert and Wilson treatment is restricted primarily to the mass transfer of odorant through space and the "active space" generated by an odor source. This active space is controlled by the geometry of the system, the concentration (K) of pheromone necessary to trigger specific behaviors in the respondent animal, and the emission rate (Q) of the source. In large, the remainder of this paper will be spent in examining those factors which control Q.

EXPERIMENTAL

A useful system for the analysis of scent marking behavior and scent mark volatilization is that of the midventral sebaceous gland of the Mongolian gerbil (Meriones unguiculatus). The gerbil concertedly marks other animals within its social group as well as objects in its immediate environment. The work of Thiessen and others has indicated a role for the scent of the midventral gland in individual and group recognition, the establishment and maintenance of territory (Thiessen, 1968), maternal behavior (Wallace, Owen and Thiessen) and other social contexts. Although there has as yet been no complete characterization of the total biochemical composition of the midventral sebum of the gerbil, Thiessen has determined that phenylacetic acid is one of its behaviorally active components (Thiessen et al. 1974). Armed with this information, an experiment was devised in which the release of phenylacetic acid from sebum and from other compounds which approximate sebum was monitored. The experimental procedure allowed for an empirical determination of the pattern of evaporation of radioactive phenylacetic acid from different classes of surfaces which had been applied with sebum or sebum-like materials. The effect of relative humidity on the observed evaporation curves was also investigated.

Methods and Materials

100 μCi of C^{14}-labeled phenylacetic acid obtained from Amersham/Searle was dissolved in 100 ml of reagent grade diethyl ether. A 5 μl aliquot of this ether solution is equal to about 20 ng of phenylacetic acid and yields a consistent count level of 15,000 ± 300 counts per minute in a liquid scintillation counter. The usual experimental procedure was to place 108 "plates" of a given type of material such as glass, stainless steel or cellulose on a 1/4" piece of plywood cut to fit the inside of a glass chromatography tank measuring 27.5 x 7.5 x 24 cm. Each plate measured 1 cm^2. Half of the plates were then applied with a 5 μl amount of an ether solution of male gerbil sebum made by dissolving 5 mg of male ventral gland sebum in 3 ml of ether. When the ether evaporates from the plate a thin film of sebum remains on the plate.

At this point all 108 plates were then applied with 5 µl amounts of
C^{14}-labeled phenylacetic acid.   A set of six plain and six sebum
treated plates were then deposited in scintillation vials.   These
constituted the control conditions for the experiment.   The remaining
plates were immediately placed inside the chromatography tank.   A
glass cover sealed with silicon stopcock grease maintained a closed
system.   After set intervals of time, sets of six plain and six
sebum-treated plates were removed from the tank and placed in scintillation vials containing scintillation cocktail.   The samples
were counted in a Beckman LS-100 liquid scintillation counter.   The
scintillation cocktail consisted of 5 gm PPO and 100 gm Napthalene
brought to a total volume of 1 1 with dioxane.   Each vial contained
15 ml of cocktail.   Using this general procedure a plot of the
gradual loss of radioactive phenylacetic acid from sebum-treated and
non-sebum treated plates was developed.   Humidity inside the tank
was manipulated by placing Drierite (0% relative humidity) or distilled water (100% relative humidity) on the bottom of the tank.

Modulation of Odorant Release from Surfaces

It is clear, especially from many insect pheromone systems,
that a given pheromone may be deposited as a scent mark in a relatively pure form, that is, unadulterated with non-pheromonal material.   In these cases the question arises of possible surface effects
due specifically to the type of surface to which the scent mark has
been applied.   To examine this, phenylacetic acid was applied to a
spectrum of surface types and its loss at room temperature and 0%
relative humidity over a 24 hour period was monitored.   It will be
seen from Fig. 1 that within 24 hrs. most of a 20 ng amount of odorant evaporates from an inert metal surface such as platinum.   In
contrast, montmorillonite, a natural clay with a surface area of
700 m^2/g, is seen to adsorb the material quite strongly.   This is
also largely true for glass, mylar plastic, cellulose, kaolinite and
balsa wood.   The considerable affinity of these latter materials
suggests that after an initial period of rapid evaporation, a polar
odorant such as phenylacetic acid will remain immobilized for long
periods of time on many kinds of surfaces which would occur in
natural contexts.   The results of this experiment clearly demonstrates that most naturally occurring surfaces are far from inert or
neutral either due to their high surface area or their charge.   Perhaps there is also a lesson here for the design of olfactometers and
other odor testing apparatuses.

It should be noted that the very high initial evaporation rate
of the phenylacetic acid, indicated by the dotted lines of Fig. 1,
is an artifact.   Since the phenylacetic acid is applied as a spot
in an ether solution and since ether has a very high vapor pressure,
the spotting procedure is affected by a "solvent effect".   Short term
studies indicate that this effect levels off within 15-20 minutes.

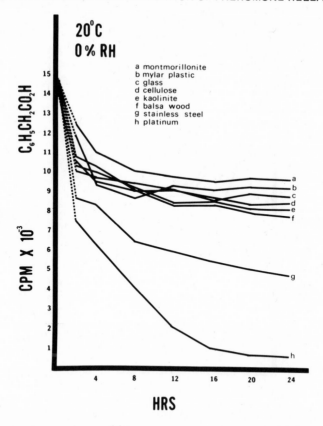

Fig. 1.   Evaporation of C^{14}-labeled phenylacetic acid at 20°C and 0% relative humidity from eight different types of surfaces.

The effect of mixing the odorant phenylacetic acid with non-volatile liquids or greases on a relatively inert surface like stainless steel is seen in Fig. 2.

The oily or greasy phases have a pronounced effect on evaporation. The polar polymeric material diethylene glycol succinate significantly retards evaporation while mineral oil has a minimal effect. Sebum and the polyisoprene lipid squalene (which is a known component of sebum) are intermediate in their effect. This data shows that non-pheromonal material can clearly have a major influence on the signal emission rate (Q) of a pheromonal compound contained within it. The large amount of di- and triglycerides

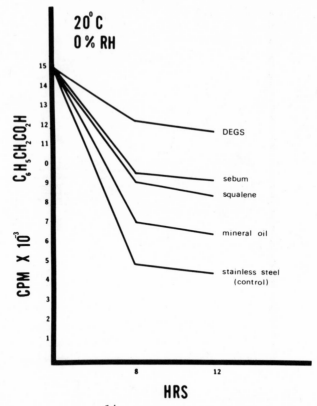

Fig. 2. Evaporation of C^{14}-labeled phenylacetic acid at 20°C and 0% relative humidity from stainless steel and from stainless steel treated with films of sebum and three other liquid materials.

and other lipid material of vertebrate scent gland secretions can thus alter the temporal emission pattern of a pheromone. A little reflection on the information presented in Figures 1 and 2 leads to a conclusion about the function of lipids and other non-pheromonal material in scent marks: on the one hand, they buffer the strong surface effects indigenous to most natural surfaces and thus insure signal emission over prolonged periods of time. Secondly, they regard the volatilization of very low molecular weight compounds and thus once again prolong signal strength. It is clear that the biochemical "junk" which we often go to great lengths to eliminate in order to identify and isolate a specific pheromonal compound has a very significant effect on the actual signal potential of a given pheromone within naturalistic contexts.

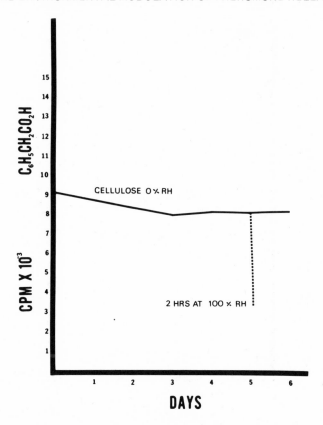

Fig. 3.   Evaporation of C^{14}-labeled phenylacetic acid from cellulose over 5 days at 20°C and 0% relative humidity. The dotted line represents the rapid loss of the compound observed after 2 hrs at 100% relative humidity.

Humidity Effect

A change in relative humidity had dramatic effects. This is clearly seen in Fig. 3. A set of plain cellulose plates were spotted and run for 5 days at 20°C and 0% relative humdidity. During the last 3 days of the experiment, bound odorant concentration decreased by less than 5%. When the plates were transferred to another tank for 2 hrs. at room temperature but at 100% relative humidity, 66% of the remaining odorant volatilized.

Further tests indicate that the same humidity phenomenon is observed when phenylacetic acid is dissolved in sebum. Fig. 4 shows the pattern of evaporation observed at 20°C and 0% humidity when phenylacetic acid was applied to plain cellulose and plain stainless

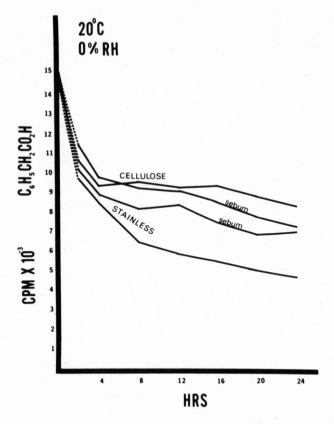

Fig. 4. Effect of sebum on evaporation of C^{14}-labeled phenylacetic acid from cellulose and stainless steel at 20°C and 0% relative humidity.

steel and to both of these surfaces after they had been pretreated with sebum. It can be seen that evaporation from cellulose was facilitated by the sebum, while evaporation from the more inert stainless steel was retarded. Figure 5 shows the effect of 100% relative humidity on the sebum-treated curves plotted in Fig. 4. High humidity significantly increases the rate of evaporation of phenylacetic acid from sebum, whether applied to cellulose or stainless steel. This effect is pronounced and has been observed throughout our experiments with other types of surfaces treated or not treated with sebum. One set of trials run on stainless steel and cellulose but at 46% relative humidity rather than 100% yielded curves very close to those shown in Fig. 5. We suspect for theoretical reasons to be discussed below, that any humidity high enough to saturate the surface with a single molecular layer of

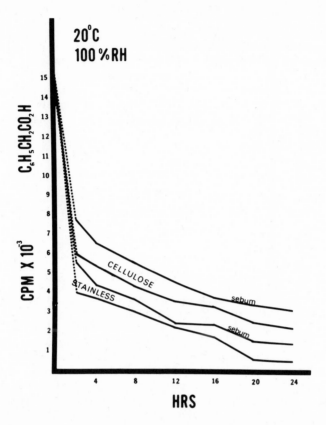

Fig. 5. Effect of 100% relative humidity on the evaporation curves shown in Fig. 4.

water will result in the dramatic humidity effect. The point is that extremely high humidity is not necessary for the effect.

From the above experimental data it may be concluded that the nature of the surface on which a scent mark is deposited, the presence and chemical nature of non-pheromonal material constituting the total make up of the mark, and relative humidity all can play a significant role in modulating the evaporation of odorants from the mark. Some possible explanations for these phenomena and their meaning in terms of pheromonal communication processes will be examined theoretically in the following sections of this paper.

EQUILIBRIUM MODEL FOR PHEROMONE RELEASE FROM A SCENT MARK

Analysis of the factors controlling pheromone emission from a scent mark are most easily handled if we consider the case of an odorant released in a small enclosed space at constant temperature and pressure. After a period of time the odorant in the surface scent mark phase (A) and the gas phase (B) above the scent mark will achieve equilibrium. When this happens the number of molecules of a substance (P) passing from phase A to B is equal to those passing from phase B to A, i.e.

$$P_A \rightleftharpoons P_B.$$

At equilibrium, the ratio of the concentration of P in these phases may be expressed by a distribution constant (K_D). This constant will be defined as

$$K_D = C_A/C_B \tag{2}$$

Those compounds with a small K_D are quite volatile while those with a large K_D are relatively non-volatile.

It is also known that when the distribution of a solute (P) is in equilibrium between two immiscible phases that the change in Gibbs free energy is zero (Purnell, 1962). This would mean that the partial molar free energy or chemical potential (μ) in the two phases is equal; $\mu_A = \mu_B$. Since $\mu = \mu° + RT \ln a$ when a is the activity of P in a given phase, $\mu°$ is the standard chemical potential, R is the gas constant, and T is the absolute temperature, it may be shown that

$$\mu_A° + RT \ln a_A = \mu_B° + RT \ln a_B. \tag{3}$$

By assuming that the concentration of P will be low in a scent mark, activity (a) may be replaced by concentration C. Making this substitution in equation (3) it is seen that

$$\ln C_A/C_B = - \frac{(\mu_A° - \mu_B°)}{RT}. \tag{4}$$

Substituting K_D from equation (2) for C_A/C_B we see that

$$\ln C_A/C_B = \ln K_D = - \Delta\mu°/RT \tag{5}$$

This thermodynamic description of pheromone volatilization indicates that the distribution constant and therefore the amount of pheromone in the gas phase that may be used for signalling is a function of temperature and the change in standard chemical potential between the two phases. It is apparent that if $\Delta\mu°$ for two pheromones is different, one pheromone will volatilize from a scent mark faster than the other. The possible significance of this differential evaporation is that animals could tell the age of a scent mark if they recognize the concentration ratio of individual components by smell.

The chromatographic process is well known to those who have worked in pheromone purification. This process is due to differential partitioning of solutes between stationary and mobile phases in the chromatography column. Interestingly, there is a very close parallel between the processes occurring in a chromatography column and the evaporation of odorants from a scent mark. Equation 5 applies equally to chromatography. During the remainder of this discussion we will draw on knowledge of chromatographic systems to explain pheromone volatilization.

Natural scent marks are probably of two types; 1) those in which pheromones are mixed with rather large amounts of non-pheromone material and 2) those in which one or two odorants in low concentration are deposited in pure form on an environmental surface. The nature of pheromone emission in these two cases will be quite different. When μg quantities of pheromone are deposited with mg quantities of other organic compounds, the pheromone is essentially in a solution of the non-pheromonal material. Evaporation of the odorant may be considered as occurring from a solution and the equilibration of pheromone may be treated by a gas-liquid partitioning model. When the concentration of pheromones deposited on the environmental surface in the absence of any other material is small (a μg or less), they will be adsorbed. This case is best treated by a gas-solid partitioning model. Situations involving intermediate concentrations of solutes will partition by a combination of these models and will be treated separately.

GAS-LIQUID PARTITIONING

As noted above, the gas-liquid partition model in general may only be used to describe the partitioning of a pheromone when the amount of organic compound in the scent mark is greater than a mg/cm^2. Since 0.13 μg of phenylacetic acid will theoretically produce a monolayer on a planar one cm^2 surface, a mg of material will easily saturate all of the adsorptive sites on a non-porous surface. If the surface on which the scent mark is deposited is of high porosity and surface area, considerably more material is required to saturate all adsorbant sites.

Equation (5) states that

$$\ln K_D = -\Delta\mu/RT.$$

From the standard thermodynamic relationship

$$\Delta\mu^\circ = \Delta h^\circ - T\Delta S^\circ \tag{6}$$

we see that the change in chemical potential associated with the phase change of a partitioning odorant is related to the standard molar heat (h°) and entropy (S°) respectively. Dal Nogare and Juvet (1962) have shown that equations (5) and (6) may be combined to produce the equation

$$\ln K_D = -\frac{\Delta Hv}{RT} + \ln \underline{C} \tag{7}$$

where ΔHv is the molar heat of vaporization and $\underline{C}$ is a constant related to the particular pheromone and partitioning system. This equation shows that as temperature increases K_D decreases and the volatility of an odorant increases. In converse, as the heat of vaporization (ΔHv) of an odorant increases its volatility decreases.

It is to be expected that equation (7) will be of limited utility in predicting the absolute volatility or vapor pressure of a substance in an incompletely defined system. The $\underline{C}$ term in equation (7) is specific for each pheromone and the organic compounds with which it is associated.

Raoult's Law

If we assume as in equations (3) and (4) that $\underline{a} = 1$ and the system is behaving as an ideal system, then we may apply Raoult's law (Glasstone and Lewis, 1960) to predict the vapor pressure of odorants. This law states that the partial pressure (p_1) of a volatile component in a liquid mixture is proportional to its mole fraction (X_1) in the solution and its saturation vapor pressure (p_1°).

$$P_1 = X_1 \, p_1^\circ \tag{8}$$

The mole fraction is defined as

$$X_1 = \frac{N_1}{N_1 + N_2} \tag{9}$$

where N_1 and N_2 are the number of moles of odorant and solvent respectively. Since the relative amount of pheromone in a scent mark is usually very small (approximately 1 µg of phenylacetic acid/mg of sebum), equation (9) reduces to $X_1 = N_1/N_2$. Therefore

$$P_1 = {}^{N}1/N_2 \, p_1^\circ. \tag{10}$$

From this equation it may be deduced that as the mole fraction of an odorant in a solution decreases, the vapor pressure of the odorant decreases. Another way of stating this is that if the amount of an odorant stays constant and a diluent is added, the vapor pressure and the amount of the material in the gas phase will decrease.

The significance of this in pheromonal communication is that the mixing of pheromones with non-pheromone material will decrease the volatility of the odorant and therefore the active space generated by the scent mark. By decreasing the rate of evaporation, a scent mark will be more persistent. This may also be a way of extending the life time of odor marks containing very volatile pheromones.

Experimental verification of this phenomenon is seen in Fig. 2. The addition of mineral oil or one of a series of other greases to phenylacetic acid reduces its volatility and therefore its rate of evaporation from a metal surface.

Henry's Law

Many cases are known where through molecular associations solutions do not behave ideally and Raoult's law does not apply. These non-ideal systems are better described by Henry's law:

$$W/V_s = k \, p \tag{11}$$

where W is the mass of a volatile odorant dissolved in a scent mark of volume V_s, p is the partial pressure of the pheromone and k is Henry's constant. For a more detailed discussion of Henry's law the reader is referred to Sheehan (1961). Again we have an equation of limited use for incompletely defined pheromone systems because Henry's constant is required.

Evidence that all solutions do not behave ideally is seen in Figure 2. Although approximately the same amount of mineral oil, squalene, sebum, and diethylene glycol succinate (DEGS) was used in all cases, they exhibit wide differences in their ability to modulate the release of phenylacetic acid from a surface. The polar DEGS is the most retentive while the non-polar mineral oil is the least. The significance of this in natural systems is that the polarity of the non-pheromonal material in a scent mark can radically alter the volatility of pheromones.

Odorant-Solvent Interactions

The deviations from ideality discussed above are in general the result of odorant-solvent interactions. The interactions are usually

of a cohesive nature but may occasionally be repulsive. Again we
may make a direct comparison to chromatography. Numerous studies have
shown that in gas chromatography columns cohesive solute-solvent
interactions are responsible for the differences in volatility of
compounds with the same saturation vapor pressure (p°).

These cohesion forces are of three types (Miller, 1975); 1)
dipole-dipole interactions, 2) inductive forces, and 3) dispersion
forces. The intermolecular interaction of permanent dipoles may produce a cohesive force that decreases volatility of odorants. This
interaction is the result of the orientation of two permanent dipoles
such that the positive head of one dipole positions itself beside the
negative head of the other. Hydrogen bonding is one of the important
cohesion forces that may be treated as a dipole-dipole interaction.
Odorant volatility would be reduced substantially by the interaction
of a small amount of polar pheromone with a large amount of polar
liquid phase.

Inductive forces result from the interaction of a permanent dipole
with an adjacent molecule in which a dipole may be induced. The size
of the induced dipole would be proportional to the size of the permanent dipole.

Cohesion forces also exist between molecules that are non-polar.
Due to the oscillation of positively charged nuclei and negatively
charged electrons, every molecule behaves as an oscillating dipole.

The cohesion forces described above explain why in Figure 2
phenylacetic acid is tightly retained by the polar diethylene glycol
succinate and weakly bound by non-polar mineral oil. Sebum and squalene are intermediates. Application of the concept of odorant-solvent
interactions would further suggest that in natural systems the nature
of the functional group(s) on both pheromones and sebum components
substantially influence pheromone volatility in a scent mark. For
example, one would expect the evaporation of a non-polar hydrocarbon
from sebum to be faster than that of a carboxylic acid.

GAS-SOLID PARTITIONING

The adsorption of an odorant directly onto a surface during
scent marking and subsequent desorption into the surrounding air is
an example of gas-solid partitioning. Through the work of Cremer
and Prior (1951) it is known that the distribution coefficient (K_D)
of an odorant is related to its heat of adsorption (ΔH_a) by the
equation

$$\ln K_D = - \Delta H_a/RT + \ln C_a. \tag{12}$$

Unfortunately, K_D often varies with odorant concentration. This is primarily because: 1) the energy of odorant adsorption is not constant and depends on the extent of surface coverage, 2) adsorption may occur at multiple surface sites and there may be interaction between adsorbed odorant molecules, and 3) more than a monolayer of odorant molecules may be adsorbed in some systems. Multilayer physical adsorption on a surface is more adequately treated by the BET equation (Brunauer et al., 1938),

$$\frac{p}{V\,(p_o - p)} = \frac{1}{V_m c} + \frac{p}{p_o} \cdot \frac{(c-1)}{V_m c}$$

where V is the volume of gas adsorbed at pressure p, V_m is the volume of a surface monolayer, and c is a constant approximately equal to $_e(E_a - E_1/RT)$. E_a is the heat evolved during the adsorption of the first layer of molecules and E_1 is the heat of liquifaction.

The chemical and physical properties of both the odorant and adsorbent influence adsorption-desorption equilibria and the mechanism of adsorption. In general the adsorption of an odorant to a surface is due to one or a combination of mechanisms composed of: 1) physical adsorption, 2) hydrogen bonding, and 3) chemical adsorption.

Physical Adsorption

Van der Waals forces resulting from dipole-dipole interaction are responsible for physical adsorption of odorants (Brunauer et al., 1938). Since the major types of dipole-dipole interactions resulting in odorant-surface association have been reviewed above, they will not be discussed here. The Van der Waals forces are responsible for the adsorption of neutral and weakly polar substances to surfaces. It is easily seen that molecules having the same vapor pressure in the pure state but different permanent or inducible dipoles will experience different rates of release from a surface.

Hydrogen Bonding

The dipole moment of a molecule may not always be an indication of its "polarity" or tendency to interact with other "polar" molecules (Kargar et al., 1973). The dipole moments of molecules that hydrogen bond are often modest while intramolecular association is quite strong. Proton-donor groups such as -OH, -NH, and -SH interact with proton-acceptors that contain unpaired electrons (e.g., -O-, =O, =N-, -C=C-) in hydrogen bonding. Volatilization of odorants that are hydrogen bonded must accompany disruption of hydrogen bonding.

The very tight association of phenylacetic acid to cellulose, glass, and montmorillonite, all of which are rich in surface hydroxyls, is due partially to hydrogen bonding. In general, the acidic protons of a carboxylic acid hydrogen bond more strongly than simple alcohols.

Chemical Adsorption

At least four different mechanisms for chemical adsorption on inorganic materials may occur (Bailey and White, 1970): 1) ion exchange, 2) protonation at the surface by reaction of an odorant base with the hydronium ion on the exchange site, 3) protonation of the organic odorant with subsequent ion exchange, and 4) in systems having water of hydration, protonation by reaction with the dissociated proton of residual water left on the surface. It is obvious that the above chemical adsorption reactions will be limited to organic cations and anions and surfaces of opposite charge.

Within the framework of the above gas-solid partitioning mechanisms there are a series of factors that influence equilibria. These factors may be divided into two major categories; 1) those dealing with the physio-chemical characteristics of the adsorbent and 2) those relating to the physico-chemical characteristics of the odorant. Since this paper is only intended to present a general model of the modulation of odorant release from a scent mark, detailed analyses of these factors will not be presented. We will simply point out that the properties of the adsorbent such as surface area, porosity, pore diameter, surface configuration, and surface polarity all influence odorant adsorption. Additionally it will be noted that in odorants, the major contributing factor in adsorption is the nature of functional groups. Secondary factors such as molecular geometry, nature and position of substituents that influence electron density in functional groups, and polarizability of odorants also contribute to adsorption.

INTERMEDIATE GAS-LIQUID/GAS-SOLID PARTITION SYSTEMS

The concepts of gas-liquid and gas-solid partitioning were used to explain chromatography in the decade beginning in 1950. Years of experience with both gas and liquid chromatography systems have taught scientists that the demarcation between these two types of partitioning is not distinct. For example, interaction of polar molecules with the support surface is often observed even on well coated gas-liquid chromatography supports. On the other hand, trace amounts of water or other polar molecules may deactivate a support by adsorbing to active sites. Presumably partitioning then occurs by association of the odorant with an adsorbed layer of organic compound or water.

Odorant partitioning in scent marks must behave in a similar manner. We will illustrate these intermediate cases by again examining those factors that are responsible for partitioning. These factors are; 1) the available surface area and polarity of the environmental surface, 2) the total volume of compounds located on this surface, 3) the polarity of compounds placed on the surface, and 4) the polarity of odorants relative to other compounds on the surface.

Adsorption of an odorant to a surface will occur anytime that the odorant has an affinity for support material and there are available binding sites on the surface. The availability of binding sites on the surface depends on the number of surface binding sites and the number of molecules deposited on that surface that have an affinity for those binding sites. When the number of molecules of odorant placed on an environmental surface in the absence of sebum is smaller than the number of active sites on the surface; gas-solid adsorption is occurring.

The evaporation of odorant from a surface as a function of time is described by the equation

$$\frac{dC}{dt} = - k_e C \tag{13}$$

where C is concentration of odorant, k_e is a rate constant, and t is time. When the quantity of odorant in a scent mark exceeds the number of surface active sites, the odorant will partition by multiple mechanisms and the constant k_e in the above equation will change during odorant evaporation. It should be noted here that the distribution coefficient (K_D) discussed previously will also change. The first surface monolayer will probably be more tightly bound than successive layers. This should generally result in an evaporation curve with several linear regions.

Addition of a large quantity (mg/cm^2) of non polar sebum to a surface will diminish the rate of volatilization because odorants must first be desorbed from the surface and then escape through the sebum for evaporation to occur. If the sebum itself is polar or contains a polar constituent it will significantly modify surface effects on odorant adsorption and desorption. Adsorption of a polar odorant ($0^{\delta+}$) on a silaceous surface ($S^{\delta-}$) is represented by the equation

$$S^{\delta-} + 0^{\delta+} \rightleftharpoons S^{\delta-}0^{\delta+}. \tag{14}$$

Polar modifiers ($M^{\delta+}$) in sebum will also adsorb to the silica surface in much the same way,

$$S^{\delta-} + M^{\delta+} \rightleftharpoons S^{\delta-}M^{\delta+}. \tag{15}$$

These equations show that both polar odorants and modifiers are competing for the same surface active sites ($S^{\delta-}$). If the quantity of polar surface modifier (M) is sufficiently large to convert all of the surface active sites to SM complexes and M is more polar than P, then the odorant is free to partition with only the sebum for volatilization.

As noted in Figure 3, water is an important modifier of odorant release. Figures 4 and 5 indicate that after odorant is deposited on the surface in both the presence or absence of sebum, high humidity initiates a rapid release of odorant. This is probably the result of water vapor acting as a modifier (M) having a higher affinity for surface active sites then the odorant and being able to break up hydrogen bonding in sebum. The environmental significance of this phenomenon is rather large. It suggests that by exhaling on a surface, licking, or in some way depositing water vapor on a scent mark, a sudden rapid release of odorant may be obtained. It also suggests that the useful lifetime of a scent mark may be strongly influenced by humidity or rain. The importance of water vapor on odorant release is an environmental factor that merits further investigation.

SUMMARY

This paper shows experimentally and theoretically that the release of odorants from a scent mark are due to at least 4 factors; 1) the nature of the surface on which the odorant is deposited, 2) the quantitative and qualitative nature of other organic compounds such as sebum that are deposited on the surface, 3) the polarity of the odorant, and 4) humidity. All of these factors determine the type and extent of interaction of odorants with the non volatile constituents in the scent mark. In general, non polar odorants will show little interaction with polar materials (both surface and sebum) and their evaporation will not be greatly influenced by humidity. On the other hand, polar odorants will interact with polar surfaces and sebum and their volatility will be substantially reduced. Competition of polar odorants and water vapor for the same polar sites causes enhanced evaporation at high relative humidity. These results appear to correlate quite well with current explanations of chromatographic processes.

This work was supported by NSF grant #BMS-74-14103.

REFERENCES

Bailey, G. W., and White, J. L.   1970.   Factors influencing the adsorption, desorption and movement of pesticides in soil.   Residue Rev. 32:29-83.

Bossert, W. H., and Wilson, E. O.   1963.   The analysis of olfactory communication in animals.   J. Theor. Biol. 5:443-69.

Brunauer, S., Emmett, P. H., and Teller, E.   1938.   Adsorption of gases in multimolecular layers.   J. Am. Chem. Soc. 60:309.

Cremer, E., and Prior, F.   1951.   Separation of gases and determination of adsorption energies.   Z. Electrochem. 55:66.

Del Nogare, S., and Juvet, R. S.   1962.   Gas-Liquid Chromatography. Interscience Publishers, New York, p. 16.

Glasstone, S., and Lewis, D.   1960.   Elements of Physical Chemistry, D. VanNorst and Company, Inc. p. 358.

Kargar, B. L., Snyder, L. R., and Horvath, C.   1973.   An Introduction to Separation Science.   Wiley-Interscience, New York, p. 41.

Miller, J. M.   1975.   Separation Methods in Chemical Analysis.   Wiley-Interscience, New York, p. 116.

Purnell, H.   1962.   Gas Chromatography.   John Wiley and Sons, Inc., New York, p. 9.

Sheehan, W. H.   1961.   Physical Chemistry.   Allyn and Bacon, Boston, p. 288.

Silverstein, R. M., and Young, J. C.   1976.   Insects generally use multicomponent pheromones, pp. 1-29 in M. Beroza (ed.), Pest Management with Insect Sex Attractants.   A.C.S. Symposium Series. No. 23.

Thiessen, D. D. 1968.   The roots of territorial marking in the Mongolian gerbil: a problem of species - common topography. Beh. Res. Meth. & Instru. 1:70-76.

Thiessen, D. D., Regnier, F. E., Rice, M., Goodwin, M., Isaacs, N., and Lawson, N.   1974.   Identification of a ventral scent marking pheromone in the male Mongolian gerbil (Meriones unguiculatus). Science 184:83-85.

Wallace, P., Owen, K., and Thiessen, D. D.   1973.   The control and function of maternal scent marking in the Mongolian gerbil (Meriones unguiculatus).   Physiol. & Beh.

CHEMICAL COMMUNICATION IN AMPHIBIANS AND REPTILES

Dale M. Madison

[1]Biology Department
McGill University
Montreal, Quebec, Canada   H3A 1B1

INTRODUCTION

Chemical communication is a relatively new field of study in which contemporary interest in the form of reviews, papers, symposia, and books is disproportionately large compared to the actual number of experimental studies on the subject. But such reviews are constructive in that they provide a broad perspective of the problem areas in the field before a more restricted experimental focus begins or resumes. This review on chemical communication in amphibians and reptiles provides an evaluation of the available information, and attempts to formulate trends and ideas for future study. Although experimental indication of chemical communication in amphibians and reptiles is generally lacking, a fair number of morphological and behavioral studies exist, which collectively indicate a distinct potential for a variety of chemical systems in several amphibian and reptilian groups. The scope of this review is restricted to intraspecific interactions in the major living taxa. Omitted from consideration are feeding interactions, defense secretions, and alarm substances. However, the scope remains sufficiently broad and insures a more detailed examination of the potential production, reception, and functional significance of sociochemical signals and cues. This review is best considered complementary to excellent reviews on the omitted subjects above and in related subject areas (Bellairs, 1970; Blair, 1968; Burghardt, 1970; Noble, 1931a; Parsons, 1967; Pfeiffer, 1974; Porter, 1972; Quay, 1972).

[1]Address after Aug. 1977 is: Biology Department, State University
    of New York, Binghamton, N.Y. U.S.A.   13901

MORPHOLOGICAL REQUISITES

Chemical communication requires scent production and scent reception. These generally peripheral functions usually have distinct anatomical and histological bases, and knowing something about this structure - even without the associated function - gives a good first reading of the potential for chemical communication in the species concerned.

Scent Production

Scent production frequently involves the presence of specialized integumentary glands. However, odors in excretory products, such as those resulting from diet or species-typical digestive processes, could be utilized in chemical communication without having an integumentary production site. Excreted hormones, especially by the female during estrus, frequently serve as sociochemical signals, but these have no integumentary production areas. For the purpose of brevity, and because virtually nothing is known about these odor sources in the amphibians and reptiles, this section will be limited primarily to the glandular basis of scent production, and generally to integumentary glands. In most cases some scent producing function of the specific glands has been suggested, but rarely demonstrated. All of these glandular areas will be reviewed at least briefly in order not to omit a glandular type that may ultimately be shown to have a role in scent production.

I. Amphibia

Amphibians generally have a rich glandular field associated with the maintenance of a flexible, moist integument. These glands, mostly alveolar in design, are of two types: the granular glands and the mucous glands (Breckenridge & Murugapillai, 1974; Dawson, 1920; Elias & Shapiro, 1957; Miscalencu et al., 1971; Noble, 1931a; Porter, 1972; Quay, 1972). The granular glands secrete alkaloid toxins, which appear viscous and milky, and usually are somewhat localized in distribution. In Ambystoma gracile the glands are found on almost all dorsal surfaces, although the main concentrations are on the dorsal tail ridge and the parotid region (Brodie & Gibson, 1969). In Taricha and Notopthalmus the dorsal trunk region contains heavy concentrations of granular glands (Brodie, 1968a,b), and similar concentrations are observed in some frogs and toads (Dodd et al., 1974; Hensel & Brodie, 1976). Proteus lacks these glands (Noble, 1931a), while some toads possess only granular glands (Muhse, 1909; Noble, 1931a). Although some granular gland secretions have peculiar odors (Phisalix, 1918), no communicatory functions have been demonstrated.

The mucous glands are more evenly distributed on the body, and are associated with clear secretions that serve to maintain a moist integument. The secretion facilitates osmoregulation, thermoregulation, respiration, and provides some protection. The mucous glands can produce irritating secretions, but these are not as toxic nor as common as the granular gland secretions (Noble, 1931a). The mucous glands, or the glandular structures derived from them, are more directly implicated in scent production than are the granular glands.

Caudata. Superimposed on the above general integumentary pattern, some salamanders show areas of modified mucous glands called the hedonic glands (Brodie, 1968b; Hays, 1966; Lanza, 1959; Noble, 1929, 1931a,b; Sever, 1976; Truffelli, 1954). These glands are tubular with tall columnar cells and a modified central lumen. They are common but not restricted to the plethodontid and desmognathid salamanders (Rogoff, 1927; Salthe, 1967; Smith, 1941), occur on adult males and some females (Aneides), and are variously seasonal in activity (Noble, 1931a). A pure cluster of hedonic glands is conspicuous as a disc-shaped swelling on the chin of eastern large Plethodon (Fig. 1,a,b), but is rather inconspicuous although just as active in the western species of the genus (Brodie, 1968b). Hedonic glands also occur in somewhat lesser concentrations on the eyelids, temporal head region, and dorsal tail base in Eurycea (Noble, 1929) and Desmognathus (Hays, 1966), and occur as isolated units over the dorsal surface of most plethodontids and desmognathids (Noble, 1931a). The hedonic gland secretions have been directly implicated in signal production, especially during courtship.

The plethodontid salamanders also possess the rather unique structures known as the nasolabial grooves (Fig. 1,c) and the nasolabial glands (Fig. 1,d) (Brown, 1968; Brown & Martof, 1966; Sever, 1975; Whipple, 1906). The large nasolabial gland complex secretes unsaturated lipids that tend to keep the olfactory chamber and nares free of excess fluids and debris, but also may be involved in scent production (Brown & Martof, 1966; Tristram, 1976). The glands are of the branched tubular type, as opposed to the generalized alveolar or acinous type. The glands serving each nasolabial groove exist in two long convoluted tubule systems and in a collection of short, more simple systems. The larger systems have ducts at the level of the external naris (at the uppermost section of the nasolabial groove). The smaller glands open in close proximity to, but not within, the remaining more ventral portion of the nasolabial groove. These dorsal, subdermal glands collectively occupy much of the dorsolateral area from the snout to the eye orbits, as well as the areas medial and ventral to the anterior portion of the nasal chambers (Fig. 1,d).

The seasonal hypertrophy of parts of this glandular system
in the lip area in male plethodontids (Sever, 1975, 1976), and the
associated substrate "nose-tapping" and "chin-touching" involving
the nasolabial grooves and the associated labial cirri in Plethodon
cinereus (Tristram, 1976), support the suggestion of Brown & Martof
(1966) that the nasolabial secretions may have hedonic functions
or other signal significance.

Another group of glands, more common to all salamanders than
the previous two gland types, are the cloacal glands. These glands
are of three different types: the cloacal, pelvic, and abdominal
glands (Dawson, 1922; Dieckmann, 1927; Heidenhain, 1890; Noble &
Weber, 1929; Williams, pers. comm.). The abdominal gland is
believed to have hedonic-like secretions (Dawson, 1922; Noble,
1931a), and is the only gland of the cloacal complex known to have
ducts that empty outside or near the outside of the cloaca, usually
on small papillae (Heidenhain, 1892; Noble, 1931a). These abdominal
glands are often more rudimentary or non-functional in female
salamanders (Noble, 1931a). The pelvic and cloacal gland types
are involved in the formation of the spermatophore in the male or
in egg capsule formation in the female. These glands are usually
less elaborated in the female (Kingsbury & Defiance, 1895).

Anura. Compared to the salamanders, the indication of scent
production in frogs and toads is almost non-existent. Departure
from the general mucous and/or granular integumentary structures
is the exception. Conaway & Metter (1967) describe specialized
"breeding glands" in the sternal region of males that function
in adhesion of the sexes during amplexus. Metter & Conaway (1969)
further demonstrated that the glands were derived from mucous
glands and were under the control of male sex hormones. No
indication of hedonic action was mentioned. Nuptial thumb pads
on frogs and toads also secrete adhesive substances that aid the
male grip during amplexus (Porter, 1972). Here again, no hedonic
functions were suggested.

Several anurans, such as Rana septentrionalis, Gastrotheca
monticola, Bufo vulgaris, and some pelobatids, have peculiar integumentary odors (Noble 1931a; Müller-Schwarze, pers. comm.). The
source of these odors appears to be integumentary, either from
the granular or mucous glands (Noble, 1931a). Although the odors
may prove to have defense or alarm functions, further study is
necessary.

Recently, Blaylock et al. (1976) demonstrated the presence of
lipid integumentary glands in the phyllomedusine frogs. The lipid
secretion is spread over the skin by "wiping" movements as part of
a mechanism for preventing cutaneous water loss. Although lipids
are commonly used in scent marking, such a function in the

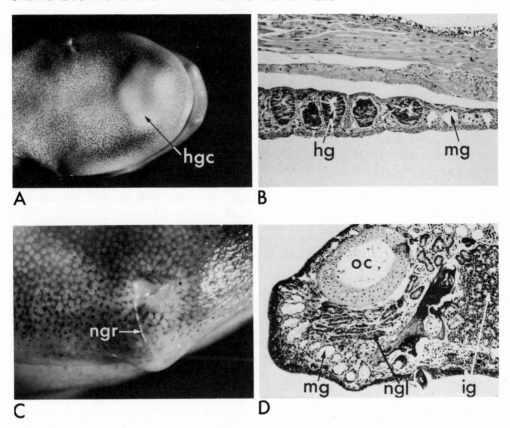

Fig. 1. Structures associated with scent production and scent reception in Plethodon salamanders. Represented are (A) the hedonic gland complex (hgc) on the chin of the male P. jordani (head width is 10 mm); (B) a longitudinal section through the lower jaw of a male P. glutinosus at the edge of the hedonic gland complex showing the tubular hedonic glands (hg) and the adjacent mucous glands (mg); (C) the nasolabial groove (ngr) on a male P. jordani extending on the right side from the external naris ventrally to the large labial cirrus; (D) a cross section of the upper jaw of a male P. glutinosus 1 mm from the snout tip at the anterior-most level of the olfactory chamber (oc) showing the nasolabial glands (ngl), the enlarged mucous glands (mg), and the intermaxillary glands (ig).

Phyllomedusa seems unlikely.

The cloaca structure of anurans is also quite simplified

compared to the caudates (Kerr, 1939). In fact, there is a complete
lack of cloaca glands, internal papillae and ridges, and ciliated
tracts in the frog (Rana temporaria). The only seasonal change in
the cloaca of the frog is the enlargement of the endothelial epithelium along the upper cloacal chamber during the breeding season. No
multicellular glandular structures appear, although the tall glandular cells making up the epithelial lining produce fluids that are
thought to flush out and lubricate the cloaca for the passage of eggs
and sperm (Kerr, 1939). Such fluids could acquire some communicative
significance in the few anurans that have elaborate courtship patterns in an aquatic environment (e.g., Pipa; Rabb & Rabb, 1963).

As a final note, Noble (1931a) observed conspicuous glandular
hypertrophies in the ventral femoral region of male Mantidactylus
luteolus, and in the inguinal region of male Cycloramphus asper
and male and female Pseudophyrne guentheri. The function of these
glandular areas is not known.

II. Reptilia

The reptilian integument differs externally from that of the
Amphibia in being scaled and less permeable to water (Bellairs,
1970; Bentley & Schmidt-Nielsen, 1966). It has few glands of the
mucous and granular variety. On the other hand, the reptiles have
many glandular structures that are affiliated with the integument.

Lacertilia. The general integumentary glands of the lizards,
especially gekkonine species, have received extensive study by
Maderson and co-workers (e.g., Chiu et al., 1975; Chiu & Maderson,
1975; Maderson, 1968, 1970, 1972; Maderson & Chiu, 1970; Menchel &
Maderson, 1975). These integumentary glands consist of two general
types: the "generation" glands and the "preanal organs". The
generation glands are found at the epidermal splitting zone throughout the body surface, and are believed to function in association
with the cycle of skin shedding. These glands may produce secretions
of odoriferous significance (Maderson, 1970). The preanal organs are
epidermal follicular structures found in the ventral abdominal and
femoral regions. These glands function independently of the epidermal shedding process, are sensitive to reproductive endocrine cycles,
and are generally implicated in chemical communication. The preanal
organs in the ventral abdominal integument are just anterior to the
cloaca and are termed the preanal glands. These glands usually have
a general transverse occurrence across the abdominal region, but may
be enclosed in a distinct longitudinal groove along the mid-ventral
axis, as in the gecko, Gymnodactylus rubidus (Smith, 1935). The
preanal organs in the femoral region occur along the mid-ventral
axis of each thigh and are called the femoral glands. Cole (1966a)
has made extensive studies of the anatomy of the femoral glands in

an iguanid lizard, and has reviewed the literature on femoral glands (Cole, 1966b). The femoral glands are usually either smaller, less complex, or absent in the females of the species studied. The glands seem to be more active during the breeding season, and their action appears to be testosterone dependent. The secretions appear to be keratin and contain protein (Hunsaker, cited by Cole, 1966b). Cole (1966b) presents five hypotheses regarding the function of the femoral glands, two of which concern chemical communication. Perhaps the best observational evidence on function is provided by Noble & Bradley (1933) and Greenberg (1943), who observed that the waxy protrusions from these glands facilitate grip or provide tactile stimulation during courtship activities.

Another epidermal gland in the cloacal region of lizards was just recently described in male Sceloporus (Burkholder & Tanner, 1974). This gland is apocrine and subcutaneous and is found immediately anterior to the ventral lip of the cloaca. Secretions which emerge from ducts just inside the anterior cloacal lip are noticeable on the cloacal margins throughout the mating season. Whether this gland would fit into the classification of Whiting (1969) is not known.

Scincid lizards (e.g., Tropidophorus) have post anal organs that have integumentary ducts (Bellairs, 1970; Kluge, 1967), but more complete information on these glands could not be obtained.

Gekkonid lizards of both sexes possess bilateral cloacal sacs (or post anal sacs) that open on the surface of the integument just posterior to the cloaca (Smith, 1935). These are believed to offer physical support by eversion during copulation (Greenberg, 1943). However, chemical communication functions have been suggested, and Smith (1935) considers them possibly homologous to the scent glands of crocodiles and snakes. Kluge (1967) favors the interpretation of Greenberg (1943).

The lizards have distinct glandular specializations associated with the cloaca, but the scent glands that will be discussed for the snakes are absent in the lizards (Gabe & Saint-Girons, 1965, 1967; Whiting, 1967, 1969). Lizards have a single, medial dorsal gland, but this gland probably functions during sperm transfer or sperm reception (Kroll & Reno, 1971; Whiting, 1969). Dorsal and ventral cloacal glands are also described (Gabe & Saint-Girons, 1965). These contain serous and/or mucous cells and secrete muco- or glycoprotein, but the function of these glands is unknown. The remaining cloacal glands seem to function directly in association with the mechanics of mating.

Ophidia. The snakes have a variety of glands implicated in scent production. Evidence by Noble & Clausen (1936) suggests that diffuse epidermal glands, perhaps similar to but not as conspicuous as the generation glands of lizards, may be involved in scent production. However, more behavioral and histological work is obviously needed.

In the cloacal region, the preanal organs (and the associated ducts or pores) that occurred widely in the lizards appear to be absent in the snakes (Whiting, 1969). However, the glandular structures associated with the cloaca are more elaborated. Snakes have paired "scent glands" (Whiting, 1969), which open into the posterolateral aspect of the cloaca and extend posteriorly in the ventrolateral subdermal space of the tail. In Leptothyphlops, these glands have been called cloacal sacs and have been described as receiving some of the waste products and glandular secretions passing into and through the cloaca, and as well some of the cornified epithelial cells sloughed from the inner surface of the "cloacal sac" (Kroll & Reno, 1971). Several authors consider the cells of the scent gland walls to be holocrine, thus contributing cellular material to the glandular secretions (Gabe & Saint-Girons, 1965; Sabnis, 1967; Volsøe, 1944; Whiting, 1969). In many snake species the glandular contents are excreted, especially during stress, as a thick, odiferous substance. In a comparative study of the lipids from the scent gland of 25 snake species (from Colubridae, Boidae, Viperidae), Oldak (1976) showed that each species could be identified by the thin layer chromatograms produced by the respective lipid samples. In a few species, sexual, individual and developmental differences occurred in the lipid fraction. These results clearly indicate the potential for the use of scent gland secretions in communication.

Also affiliated with the cloaca is the median cloacal gland (Whiting, 1969), or the "cloacal gland" in Leptotyphlopidae (Kroll & Reno, 1971). In Leptotyphlopidae, this gland opens through a duct into the dorsal wall of the cloacal chamber, but lies mid-ventrally in the subdermal space posterior to the cloacal orifice. This gland secretes lipids and is active during the reproductive period (Fox, 1965; Kroll & Reno, 1971). It probably functions only to facilitate sperm transfer and reception (Kroll & Reno, 1971).

Two glandular structures appear to be peculiar to the Leptotyphlopidae: the "margino-cloacal glands" (Gabe & Saint-Girons, 1965; Kroll & Reno, 1971; Whiting, 1969) and the gland at the base of the tail spine in L. dulcis (Kroll & Reno, 1971). No communicatory significance has been suggested for these glands.

One group of snakes, comprising many species of Natrix and Macropisthodon, has been found to contain an extensive holocrine gland system under the skin in the dorsal neck region (Smith, 1938).

With only one exception, these "nucho-dorsal" glands are different
for the two genera. They extend the entire length of the body in
a few species, occur in both sexes, are less developed in immature
individuals, and vary in their size and occurrence within conspecifics of the same body size. The secretion of these glands
are thought to have a defensive function (Smith, 1938), but their
location also suggests possible use during courtship activity.

Chelonia. Possible scent producing glands in the turtles
occur under the chin (Winokur & Legler, 1975), along the lateral
margin of the shell (Ehrenfeld & Ehrenfeld, 1973), and in association with the cloaca. The chin or "mental glands" have received the
greatest amount of attention in connection with scent production
(Winokur & Legler, 1975). These glands exist as either large,
complex, multilobed, holocrine epidermal glands (Class I), or small,
cryptic epidermal glands with keratinized or holocrine cells (Class
II) (Winokur & Legler, 1975). Class I mental glands occur among
the Testudinidae; Class II glands occur widely in the families
Emydidae and Platysternidae. The glands secrete both saturated
and unsaturated fatty acids, and have demonstrated communicatory
significance (Rose, 1970).

All families of turtles except the Testudinidae have been
reported as having the inguinal and axillary glands, collectively
referred to as Rathke's glands (Ehrenfeld & Ehrenfeld, 1973). The
ducts of these holocrine glands occur close to the junction of the
plastron and the carapace just posterior to the forelimbs and just
anterior to the hind limbs. Smaller "supernumerary" glands and
their ducts may occur along the same line but between the axillary
and inguinal pores. The glands are believed to have defensive
functions (Ehrenfeld & Ehrenfeld, 1973), although possible communicatory functions have been implicated (Mahmoud, 1967).

Not much appears to be published on the cloacal glands of
turtles. However, the functioning of these glands in chemical
communication is indicated (e.g., Auffenberg, 1969; Jackson &
Davis, 1972a,b; Weaver, 1967).

Crocodilia. Very little is known about scent production in the
Crocodilia. There appear to be four musk glands: two mandibular or
throat musk glands, and two cloacal glands (Ditmars, 1910; Guggisberg,
1972; Reese, 1920, 1921). Reese (1921) further mentions two rows of
integumentary glands under the second and third row of dorsal scales
on the back of crocodiles.

The mandibular glands occur in both sexes, are most active
during the breeding season, and are everted to expel their contents
(Reese, 1921). The ducts of these large glands open on either side
of the midline in the throat region. The scent is released during

excitement and courtship activities. The cloacal glands are similar
to the mandibular glands in being invaginated epidermal specializations for scent production (Reese, 1921). The ducts of the cloacal
glands open into the cloacal chamber. The glands secrete lipids
and are holocrine or apocrine. The dorsal glands, because of their
abundance, their apparent absence of odor compared to the musk
glands, and their lipid secretions, have been interpreted to have
a lubricative function (Reese, 1921).

Scent Reception

In contrast to the available information on scent production in
amphibians and reptiles, the anatomy and physiology of scent reception is relatively well known, or at least well reviewed (e.g.,
Allison, 1953; Bertmar, 1969; Burghardt, 1970; Heimer, 1969; Moulton
& Beidler, 1967; Parsons, 1967; Tucker, 1971; Tucker & Smith, 1969).
Thus, I see no reason to review the information again. What I would
like to introduce here are a few anatomical features of amphibians
that serve as auxillary channels of chemoreception. These features
are the amphibian skin and the integumentary grooves.

As mentioned previously, the amphibian integument is generally
thin, moist, and non-keratinized. As such, although not necessarily,
it has the potential for dissolving substrate or airborne substances
and allowing their trans-cutaneous diffusion and eventual absorption
or passage into the blood stream. This is especially true in the
lungless salamanders (Plethodontidae) that have superficially vascularized skin for cutaneous gas exchange. Mucous secretion, of
course, is one agent that protects the internal osmotic and chemical
environment, but it may also enhance the inward diffusion of certain
chemical substances. Rie (1973) exploited the somewhat permeable
qualities of the amphibian skin in a successful study of the dermal
application of a variety of drugs in the relatively thick-skinned
newt, Taricha granulosa. This result suggests that pheromones could
enter by similar processes without the necessity of peripheral
sensory pathways. An experiment demonstrating whether the hedonic
gland secretions can pass directly through the integument of the
salamander would be most interesting from the point of view of
mechanism.

The likelihood of the above process is further enhanced by the
implications of the findings of Lillywhite & Licht (1974). They
found that integumentary grooves and sculpturing provide a physical
mechanism for the passive movement of water (and possibly substances
in solution) from the substrate onto the skin in several species of
toad. The passive transport of moisture from the substrate, in part
through "capillarity", has been demonstrated in the case of the
nasolabial grooves of salamanders (Brown, 1968). A similar mechanism

would seem possible in other cases where epidermal sculpturing occurs, for example the costal grooves of salamanders that correspond to vertebrae number (Highton, 1957). These costal grooves are vertical furrows that extend essentially from ventral to dorsal midlines at regular intervals along the axis of the body of a wide variety of caudates. Although the direct application of hedonic secretions would seem more likely than a mechanism involving cutaneous absorption from an intermediate substrate, the latter remains a possibility for consideration.

BEHAVIORAL EVIDENCE

The behavioral evidence supporting chemical communication in amphibians and reptiles is such that it would be difficult to organize the material in functional categories (e.g., species recognition, sex recognition, space ownership) and yet have contributions from each major taxon in each subject. Instead, the available information will be viewed from a taxonomic perspective, and in this way will highlight morphological and ecological differences between the various groups.

I. Amphibia

Any consideration of amphibian biology, and this includes their means of communication, must recognize that this group is quite susceptible to dehydration. The integument is generally thin, moist, and highly glandular, and in some groups serves as the primary surface of gas exchange in the adult. Correlated with this vulnerability is the wide-spread occurrence of nocturnality (a time of greater relative humidity), the use of subterranean retreats, and/or the tendency to remain associated with aquatic seepage areas, waterways, or other bodies of water. These habits do not generally enhance the evolution of visual communication systems, and one sees the predominance of vocal-tactile signals in the Anura and chemotactile signals in the caudata. Entirely aquatic species or aquatic breeders are somewhat released from the constraint of nocturnality, and it is in these species that one occasionally observes visual displays complementing the signal repertoire of the organism (e.g., Triturus).

Caudata. The salamanders are essentially avocal and rely on visual, chemical and tactile cues for species, sex, and perhaps kin and individual recognition (Madison, 1975; Noble, 1931a; Porter, 1972). Although the number of verified cases of chemical communication in the caudates is few, it would be surprising if the conspicuous linkage between courtship movements and body contacts on the one hand, and the distribution and activity of the special-

ized glandular tissue on the other, was not related to the exchange
of chemical information.

One association that appears to be evident in the caudates is
that the non-migratory, mostly terrestrial groups (Plethodontidae,
Desmognathidae) appear to rely extensively on the hedonic glands in
the organization and stimulation of courtship activity. The migratory, mainly semi-terrestrial groups (Ambystomatidae, Salamandridae)
rely to a greater extent on cloacal odors in courtship activities.
The remaining non-migratory, mostly aquatic groups, as far as can
be determined, resemble the ambystomatids and salamandrids.

In the Plethodontidae, and specifically in Plethodon, the
hedonic glands on the chin of the male become conspicuously swollen
and vascularized during the breeding season. In virtually all male
plethodontids and desmognathids, this or another region of hedonic
gland tissue is rubbed, tapped or pressed against the body or snout
of the female during an early phase of courtship (e.g., Cupp, 1971;
Noble, 1929, 1931; Noble & Brady, 1930; Organ, 1958, 1960a,b, 1961;
Salthe, 1967; Salthe & Salthe, 1964; Thorn, 1963). The movements
of the female become noticeably slowed and "tuned" to the subsequent
movements of the male. In a later stage of courtship, the female
walks immediately behind the male with her chin resting on the dorsal
surface of the base of his tail, where another major region of
hedonic glands occurs.

The same correlation between glandular presence and courtship
behavior occurs in Salamandridae and Ambystomatidae, except that in
these groups the cloaca region, and quite likely the abdominal
glands, appear to be the focus of much premating activity. Smith
(1941) considers this region to have functions analogous to the
hedonic glands of the Plethodontidae. Although a few species are
reported to have hedonic glands and some have nose or head rubbing
components in their courtship patterns (Rogoff, 1927; Smith, 1941;
Salthe, 1967), the general shift in courtship activity favoring the
cloacal region is clear. In many Ambystoma, Taricha, Triturus,
Notophthalmus, and Cynops, the male rubs his cloacal orifice against
the back of the female (Davis & Twitty, 1964; Licht, 1969) or against
the substrate (Kumpf & Yeaton, 1932). Frequently, one sex, usually
the female, probes and follows closely the cloacal area of the other
sex (Garton, 1972; Halliday, 1975; Kumpf & Yeaton, 1932; Noble &
Brady, 1933; Tsutsui, 1931; Wyman, 1971). The cloacal orifice of
the male may also be open with papillae and/or glandular swelling
apparent (Halliday, 1974, 1975; Licht, 1969; Tsutsui, 1931). Tail
"fanning", "waving", or "whipping" occur in many courtship sequences
and are thought to enhance the spread of glandular secretions from
the cloacal region of the male toward the female. Such generated
current flows in an otherwise still pond environment would have
distinct close range advantages. The remaining caudate families,

especially Cryptobranchidae, Proteidae and Hynobiidae, may tentatively be considered to have a cloacal odor orientation in their
courtship activities similar to that found in the Salamandridae
and Ambystomatidae (Salthe, 1967), but further study of these groups
is needed.

One of the most dramatic indications of sex, possibly species,
odor recognition in the Salamandridae was recorded by Twitty (1955).
Sponges soaked with the odors of female _Taricha rivularis_ and placed
in the breeding stream immediately stimulated the approach of males
downstream from the sponge. Sponges that were not soaked with female
odors failed to elicit any response. The upstream movements of the
males were often in long lines or "queues" immediately downstream
from the scented sponge. This result is somewhat similar to the
findings of Zippelius (1948), who stimulated courtship behavior in
male _Triturus vulgaris_ by exposing them to cotton balls scented with
female odors. No courtship, i.e., "tail-wagging", was observed when
unscented cotton balls were presented.

In addition to having functions during courtship, chemical
signals may be involved in territorial maintenance. This function
is suggested in the Plethodontidae and Desmognathidae, where
abdominal glands and hedonic glands both occur. Recent studies with
Plethodon cinereus demonstrate that individuals can discriminate
their own substrate odors from those of conspecifics during the non-
breeding season (Tristram, 1976), and that fecal pellets on the
substrate seem to be points of interest (Jaeger, personal communication). The abdominal glands at the posterior margin of the
cloacal orifice could easily scent the fecal pellets during
elimination. Tristram (1976) suggests that the avoidance of
conspecific odors, the apparent scent marking behavior called
"chin-touching", and the chemoreceptive behavior called "nose-
touching", are part of a territorial maintenance mechanism that
spaces out members of the species. The discrimination of odors of
neighbors and non-neighbors in _P. jordani_ (Madison, 1975) further
supports a home area, or possible "territorial", scent marking
hypothesis. Since the plethodontids and desmognathids are non-
migratory, it is somewhat more important for these species to
maintain a predictable location in space to compensate for the
lack of high density breeding aggregations, and to defend the
associated nesting retreats and food supplies. This reasoning is
supported by the well defined home range and homing ability in _P.
jordani_ throughout the season of activity (Madison, 1972; Madison
& Shoop, 1970). The use of the abdominal glands, nasolabial glands,
or even more subtle combinations of mucous gland, skin odors and
excretory products could serve, in part, to identify different
conspecifics, their home areas, and even their reproductive states.
That possible scent marking is not exclusive to the plethodontids
or desmognathids is indicated in a study by Gauss (1961) on several

<u>Triturus</u> species. He observed males scent marking their area of
residence with cloacal secretions, noted a peculiar change in
behavior of the females when they encountered these marks, and
observed that this change in behavior was attractive to the resident
male. This observation by Gauss supports multiple functions of the
odors associated with the cloaca of <u>Triturus</u>.

Another possible role of chemical signals may be in the
recognition of the eggs or young. Early accounts of brooding
behavior in salamanders suggested that parental recognition of
odors may be involved in the affiliation of the female (or male
in <u>Cryptobranchus</u>) with their respective clutches. Noble & Evans
(1932) state instead that female <u>Desmognathus</u> <u>fuscus</u> show a marked
site memory, and that living eggs of conspecifics and even artificial eggs will be brooded so long as these are placed in the
original nest location. Similar findings were reported for the
same species (Dennis, 1962) and for <u>Hydromantes</u> (Gorman, 1956).
It should be added, however, that these studies were not extended
throughout the brooding period, and that brood recognition may
develop slowly up to and immediately following hatching. The
recognition of sibling odors may be important to the identification
and defense of the young, to the avoidance of cannibalism, or to
the prevention of inbreeding once the young grow to maturity
(Madison, 1975).

<u>Anura</u>. Frogs and toads have largely exploited vocal, tactile,
and, in some cases, visual signals for communication (e.g.,
Duellman, 1967; Emlen, 1968; Jameson, 1955; Rosen & Lemon, 1974;
Weiwandt, 1969). However, some anurans do not use vocal signals,
as seen in the ascaphids (Noble & Putnam, 1931; Slater, 1931).
Tactile signals become progressively important once potential
mates have come together (Noble & Farris, 1929).

Olfactory cues have been implicated in sex recognition in
wood frogs, <u>Rana</u> <u>sylvatica</u> (Banta, 1914). However, evidence
collected by Noble & Farris (1929) indicates an alternative
explanation. In a particularly revealing experiment, Noble &
Farris anesthetized females, stripped them of eggs, and then
increased some of the females to their former size by injecting
water into their coelom. Males actively gripped both the injected
and non-injected females, but maintained their hold only on the
injected females. Similar results were obtained without anesthesia.
Males also seized and immediately released anesthetized males.
These results indicate that it is not the scent of the female that
controls clasping behavior, rather it is the physical size or girth
of the object seized.

A case of apparent chemical communication in anurans was
reported by Rabb & Rabb (1963). They found that the odor of a

"ready" female toad, Pipa pipa, stimulated and attracted males.
Mature males and "unready" females became quiescent when removed
from the presence of the receptive female. When a gallon of water
from a tank containing the receptive female was put into a tank of
quiescent males and non-receptive females, the males became agitated,
called, and attempted clasping. Although several variables were not
controlled for in this study, the observation remains as one of the
few potential cases of chemical communication in the Anura.

Mention should also be made of the studies of Grubb (1970,
1973a,b, 1975). He demonstrated that in several species of anurans,
both males and females prefer the odor of water from their own
breeding pond compared to that from another breeding pond used by
the same species. Although this is likely a case of a conditioned
or imprinted preference for the odors of the natal environment, as
demonstrated in salmon (e.g., Cooper & Hasler, 1973; Cooper et al.,
1976; Hasler, 1966, 1971; Madison et al., 1973; Scholz et al.,
1976), the possibility exists that intraspecific odor preferences
are involved. In a subsequent paper, Grubb (1976) showed that a
toad, Bufo valliceps, could be conditioned to orient with reference
to an odoriferous landscape. Thus, some anurans have the ability
to detect, learn, and use chemical information about their environment. It remains to be demonstrated whether this ability is
exploited for communicatory purposes.

II. Reptilia

A dominant feature of the reptiles differing from the amphibians
is their relatively greater freedom from the requirement of ambient
moisture or humidity within the environment. This freedom is
associated with the scaled integument, the amniote egg, and the
occurrence of internal fertilization with the use of intromittent
organs. Manifestations of these factors are the elimination of the
seasonal breeding migration to an aquatic habitat, the greater
frequency of diurnal, terrestrial species, and an increased potential
for social stability during the season of growth and reproduction.
As in the caudate amphibians, the use of chemical signals appears
to be associated with terrestrial or semi-terrestrial habits,
suggesting perhaps that scent marking is more feasible in a terrestrial environment, and/or that residential patterns, as opposed to
those requiring a seasonal breeding migration, promote sophistication
in the use of chemical cues. The occurrence of only a few reptiles
that have exploited audition to some degree (e.g., Crocodilia,
Gekkonidae) emphasize that this is not the dominant sensory mode
for communication in the reptiles. Rather, vision seems to be the
primary sensory mode for the exchange of information within the
Reptilia.

Lacertilia. An excellent review by Evans (1961), and a
subsequent update (Evans, 1965), form a good starting point for a
consideration of chemical communication in the lizards. Evans
makes the generalization that the lizards of the families Iguanidae,
Agamidae, and Chamaeleontidae rely primarily on visual releasers in
courtship and territorial interactions. These lizards are diurnal
and conspicuous for the most part, and dichromatism is prominent
among the species. In contrast, families in the Autarchoglossid
Division (Scincidae, Lacertidae, Teiidae, and others) depend more
on a balance of tactile, olfactory, and vocal stimuli. The species
in this Division are generally secretive, even if diurnal; many
representatives are crepuscular or nocturnal; and some are completely subterranean. The Gekkonidae generally fit the characterization
of the autarchoglossids.

Recent information on communication in lizards supports the
dichotomy above. The iguanid lizards have shown a rich variety of
movements and color patterns in their social interactions (see
Carpenter, 1967; Crews, 1975; Purdue & Carpenter, 1972), as have
the agamids (Bellairs, 1970; Blanc & Carpenter, 1969; Harris, 1964)
and the chamaeleontids (Parcher, 1974). Ferguson (1966) made an
experimental attempt to determine the possible use of chemical signals in the communication of Uta stansburiana. Blind-folded males
failed to respond to receptive females, but responded normally when
the blind folds were removed. Males courted receptive females that
were covered with vaseline and even courted males of different
species. The males showed no response to models covered with solutions from the cloaca, skin, or whole animal. Ferguson concluded
that odor was unimportant in the release of male social behavior.

Even though visual signals dominate in the infraorder Iguania,
some indication of the use of chemical signals exists. Blanc &
Carpenter (1967) observed that female Chalarodon madagascariensis
hold their cloaca against the ground while remaining in the alert
position. When the male approaches, the female often assumes a
position with her brightly colored cloaca oriented toward the male
and her tail held up and arched forward. The male may subsequently
bring his nose close to her cloaca. In the chuckwalla, Sauromalus
obesus, Berry (1974) observed individuals licking the substrate,
licking the cloaca, rubbing their jaws on the substrate, and rubbing
the femoral pores on the substrate. Carpenter (1975) suggests that
defecation by male chuckwallas at lookout prominences may generate
chemical signals that assert the male's presence. In a review of
Berry's work, Carpenter (1975) considered the licking and rubbing
movements and other behavior patterns typical in most respects to
what is observed in other iguanid lizards. This suggests that many
iguanids may use certain chemical signals to enhance or supplement
the effect of their visual signals. Burghardt (1977) and Burghardt
et al. (in press) have observed well coordinated rubbing movements

with the chin, venter, and thighs in hatchling green iguanas (Iguana iguana) as well as frequent tongue flicking directed at conspecifics and the substrate. Finally, in Chamaeleo, all individuals "tongue test", during which the tongue is touched against the substrate (Parcher, 1974). This behavior along with cloacal wiping after defaecation could have communicatory significance.

In Gekkonidae, the information on integumentary glands (e.g., Chiu & Maderson, 1975; Maderson, 1972) and post anal or cloacal sacs (Kluge, 1967; Smith, 1935) would seem to suggest chemical communication in this family. The experiments by Greenberg (1943) suggest sex recognition based on integumentary odors in the gecko, Coleonyx variegatus. Courting males were exposed to partially anesthetized male and female lizards that had their tails surgically exchanged. The normal courtship response of the male is to bite and hold the tail of the female. In the experiment, the courting males gripped the female's body or tail parts, and only rarely gripped the male parts. Three mating attempts were observed during which the male gripped and held the female tail on the male individual.

In the Division Autarchoglossida, Carpenter (1962) describes an apparent case of substrate marking in the teiid, Cnemidophorus sexlineatus. In association with space defence, the male wags his posterior trunk region while rubbing his cloaca on the substrate. As the male moves forward, the hind legs are dragged flat against the ground. These movements are repeated at key loci within the habitat. Noble & Teale (1930) compared the courtship of several teiid and iguanid species and found the teiid, Ameiva chrysolaema, to have several courthsip patterns suggesting chemical communication. The beginning of courtship in this species always began with the male rubbing its cloaca laterally over substrate objects. Prior to insertion of the hemipenis, the male rubs his protruding "combs" of the femoral gland along the ventral pelvic surface of the female. In the context of maternal care, several workers have noticed nosing and licking of eggs and young in Scincidae, suggestive of chemical sensing and identification (Evans, 1959; Noble & Kumpf, 1936; Noble & Mason, 1933).

Ophidia. Intraspecific communication in the snakes is dominated by the visual and olfactory senses (Bellairs, 1970; Porter, 1972). Visual displays, although conspicuous in many snake species, do not reach the diversity and complexity of the displays characterizing the iguanid and agamid lizards. Instead, snakes as a group seem to rely more on chemical cues, especially for species and sex recognition. The cloacal "scent glands" peculiar to the ophidians (Whiting, 1969) and the interspecific and intraspecific differences in lipid secretion from these glands (Oldak, 1976) correlate with the difference in emphasis in signal production between the lizards and snakes.

The relatively greater reliance on chemical signals in snakes is probably affected by their evolution from autarchoglossid ancestors (Evans, 1961); their secretive, nocturnal, or subterranean activity patterns; and, in the higher latitudes, their seasonal migration to and from underground hibernation dens (Bellairs, 1970; Burghardt, 1970; Evans, 1961). Information on chemical communication in snakes is generally limited to the Colubridae, particularly because more species occur in this family than in all other families combined (Porter, 1972).

In the Colubridae, odors from the cloacal glands and possibly from the integument appear to function in the "recognition" of sex, reproductive state, individual identity, and species identity. The identification of sex, reproductive state and species is manifest in the trail following behavior of males in response to the odor trails of estrous females (Carpenter, 1955; Gehlbach et al., 1971; Noble, 1937; Noble & Clausen, 1936). In an extensive series of studies on the garter snake, Thamnophis sirtalis, Noble (1937) showed that integumentary odors (from the "back" and "lateral body" skin) of estrous females were attractive to conspecific males during the breeding season, but that the integumentary odors of males and non-estrous females of the same species were non-attractive during the same period. Similar results were recorded for Storeria dekayi (Noble & Clausen, 1936). In both sets of experiments, the snakes seemed to avoid or be indifferent to the cloacal odors of the estrous female, thus suggesting the presence of an unidentified, possibly integumentary, production site. The response to female odors has as its normal outcome the formation of courtship "balls" or mating groups of a single female and multiple males (Aleksiuk & Gregory, 1974; Davis, 1936; Fitch, 1965; Gardner, 1957). In species where the males are less tolerant of subsequent males in the vicinity of the estrous female (e.g., Masticophis), such mating aggregations do not occur (Bennion & Parker, 1976).

With the demonstration that males are attracted to the odors of the dorsal and lateral skin of estrous females, the courtship patterns of the colubrids become conspicuously suggestive of chemical communication processes. A common behavior pattern among the Colubridae is for the male to "chin-rub" or "tongue-flick" the female on the back, neck or head during courtship (Bennion & Parker, 1976; Carpenter, 1955; Fitch, 1963; List, 1950; Noble, 1937; Noble & Clausen, 1936; Pisani, 1967). In species of Natrix that are known to have the nucho-dorsal glands, the courtship patterns appear to have similar elements, except that the male adds "jerky-nodding" to chin-rubbing in the neck region of the female (Noble, 1937).

In addition to the discrimination of sex, reproductive state, and species, snakes also respond differently to the odors of themselves or of conspecifics of the same sex. Gehlbach et al. (1971)

studied trail following in five species of colubrid snakes by
presenting snakes with odor trails made by themselves, by conspecifics of opposite sex, and by conspecifics of the same sex.
In each species, either opposite sex odors or "self" odors elicited
the most trail following. When given a choice between compartments
that were clean or scented with the cloacal sac (scent gland) odors
of conspecifics, the same colubrid species preferred the clean
compartment (Watkins et al., 1969). These results indicate that
the trail following behavior was not in response to cloacal odors,
which agrees with the findings of Noble (1937) and Noble & Clausen
(1936) reported earlier.

A concurrent series of studies with the blind snake,
Leptotyphlops dulcis, gave somewhat different results. Gehlbach
et al. (1971) observed that L. dulcis was attracted to the odor
trails of conspecifics of the opposite sex, of themselves, and of
conspecifics of the same sex, in that order. However, when given
a choice between clean compartments and compartments scented with
the cloacal sac odors of conspecifics, the snakes preferred the
latter (Watkins et al., 1969). The differing result for L. dulcis
suggests a functional difference in the cloacal odors of Colubridae
and Leptotyphlopidae. This difference may be related to the differences in cloacal gland anatomy between the two families (Kroll &
Reno, 1971; Whiting, 1969).

In another series of experiments, Porter & Czaplicki (1974)
showed that individual Natrix rhombifera preferred a clean substrate
to one soiled by a conspecific. These individuals did not show a
preference between substrates that were clean or scented with "self"
odors. In contrast, Thamnophis radix preferred the substrates
soiled by themselves or conspecifics when these were paired with a
clean substrate. Although the results for these two colubrids may
reflect differences in the ecology or social tolerance of the two
species, it is not clear whether the results are free from the
influence of uncontrolled variables. Age and sex were not controlled, although the experiments were conducted during the non-breeding
season; the Natrix were littermates reared from hatchlings in the
laboratory; the Thamnophis were wild caught adults of unknown mutual
familiarity or den affiliation.

The attraction to conspecific odors is also manifest in the
tendency of snakes to group themselves together independent of sex
differences. Neonate Thamnophis sirtalis and Storeria dekayi show
aggregative behavior and species discrimination that is dependent
on a functioning Jacobson's organ (Burghardt, 1977; pers. comm.).
Dundee & Miller (1968) show similar responses in the ringneck snake,
Diadophis punctatus. In this study, discs and the underlying
sand were experimentally manipulated showing that subtle
chemical cues from the previous snake's presence attract

subsequent snakes to the same disc or collection of sand. Noble
& Clausen (1936) conducted an extensive series of studies on the
sensory basis of the aggregative behavior of Storeria dekayi. They
showed that the olfactory sense mediated the aggregative response,
and that the attractive odors "originated" on the integument of the
conspecific. This result slightly differs from the recent findings
of Burghardt in that Noble & Clausen did not feel that Jacobson's
organ was involved in the response. The difference between the
two sets of results could be attributed to the difference in the
species and age class of the individuals studied. A recent finding
of natural summer aggregations of gravid, non-gravid, or mixed
species groups of colubrids (Gregory, 1976) suggests that snake
aggregations serve many different functions.

Before closing this section on the Ophidia, one apparent
inconsistency should be considered that emerges between the
information on scent production, or the lack of scent production,
and the response of snakes to odors from conspecifics. The pre-
vailing evidence indicates that in the Colubridae the odors from
the cloacal area of sexually inactive individuals or estrous females
(presumably odors from excretory products and/or glands affiliated
with the cloacal region) are avoided by conspecifics of either sex.
On the other hand, odor "from" the dorsal or lateral integument is
attractive to conspecifics, especially to males when the donor is
an estrous female. The inconsistency is that there are no known
specialized odor production sites in the dorsal and lateral integument, except in a few natricines, whereas the cloaca scent glands
are capable of signaling information concerning age, sex, and perhaps
individual identity (Oldak, 1976). One explanation for these
findings could be that, when excited, a snake releases feces along
with the scent gland contents, and these act in concert to signal
alarm (Oldak, pers. comm.). In other situations, the snake may
release voluntarily some of the gland contents without mixing it
with feces, and then spread the secretions on the integument through
coiling behavior. The integument thus could act as a receiving
surface for the complex lipid secretions of the scent gland. It
becomes imperative, then, for studies concerning the signal significance of the cloacal secretions to note exactly what the source
area is that is sampled. Oldak (1976) has perfected a method that
only samples the scent gland contents of live snakes. The method
produces similar results to the substances collected from excised
glands (Oldak, pers. comm.).

Chelonia. In his review of the field of herpetology, Porter
(1972) concluded that chemical communication in turtles is largely
restricted to terrestrial or semiaquatic species. The study of
Winokur & Legler (1975) on the occurrence of mental (chin) glands
in 69 of the 74 recognized genera of turtles seems to support
Porter's conclusion. Only 3 families of turtles including

Testudinidae, Emydidae, and Platysternidae had representatives
with mental glands, and these are terrestrial or semi-aquatic
families for the most part. The recent behavioral literature
further corroborates these findings, but because of the inherent
bias favoring terrestrial over aquatic studies, particularly if
the aquatic habitat is turbid or the turtles are wide-ranging, it
is still rather early to accept this generality as more than a
good starting point for further investigation. Most needed is a
comparative study of the cloacal glands comparable to Winokur &
Legler's study of the mental glands and Ehrenfeld & Ehrenfeld's
study of Rathke's glands.

The most extensive investigation of chemical communication in
turtles concerns the species of the genus Geochelone (Auffenberg,
1965, 1969; Carpenter, 1966) and Gopherus (Auffenberg, 1966;
Patterson, 1971; Rose, 1970; Weaver, 1967), both being land
tortoises in the family Testudinidae. A comparison of these
genera is particularly interesting because Gopherus have mental
glands and Geochelone do not (Winokur & Legler, 1975).

In an excellent experimental study of Gopherus belandieri,
Rose (1970) identified the fatty acid composition of the secretion
of the mental gland from male tortoises (the glands of the females
were too small to permit sample collection) and then proceeded to
determine what effect this lipid complex had on the behavior of
adult male and female tortoises. For the behavioral studies, Rose
made a plaster model of a tortoise and then painted the head with
either carbon disulfide (control) or carbon disulfide plus a
solution of fatty acids from laboratory standards whose composition
simulated the glandular secretions analysed. In a natural encounter
situation, a total of 68 tortoises were exposed to the control
treatment, both groups approximating a 1:1 sex ratio. Of the control
encounters, 6 males responded by head-bobbing. Two of these 6 males
also rammed the model. None of the females responded. When the
model carried the lipid treatment, 23 males responded, all with
head-ramming, and 16 females responded, all with head-bobbing. Six
of the male responses actually turned the heavy model over. Since
the courtship pattern of biting and mounting were not observed in
the male, the male response was interpreted to be combative. Since
the females did not ram, their behavior was interpreted to be one
of attraction. This study is conclusive evidence that the lipid
fraction of the mental gland secretion has signal significance, and
that the response to the lipid solution is sex dependent. The gland
secretion probably functions in sex recognition, and the size and
secretory activity of the gland could easily be an expression of
dominance status.

In another study on the same species, Weaver (1967) indicates
further complexity in the chemical communication system. He states

that sex recognition involves differences in behavior, mental gland
secretion, and cloacal scent; that only the females can differentiate sex by cloacal odor; and that preferences for the opposite
sex are made only when the mental glands of both sexes are secreting.
He also adds that when combative behavior involves only one aggressive tortoise, the behavior is very similar to courtship behavior.
When combined with the findings of Rose (1970), these data indicate
that the lipid secretion and maintenance of position of the plaster
model was interpreted by the male tortoise as an aggressive "male"
sign or challenge, and that the combat in effect was a bilateral
one. The failure of Rose to observe courtship biting or mounting,
or to obtain mental gland secretions from the female, supports this
interpretation. It therefore seems likely that the male would be
able to discriminate between sexually active males and females on
the basis of mental gland odors, and that failure to discriminate
could be a sign of a male (possibly subordinate) with fairly
inactive mental glands. In Rose's study, the 40% of the males
that did not respond to the fatty acid odors with ramming could
have accepted the model as a more dominant male, based on mental
gland odors alone. Weaver's other conclusions are difficult to
interpret, but they do indicate the function of the cloacal glands
as well as the mental glands in signal production.

In a related species, Gopherus polyphemus, Auffenberg (1966)
observed that the male tortoise head-bobs upon approaching the
female, while at the same time everting his enlarged mental gland.
These actions could facilitate the release of scent into the air.
At a later stage in the courtship, the male bites the female in the
throat region, primarily on the gular projection of the plastron.
Such close proximity of the nose of the male and the mental gland
of the female may have chemical as well as tactile significance.
These results and particularly those of Weaver (1967) indicate
female participation in the generation of mental gland signals.
The somewhat conflicting results of Rose (1970) show the need for
further study. In a third species, Gopherus agassizi, Patterson
(1971) introduces perhaps what is a third expression of chemical
communication - the use of excretory products, especially feces,
in the regulation of spacing. Dominant males produce fecal pellets
that effectively disperse conspecifics until the pellets dry out
several days after deposition. The above studies collectively
indicate a rich repertoire of chemical signals in Gopherus, although
the chemical system within any one species remains to be clearly
described.

In Geochelone, a genus not having mental glands, the chemical
signals do not appear to be as complex. Carpenter (1966) observed
that male G. elephantopus sniffs primarily the tail region of the
female before mounting. Auffenberg (1965, 1969) indicates the
involvement of the cloacal region in the production of chemical

signals in G. carbonaria and G. denticulata. During courtship, the male challenges the intruder with visual signals. If the challenged tortoise does not respond, the male approaches and sniffs the cloacal region. If the tortoise is a receptive female, mounting is attempted. In some cases, Auffenberg (1965) was able to induce mounting behavior in male tortoises by rubbing cloacal secretions of a receptive female on a skeletonized shell or on other species. Thus, much of the focus of courtship activity appears to shift from the head region in Gopherus to the tail region in Geochelone, at least as far as smelling movements are concerned. This nicely correlates with the occurrence of the mental glands.

In the genera Graptemys and Chrysemys of the family Emydidae, the males bite, sniff, or nose the tail region of females during courtship (Ernst, 1974; Jackson & Davis, 1972a,b). These genera do not have mental glands, so it would not be unexpected to find cloacal scent involved in sex or species recognition.

In the mud and musk turtles of the family Kinosternidae, Mahmoud (1967) gave good descriptive evidence suggesting that species, sex, or sexual receptivity was being determined through olfactory cues. In the four species studied, the male initially approached the female from behind and touched or smelled the tail region of the female. Next, the male "nudged" the side (bridge) of the female near the "musk" (Rathke's) glands. Nudging proceeded anterior to the head region after which mounting would take place. The female would move off during side-nudging if not receptive. These observations suggest that cloacal odors may contribute to sex or species identity, but that receptivity is determined by the response of the female alone.

Crocodilia. Information on communication in Crocodilia is limited to the Nile crocodile (Crocodylus niloticus) and the American alligator (Alligator mississippiensis). There is little doubt that the communication system involves the generation of visual, auditory, and chemical signals (Burrage, 1965; Cott, 1961; Ditmars, 1910; Guggisberg, 1972; McIlhenny, 1934; Pooley & Gans, 1976).

Chemical signals appear to be involved in territoriality and courtship displays. According to McIlhenny (1934), when the bull alligator raises his head to give the territorial roar, the paired musk glands under the jaw evert and a large quantity of pungent smelling musk drips or is rubbed from the gland. This fluid notice- ably scents the air and also lays on the water surface (lipid?). The female alligator rarely releases musk from these glands, but instead produces large amounts of cloacal musk odor during the mating season, presumably to attract the territorial male. Ditmars (1910) suggests that both the male and the female use the jaw gland for mutual attraction. Burrage (1965) noticed that at one point in the alligator courtship sequence the male placed his lower jaw over the

snout of the female and gently rubbed the area. Odors from the
musk gland of the male could well be meaningful at this time, for
the female soon becomes docile and quite receptive. The account
of Guggisberg (1972) and the description of social behavior by
Pooley & Gans (1976) suggests that similar chemical signals exist
for the Nile crocodile.

CONCLUSIONS

This review attempted to bring information together on odor
production, odor reception, and social behavior in amphibians and
reptiles, and thereby provide an evaluation of chemical communication
within these taxonomic groups. Many of the statements were inferen-
tial, and much of the supporting evidence was circumstantial.
Despite these weaknesses, the information strongly suggests that
research into several areas would yield interesting results. In
particular, the hedonic and abdominal glands of salamanders, the
scent glands of snakes, the mental glands of turtles, and the "musk"
glands of the crocodilians are glandular structures that almost
certainly produce a rich variety of sociochemical signals, but
experimental evidence to support this action is critically needed.

The existing evidence of chemical communication in amphibians
and reptiles is likely biased by the selection of animals that are
easily "observed". A critical need is a study of a phylogenetically
related group of amphibians or reptiles whose species live under a
range of different environmental conditions, e.g. subterranean, open-
terrain, and arboreal conditions. The inspection of these organisms
must be sensitive to the generation of chemical and visual signals,
and also to signals in other sensory modes. Only then can we begin
to view objectively some of the more interesting questions regarding
chemical communication as an ecological adaptation.

SUMMARY

The possibility of chemical communication in amphibians and
reptiles was considered from the point of view of scent production,
scent reception, and the associated patterns of behavior. The major
findings of this review are as follows:

1) On the basis of the association between glandular structure and
   general patterns of behavior, the caudate amphibia, and especially the Plethodontidae and Desmognathidae, appear to have a
   distinct potential for a complex system of chemical signals.
2) In the Caudata, the potential for complexity in chemical signals
   is associated with organisms that are non-migratory, nocturnal,
   and terrestrial or semi-terrestrial.

3) Chemical communication does not emerge as a significant trend in any of the anuran Amphibia surveyed.
4) Although the reptiles as a group rely predominantly on visual signals for the exchange of information, many snakes, turtles, and crocodilians have distinct scent producing glands and conspicuous patterns of behavior that implicate the importance of chemical signals in the coordination of social behavior.
5) Some groups of lacertilian reptiles, especially the Autarchoglossida, appear to use chemical communication as a complementary means of information exchange between conspecifics.
6) In the reptiles the more elaborate cases of chemical communication appear to be associated with species that are terrestrial or subterranean in habit.

ACKNOWLEDGEMENTS

I am especially grateful to the following persons for providing references, reprints, and their own observations covering many aspects of chemical communication in amphibians and reptiles: M. Aleksiuk, W. Auffenberg, R. Brandon, B.H. Brattstrom, N. Bray, G. Burghardt, C.C. Carpenter, A. Carr, J. Grubb, H. Heusser, J. Huheey, R. Jaeger, P.F.A. Maderson, P. Oldak, A.S. Rand, R. Ruibal, R. Shoop, S. Tilley, D. Tinkle, and A. Williams. I thank Eva, Kathy and Louissa for extensive library and clerical assistance, and Robert LaMarche for excellent photographic services. This study received financial assistance from the National Research Council of Canada.

REFERENCES

Aleksiuk, M. & P.T. Gregory. 1974. Regulation of seasonal mating behavior in Thamnophis sirtalis parietalis. Copeia 1974: 681–689.

Allison, A.C. 1953. The morphology of the olfactory system in the vertebrates. Biol. Rev. 28: 195–244.

Auffenberg, W. 1965. Sex and species discrimination in two sympatric South American tortoises. Copeia 1965: 335–342.

_____. 1966. On the courtship of Gopherus polyphemus. Herp. 22: 113–117.

_____. 1969. Tortoise Behavior and Survival. Rand McNally & Co., Chicago. 38 p.

Banta, A.M. 1914. Sex recognition and the mating behavior of the wood frog, Rana sylvatica. Biol. Bull. 26: 171–183.

Bellairs, A. 1970. The Life of Reptiles. Vol. II. The Universe Natural History Series. Universe Books, New York.

Bennion, R.S. & W.S. Parker. 1976. Field observations on courtship and aggressive behavior in desert striped whipsnakes, Masticophis t. taeniatus. Herp. 32: 30–35.

Bentley, P.J. & K. Schmidt-Nielsen. 1966. Cutaneous water loss in
    reptiles. Science 151: 1547-1549.
Berry, K. 1974. The ecology and social behavior of the chuckwalla,
    Sauromalus obesus obesus Baird. U.C. Publ. Zool. 101: 1-60.
Bertmar, G. 1969. The vertebrate nose, remarks on its structural
    and functional adaptation and evolution. Evol. 23: 131-152.
Blair, W.F. 1968. Amphibians and reptiles. Pp. 289-310 in T.A.
    Sebeok (ed.), Animal Communication, Indiana University Press,
    Bloomington, Indiana.
Blanc, C.P. & C.C. Carpenter. 1969. Studies on the Iguanidae of
    Madagascar. III. Social and reproductive behavior of
    Chalarodon madagascariensis. J. Herp. 3: 125-134.
Blaylock, L.A., R. Ruibal & K. Platt-Aloia. 1976. Skin structure
    and wiping behavior of phyllomedusine frogs. Copeia 1976:
    283-295.
Breckenridge, W. & R. Murugapillai. 1974. Mucous glands in the
    skin of Ichthyophis glutinosus (Amphibia: Gymnophiona).
    Ceylon J. Sci. (Bio. Sci.) 11: 43-52.
Brodie, E.D., Jr. 1968a. Investigations on the skin toxin of the
    adult rough-skinned newt, Taricha granulosa. Copeia 1968:
    307-313.
_____. 1968b. Observations on the mental hedonic gland-clusters
    of western salamanders of the genus Plethodon. Herp. 24:
    248-250.
Brodie, E.D., Jr. & L.S. Gibson. 1969. Defensive behavior and
    skin glands of the northwestern salamander, Ambystoma gracile.
    Herp. 3: 187-194.
Brown, C.E. & B.S. Martof. 1966. The function of the naso-labial
    groove of plethodontid salamanders. Phys. Zool. 4: 357-367.
Brown, C.W. 1968. Additional observations on the function of the
    nasolabial grooves of plethodontid salamanders. Copeia 1968:
    728-731.
Burghardt, G. 1970. Chemical perception in reptiles. Pp. 241-308
    in J.W. Johnston, Jr., D.G. Moulton and A. Turk (eds.). Advances
    in Chemoreception, Vol. I., Meredith Corp., New York.
_____. 1977. Of iguanas and dinosaurs: Social behavior and
    communication in neonate reptiles. Amer. Zool. in press.
Burghardt, G.M., H.W. Greene, & A.S. Rand. in press. Social
    behavior in hatchling green iguanas. Science, in press.
Burkholder, G.L. & W.W. Tanner. 1974. A new gland in Sceloporus
    graciosus males. Herp. 30: 368-371.
Burrage, B.R. 1965. Copulation in a pair of Alligator
    mississipiensis. Brit. J. Herp. 3: 207-208.
Carpenter, C.C. 1955. The garter snake. The Scientific Monthly
    81: 248-252.
_____. 1962. Patterns of behavior in two Oklahoma lizards. Am.
    Midl. Natur. 67: 132-151.
_____. 1966. Notes on the behavior and ecology of the Galapagos

tortoise on Santa Cruz Island. Proc. Okla. Acad. Sci. 46: 28-32.

_____. 1967. Aggression and social structure in iguanid lizards. Pp. 87-105 in W.W. Milstead (ed.), Lizard Ecology: A Symposium, Univ. Missouri Press, Columbia, Missouri.

_____. 1975. Review. Copeia 1975: 388-389.

Chiu, K. & P.F.A. Maderson. 1975. The microscopic anatomy of epidermal glands in two species of gekkonine lizards, with some observations on testicular activity. J. Morph. 147: 23-40.

Chiu, K., P.F.A. Maderson, S.A. Alexander & K.L. Wong. 1975. Sex steroids and epidermal glands in two species of gekkonid lizards. J. Morph. 147: 9-22.

Cole, C.J. 1966a. Femoral glands of the lizard, Crotaphytus collaris. J. Morph. 118: 119-135.

_____. 1966b. Femoral glands in lizards: a review. Herp. 22: 199-206.

Conaway, C.H. & D.E. Metter. 1967. Skin glands associated with breeding in Microhyla carolinensis. Copeia 1967: 672-673.

Cooper, J.C. & A.D. Hasler. 1973. II. An electrophysiological approach to salmon homing. Fish. Res. Bd. Can. Tech. Rep. 415: 1-44.

Cooper, J., A. Scholz, R. Horrall, A. Hasler & D. Madison. 1976. Experimental confirmation of the olfactory hypothesis with homing, artificially imprinted coho salmon (Oncorhynchus Kisutch). J. Fish. Res. Bd. Can. 33: 703-710.

Cott, H.B. 1961. Scientific results of an enquiry into the ecology and economic status of the Nile crocodile (Crocodilus niloticus) in Uganda and Northern Rhodesia. Trans. Zool. Soc. London 29: 211-356.

Crews, D. 1975. Inter- and intra-individual variation in display patterns in the lizard, Anolis carolinensis. Herp. 31: 37-47.

Cupp, P.V., Jr. 1971. Fall courtship of the green salamander, Aneides aeneus. Herp. 27: 308-310.

Davis, D.D. 1936. Courtship and mating behavior in snakes. Zool. Ser. Field Mus. Nat. Hist. 20: 257-290.

Davis, W.C. & V.C. Twitty. 1964. Courtship behavior and reproductive isolation in the species of Taricha (Amphibia, Caudata). Copeia 1964: 601-610.

Dawson, A.B. 1920. The integument of Necturus maculosus. J. Morph. 34: 487-589.

_____. 1922. The cloaca and cloacal glands of the male Necturus. J. Morph. 36: 447-466.

Dennis, D. 1962. Notes on the nesting habits of Desmognathus fuscus fuscus (Raf.) in Licking County, Ohio. J. Ohio. Herp. Soc. 3: 28-35.

Dieckmann, J.H. 1927. The cloaca and spermatheca of Hemidactylium scutatum. Biol. Bull. 53: 281-285.

Ditmars, R.L. 1910. Reptiles of the World. Sturgis and Walton Co., New York.

Dodd, C.K., Jr., J.A. Johnson & E.D. Brodie, Jr. 1974. Noxious skin secretions of an eastern small Plethodon, P. nettingi hubrichti. J. Herp. 8: 89-92.

Duellman, W.E. 1967. Social organization in the mating calls of some Neotropical anurans. Am. Midl. Natur. 77: 156-163.

Dundee, H.A. & M.C. Miller III. 1968. Aggregative behavior and habitat conditioning by the prairie ringneck snake Diadophis punctatus arnyi. Tulane Stud. Zool. Bot. 15: 41-58.

Ehrenfeld, J.G. & D.W. Ehrenfeld. 1973. Externally secreting glands of freshwater and sea turtles. Copeia 1973: 305-314.

Elias, H. & J. Shapiro. 1957. The histology of the skin of some frogs and toads. Am. Mus. Novit. No. 1819, 27 pp.

Emlen, S.T. 1968. Territorality in the bullfrog, Rana catesbeiana. Copeia 1968: 240-243.

Ernst, C. 1974. Observations on the courtship of male Graptemys pseudogeographica. J. Herp. 8: 377-378.

Evans, L.T. 1959. A motion picture study of maternal behavior of the lizard. Eumeces obsoletus Baird & Girard. Copeia 1959: 103-110.

_____. 1961. Structure as related to behavior in the organization of populations in reptiles. Pp. 148-178 in W.F. Blair (ed.), Vertebrate Speciation, University of Texas Press, Austin, Texas.

_____. 1965. Introduction. Pp. 83-86 in W.M. Milstead (ed.), Lizard Ecology: A Symposium, University of Missouri Press, Columbia, Missouri.

Ferguson, G.W. 1966. Releasers of courtship and territorial behavior in the side-blotched lizard Uta stansburiana. Anim. Behav. 14: 89-92.

Fitch, H.S. 1963. Natural history of the racer Coluber constrictor. Univ. Kans. Publ. Museum Nat. Hist. 15: 351-468.

_____. 1965. An ecological study of the garter snake Thamnophis sirtalis. Univ. Kans. Publ. Museum Nat. Hist. 15: 493-564.

Fox, W. 1965. A comparison of the male urogenital systems of blind snakes, Leptotyphlopidae and Typhlopidae. Herp. 21: 241-256.

Gabe, M. & H. Saint-Girons. 1965. Contribution à la morphologie comparée du cloaque et des glandes épidermoides de la région cloacale et des glandes épidermoides de la région cloacale chez les Lépidosauriens. Mem. Mus. Nat. Hist. Natur. (Paris). Ser. A. (Zool.) 33: 150-292.

_____. 1967. Données histologiques sur le tégument et les glandes épidermoides céphaliques des Lépidosauriens. Acta Anat. 67: 571-594.

Gardner, J.B. 1957. A garter snake "ball". Copeia 1957: 48.

Garton, J.S. 1972. Courtship of the small-mouthed salamander, Ambystoma texanum, in southern Illinois. Herp. 28: 41-45.

Gauss, G.H. 1961. Ein Beitrag zur Kenntnis des Balzverhaltens einheimischer Molche. Z. Tierpsychol. 18: 60-66.

Gehlbach, F.R., J. Watkins II and J. Kroll. 1971. Pheromone trail-following studies of Typhlopid, Leplotyphlopid, and Colubrid snakes. Behaviour 40: 282-294.

Gorman, J. 1956. Reproduction in the plethodontid salamanders of the genus Hydromantes. Herp. 12: 249-259.

Greenberg, B. 1943. Social behavior of the western banded gecko, Coleonyx variegatus Baird. Physiol. Zool. 16: 110-122.

Gregory, P.T. 1975. Aggregations of gravid snakes in Manitoba, Canada. Copeia 1975: 185-186.

Grubb, J.C. 1970. Orientation in post-reproductive Mexican toads, Bufo valliceps. Copeia 674-680.

_____. 1973a. Olfactory orientation in breeding Mexican toads, Bufo valliceps. Copeia 1973: 490-497.

_____. 1973b. Olfactory orientation in Bufo woodhousei fowleri, Pseudacris clarki and Pseudacris streckeri. Anim. Behav. 21: 726-732.

_____. 1975. Olfactory orientation in southern leopard frogs, Rana utricularia. Herp. 31: 219-221.

_____. 1976. Maze orientation by Mexican toads, Bufo valliceps, using olfactory and configurational cues. J. Herp. 10(2): 97-104.

Guggisberg, C.A.W. 1972. Crocodiles. Wren Publishing, Victoria, 195 p.

Halliday, T.R. 1974. Sexual behavior of the smooth newt. J. Herp. 8: 277-292.

_____. 1975. An observational and experimental study of sexual behaviour in the smooth newt, Triturus vulgaris. Anim. Behav. 23: 291-322.

Harris, V. 1964. The Life of the Rainbow Lizard. Hutchinson and Co., London, 174 p.

Hasler, A.D. 1966. Underwater Guideposts. Univ. Wisconsin Press, Madison, Wis., 155 p.

_____. 1971. Orientation and fish migration. Pp. 429-510, Hoar & Randall (eds.). Fish Physiology, Vol. 6, Academic Press, New York.

Hays, R.M. 1966. The mental hedonic gland-cluster of the male salamander, Desmognathus fuscus. Diss. Abst. 27B: 2187-B.

Heidenhain, M. 1890. Beiträge zur Kenntnis der Topographie und Histologie der Kloake und ihrer drüsigen Adnexa bei den einheimischen Tritonen. Arch. f. mikr. Anat. 35: 173-266.

Heimer, L. 1969. The secondary olfactory connections in mammals, reptiles and sharks. Ann. N.Y. Acad. Sci. 167: 129-146.

Hensel, J.L., Jr. & E.D. Brodie, Jr. 1976. An experimental study of aposematic coloration in the salamander Plethodon jordani. Copeia 1976(1): 59-65.

Highton, R. 1957. Correlating costal grooves with trunk vertebrae in salamanders. Copeia 1957: 107-109.

Jackson, G., Jr. & J.D. Davis. 1972a. Courtship display behavior of Chrysemys concinna suwanniensis. Copeia 1972(2): 385-387.

_____. 1972b. A quantitative study of the courtship display of the red-eared turtle, Chrysemys scripta elegans. Herp. 28: 58-64.

Jameson, D.L. 1955. Evolutionary trends in the courtship and
    mating behavior of Salientia. Syst. Zool. 4: 105-119.
Kerr, T. 1939. On the structure and function of the cloaca of
    the common frog (Rana t. temporaria). Proc. Zool. Soc.
    London 1939: 63-73.
Kingsbury, B.F. & O. Defiance. 1895. The spermatheca and methods
    of fertilization in some American newts and salamanders.
    Trans. Amer. Micro. Soc. 17: 261-304.
Kluge, A.G. 1967. Higher taxonomic categories of gekkonid lizards
    and their evolution. Bull. Am. Mus. Nat. Hist. 135: 1-60.
Kroll, J.C. & H.W. Reno. 1971. A re-examination of the cloacal
    sacs and gland of the blind snake, Leptotyphlops dulcis
    (Reptilia: Leptotyphlopidae). J. Morph. 133: 273-280.
Kumpf, K.F. & S.C. Yeaton, Jr. 1932. Observations of the courtship
    behavior of Ambystoma jeffersonianum. Am. Mus. Novit. 546: 1-6.
Lanza, B. 1959. Il corpo ghiandolare mentoniero dei
    "Plethodontidae" (Amphibia, Caudata). Monit. Zool. Ital. 67:
    15-53.
Licht, L. 1969. Observations on the courtship behavior of
    Ambystoma gracile. Herp. 25: 49-52.
Lillywhite, H.B. & P. Licht. 1974. Movement of water over toad
    skin: functional role of epidermal sculpturing. Copeia 1974:
    165-170.
List, J.C., Jr. 1950. Observations on the courtship behavior of
    Thamnophis s. sirtalis. Herp. 6: 71-74.
Maderson, P.F.A. 1968. The epidermal glands of Lygodactylus
    (Gekkonidae, Lacertilia). Breviora No. 288: 1-35.
_____. 1970. Lizard glands and lizard hands: models for
    evolutionary study. Forma et Functio 3: 179-204.
_____. 1972. The structure and evolution of holocrine epidermal
    glands in Sphaerodactyline and Eublepharine gekkonid lizards.
    Copeia 1972(3): 559-571.
Maderson, P.F.A. & K.W. Chiu. 1970. Epidermal glands in gekkonid
    lizards: evolution and phylogeny. Herp. 26: 233-238.
Madison, D.M. 1972. Homing orientation in salamanders: a
    mechanism involving chemical cues. Pp. 485-498, S.R. Galler,
    K. Schmidt-Koenig, G.J. Jacobs & R.E. Belleville (eds.).
    Animal Orientation and Navigation. Special Publication 262,
    North American Space Administration, Washington, D.C.
_____. 1975. Intraspecific odor preferences between salamanders
    of the same sex: dependence on season and proximity of
    residence. Can. J. Zool. 53: 1356-1361.
Madison, D.M., A. Scholz, J. Cooper, R. Horrall, A.D. Hasler & A.
    Dizon. 1973. I. Olfactory hypotheses and salmon migration:
    a synopsis of recent findings. Fish. Res. Bd. Canada Tech.
    Report 414, pp. 1-35.
Madison, D.M. & C.R. Shoop. 1970. Homing behavior, orientation,
    and home range of salamanders tagged with Tantalum-182. Science
    168: 1484-1487.

Mahmoud, I.Y. 1967. Courtship behavior and sexual maturity in four species of kinosternid turtles. Copeia 1967(2): 314-319.

McIlhenny, E.A. 1934. The Alligator's Life History. Christopher Publ. House, Boston.

Menchel, S. & P.F.A. Maderson. 1975. The post-natal development of holocrine epidermal specializations in gekkonid lizards. J. Morph. 147: 1-8.

Metter, D. & C. Conaway. 1969. The influence of hormones on the development of breeding glands in Microhyla. Copeia 1969: 621-622.

Miscalencu, D., M.D. Ionescu & A. Petrovici. 1971. Fine structure of integumentary glands in Triturus cristatus. Anat. Anz. 129: 225-232.

Moulton, D.G. & L.M. Beidler. 1967. Structure and function in the peripheral olfactory system. Physiol. Rev. 47: 1-52.

Muhse, E.F. 1909. The cutaneous glands of the common toads. Am. J. Anat. 9: 322-359.

Noble, G.K. 1929. The relation of courtship to the secondary sexual characters of the two-lined salamander, Eurycea bislineata (Green). Amer. Mus. Novit. No. 362: 1-5.

_____. 1931a. Biology of the Amphibia. McGraw-Hill, New York, 577 p.

_____. 1931b. The hedonic glands of the plethodontid salamanders and their relation to sex hormones. Anat. Record (suppl.) 48: 57-58 (abstract).

_____. 1937. The sense organs involved in the courtship of Storeria, Thamnophis and other snakes. Bull. Amer. Mus. Nat. Hist. 73: 673-726.

Noble, G.K. & H.T. Bradley. 1933. The mating behavior of lizards; its bearing on the theory of sexual selection. Ann. N.Y. Acad. Sci. 35: 25-100.

Noble, G.K. & M.K. Brady. 1930. The courtship of the plethodontid salamanders. Copeia 1930: 52-54.

_____. 1933. Observations on the life history of the marbled salamander Ambystoma opacum. Zoologica 11: 89-132.

Noble, G.K. & H.J. Clausen. 1936. The aggregation behaviour of Storeria dekayi and other snakes with special reference to the sense organs involved. Ecol. Monog. 6: 269-316.

Noble, G.K. & G. Evans. 1932. Observations and experiments on the life history of the salamander, Desmognathus fuscus fuscus (Rafinesque). Amer. Mus. Novit. 533: 1-16.

Noble, G.K. & E.J. Farris. 1929. The method of sex recognition in the woodfrog, Rana sylvatica Le Conte. Amer. Mus. Novit. 363: 1-17.

Noble, G.K. & K.F. Kumpf. 1936. The function of Jacobson's organ in lizards. J. Genet. Psychol. 48: 371-382.

Noble, G.K. & E.R. Mason. 1933. Experiments on the brooding habits of the lizards Eumeces and Ophisaurus. Amer. Mus. Novit. 619: 1-29.

Noble, G.K. & P.G. Putnam. 1931. Observations on the life history of *Ascaphus truei* Stejneger. Copeia 1931: 97-101.

Noble, G.K. & H.K. Teale. 1930. The courtship of some iguanid and teiid lizards. Copeia 1930: 54-56.

Noble, G.K. & J.A. Weber. 1929. The spermatophores of *Desmognathus* and other plethodontid salamanders. Amer. Mus. Novit. No. 351: 1-15.

Oldak, P. 1976. Comparison of the scent gland secretion lipids of twenty-five species of snakes: implications for biochemical systematics. Copeia 1976: 320-326.

Organ, J.A. 1958. Courtship and spermatophore of *Plethodon jordani metcalfi*. Copeia 1958: 251-259.

_____. 1960a. The courtship and spermatophore of the salamander *Plethodon glutinosus*. Copeia 1960: 34-40.

_____. 1960b. Studies on the life history of the salamander, *Plethodon welleri*. Copeia 1960: 287-297.

_____. 1961. Life history of the pygmy salamander, *Desmognathus wrighti* in Virginia. Am. Midl. Natur. 66: 384-390.

Parcher, S. 1974. Observation on the natural histories of six Malagasy Chamaeleontidae. Z. Tierpsychol. 34: 500-535.

Parsons, T.S. 1967. Evolution of the nasal structure in the lower tetrapods. Am. Zool. 7: 397-413.

Patterson, R. 1971. Aggregation and dispersal behavior in captive *Gopherus agassizi*. J. Herp. 5: 214-216.

Pfeiffer, W. 1974. Pheromones in fish and amphibia. Pp. 269-296 in M.C. Birch (ed.). Pheromones. Frontiers of Biology, Vol. 32, North-Holland Publ. Co., Amsterdam.

Phisalix, M. 1918. Les venins cutanés dur *Spelerpes fuscus* Gray. Bull. Mus. Nat. Paris 24: 92-96.

Pisani, G.R. 1967. Notes on the courtship and mating behavior of *Thamnophis brachystoma*. Herp. 23: 112-115.

Pooley, A.C. & C. Gans. 1976. The Nile crocodile. Sci. Amer. 234: 114-124.

Porter, K.R. 1972. Herpetology. W.B. Saunders Co., Philadelphia. pp. 524.

Porter, R.H. & J.A. Czaplicki. 1974. Responses of water snakes (*Natrix r. rhombifera*) and garter snakes (*Thamnophis sirtalis*) to chemical cues. Anim. Learn. Behav. 2: 129-132.

Purdue, J.R. & C.C. Carpenter. 1972. A comparative study of the body movements of displaying males of the lizard genus *Sceloporus* (Iguanidae). Behav. 41: 68-81.

Quay, W.B. 1972. Integument and the environment: glandular composition, function and evolution. Amer. Zool. 12: 95-108.

Rabb, G.B. & M.S. Rabb. 1963. Additional observations on breeding behavior of the Surinam toad, *Pipa pipa*. Copeia 1963: 636-642.

Reese, A.M. 1920. The integumental glands of *Alligator mississipiensis*. Anat. Record 20: 203.

_____. 1921. The structure and development of the integumental glands of the Crocodilia. J. Morph. 35: 581-611.

Rie, I.P. 1973. Application of drugs to the skin of salamanders. Herp. 29: 55-59.

Rogoff, J.L. 1927. The hedonic glands of Triturus viridescens; a structural and morphological study. Anat. Rec. 34: 132-133.

Rose, F.L. 1970. Tortoise chin gland fatty acid composition: behavioral significance. Comp. Biochem. Physiol. 32: 577-580.

Rosen, M. & R. Lemon. 1974. The vocal behavior of spring peepers, Hyla crucifer. Copeia 1974: 940-950.

Sabnis, J.H. 1967. Anatomy and histology of the hemipenis and associated structures in Natrix piscator piscator (Schneider). Brit. J. Herp. 4: 51-54.

Salthe, S.N. 1967. Courtship patterns and phylogeny of the urodeles. Copeia 1967 (1): 100-117.

Salthe, S. & B. Salthe. 1964. Induced courtship in the salamander Pseudoeurycea belli. Copeia 1964: 574-576.

Scholz, A., R. Horrall, J. Cooper & A.D. Hasler. 1976. Imprinting to chemical cues: the basis for home stream selection in salmon. Science 192: 1247-1249.

Sever, D.M. 1975. Morphology and seasonal variation of the nasolabial glands of Eurycea quadridigata. J. Herp. 9: 337-348.

_____. 1976. Morphology of the mental hedonic gland clusters of plethodontid salamanders (Amphibia, Urodela, Plethodontidae). J. Herp. 10: 227-239.

Slater, J.R. 1931. The mating behavior of Ascaphus truei Stejneger. Copeia 1931: 62-63.

Smith, M.A. 1935. The Fauna of British India, including Ceylon and Burma. Reptilia and Amphibia, Vol. II. Sauria. Taylor & Francis Ltd., London. 440 pp.

_____. 1938. The nucho-dorsal glands of snakes. Proc. Zool. Soc. London, B. 108: 575-583.

Smith, R.E. 1941. Mating behavior in Triturus torosus and related newts. Copeia 1941(4): 255-262.

Thorn, R. 1963. Contribution à l'étude d'une salamander Japonaise, l'Hynobius nebulosus (Schlegel). Comportement et reproduction en captivité. Arch. Inst. Gr-Duc. Luxembourg N.S. 29: 201-215.

Tristram, D.A. 1976. Intraspecific olfactory communication in the terrestrial salamander, Plethodon cinereus. Copeia (In press).

Truffelli, G.T. 1954. A macroscopic and microscopic study of the mental hedonic gland-clusters of some plethodontid salamanders. Univ. Kans. Sci. Bull. 36: 3-39.

Tsutsui, Y. 1931. Notes on the behavior of the common Japanese newt Cynops pyrrhogaster Boie. 1) Breeding habit. Mem. Coll. Sci. Kyoto Imp. Univ., Ser. B., 7: 159-167.

Tucker, D. 1971. Nonolfactory responses from the nasal cavity: Jacobson's Organ and the trigeminal system. Pp. 151-181, L.M. Beidler (ed.), Handbook of Sensory Physiology IV, Springer-Verlag, New York.

Tucker, D. & J.C. Smith. 1969. The chemical senses. Ann. Rev. Psych. 20: 129-158.

Twitty, V.C. 1955. Field experiments on the biology and genetic relationships of the California species of Triturus. J. Exp. Zool. 129: 129-148.

Volsøe, H. 1944. Structure and seasonal variation of the male
     reproductive organs of Vipera berus (L.). Spolia Zoolog.
     Musei Haunieneis 5: 9-157.
Watkins II, J.F., F.R. Gehlbach & J.C. Kroll. 1969. Attractant-
     repellent secretions of blind snakes (Leptotyphlops dulcis)
     and their army ant prey (Neivamyrmex nigrescens). Ecol. 50:
     1098-1102.
Weaver, W.G., Jr. 1967. Courtship and combat behavior in Gopherus
     berlandieri. Ph.D. Dissertation, Univ. Florida, Gainesville.
Whipple, I. 1906. The naso-labial groove of lungless salamanders.
     Biol. Bull. 11: 1-26.
Whiting, A.M. 1967. Amphisbaenian cloacal glands. Am. Zool. 7:
     776 (abstract).
_____. 1969. Squamate cloacal glands: morphology, histology and
     histochemistry. Ph.D. Dissertation, Pennsylvania State
     University.
Wiewandt, T.A. 1969. Vocalization, aggressive behavior, and
     territoriality in the bullfrog, Rana catesbeiana. Copeia
     1969: 276-285.
Winokur, R.M. & J.M. Legler. 1975. Chelonian mental glands. J.
     Morph. 147: 275-292.
Wyman, R.L. 1971. The courtship behavior of the small-mouthed
     salamander, Ambystoma texanum. Herp. 27: 491-498.
Zippelius, H.M. 1948. Untersuchungen uber das Balzverhalten
     einheimischer Molche. Verh. dt. Zool. Kiel 1948: 127.

CHEMICAL SIGNALS IN AGONISTIC AND SOCIAL BEHAVIOR OF RODENTS

Ph. Ropartz

Laboratoire de Psychophysiologie (U.L.P.)

67000 Strasbourg, France

Ten years ago, a review regarding the role of chemical signals in the social behavior of rodents would have been a simple task ; there were only a few works showing the involvement of olfactory cues in the social relations of mice and rats. Today, it seems impossible to compile an exhaustive list of the results of all such studies.

As Stoddart (1974) said "if odors are used for social purposes, one might expect that the range of messages to be exchanged would be as large as the range of visual signals observed in social groups of large mammals." Perhaps this explains the fact that the list of pheromones involved in the social behavior of rodents continues to increase.

"Social behavior" may be defined as all types of relations which can occur between at least two members of the same species apart from the sexual relations. However, as several excellent reviews have been published recently on maternal pheromones (Leon, 1974), spacing behavior (Mykytowycz, 1974), and olfactory marking (Ralls, 1971 ; Johnson, 1973) in mammals, this paper about social behavior will exclude these aspects. Some precise examples will emphasize other recent trends in this type of research.

Even in mice, the most studied of the rodent species, the exact social life is not well known. Crowcroft (1966) showed that a population of wild mice is divided in numerous subunits, each one involving one adult male, some adult females, and their young. Such a social structure is probably very common among rodents. Barnett (1958) described the same relations among domestic or wild rats. Such a social structure does not necessitate as many social signals and means of communication as in true animal societies such as in some birds and primates or even in insect societies. Nevertheless rodents seem to use chemical communication preferentially and almost

all social information is transmitted through the olfactory
channel. It is probably the reason why so many pheromones, for
example, have been found in mice. But, as we don't know the chemical nature of these components, it remains possible that the
same pheromone possesses two functions according to the context,
or that a sum of pheromones does not correspond to a simple addition of the meanings of each of them.

The list of pheromones one can establish today is certainly
not complete and it would be better to present some questions implied by the chemical signals in the social life of rodents.

OLFACTORY RECOGNITION : INDIVIDUAL, GROUP, AND SPECIES

Ten years ago, few authors assumed that small mammals could
use chemical cues for the recognition of members of their species.
Nevertheless, it seemed likely that even in the very simple social
structure of rats and mice there was an elementary ability. To
recognize members of the species and even members of the social
subunits. One cannot understand how these sorts of social subunits
could exist and maintain relationships between them without the
ability of group recognition.

Several authors attempted to give evidence of this ability in
gerbils (Halpin, 1974), guinea pigs (Berüter et al., 1974), voles
(Moncrieff, 1962 and Doty and Kart, 1972), rats (Krames, 1970),
and mice (Bruce, 1960). For example, Halpin discovered that gerbils
have to learn the olfactory cues of members of their social group ;
voles were found to recognize members of their own species (Doty,
1972) in a sexual context or other social contexts (Moncrieff,
1962). In mice, the pregnancy-block discovered by Bruce (1960) and
further studied by Dominic (1969) is an indirect evidence of a recognition of different male mice by a female.

It is not easy to prove a true recognition through chemical
messages. For example, the works of Bruce and Dominic showed that
a female mouse can discriminate between two males using only olfactory cues ; but we have to notice that the more significant results,
that is the higher frequency of olfactory pregnancy-block, were obtained when the two males belonged to different strains. This is certainly an olfactory recognition between strains but not necessarily
between individuals.

In other words , in most of these cases we need better experimental proofs, as in the work of Bowers and Alexander (1967) in
mice. This work was improved by Hahn and Simmel (1968), who showed
very clearly that mice are able to recognize other members of their
species by olfactory cues.

Kalkowski (1967) showed that the olfactory recognition of
species-partners in mice remains possible at a distance of 17.5 cm
and he emphasized the advantage of such a capacity in the communication between members of the same social unit or between small
groups belonging to the same population.

While studying the social behavior of wild and domestic rats, Barnett (1963) attempted to explain some aspects of the aggressive patterns of this rodent. He came to the conclusion that two sorts of smell could explain the relations between members of the same group and between strange groups : a "type odour rats", allowing a recognition of the individual, its age, its sex, and perhaps its physiological state, and a "colony odour rats", which enables an animal to distinguish a member of its social unit from strange rats. But, at that time, there was no evidence to test this hypothesis.

Using a very simple actographic method, Ropartz (1968b) showed that between small groups of male mice (C 57 Bl) there existed two sorts of odors. The first one, named "plantar factor mice" is produced by the sudoriferous secretion in the paws of male mice when isolated or grouped. The second one, named "urinary factor mice" is produced by the coagulating glands of the genital tract of the male and is only secreted by grouped mice when they can exchange tactile stimuli.

These results show, at least in male mice, that Barnett's "colony odour" is not the sum of the "type odours" of the members of the group and that it is necessary for males to exchange some kind of social stimuli to elicit the secretion of this odor.

These two types of odor are used by male mice in their relations with intruders. Owing to both these messages, a group of male mice which is receiving the chemical signal, can determine whether the odor comes from a group of strange males or from a single male mouse.

It is not known yet if exactly the same mechanism of recognition exists in other species of rodents, but it is likely. The "plantar factor" is perhaps only one of the "type odours" known in mice ; one can assume that other odors could indicate the age or the physiological state of the subjects as we shall see when we investigate their social status.

SOCIAL STATUS AND CHEMICAL COMMUNICATION

Even if the social structure of most species of rodents is not as sophisticated as in higher mammals, one cannot conceive how even such simple structures could maintain their level of stability without some means of social recognition. Grant and Mackintosh (1963) showed the role of social postures in several laboratory rodents in this context. After these results appeared, numerous studies showed that chemical signals are involved in this kind of social communication.

First of all, anosmia is known to reduce the social reactivity of gerbils (Christenson et al., 1973 ; Hull et al., 1974), hamsters (Devor and Murphy, 1973) and mice (Edwards, 1974). The same results could probably be found in other species of macrosmatic rodents. But we have to notice that the results are linked to the method used. For example, bilateral removal of the olfactory bulbs provokes a decrease in the social behavior of mice while an intranasal application of a zinc sulfate solution does not produce the same effect.

The ontogeny of chemical communication in rodents is unknown, but it is sure that the age of the animal when the bulbectomy is performed, is a very important factor (Leonard, 1972).

In some cases dominant individuals produce odors different from those of subordinate ones (Carr and Martorano, 1967). The urine of dominant male mice contains an aversive factor mice for the other males (Jones and Nowell, 1974) but the urine of subordinate males does not contain this compound : the same phenomenon has been shown in rats (Krames et al., 1969).

In other cases, the frequency of olfactory marking depends upon the social status of the animals. Male rabbits of high rank deposit the secretion of their anal glands more frequently than subordinates (Mykytowycz, 1967). The same observation was made in male hamsters (Johnston, 1972) and in mice, owing to a very interesting technique (Desjardins et al., 1973) ; by ultraviolet visualisation of urine marks, it was possible to show that dominant male mice mark their entire cages, whereas subordinate ones deposit their urine in only two to four places within their cages. Using another method , Reynolds (1971) found a similar result.

Besides these effects of social status upon the quality and the frequency of olfactory marking, other studies have demonstrated morphological changes in glands as a function of social rank. This phenomenon has been well studied by Mykytowycz and Dudzinski (1966) in the rabbit. In mice, Bronson and Marsden (1973) showed that the males' preputial glands are heavier in dominant individuals than in subordinate ones. Quantitative differences in the secretion of some kinds of glands between dominant and subordinate males were found in mice (Mugford and Nowell, 1970), hamsters (Beauchamp, 1974 ; Beruter et al., 1974), and gerbils (Thiessen et al., 1974).

As indicated by most of the studies above, the modification of a skin gland or of its secretion seem correlated both to social rank and to the level of androgens   (Mugford and Nowell, 1971 ; Mykytowycz and Dudzinsky, 1966 ; Doty and Kart, 1972 ; Thiessen et al., 1973).

When considering these qualitative or quantitative differences between dominant and subordinate animals, it should be remembered that a stressed rodent emits a specific odor. This has been documented in mice (Muller-Velten, 1966 ; Carr et al., 1970 ; Rottman and Snowdon, 1972) and rats (Valenta and Rigby, 1968). In learning experiments with mice (Sprott, 1969, Davis, 1970) or rats (Wasserman and Jensen , 1969 ; Ludvigson and Morrison, 1970 ; Ludvigson and Collerain, 1972), different authors found that some kinds of odors linked to extinction or frustration are perhaps of the same nature. However, as King and Pfister (1975) emphasized, an "alarm pheromone" does not correspond necessarily to the odor of a stressed animal. Moreover, the "alarm pheromone" and the "frustration phe-

romone" possess aversive effects while a subordinate animal produces an attractive odor.

PRIMING PHEROMONES

Generally speaking, priming pheromones are known to be involved in the sexual life of rodents (Bronson, 1974). Those
chemical messages which prime a physiological or a behavioral response with some delay, influence the social life of rodents in the
same manner.

In several species, for example, it was possible to show a
kind of "olfactory imprinting" (Marr and Gardner, 1965 ; Marr and
Lilliston, 1969 ; Mainardi et al., 1965 ; Carter and Marr, 1970 ;
Carter, 1972 ; Porter and Etscorn, 1974, 1975) in the same manner
as in birds. An early exposure to visual stimuli provokes the
establishment of a temporary link between the young bird and the
stimulus ; an exposure of young rodents to a novel and unknown
odor could produce the same effect. Mainardi et al. (1965) raised
young mice with their mother in an intact nest or in a perfumed
nest ; in this last case, the mother and the nest were scented
every day with an artificial perfume. Between weaning and the age
of 75 days, all the mice were kept without the smell of the perfume. When 75 days old, the female mice had to choose between
unperfumed or perfumed sexual partners ; the females which had
been exposed early to a scented environment prefered the perfumed
males. Similar results were obtained in rats, guinea pigs and
Acomys.

It is not certain whether the scents used in these experiments
possess some attractive or aversive effects on the animals and
caution must be taken in interpreting the results of such studies.
But, if this imprinting phenomenon actually occurs, it is certainly
involved in the ontogeny of the social life of rodents. As in
birds, this kind of imprinting should permit an olfactory knowledge
of the species characters for the young. This hypothesis could
perhaps be confirmed by experiments of crossfosterings between rats
and mice (Denenberg et al., 1964, 1969 ; Hudgens et al., 1968).

Similar results were obtained in another context by Mugford
and Nowell (1972). They showed that if an adult male is maintained
with the female mouse and its young the male mice coming from this
litter are more aggressive than ones which were reared by the female
temple alone.

Another priming effect was found by Fullerton and Cowley (1971).
According to the physiological state of a female mouse, its urine
can increase or decrease the growth rate of young mice exposed to
these olfactory stimuli. This particular effect could facilitate
the regulation of population growth in rodents.

This last question, the study of population dynamics of rodents
and its regulation, has been studied by numerous researchers. It
is well known that the curve of population growth sooner or later
attains a maximum which is followed by a rapid decrease. There

were numerous hypotheses for explaining the decrease in the curve, some of them involving behavioral and physiological factors. For example, Christian (1963, 1965, 1968) showed that an increase in population density produced an increase in the weight and activity of the adrenal glands of mice and voles and a decrease in the weight of genital tracts. It was assumed that an increase of the adreno-cortical secretion would increase mortality through a reduced resistance to disease and decreased natality through a decline of sexual maturation, an increased intrauterine mortality, and inda-dequate lactation. These studies of the physiological consequences of crowding in mice and voles have neglected the behavioral causes of this phenomenon. At first Christian and his team assumed that the density per se was responsible of these effects. Then Davis and Christian (1957) and Louch and Higginbotham (1967) revealed that submissive or defeated animals were more affected by these physiological effects than the dominant ones ; the increase in aggressive behavior that can be correlated with an increase in population density had to be responsible for the physiological or behavioral responses of the animals.

All these investigations neglected the actual stimuli exchan-ged by the animals which were responsible for the physiological changes. Ropartz (1966) showed that the odors of a group of male mice sufficed to provoke an increase in the weight of the adrenal glands in isolated male mice exposed to these olfactory stimuli for one week. Different experiments proved (Ropartz, 1968a) that the "urinary factor" produced by the coagulating glands of male mice was the stimulus responsible for the increase.

Using another method, Archer (1969) confirmed this result and found an increase in adrenocortical activity. Sattler (1972) found the same results.

These findings confirm the involvement of olfactory cues in the social regulation and in the population dynamics of rodents.

CHEMICAL SIGNALS AND AGONISTIC BEHAVIOR

In social species it is evident that aggressive trends have not disappeared but that they have come to be controlled by means of behavioral patterns such as displays or ritualized behaviors. One could assume that in macrosmatic rodents chemical signals were used in this kind of regulation.

Indeed, premiliminary studies indirectly demonstrated that pheromones influenced the release or even the control of the aggressive responses of male mice (Mackintosh and Grant, 1966 ; Archer, 1968). It was easy to observe that two males which meet for the first time begin to sniff each other. This behavior is used for olfactory recognition : is the opponent known or unknown ? The urine seems to be an important factor in this recognition.

Ropartz (1968b) noticed a reduction of aggression in male mice when the animals' natural odors were masked by an artificial perfume. Then, by removal of the olfactory bulbs, it was possible to obtain

a complete elimination of aggression. The first findings were confirmed and precisely determined. To clarify the results, studies
were made of males and females.

Male Aggression

It is not possible to present all the results on aggressive
behavior and olfactory signals of male mice. We prefer to stress
the questions still not solved and the remaining contradictions.
The removal of olfactory bulbs in rats provokes an increase of the
interspecific aggression between rats and mice (Didiergeorges
and Karli, 1966). In mice, the same surgical operation results in
the elimination of aggression. But Edwards et al. (1972) showed
that an intranasal application of a zinc sulfate solution does not
produce an elimination of aggression but only a decrease of this
behavioral pattern.

It is absolutely necessary to distinguish between an anosmia
produced by removal of the olfactory bulbs which may lead to some
definitive changes in the central nervous system and an anosmia
produced by an anaesthesia of the olfactory epithelium (Hull et al.,
1974). In the same way it is necessary to distinguish between inter-
and intraspecific aggression (Jeffrey and Bennett, 1973).

The results recently obtained on the release of the aggressive
behavior in mice, are very clear. Archer (1968) found that male
mice which occupied the same cage for some days emitted an odor
which stimulated the aggression of their successors,Haug (1970)
confirmed this result and showed that if two males had just fought
vigorously in a cage, they deposited an odor which actually stimulated the aggression of others. On the other hand, if these two males
lived peacefully, for one week, in the cage, they deposited another
odor which decreased the aggression of their successors.

Mugford (1973) attempted to desmonstrate that in all cases the
odor deposited by one or more male mice in a cage possessed a stimulating effect on the agression of other males. One can assume
that these contradictions are due to different strains of mice and
different experimental methods, but the problem is not yet solved.
A study by Harmatz et al. (1974) seemed to confirm Haug's results
partially but Archer (1975) presented some severe criticisms of
this experimental work.

Similar contradictions exist in regard to the function of the
secretion of preputial glands of male mice. According to Bronson
and Marsden (1973), this secretion could possess an attractive
effect and a sexually stimulating effect for female mice while for
other males, this secretion could have stimulating properties on
aggression. This would also apply to the preputial glands of spayed
females treated with androgens (Mugford and Nowell, 1971).

But, in 1973, Mugford contended that the pheromone left in a
cage by one or more males, which stimulated the aggression of their
successors does not come from the preputial glands !

Haug (1971), using another method involving gland ablation,
showed that the preputial glands do not interfere with the aggressive

behavior of male mice.

One could relate such uncertain results about the secretion of coagulating glands and their role in the regulation of aggressive behavior in mice (Haug, 1971 ; Jones and Nowell, 1973). According to Haug, the coagulating glands of male mice would produce an aggression-inhibiting pheromone since the removal of these glands and seminal vesicles reduces the aggression of these animals. Jones and Nowell contend however that the same coagulating glands produce an aggression-inhibiting pheromone and that another pheromone (or the same ?) has an aversive effect on the olfactory investigation of other males, and, furthermore, that grouped males would not be able to emit this pheromone. However, Ropartz (1968b) had shown that the "urinary factor" coming from the same coagulating glands is only produced when male mice are grouped and can exchange tactile stimuli.

Despite the numerous unresolved contradictions in these results, it is likely that some chemical signals are involved in the release and the control of aggression between rodents (see also Lee and Brake, 1971 ; Lee and Griffo, 1972 ; Kielman and Lubow, 1974 ; Wechkin and Breuer, 1974 ; Alderson, 1975 ; Johnston, 1975).

Female Aggression

When one observes an encounter between a male and a female, there is generally no aggressive behavior. After some time devoted to mutual olfactory investigation, sexual patterns appear. Most authors admit that the female mice's urine contains an aggression-inhibiting pheromone (Mugford, 1973 ; Dixon and Mackintosh, 1971, 1975).

It is well known that females generally present some aggressive responses towards intruders when lactating. By chance, Haug (1972a) found a new pattern of aggressive behavior in female mice. When a female is introduced in a small group of adult females (3 to 4), she is attacked by the group.

Haug noticed that the aggressive response of the grouped female mice is quicker and stronger when the strange female is a lactating one. It was possible to show that this agonistic behavior of grouped females varies as a function of the stage of lactation of the intruder. It is when the female is in its second week of nursing that it provokes the strongest aggressive response.

In another study, Haug tried to investigate the determinism of this phenomenon ; by surgical removal of their olfactory bulbs, the grouped females were rendered anosmic. Such anosmic female mice never again attacked an intact or lactating intruder. Thus, it was likely that chemical signals were involved in this agonistic pattern. But as it is well known that the ablation of olfactory bulbs produces aversive effects on aggressive behavior, Haug preferred to use another method.

He compared (1972b) the aggressive response of a group of females when an intact female mouse, a spayed one or a spayed female coated with urine collected from intact females was introduced in their cage. He found that the scent of the urine of a strange female is a sufficient stimulus to release the attacks from grouped females. Moreover this effect disappears after removal of the ovaries of the mouse which gives urine.

In another experiment (Haug, 1973a)it was possible to show that a virgin female which was placed for several days in the cage of a lactating female was attacked more by grouped females than another female which spent the same time in the cage of a virgin female.

There is certainly at least one chemical message which is coming from the urine of the female mouse and which stimulates aggression in grouped females. The second week of nursing increases the stimulating effect of this compound (Ropartz and Haug, 1975).

Currently, a genetic study of this behavioral pattern and attempts at identifying the chemical nature of this compound have been started.

CONCLUSION

It seems that the only method to determine accurately the chemical messages involved in the social behavior of rodents and to explain the apparent contradictions which remain, would be to determine the exact chemical nature of these olfactory substances. In this context rodent studies are far behind when compared the work accomplished with invertebrates and particularly insects. We are just trying to establish the list of pheromones involved in the social or sexual life of some species of rodents, using indirect methods, while entomologists are studying the natural origin of insects' pheromones and the role of these chemical signals in the evolution of the species.

There are perhaps two reasons for this delay between the two kinds of research : the social relations of mammals are far more sophisticated and their phylogenetic level is far above that of insects ; moreover, the chemical nature of rodent pheromones is likely to be more complex than that of insects.

Several authors attempted to learn the chemical nature of olfactory signals used in the social or sexual relations of rodents (see, for example, Stoddart, 1974). One contended that pheromones of mammals might be complex molecules or combinations of molecules. The first results showed that there exists a great variety of these components ; some of them being very simple, others very complex ; some of them are by-products of a bacterial metabolism, others are the secretions of the skin glands.

Stoddart (1973) began a very fine study of the secretion of the caudal gland in Apodemus flavicollis ; he found at least 26 components. Other attempts for isolating chemical messages were made by Beruter and associates (1973, 1974) and Gaunt (1967) in guinea pigs and mice respectively. In other cases, only one active component is known, but not the entire combination (Thiessen and Regnier, 1974).

It is hoped that the advance of biochemical studies will permit the establisment of a more accurate list of chemical signals involved in the social behavior of rodents.

REFERENCES

Alderson, J., and Johnston , R.E., 1975. Responses of male golden Hamster (Mesocricetus auratus) to clean and male scented areas. Behav. Biol., 15 : 505-510.

Archer, J., 1968. The effect of strange male odour on aggressive behavior in male mice. J. Mammal., 49 : 572-575.

Archer, J., 1969. Effects of social stimuli on the adrenal cortex in male mice. Psychon. Sci., 14 : 17-18.

Archer, J., 1975. Comment on Harmatz, Boelkins and Kessler (1975) : habituation not primer pheromone reduces attack in odour-exposed mice. Behav. Biol., 15 : 519-520.

Barnett, S.A., 1958. An analysis of social behaviour in wild rats. Proc. zool. Soc. London, 130 : 107-152.

Barnett, S.A., 1963. A study in behaviour. Methuen & Co, London.

Beauchamp, G.K., and Beruter, J., 1973. Source and stability of attractive components in guinea pig (Cavia porcellus) urine. Behav. Biol., 9 : 43-47.

Beauchamp, G.K., 1974. The perineal scent gland and social dominance in the male guinea pig. Physiol. Behav., 13 : 669-673.

Beruter, J., Beauchamp, G.K., and Muetterties, E.L., 1973. Complexity of chemical communication in mammals : urinary components mediating sex discrimination by male guinea pigs. Biochem. biophys. Res. Commun. 53 : 264-271.

Beruter, J., Beauchamp, G.K., and Muetterties, E.L., 1974. Mammalian chemical communication : perineal gland secretion of the guinea pig. Physiol. zool., 47 : 130-136.

Bowers, J.M., and Alexander, B.K., 1967. Mice : individual recognition by olfactory cues. Science, 158 : 1208-1210.

Bronson, F.H., and Marsden, H.M., 1973. The preputial gland as an indicator of social dominance in male mice. Behav. Biol., 9 : 625-628.

Bronson, F.H., 1974. In"Pheromones", M.C. Birch ; Elsevier, New York.

Bruce, H.M., 1960. A block to pregnancy in the mouse caused by proximity of strange males. J. Reprod. Fert., 1 : 96-103.

Carr, W.J., and Martorano, R.D., 1967. The response of mice to odors from trained fighters versus submissive animals. East. Psychol. Assoc., Boston.

Carr, W.J., Martorano, R.D., and Krames, L., 1970. Responses of mice to odors associated with stress. J. Comp. Physiol. Psychol., 71 : 223-228.

Carter, C.S., and Marr, J.N., 1970. Olfactory imprinting and age variables in the guinea pig, Cavia porcellus. Animal Behav., 18 : 238-244.

Carter, C.S., 1972. Effects of olfactory experience in the behaviour of the guinea pig (Cavia porcellus). Animal Behav., 20 : 54-60.

Christian, J.J., 1963. Endocrine adaptive mechanisms and the physiologic regulation of population growth. In Physiological Mammalogy, edited by W.V. Mayer and R.G. van Gelder. Academic Press, New York.

Christian, J.J., Llyod, J.A., and Davis, D.E., 1965. The role of endocrines in the self-regulation of mammalian populations. Recent Progr. Hormone Res., 21 : 501-578.

Christian, J.J., 1968. Endocrine-behavioral negative feed-back responses to increased population density. In L'effet de groupe chez les animaux. Coll. int. C.N.R.S., Paris.

Christenson, T., Wallen, K., Brown, B.A., and Glickman, S.E. 1973. Effect of castration, blindness and anosmia on social reactivity in the male mongolian gerbil (Meriones unguiculatus). Physiol. Behav., 10 : 989-994.

Crowcroft, P., 1966. Mice all over. London.

Davis, D.E., and Christian, J.J., 1957. Relation of adrenal weight to social rank of mice. Proc. Soc. exp. Biol. Med., 94 : 728-731.

Davis, S.F., 1970. Conspecific odors as cues for runway behavior in mice. Psychon. Sci., 19 : 169-170.

Denenberg, V.H., Hudgens, G.A., and Zarrow, M.X., 1964. Mice reared with rats : modification of behavior by early experience with another species. Science, 143 : 380-381.

Denenberg, V.H., Rosenberg, K.M., and Zarrow, M.X., 1969. Mice reared with rats' aunts : effects in adulthood upon plasma corticosterone and open-field activity. Physiol. Behav., 4 : 705-707.

Desjardins, C., Maruniak, J.A., and Bronson, F.H., 1973. Social rank in house mice : differentiation revealed by ulraviolet visualization of urinary patterns. Science, 182 : 939-941.

Devor, M., and Murphy, M.R., 1973. The effect of peripheral olfactory blockage on the social behavior of the male golden hamster. Behav. Biol., 9 : 31-42.

Didiergeorges, F., and Karli, P., 1966. Stimulations "sociales" et inhibition de l'agressivité interspécifique chez le rat privé de ses afférences olfactives. C.R. Soc. Biol., 160 : 244-245.

Dixon, A.K., and Mackintosh, J.H., 1971. Effects of female urine upon the social behaviour of adult male mice. Animal Behav., 19 : 138-140.

Dixon, A.K., and Mackintosh, J.H., 1975. The relationship between the physiological condition of female mice and the effects of their urine on the social behaviour of adult males. Animal Behav., 23 : 513-520.

Dominic, C.J., 1969. Pheromonal mechanisms regulating mammalian reproduction. Gen. comp. Endocrin., 2 : 260-267.

Doty, R.L., 19 . Odor preferences of female Peromyscus maniculatus bairdi for male mouse odors of P. maniculatus bairdi and P. leucopus noveboragensis as a function of estrous state. J. Comp. Physiol. Psychol., 81 : 191-197.

Doty, R.L., and Kart, R., 1972. A comparative and developmental analysis of the mid ventral sebaceous glands in 18 taxa of Peromyscus, with an examination of gonadal steroid influences in Peromyscus maniculatus bairdi. J. Mammal., 53 : 83-99.

Edwards, D.A., 1974. Non-sensory involvement of the olfactory bulbs in the mediation of social behavior. Behav. Biol., 11 : 287-302.

Edwards, D.A., Thompson, M.L., and Burge, K.G., 1972. Olfactory bulb removal vs peripherally induced anosmia : differential effects on the aggressive behavior of male mice. Behav. Biol., 7 : 823-828.

Fullerton, C., and Cowley, J.J., 1971. The differential effect of the presence of adult male and female mice on the growth and development of the young. J. Gen. Psychol., 119 : 89-98.

Gaunt, S.L.L., 1967. Classification and effect of the preputial pheromone in the mouse. Am. Zool., 7 : 713.

Grant, E.C., and Mackintosh, J.H., 1963. A comparison of the social postures of some common laboratory rodents. Behaviour, 21 : 246-259.

Hahn, M.E., and Simmel, E.C., 1968. Individual recognition by natural concentrations of olfactory cues in mice. Psychon. Sci., 12 : 183-184.

Halpin, Z.T., 1974. Individual difference in the biological odors of the mongolian gerbil (Meriones unguiculatus). Behav. Biol., 11 : 253-259.

Harmatz, P., Boelkins, R.C., and Kessler, S., 1974. Postisolation aggression and olfactory cues. Behav. Biol., 13 : 219-224.

Haug, M., 1970. Mise en évidence de deux odeurs aux effets opposés de facilitation et d'inhibition des conduites agressives chez la souris mâle. C.R. Acad. Sci. Paris, 271 : 1567-1570.

Haug, M., 1971. Rôle probable des vésicules séminales et des glandes coagulantes dans la production d'une phéromone inhibitrice du comportement agressif chez la souris. C.R. Acad. Sci. Paris, 273 :

1509-1510.

Haug, M., 1972. Phénomènes d'agression liés à l'introduction d'une femelle étrangère vierge ou allaitante au sein d'un groupe de souris femelles. C.R. Acad. Sci. Paris, 275 : 2729-2732.

Haug, M., 1972. Effet de l'urine d'une femelle étrangère sur le comportement agressif d'un groupe de souris femelles. C.R. Acad. Sci. Paris, 275 : 995-998.

Haug, M., 1973. Mise en évidence d'une odeur liée à l'allaitement et stimulant l'agressivité d'un groupe de souris femelles. C.R. Acad. Sci. Paris, 276 : 3457-3460.

Haug, M., 1973. L'urine d'une femelle allaitante contient une phéromone stimulant l'agressivité de petits groupes de souris femelles. C.R. Acad. Sci. Paris, 277 : 2053-2056.

Hudgens, G.A., Denenberg, V.H., and Zarrow, M.X., 1968. Mice reared with rats : effects of preweaning and post-weaning social interactions upon adult behaviour. Behaviour, 30 : 259-274.

Hull, E.M., Hamilton, K.L., Engwall, D.B., and Rosselli, L., 1974. Effects of olfactory bulbectomy and peripheral deafferentation on reactions to crowding in gerbils (Meriones unguiculatus). J. Comp. Physiol. Psychol., 86 : 247-254.

Jeffrey, R.A., and Bennett, G.G., 1973. Olfactory cues and movement : stimuli mediating intraspecific aggression in the wild norway rat. J. Comp. Physiol. Psychol., 85 : 233-242.

Johnson, R.P., 1973. Scent marking in mammals. Animal Behav., 21 : 521-535.

Johnston, R.E., 1972. Scent marking, olfactory communication and social behavior in the golden hamster, Mesocricetus auratus. Dissertation Abstr. int., 32 : 10.

Johnston, R.E., 1975. Scent marking by male golden hamsters (Mesocricetus auratus). I. Effects of odors and social encounters. Z. Tierpsychol., 37 : 75-98.

Jones, R.B., and Nowell, N.W., 1973. The coagulating glands as a source of aversive and aggression-inhibiting pheromone (s) in the male albino mouse. Physiol. Behav., 11 : 455-462.

Jones, R.B., and Nowell, N.W., 1974. A comparison of the aversive and female attractant properties of urine from dominant and subordinate male mice. Animal Learn. Behav., 2 : 141-144.

Kalkowski, W., 1967. Olfactory bases of social orientation in the white mouse. Folia Biol. (Prague - Warsaw), 15 : 69-86.

Kielman, B.R., and Lubow, R.E., 1974. The inhibitory effect of preexposed olfactory cues on intermale aggression in mice. Physiol. Behav., 12 : 919-922.

King, M.G., and Pfister, H.P., 1975. Differential preference for an activation by the odoriferous compartment of a shuttlebox in fear-conditioned and naive rats. Behav. Biol., 13 : 175-181.

Krames, L., 1970. Responses of female rats to the individual body odors of male rats. Psychon. Sci., 20 : 274-275.

Krames, L., Carr, W.J., and Bergman, B., 1969. A pheromone associated with social dominance among male rats. Psychon. Sci., 16 ; 11-12.

Lee, C.T., and Brake, S.C., 1971. Reactions of male fighters to male and female mice, untreated or deodorized. Psychon. Sci., 24 : 209-211.

Lee, C.T., and Griffo, W., 1972. Early androgenization and aggression pheromone in inbred mice. Am. Zool., 12 : 79.

Leon, M., 1974. Maternal pheromone. Physiol. Behav., 13 : 441-453.

Leonard, C.M., 1972. Effects of neonatal (day 10) olfactory bulb lesions on social behavior of female golden hamsters (Mesocricetus auratus). J. Comp. Physiol. Psychol., 80 : 208-215.

Louch, C.D., and Higginbotham, M., 1967. The relation between social rank and plasma corticosterone in mice. Gen. Comp. Endocrin., 8 : 441-445.

Ludvigson, H.W., and Morrison, R.R., 1970. Discrimination by rats of conspecific odors of reward and nonreward. Science, 167 : 904-905.

Ludvigson, H.W., and Gollerain, I., 1972. Aversion of conspecific odor of frustrative nonreward in rats. Psychon. Sci., 27 : 54-56.

Mackintosh, J.H., and Grant, E.C., 1966. The effect of olfactory stimuli on the agonistic behaviour of laboratory mice. Z. Tierpsychol., 23 : 584-587.

Mainardi, D., Marsan, M., and Pasquali, A., 1965. Causation of sexual preferences cf the house mouse. The behaviour of mice reared by parents whose odour was artificially altered. Atti. Soc. Ital., Sc. Nat. Mus. Civ. Milano, 104 : 325-338.

Marr, J.N., and Gardner, L.E., 1965. Early olfactory experience, and later social behavior in the rat : preference, sexual responsiveness, care of young. J. Genet. Psychol., 107 : 167-174.

Marr, J.N., and Lilliston, L.G., 1969. Social attachment in rats by odor and age. Behaviour, 33 : 277-282.

Moncrieff, R.W., 1962. Mice and voles. The effect of olfaction on behavior. Am. Perfumer, 77 : 52-55.

Mugford, R.A., and Nowell, N.W., 1970. Pheromones and their effect on aggression in mice. Nature, 226 : 967-968.

Mugford, R.A., and Nowell, N.W., 1971. The preputial glands as a source of aggression-promoting odors in mice. Physiol. Behav., 6 : 247-249.

Mugford, R.A., and Nowell, N.W., 1972. Paternal stimulation during infancy : effects upon aggression and openfield performance of mice. J. Comp. Physiol. Psychol., 79 : 30-36.

Mugford, R.A., and Nowell, N.W., 1974. Intermale fighting affected by homecage odors of male and female mice. J. Comp. Physiol. Psychol., 84 : 289-295.

Müller-Velten, H., 1966. Über den Angstgeruch bei der Hausmaus (Mus musculus). Z. Physiol., 52 : 401-429.

Mykytowycz, R., and Dudzinski, M.L., 1966 . A study of the weight of odoriferous and other glands in relation to social status and degree of sexual activity in the wild rabbit, Oryctolagus cuniculus (L.). C.S.I.R.O. Wildlife Res., 11 : 31-47.

Mykytowycz, R., 1967. Communication by smell in the wild rabbit. Proc. Ecol. Soc., 2 : 125-131.

Mykytowycz, R., 1974. Odor in the spacing behavior of Mammals. In "Pheromones",   M.C. Birch., Elsevier, New York.

Porter, R.H., and Etscorn, F., 1974. Olfactory imprinting resulting from brief exposure in Acomys cahirinus. Nature, 250 : 732-733.

Porter, R.H., and Etscorn, F., 1975. A primacy effect for olfactory imprinting in spiny mice (Acomys cahirinus). Behav. Biol., 15 : 511-517.

Ralls, K., 1971. Mammalian scent marking. Science, 171 : 443-449.

Reynolds,E., 1971. Urination as a social response in mice. Nature, 234 : 481.

Ropartz, Ph., 1966. Contribution à l'étude du déterminisme d'un effet de groupe chez les souris. C.R. Acad. Sci. Paris, 262 : 2070-2072.

Ropartz, Ph., 1968. The relation between olfactory perception and aggressive behaviour in mice. Animal Behav., 16 : 97-100.

Ropartz, Ph., 1968. Le rôle de l'olfaction dans le comportement social des souris mâles. Rev. Comp. animal., 2 : 1-39.

Ropartz, Ph., and Haug, M., 1975. Olfaction and aggressive behaviour in female mice. In Olfaction and Taste, D.A. Denton and J.P. Coghlan ; Academic Press, New York.

Rottman, S.J., and Snowdon, Ch.T., 1972. Demonstration and analysis of an alarm pheromone in mice. J. Comp. Physiol. Psychol., 81 : 483-490.

Sattler, K.M., 1972. Olfactory and auditory stress in mice. Psychon. Sci., 29 : 294-296.

Sprott, R.L., 1969. "Fear communication" via odor in inbred mice. Psychol. Rep., 25 : 263-268.

Stoddart, D.M., 1973. Preliminary characterisation of the caudal organ secretion of Apodemus flavicollis. Nature, 246 : 501-503.

Stoddart, D.M., 1974. The role of odor in the social biology of small mammals. In Pheromones, edited by M.C. Birch ; Elsevier, New York.

Thiessen D.D., Wallace P., and Yahr, P., 1973. Comparative studies of glandular scent marking in Meriones Tristani, an Israeli gerbil. Horm. Behav., 4 : 143-147.

Thiessen, D.D., and Regnier, F.E., 1974. Identification of a ventral scent marking pheromone in the male Mongolian gerbil (Meriones unguiculatus). Science, 184 : 83-85.

Valenta, J.G., and Rigby, M.K., 1968. Discrimination of the odor of stressed rats. Science, 161 : 599-600.

Wasserman, E.A., and Jensen, D.D., 1969. Olfactory stimuli and the "pseudo-extinction" effect. Science, 166 : 1307-1309.

Wechkin, S., and Breuer, L.F., 1974. Effect of isolation on aggression in the mongolian gerbil. Psychol. Rep., 35 : 415-421.

PHEROMONAL INFLUENCES ON RODENT AGONISTIC BEHAVIOR

Barry Fass and David A. Stevens

Clark University

Worcester, Massachusetts   01610

The effects of odors on rodent agonistic behavior were un-
noticed for many years. Attention was focused instead on examining
the role of hormones in the regulation of attack and defensive
responses. The impetus for study of the area was provided by
Beeman's (1947) discovery that androgen injections restored intraspecific sparring in castrated male rodents. Subsequent investigations demonstrated that female rodents exhibited fighting when
treated with testosterone (Edwards, 1968) and males fought less
when treated with estrogen or progesterone (Suchowski, Pegrassi, &
Bonsignori, 1969; Erpino & Chappelle, 1971). These changes in
agonistic behavior were thought to be mediated by an interaction
between the hormones and a neural substrate for aggression (Edwards,
1969; Leshner, 1975).

Olfaction also serves to regulate rodent agonistic behavior.
However, the important degree to which odors play a role has only
recently become apparent. Now it is known that the members of
several rodent species produce pheromones which elicit or inhibit
agonistic behavior. To date, no one has summarized all the data
which indicate that olfactory cues are involved (Cheal & Sprott,
1971; Schultz & Tapp, 1973; and Bronson, 1974, among others, have
provided brief reviews). Therefore, the present paper is a review
of studies which demonstrate pheromonal influences on rodent
agonistic behavior. The topic of scent marking is not addressed
since it has already been thoroughly reviewed (most recently by
Thiessen & Rice, 1976). Operational definitions are given for the
terms "pheromone" and "agonistic behavior." Then the findings of
behavioral and endocrinological experiments using mice, rats, or
hamsters as subjects are discussed. Finally, the results and in-

terpretations of experiments in which rodents were rendered anosmic
are evaluated, and suggestions for future research are made.

OPERATIONAL DEFINITIONS

Pheromones are "active substances . . . which are secreted to
the outside by an individual and received by a second individual
of the same species, in which they release a specific reaction,
for example, a definite behaviour or developmental process"
(Karlson & Lüscher, 1959, p. 55). These secretions are characterized as "active" because they stimulate olfactory receptors which,
in turn, trigger various mechanisms and thereby produce behavioral
and/or physiological effects. (Vomeronasal reception is included
here as olfactory since the vomeronasal organ has been viewed as
part of the odor detection system: McCotter, 1912; Tucker, 1963;
Powers & Winans, 1973; 1975.) The type of effect depends upon the
type of pheromone; a signaller (or releaser) pheromone brings about
a prompt behavioral reaction whereas a primer pheromone serves to
physiologically alter the endocrine and/or reproductive systems
(Wilson & Bossert, 1963; Bronson, 1974). The findings to be discussed here indicate that several rodent species produce signaller
pheromones which effect agonistic behavior.

Agonistic behavior is a general category which includes "any
aggressive, defensive, submissive, flight, or ambivalent component
of behavior" (Payne & Swanson, 1972, p. 688). Offensive posturing,
attacking, biting, fighting, and chasing are also agonistic behaviors
because they are all aggressive acts. In other words, any "behavior
pattern which may provide some measure of adjustment when two organisms come into conflict" (Scott & Fredericson, 1951, p. 273) is
agonistic. The present paper is concerned with intraspecific
agonistic behavior, which is elicited by "the presence of a . . .
conspecific to which the attacker has not become habituated" (Moyer,
1968, p. 67), and territorial agonistic behavior, which is elicited
"simply by the presence of another animal in the territory" (Moyer,
1968, p. 67).

MICE

Attack-Eliciting and Attack-Inhibiting Pheromones

Male mice readily attack other male mice, but not females.
Although a male territory-defender vigorously fights and pursues
strange male intruders, he allows unfamiliar female conspecifics
to go unmolested (Mackintosh, 1970).

Evidently males have some quality which elicits attack from

other males, and females have some quality which does not elicit
attack from males. Several lines of evidence suggest that the
quality is probably olfactory. Mice sniff a conspecific's anogenital region prior to engaging in agonistic activity (Lee & Brake,
1971). Grouped males exhibit more intragroup aggression following
exposure to the odor of a group of strange males than following
exposure to odorless air (Ropartz, 1966; 1968a). Fighting is infrequently displayed by a pair of males when one is coated with a
masking agent such as perfume (Ropartz, 1968b) or commercial
deodorant (Lee & Brake, 1971). Conversely, a female treated with
commercial deodorant is attacked as infrequently as an untreated
one (Lee & Brake, 1971). Taken together, these findings indicate
that male mice produce an attack-eliciting pheromone, whereas
females do not. Other results (Mugford & Nowell, 1970a; Dixon &
Mackintosh, 1971) suggest that female mice produce a pheromone
which is inhibitory (see below).

The aggression-eliciting and -inhibiting pheromones are
thought to be present in the urine of male and female mice, respectively. Intermale fighting is increased when the opponents'
fur has been painted with the urine of a strange male, but not
when coated with water (Mackintosh & Grant, 1966; Mugford &
Nowell, 1970a). Furthermore, a pair of uncoated males placed in a
cage containing bedding soiled by another male engage in more bouts
than those placed in a cage with clean shavings (Archer, 1968).
While male urine is effective in promoting the incidence of intermale agonistic behavior, female urine reduces it. Fighter mice
rarely attack an opponent whose fur has been treated with female
urine (Mugford & Nowell, 1970a; Dixon & Mackintosh, 1971). Also,
a pair of untreated males placed in a cage with shavings soiled by
a female engage in less agonistic behavior than fighters put in a
clean cage (Mugford, 1973). Whereas female mouse pheromone inhibits aggression in male mice, it promotes aggression in female
mice. A female is not attacked when placed in a cage with males.
However, she is attacked when placed in a cage with females,
especially if she is lactating (Ropartz, 1975; Ropartz & Haug, 1975).

The role of nonolfactory variables must be determined in order
to be certain that the effects just described are, in fact, mediated
by urinary pheromones. For example, the possibility exists that the
urine-painted opponent provides an aggression-eliciting visual cue
while the water-coated one does not. This possibility has been
ruled out, however. Genetically blind fighters do not differ from
sighted fighters. Both engage in anogenital sniffing during intraspecific encounters, promptly spar with other male mice, and attack
female mice infrequently (Lee & Brake, 1971). Alternatively, the
effects of urine on agonistic behavior may be secondary to changes
in other forms of social behavior. Perhaps males do not frequently
attack female opponents because the latter elicit competing sexual

behaviors. Perhaps males frequently fight male opponents because
the latter are not sexually attractive and consequently do not
elicit competing copulatory responses. However, several studies
have shown that changes in attack frequency are not attributable
to changes in the frequency of competing sexual responses (Dixon &
Mackintosh, 1971; Connor, 1972; Davies & Bellamy, 1974). Furthermore, the elicitation and inhibition of agonistic behavior by male
and female urine cannot be due to some special characteristic of
domesticated animals; the phenomena are observed in both laboratory
bred and wild mice (Connor, 1972).

The behavioral effects of urinary pheromones are experience
dependent. Unfortunately, the data bearing on this point are not
totally consistent. On the one hand, familiarity with olfactory
cues has been shown to facilitate agonistic behavior. Intermale
fighting is enhanced in a cage containing soiled bedding from the
home (i.e., familiar) cage compared to that in a cage containing
strange soiled bedding (Jones & Nowell, 1973a; 1975). Peromyscus
and Mus mice exposed to each other during development win more
interspecific battles than do those exposed only to conspecifics
(Stark & Hazlett, 1972). Finally, the latency to attack an opponent is considerably reduced when the encounter takes place in a
cage previously occupied by a group of males, rather than in a cage
previously inhabited by only one mouse (Mugford, 1973; Jones &
Nowell, 1975). These findings are consistent with the notion that
familiarity with urinary pheromones facilitates agonistic behavior
by reducing the incidence of competing investigatory and freezing
responses.

On the other hand, familiarity with olfactory stimuli has also
been demonstrated to attentuate agonistic behavior. Fighters exposed to soiled bedding from the cages of their future opponents
ultimately exhibit less attack behavior than fighters exposed to
the bedding of a strange conspecific (i.e., one not serving as a
future opponent) or to clean bedding. Furthermore, the fighters
pre-exposed to their opponents' soiled bedding never initiate a
bout, whereas those fighters previously exposed to clean bedding
initiate an encounter 66% of the time (Kimelman & Lubow, 1974).
Male mice housed individually in cages containing shavings soiled
by group-housed mice are rarely attacked when paired with a
fighter. In contrast, rapid and frequent attacks are sustained
by males housed individually in cages with bedding soiled by a
single male (Harmatz, Boelkins, & Kessler, 1975). The above
findings support the contention that familiarity with an opponent's
aggression-eliciting pheromone results in habituation and, therefore, a reduction in agonistic behaviors. According to this line
of reasoning, the phenomenon "isolated-induced aggression" (first
reported by Scott & Fredericson, 1951) is attributable to the pre-
clusion of habituation to pheromones. Alternatively, the

possibility exists that olfactory experience brings about physiological changes which result in decreased production of eliciting
pheromone and, therefore, less agonistic behavior.

Production of aggression-eliciting pheromone appears to be
age dependent. Juvenile male mice up to 8 weeks old traverse the
territory of an adult conspecific unharmed, whereas adult male
intruders are attacked (Mackintosh, 1970). Young males are immune
from attack, conceivably because they do not produce the eliciting
pheromone. Support for this notion is provided by the finding
that the urine of 35 day old males is ineffective in enhancing
intraspecific fighting (Svare & Gandelman, 1975). Similarly,
several results indicate that production of inhibiting pheromone
is age dependent. An adult female shows little agonistic behavior
toward another adult female, but she frequently attacks a 25-30 day
old female (White, Mayo, & Edwards, 1969). The young female is
subjected to attack, perhaps because she does not produce inhibiting pheromone. However, pheromonal production must commence by
30-35 days since the urine of 5 week-old females is effective in
reducing intraspecific fighting (Svare & Gandelman, 1975).

Aversive Pheromones

Male mice do not invariably attack each other. For instance,
male albinos typically form a social hierarchy in which one member
is exclusively dominant, with no fighting exhibited by the subordinates (Uhrich, 1938). Aggression is not displayed by all pairs
of males exposed to the urine of a strange male; usually, only 80%
of the pairs actually fight (e.g., Archer, 1968). Perhaps the
failure of strange male urine to elicit attack in all males is due
to the presence of an additional, aversive or repellent pheromone
which precludes the occurrence of an agonistic encounter. The
finding that male mice avoid the side of a shuttlebox containing
air which passed over a male conspecific (Chanel & Vernet-Maury,
1963) suggests that male mouse urine does have an aversive (in
addition to aggression-eliciting) property.

The presence of an aversive pheromone in mouse urine was
probably first demonstrated by Müller-Velten (1966). Male house-
mice avoided an alley containing the odor of stressed conspecifics
and approached an alley containing the odor of unstressed conspecifics. Although there are methodological problems with Müller-
Velten's experiment (Stevens & Gerzog-Thomas, 1977), more recent
data confirm his notion that male mouse urine has an aversive
quality. Male mice prefer the odor of conspecifics which lost an
agonistic encounter over that of victorious conspecifics (Carr,
Martorano, & Krames, 1970). Furthermore, males prefer the odor of
nonshocked conspecifics over that of conspecifics which have been
given a series of electric shocks (Carr, Roth, & Amore, 1971).

Jones and Nowell (1973b) demonstrated the presence of an aversive pheromone in the urine of unstressed mice. Male mice spent less time in the half of an open-field where drops of strange male urine had been placed than on the side where drops of water had been deposited. Further, the subjects traversed fewer squares on the urine-treated side than on the water-treated side. The aversive property of male urine was most pronounced when the donor had been housed in isolation (Jones & Nowell, 1974b) (but see also Whittier & McReynolds, 1965). These results could not be attributed to any visual or humidity cues in the apparatus or surroundings (Jones & Nowell, 1973b; 1973c). No differences in time spent or squares traversed were observed when both sides were clean nor when one side was clean and the other spotted with water. Hence, the investigatory behavior of a male mouse can be attentuated by the presence of a strange male's urine.

The findings described so far indicate that male mouse urine has aggression-eliciting and aversive properties. Two remaining questions are: 1) what factors determine whether a mouse will react aggressively to, avoid, or approach a stranger's urine, and 2) are the behavioral effects of male mouse urine mediated by a single pheromone or two separate ones?

The first question has been examined by Jones and Nowell. They found that social rank, success in fighting, species, strain, and sex of both the urine donor and test subject all are critical variables. The urines of socially dominant and submissive males are aversive, but not equally so. Both dominant and submissive male test subjects stay on the clean side of an open-field for a longer period of time than on the side treated with dominant male urine. While both dominant and submissive subjects also spend more time on the clean side than on the side treated with subordinate male urine, the clean side vs. dominant urine side difference is greater than the clean side vs. submissive urine side difference (Jones & Nowell, 1973d; 1974a). Also, test males spend far less time on the male urine treated side of an open-field after losing an aggressive encounter with a male urine donor than prior to the encounter (Jones & Nowell, 1973c). These findings suggest that to a male, dominant male urine is more aversive than subordinate male urine. However, the former is also more likely to elicit aggression than the latter. Opponents painted with dominant urine are attacked more frequently by fighters than those coated with subordinate urine (Mugford & Nowell, 1970a).

Factors affecting responses of female mice to male urine have also been examined experimentally. In contrast to the findings with males, female test subjects stay on a side treated with dominant urine longer than on a clean side, but stay on a subordinate urine treated side as long as on a clean side (Jones & Nowell, 1974a). Therefore, to a female, dominant male urine has an attractive quality. However, the urine donor and test subject must be members of the same species and strain for the effects described above to obtain (Jones & Nowell, 1974b; Doty, 1973).

Unfortunately, there are no data available which pertain to the question of whether the effects of male mouse urine on agonistic behavior are due to one and the same pheromone or two different ones.

Hormones and Pheromones in Males

As mentioned earlier, androgens are essential for the expression of attack behavior in mice. Intermale aggression is eliminated by castration and restored by testosterone replacement therapy (Beeman, 1947; Tollman & King, 1956). In addition, females given testosterone exhibit fighting whereas those which are untreated do not (Edwards, 1968).

These effects of hormonal manipulations on agonistic behavior could conceivably be due to alterations in the production and/or release of aggression-eliciting pheromone. Whereas males painted with the urine of another intact male are frequently attacked by a fighter, those coated with a castrate's urine are attacked no more often than those coated with water (Mugford & Nowell, 1970a). Fighters also exhibit higher levels of aggression toward opponents coated with the urine of an androgenized female than toward those treated with nonandrogenized female urine (Mugford & Nowell, 1970b; 1971a; 1971b; Lee & Griffo, 1973). Procedures other than urine painting have been used to demonstrate the androgen dependency of male aggression-eliciting pheromone. Testosterone-injected and non-injected castrates have been paired with fighters in order to compare the number of attacks each sustains. The former group is attacked more frequently and longer than the latter, although the injected animals never initiate an attack (Lee & Brake, 1972). Male fighters attack females given androgen at birth or in adulthood more frequently than untreated females, and, again, androgenized mice never initiate an encounter (Lee & Griffo, 1973; Mugford, 1974).

Production and/or release of the male aversive pheromone is also androgen dependent. A male mouse spends less time on the portion of an open-field treated with a castrated male's urine than on the clean portion; however, the aversiveness is not as pronounced as that elicited by an intact male's urine (Jones & Nowell, 1973b). Conversely, the urine of a castrate given testosterone produces as great an aversion as that which a normal's urine produces. Hence, castration eliminates aversive pheromone production and/or release, and androgen therapy restores it.

Hormones and Pheromones in Females

Female mice are rarely attacked by male fighters or other females (Kahn, 1954; Mackintosh, 1970; Tollman & King, 1956). The
female's immunity from attack has been attributed to an aggression-
inhibiting pheromone thought to be present in the mouse's urine,
as the evidence described earlier indicates. Perhaps the production
and/or release of this pheromone is dependent upon ovarian hormones,
just as the male pheromones are dependent upon androgenic hormones.

If estrogen does play an important role in anti-aggression
pheromone production and/or release, one would expect ovariectomy
to render females susceptible to attack. However, this is not the
case; gonadectomized females are attacked less frequently than intact females (Mugford & Nowell, 1971a). Furthermore, the urine of
ovariectomized females injected with estrogen does not suppress
aggression as effectively as intact female urine does (Mugford &
Nowell, 1971b). While these findings cast doubt upon the notion
that the inhibiting pheromone is estrogen-dependent, there are
other data which support this notion. Estrogen treatment has been
found effective in reducing intermale aggression (Suchowski, et al.,
1969; Erpino & Chappelle, 1971), and progesterone has been shown to
counteract the effects of androgens on fighting behavior (Lee &
Griffo, 1974). These results suggest that the ovarian hormones do
stimulate production of an inhibitory pheromone. More importantly,
estrogen applied directly to the fur of a male protects the mouse
from attack (Haag, Jerhoff, & Kirkpatrick, 1974). Therefore, the
possibility exists that this female hormone is the aggression-
inhibiting pheromone.

Source of the Pheromones

Aggression-Eliciting. There is reason to believe that the
preputials (sebaceous glands located anterior to the penis between
the skin and body wall) are the source of male aggression-eliciting
pheromone. Dominant males are knwon to have heavier preputials than
subordinates (Davis & Christian, cited in McKinney & Christian, 1970;
Bronson & Marsden, 1973). Furthermore, isolation-housed males have
heavier preputials containing smaller amounts of stored sebum than
do group-housed males (Hucklebridge, Nowell, & Wouters, 1972). Since
dominant and isolated males both exhibit more fighting than subordinate and group-housed males, preputial secretions might have an influence on aggressive behavior. This notion has been confirmed;
preputialectomy results in increased fighting when the preputialectomized male is paired with an intact male and decreased fighting
when two preputialectomized males are paired (McKinney & Christian,
1970).

Other findings lead one to believe that the preputials are not

the source of the aggression-eliciting pheromone. For example, the
urine-painting technique has been used to assess more precisely the
way in which preputial secretions affect aggression. Fighters attack an opponent coated with water plus preputial secretion, or
bladder urine plus preputial secretion more frequently than one with
bladder urine or water alone (Jones & Nowell, 1973e). This finding
indicates that preputials are, in part, important in the production
of aggression-eliciting pheromone. In contrast, intact fighter-
opponent pairs introduced into a cage previously occupied by preputialectomized males fight for a longer period than those introduced into a clean cage (Mugford, 1973). Evidently, then, preputialectomized mice do produce aggression-eliciting pheromone, and
hence the preputials must not play a role in the production of this
pheromone.

The female mouse's preputials are vestigial (since they are
androgen-dependent) and known as the clitoral glands. Whereas
intact females are immune from attack (e.g., Kahn, 1954), those
treated with testosterone are not (Mugford & Nowell, 1971c). Perhaps the latter's susceptibility to attack is due to production of
aggression-eliciting pheromone by the clitorals, which are rendered
functional by testosterone. This possibility has been examined, but
the available data are not consistent. On the one hand, male
fighters attack clitoralectomized females given testosterone less
frequently than intact females given testosterone, but more frequently than untreated intact females (Mugford & Nowell, 1971c).
This finding indicates that females without clitorals do elicit
aggression. Furthermore, the urine of clitoralectomized females
given testosterone is as effective in promoting attack in fighters
as that of intact females given testosterone (Mugford & Nowell,
1971c). Hence, testosterone induces production of aggression-
eliciting pheromone in females, but not by their clitoral glands.
On the other hand, the same investigators have shown that the urine
of clitoralectomized-ovariectomized females given testosterone is
less effective in inducing fighters to attack than that of intact
females (Mugford & Nowell, 1972). This would seem to indicate that
the clitorals and/or ovaries are responsible for the elicitation of
aggression, although the issue remains unresolved.

Aversive and Aggression-Inhibiting. The coagulating glands
(which are located along the urino-genital tract and are responsible
for producing semen coagulating agent) have been implicated in the
production of the aversive and aggression-inhibiting pheromones. As
previously described, male mice stay away from an area treated with
the urine of a testosterone-injected male castrate. However, they
spend as much time in an area treated with the urine of a coagulating
glandectomized castrate given testosterone as in a clean area (Jones
& Nowell, 1973f). This implies that the animal without coagulating
glands does not produce the aversive urinary pheromone. The coagulating gland secretion is not the pheromone; mice find a mixture of

coagulating gland secretion plus bladder urine aversive, but do not avoid either of the two alone (Jones & Nowell, 1973f). Evidently, then, some sort of reaction which yields the aversive pheromone takes place when the contents of the coagulating glands are combined with bladder urine. It is not clear, however, how the mixing of gland secretion and bladder urine might take place in vivo.

The coagulating glands are also involved in the production of the aggression-inhibiting pheromone. Fighters attack an opponent coated with the urine of a testosterone-injected castrate from which the coagulating glands were excised more often than one painted with the urine of a testosterone-injected castrate with the coagulating gland intact (Jones & Nowell, 1973f). This finding implies that removal of the glands renders a mouse's urine more effective in promoting attack. Furthermore, a mixture of coagulating gland secretion plus water or coagulating gland secretion plus bladder urine applied to the fur of an opponent mouse renders the animal less susceptible to attack than water alone (Jones & Nowell, 1973e).

Further research is needed to determine whether the coagulating glands produce one pheromone which can elicit aggression or avoidance (depending upon certain conditions), or two distinct pheromones.

RATS

Very little attention has been paid to the role pheromones might play in the agonistic behavior of rats. There is reason to believe that olfactory cues are involved, however. A male territory-defender fights with male intruders (Grant & Chance, 1958), initiates the bout (whereas the stranger reacts submissively and defensively) (Barfield, Busch, & Wallen, 1972), and almost invariably emerges as the winner (Davis, 1933). Conversely, a male territory-defender does not attack a female conspecific intruder (Grant & Chance, 1958). Evidently, male rats (like male mice) have an aggression-eliciting quality whereas female rats (like female mice) have an aggression-inhibiting quality. That quality may be olfactory; rats (like mice) sniff their conspecifics' anogenital regions prior to behaving agonistically (Barnett, 1963; Adams, 1976).

In order to conclude that pheromones do influence agonistic behavior in rats, the effects of nonolfactory variables must be studied. For example, changes in attack frequency may be secondary to altered frequencies of competing sexual behaviors. An experiment analogous to that of Dixon and Mackintosh (1971) would help in resolving the issue.

To our knowledge, no one has examined the possibility that male rats produce an aggression-eliciting pheromone and female rats an aggression-inhibiting pheromone using the techniques devised by

Nowell and colleagues. In fact, we know of only one study which relates to this issue. Alberts and Galef (1973) introduced an anesthetized rat in a plastic bag into the home cage of another male. The anesthetized intruder was attacked only when a perforated bag was used. These findings suggest that an aggression-eliciting pheromone was involved.

Male rats, unlike mice, spend more time on surfaces which strange male rats have previously occupied (Stevens & Köster, 1972) or on which a strange male's urine was deposited (Richards & Stevens, 1974) than on clean surfaces. However, male rats, like mice, approach areas previously occupied by a strange subordinate male more often than those previously occupied by a strange dominant male rat (Krames, Carr, & Bergman, 1969). Taken together, these results suggest that only in certain situations do rats produce an investigation-inhibiting pheromone, but supporting evidence is lacking.

Although hormones are known to influence the rat's ability to detect odors (Pietras & Moulton, 1974), no one knows if and how hormones interact with pheromones to affect agonistic behavior in rats. Studies similar to those of Mugford and Nowell, using rats, would be most informative with regard to this issue.

HAMSTERS

Of the three species considered in this paper, the hamster is perhaps the most interesting. Hamsters are similar to mice and rats; they sniff their opponent's anogenital region prior to initiating a fight (Payne & Swanson, 1970; Lerwill & Makings, 1971). However, they are dissimilar since females fight with each other and with males. Although males also fight with each other, they attack females only rarely and are less aggressive than females (Payne & Swanson, 1971). In an intersexual encounter, the female initiates the fighting and almost always wins (Dieterlen, 1959), perhaps due to her size advantage over the male (Phoenix, Goy, & Resko, 1968; Payne & Swanson, 1970). Also, the female may produce a pheromone which inhibits the male's aggressiveness and the male may produce one which elicits attack from his opponent. Male hamsters rarely attack another male which has been coated with female hamster vaginal discharge (Murphy, 1973). Further, male hamsters treated with female urine are attacked as frequently as males treated with castrated male urine (Payne, 1974b). These effects are probably attributable to a sex-attractant rather than an aggression-inhibiting pheromone (Murphy, 1973). Aggression-eliciting pheromone is present in the urine of male hamsters since a castrated male painted with intact male urine is subjected to frequent attacks (Payne, 1974b). Further investigation is necessary to determine whether male hamsters also produce an aversive urinary pheromone.

The effects of hormonal manipulations on hamster aggressiveness have been examined in some detail. Several findings are consistent with the notion that pheromones also play a role. Male
hamsters show greater aggression toward intact males than toward
castrates, and show more aggression toward androgen-treated or
ovariectomized females than toward intact females (Payne, 1974a).
Gonadectomy diminishes the aggressiveness of both sexes, while
testosterone treatment restores it. Ovarian hormones are the more
effective (Vandenbergh, 1971; Payne & Swanson, 1972) with progesterone the most effective (Payne & Swanson, 1971). However, in an
intersexual encounter, the castrated male loses no matter which
hormone is administered (Payne & Swanson, 1971). The observation
that castrates coated with vaginal secretion elicit aggression less
frequently than untreated castrates (Johnston, 1975) indicates that
pheromones affect agonistic behavior in hamsters.

EXPERIMENTALLY-INDUCED ANOSMIA AND AGONISTIC BEHAVIOR

Procedures which interfere with the ability to smell have been
employed extensively in studying pheromonal influences on agonistic
behavior. According to the rationale of this approach, the importance of olfactory cues can be revealed by determining the extent
to which anosmia blocks aggressiveness. The two techniques which
have been used to induce anosmia are olfactory bulbectomy and intranasal application of zinc sulphate solution.

The approach of surgically ablating the olfactory bulbs, as
Cain (1974a) noted in his excellent review, is based upon "the
assumption that the olfactory bulb is exclusively a sensory relay
structure whose only action is to transmit neuronal activity from
the olfactory receptors to higher olfactory centers" (p. 654).
Given this assumption, one would expect that removal of this central nervous structure diminishes aggressive behavior by eliminating the ability to detect aggression-eliciting or -inhbiting
pheromones. Many investigators, in fact, have found that bulbectomized rodents show low levels of aggressiveness.

Animals have also been rendered anosmic by intranasally ir-
rigating a zinc sulphate solution. This chemical damages the
olfactory epithelium, but only temporarily; neuronal regeneration
takes place 2 to 3 days later (Matulionis, 1976). Furthermore, a
single application of the solution temporarily induces anosmia
(Alberts & Galef, 1971), and multiple applications prolong the
inability to smell (Thor, Carty, & Flannelly, 1976). Hence, zinc
sulphate is considered a valuable tool for studying the role of
olfaction in guiding rodent behavior.

Mice

Removal of the olfactory bulbs abolishes aggressive behavior. Ropartz (1967) discovered this phenomenon and suggested that it is due to elimination of the mouse's ability to detect aggression-eliciting pheromone. His hypothesis has recently been tested by comparing the effects of bulbectomy with those of intranasal zinc sulphate application. Whereas both procedures produce an impairment in the ability to smell, only bulbectomy reduced aggressiveness (Edwards, Thompson, & Burge, 1972). Further consideration of this finding is given below.

Rats

The effects of bulbectomy on aggression in rats are not clear-cut. Bulbectomized females have been characterized as "aggressively hostile" based upon the results of a standard emotionality test (Douglas, Isaacson, & Moss, 1969). However, Douglas et al. did not examine agonistic behavior and therefore the conclusion that bulbectomy enhances intraspecific aggressiveness in rats is unwarranted. More recently, bulbectomized males have been tested for aggression (Bandler & Chi, 1972). Apparently, extent of lesion plays a critical role in the effects which are obtained; partial removal results in increased display of attack whereas total removal eliminates attack behavior. Furthermore, bulbectomy converts nearly half of the animals into mouse killers. Taken together, the above findings do not allow for any conclusive statement about the involvement of pheromones in guiding aggressive behavior of rats.

Hamsters

Olfactory bulb ablation results in decreased agonistic and sexual behavior (Murphy & Schneider, 1970). In addition, intermale aggression can be abolished by destroying the bulbs unilaterally and simultaneously blocking the contralateral nostril. Zinc sulphate treatment is also effective in eliminating agonistic behavior (Devor & Murphy, 1973).

Behavioral Changes and Anosmia

Are the behavioral changes due to anosmia? In the case of bulbectomy, it is becoming increasingly apparent that the answer is no. Bulbectomy not only impairs the ability of mice to smell, it also produces hyperactivity, increased ambulation in the open-field, circling, random "darting" movements, wall climbing, and decreased grooming (Burge & Edwards, 1976). Similarly, bulb removal has multiple behavioral consequences in rats (increased mouse killing, reactivity to handling, and emotionality) (Alberts & Friedman, 1972; Cain, 1974b) and in hamsters (elimination of mating, decreased food consumption, body weight, running activity, and lateral hypothalamic self-stimulation rates) (Murphy & Schneider, 1970; Borer, Powers, Winans, & Valenstein, 1974). None of these additional behavioral effects are observed in animals treated with zinc sulphate (Edwards et al., 1972; Alberts & Friedman, 1972; Cain & Paxinos, 1974; Burge & Edwards, 1976); the only similarity between bulbectomized and zinc sulphate treated rodents is loss of smell. These findings suggest that the decreased aggressiveness of bulbectomized animals is part and parcel of a wide-ranging behavioral syndrome caused by central nervous tissue destruction.

To what are bulbectomy-induced increases in aggressiveness attributable? Cain (1974a) suggests that the bulbs are part of an emotionality-arousal circuit. Neuroanatomical studies show that the anterior olfactory nucleus, olfactory bulbs, and structures caudal to the bulbs all receive projections from the stria terminalis(deOlmos & Ingram, 1972), a tract which has been implicated in the septal hyperreactivity syndrome (Turner, 1970). Therefore, the changes in aggressiveness seen in bulbectomized rodents may be due to the manifestation of competing (emotional) responses which result when the circuit is interrupted. Edwards (1974; Burge & Edwards, 1976) has also emphasized the nonolfactory functions of the bulbs and supported Cain's position:

"It is possible that the wide variety of social behavioral deficits seen after bulbectomy occurs as a consequence of the disruptive effects of hyperactivity. However, it seems more likely that the olfactory bulb removal in mice produces alteration in nervous system processes underlying behavioral arousal" (Burge & Edwards, 1976, p. 87).

Therefore, interference with the ability to detect aggression-eliciting pheromone seemingly has little to do with the phenomena just described.

Application of zinc sulphate impairs a rodent's ability to smell, yet it has no effect on agonistic behavior (e.g., Edwards et al., 1972). Furthermore, there is now reason to doubt the assumption that zinc sulphate's behavioral effects are attributable to anosmia alone. Sieck and Baumbach (1974) have shown that this solution is toxic and makes rodents sick when administered intraperitoneally. Because zinc sulphate can cause systemic poisoning, its usefulness as a tool for examining the role of olfaction in guiding rodent behavior must be reassessed.

CONCLUSIONS

Role of Pheromones in Rodent Agonistic Behavior

Mice. As Nowell and his coworkers have pointed out, the
pheromones which influence agonistic behavior confer several advantages upon the mouse. For one, a male territory owner can
distinguish the sex of an intruder on the basis of olfaction.
Secondly, the pheromones induce a male territory owner to react to
intruders in such a way that he either defends his territory or ends
up with a mate. A male intruder, which potentially "threatens" to
evict the territory-defender, produces attack-eliciting pheromone;
consequently, the resident mouse protects his home region by warding
off the intruder. Conversely, a strange female, which represents a
potential mate, produces attack-inhibiting pheromone; consequently,
the resident male courts the intruder and propagates the species.
Furthermore, since the aggression-eliciting pheromone of dominant
males is particularly effective, only the "fittest" mice engage in
agonistic encounters and thereby maintain territories. Subordinate
mice, which find male conspecific urine aversive, stay away from
"dangerous" territory saturated with investigation-inhibiting pheromone. Clearly, pheromones play a most important role in the intraspecific and territorial aggressiveness of mice.

Rats. Unfortunately, little is known about pheromonal effects
on rat agonistic behavior. More work needs to be done before general
statements can be made. The limited data available at this time do
suggest, however, that rats produce pheromones which are functionally
similar to the ones produced by mice.

Hamsters. This species is perhaps the most interesting inasmuch as the female outfights the male. If we are to completely
understand the importance of olfactory cues in guiding agonistic
behavior, more research must be conducted using hamsters. Otherwise, our generalizations derived from studies employing species in
which the male is the aggressor may be hasty.

Directions for Future Research

Although the findings described in this chapter are somewhat
convincing, we still cannot be absolutely certain that the behavioral effects are mediated by pheromones. In order to be sure, the
putative pheromones must be isolated, identified, and synthesized.
So far, no one has attempted to undertake these tasks. There are
several clues, however, which can be used to guide future investigations. For example, preputial gland secretion apparently has
something to do with the aggression-eliciting pheromone of male

mice, whereas the contents of the coagulating gland are related to the female aggression-inhibiting pheromones and male aversive phero-mones. Since urine is also effective in altering the mouse's agon-istic behavior, one might suspect that a substance common to the glands and urine is the pheromone. The means for isolation and identification of that substance are available (see Singer, Agosta, O'Connell, Pfaffman, Bowen & Field, 1976). Testosterone must be considered a potential candidate for the male pheromones in light of the finding of Haag et al. (1974).

The question of whether the male aggression-eliciting and aversive qualities are mediated by a single pheromone or two sepa-rate ones will be answered when the pheromones are identified.

It was mentioned earlier that an interaction between hormones and central nervous structures involved in agonistic behavior is known to exist, and an interaction between hormones and pheromones has been considered and examined. Now, a three-way interaction between hormones, the central nervous system, and pheromones must be investigated, if we are to have a fuller understanding of the mechanisms underlying agonistic behavior. Some progress has been made toward this goal; for example, Pfaff and Pfaffman (1969) have shown that testosterone implants in the preoptic-anterior hypo-thalamic area of the male rat produce changes in olfactory bulb electrical activity responses to odorants. As Svare and Gandelman (1975) suggested:

"androgen may produce its effect upon aggressive behavior by altering, at the neuronal level, the quality and/or quantity of olfactory signals with the result being an enhancement of the 'motivational impact' of the stimuli" (p. 413).

Acknowledgements. The authors are grateful to J.G. Valenta and M.K. Rigby for their translation of Müller-Velten (1966), to P.A. Spiers for his translation of various French articles, and to J.M. Dawley for helpful comments on the manuscript.

REFERENCES

Adams, D.B. 1976. Relation of scent-marking, olfactory investiga-tion, and specific postures in isolation-induced fighting of rats. Behaviour 56:286-298.
Alberts, J.R., and Friedman, M.I. 1972. Olfactory bulb removal but not anosmia increases emotionality and mouse killing. Nature 238:454-455.
Alberts, J.R., and Galef, B.G., Jr. 1971. Acute anosmia in the rat: A behavioral test of a peripherally-induced olfactory deficit. Physiol. Behav. 6:619-621.

Alberts, J.R., and Galef, B.G. 1973. Olfactory cues and movement:
    Stimuli mediating intraspecific aggression in the wild Norway
    rat. J. Comp. Physiol. Psychol. 85:233-242.
Archer, J. 1968. The effect of strange male odor on aggressive be-
    havior in male mice. J. Mammal. 49:572-575.
Bandler, R.J., Jr., and Chi, C.C. 1972. Effects of olfactory bulb
    removal on aggression: A reevaluation. Physiol. Behav. 8:
    207-211.
Barfield, R.J., Busch, D.E., and Wallen, K. 1972. Gonadal influen-
    ces on agonistic behavior in the male domestic rat. Horm.
    Behav. 3:247-259.
Barnett, S.A. 1963. The rat: A study in behaviour. Aldine, Chicago.
Beeman, E.A. 1947. The effect of male hormone on aggressive behavior
    in mice. Physiol. Zool. 20:373-405.
Borer, K.T., Powers, J.B., Winans, S.S., and Valenstein, E.S. 1974.
    Influence of olfactory bulb removal on ingestive behaviors,
    activity levels, and self-stimulation in hamsters. J. Comp.
    Physiol. Psychol. 86:396-406.
Bronson, F.A. 1974. Pheromonal influences on reproductive activi-
    ties in rodents. Pages 344-365 in M.C. Birch (ed.), Pheromones.
    North-Holland, Amsterdam.
Bronson, F.H., and Marsden, H.M. 1973. The preputial gland as an in-
    dicator of social dominance in male mice. Behav. Biol. 9:
    625-628.
Burge, K.G. and Edwards, D.A. 1976. Olfactory bulb removal results
    in elevated spontaneous locomotor activity in mice. Physiol.
    Behav. 16:83-89.
Cain, D.P. 1974a. The role of the olfactory bulb in limbic mecha-
    nisms. Psychol. Bull. 81:654-671.
Cain, D.P. 1974b. Olfactory bulbectomy: Neural structures involved
    in irritability and aggression in the male rat. J. Comp.
    Physiol. Pyschol. 86:213-220.
Cain, D.P., and Paxinos, G. 1974. Olfactory bulbectomy and mucosal
    damage: Effects on copulation, irritability, and interspecific
    aggression in male rats. J. Comp. Physiol. Psychol. 86:202-212.
Carr, W.J., Martorano, R.D., and Krames, L. 1970. Responses of mice
    to odors associated with stress. J. Comp. Physiol. Psychol.
    71:223-228.
Carr, W.J., Roth, P., and Amore, M. 1971. Responses of male mice
    to odors from stress vs. nonstressed males and females.
    Psychon. Sci. 25:275-276.
Chanel, J. and Vernet-Maury, E. 1963. Détermination par un test
    olfactif des interattractions chez la souris. J. Physiol.
    (Paris) 55:121-122.
Cheal, M.L., and Sprott, R.L. 1971. Social olfaction: A review of
    the role of olfaction in a variety of animal behaviors. Psy-
    chol. Rep. 29:195-243.
Connor, J. 1972. Olfactory control of aggressive and sexual beha-
    vior in the mouse. Psychon. Sci. 27:1-3.

Davies, V.J. and Bellamy, D. 1974. Effects of female urine on so-
    cial investigation in male mice. Anim. Behav. 22:239-241.
Davis, F.C. 1933. The measurement of aggressive behavior in lab-
    oratory rats. J. Genet. Psychol. 43:213-217.
deOlmos, J., and Ingram, W.R. 1972. The projection field of the
    stria terminalis in the rat brain. An experimental study.
    J. Comp. Neurol. 146:303-334.
Devor, M., and Murphy, M.R. 1973. The effect of peripheral ol-
    factory blockade on the social behavior of the male golden
    hamster. Behav. Biol. 9:31-42.
Dieterlen, F. 1959. Das Verhalten des Syrischen Goldhamsters
    Z. Tierpsychol. 16:47-103.
Dixon, A.K., and Mackintosh, J.H. 1971. Effects of female urine
    upon the social behaviour of adult male mice. Anim. Behav.
    19:138-140.
Doty, R.L. 1973. Reactions of deer mice (Peromyscus mapiculatus)
    and white-footed mice (Peromyscus leucopus) to homospecific
    and heterospecific urine odors. J. Comp. Physiol. Psychol.
    84:296-303.
Douglas, R.J., Isaacson, R.L., and Moss, C.L. 1969. Olfactory
    lesions, emotionality and activity. Physiol. Behav. 4:379-
    381.
Edwards, D.A. 1968. Fighting by neonatally androgenized females.
    Science 161:1027-1028.
Edwards, D.A. 1969. Early androgen stimulation and aggressive
    behavior in male and female mice. Physiol. Behav. 4:333-338.
Edwards, D.A. 1974. Non-sensory involvement of the olfactory bulbs
    in the mediation of social behaviors. Behav. Biol. 11:287-302.
Edwards, D.A., Thompson, M.L., and Burge, K.G. 1972. Olfactory
    bulb removal vs. peripherally induced anosmia: Differential
    effects on the aggressive behavior of male mice. Behav. Biol.
    7:823-828.
Erpino, M. J., and Chappelle, T.C. 1971. Interactions between andro-
    gens and progesterone in mediation of aggression in the mouse.
    Horm. Behav. 2:265-272.
Grant, E.C. and Chance, M.R.A. 1958. Rank order in caged rats.
    Anim. Behav. 6:183-194.
Haag, C., Jerhoff, B., and Kirkpatrick, J.F. 1974. Ovarian hormones
    and their role in aggression inhibition among male mice.
    Physiol. Behav. 13:175-177.
Harmatz, P., Boelkins, R.C., and Kessler, S. 1975. Postisolation
    aggression and olfactory cues. Behav. Biol. 13:219-224.
Hucklebridge, F.H., Nowell, N.W., and Wouters, A. 1972. A relation-
    ship between social experience and preputial gland function in
    the albino mouse. J. Endocr. 55:449-450.
Johnston, R.E. 1975. Sexual excitation function of hamster vaginal
    secretion. Anim. Learn. Behav. 3:161-166.
Jones, R.B., and Nowell, N.W. 1973a. The effect of familiar visual
    and olfactory cues on the aggressive behavior of mice. Physiol.
    Behav. 10:221-223.

Jones, R.B., and Nowell, N.W. 1973b. The effect of urine on the investigatory behaviour of male albino mice. Physiol. Behav. 11:35-38.

Jones, R.B., and Nowell, N.W. 1973c. Aversive effects of the urine of a male mouse upon the investigatory behaviour of its defeated opponent. Anim. Behav. 21:707-710.

Jones, R.B., and Nowell, N.W. 1973d. Aversive and aggression-promoting properties of urine from dominant and subordinate male mice. Anim. Learn. Behav. 1:207-210.

Jones, R.B., and Nowell, N.W. 1973e. Effects of preputial and coagulating gland secretions upon aggressive behaviour in male mice: A confirmation. J. Endocr. 59:203-204.

Jones, R.B., and Nowell, N.W. 1973f. The coagulating glands as a source of aversive and aggression-inhibiting pheromone(s) in the male albino mouse. Physiol. Behav. 11:455-462.

Jones, R.B., and Nowell, N.W. 1974a. A comparison of the aversive and female attractant properties of urine from dominant and subordinate male mice. Anim. Learn. Behav. 2:141-144.

Jones, R.B. and Nowell, N.W. 1974b. The urinary aversive pheromone of mice: Species, strain and grouping effects. Anim. Behav. 22:187-191.

Jones, R.B. and Nowell, N.W. 1975. Effects of clean and soiled sawdust substrates and of different urine types upon aggressive behavior in male mice. Aggress. Behav. 1:111-121.

Kahn, M.W. 1954. Infantile experience and mature aggressive behavior of mice; some maternal influences. J. Genet. Psychol. 84:65-76.

Karlson, P. and Lüscher, M. 1959. "Pheromones": A new term for a class of biologically active substances. Nature 183:55-56.

Kimelman, B.R., and Lubow, R.E. 1974. The inhibitory effect of preexposed olfactory cues on intermale aggression in mice. Physiol. Behav. 12:919-922.

Krames, L., Carr, W.J., and Bergman, B. 1969. A pheromone associated with social dominance among male rats. Psychon. Sci. 16:11-12.

Lee, C.T., and Brake, S.C. 1971. Reactions of male fighters to male and female mice, untreated or deodorized. Psychon. Sci. 24:209-211.

Lee, C.T., and Brake, S.C. 1972. Reaction of male mouse fighters to male castrates treated with testosterone proprionate or oil. Psychon. Sci. 27:287-288.

Lee, C.T., and Griffo, W. 1973. Early androgenization and aggression pheromone in inbred mice. Horm. Behav. 4:181-189.

Lee, C.T., and Griffo, W. 1974. Progesterone antagonism of androgen-dependent aggression-promoting pheromone in inbred mice. J. Comp. Physiol. Psychol. 87:150-155.

Lerwill, C.J. and Makings, P. 1971. The agonistic behaviour of the golden hamster Mesocricetus auratus (Waterhouse). Anim. Behav. 19:714-721.

Leshner, A.I. 1975. A model of hormones and agonistic behavior.
     Physiol. Behav. 15:225-235.
Mackintosh, J.H. 1970. Territory formation by laboratory mice.
     Anim. Behav. 18:177-183.
Mackintosh, J.H. and Grant, E.C. 1966. The effect of olfactory
     stimuli on the agonistic behaviour of laboratory mice.
     Z. Tierpsychol. 23:584-587.
Matulionis, D.H. 1976. Light and electron microscopic study of
     the degreration and early regreration of olfactory epithelium
     in the mouse. Am. J. Anat. 145:79-100.
McCotter, R.E. 1912. The connection of the vomeronasal nerves with
     the accessory olfactory bulb in the opposum and other mam-
     mals. Anat. Rec. 6:299-318.
McKinney, T.D., and Christian, J.J. 1970. Effects of preputial-
     ectomy on fighting behavior in mice. Biol. Med. 134:291-293.
Moyer, K.E. 1968. Kinds of aggression and their physiological
     basis. Commun. Behav. Biol. 2:65-87.
Mugford, R.A. 1973. Intermale fighting affected by home-cage
     odors of male and female mice. J. Comp. Physiol. Psychol.
     84:289-295.
Mugford, R.A. 1974. Androgenic stimulation of aggression-eliciting
     cues in adult opponent mice castrated at birth, weaning or
     maturity. Horm. Behav. 5:93-102.
Mugford, R.A. and Nowell, N.W. 1970a. Pheromones and their effect
     on aggression in mice. Nature 226:967-968.
Mugford, R.A. and Nowell, N.W. 1970b. The aggression of male
     mice against androgenized females. Psychon. Sci. 20:191-192.
Mugford, R.A., and Nowell, N.W. 1971a. The relationship between
     endocrine status of female opponents and aggressive behavior
     of adult mice. Anim. Behav. 19:153-155.
Mugford, R.A., and Nowell, N.W. 1971b. Endocrine control over
     production and activity of the anti-aggression pheromone
     from female mice. J. Endocr. 49:225-232.
Mugford, R.A., and Nowell, N.W. 1971c. The preputial glands as a
     source of aggression-promoting odors in mice. Physiol.
     Behav. 6:247-249.
Mugford, R.A. and Nowell, N.W. 1972. The close-response to testo-
     sterone proprionate of preputial glands, pheromones and
     aggression in mice. Horm. Beh. 3:39-46.
Müller-Velten, H. 1966. Über den Angstgeruch bei der Hausmaus
     (Mus musculus L.). Z. Vergleich. Physiol. 52:401-429.
Murphy, M.R. 1973. Effects of female hamster vaginal discharge
     on the behavior of male hamsters. Behav. Biol. 9:367-375.
Murphy, M.R., and Schneider, G.E. 1970. Olfactory bulb removal
     eliminates mating behavior in the male golden hamster.
     Science 167:302-304.
Payne, A.P. 1974a. The aggressive response of the male golden ham-
     ster towards males and females of differing hormonal status.
     Anim. Behav. 22:829-835.

Payne, A.P. 1974b. The effects of urine on aggressive responses by male golden hamsters. Aggress. Behav. 1:71-79.

Payne, A.P. and Swanson, H.H. 1970. Agonistic behavior between pairs of hamsters of the same and opposite sex in a neutral observation area. Behaviour 36:259-269.

Payne, A.P. and Swanson, H.H. 1971. Hormonal control of aggressive dominance in the female hamster. Physiol. Behav. 6:355-357.

Payne, A.P. and Swanson, H.H. 1972. The effect of sex hormones on the agonistic behavior of the male golden hamster. Physiol. Behav. 8:687-691.

Pfaff, D.W., and Pfaffman, C. 1969. Olfactory and hormonal influences on the basal forebrain of the male rat. Brain Res. 15:137-156.

Phoenix, C.H., Goy, R.W., and Resko, J.A. 1968. Psychosexual differentiation as a function of androgenic stimulation. Pages 33-49 in M. Diamond (ed.), Perspectives in reproduction and sexual behavior. Indiana Univ., Bloomington.

Pietras, R.J., and Moulton, D.G. 1974. Hormonal influences on odor detection in rats: Changes associated with the estrous cycle, pseudopregnancy, ovariectomy, and administration of testosterone proprionate. Physiol. Behav. 12:475-491.

Powers, J.B. and Winans, S.S. 1973. Sexual behavior in peripherally anosmic male hamsters. Physiol. Behav. 10:361-368.

Powers, J.B. and Winans, S.S. 1975. Vomeronasal organ: Critical role in mediating sexual bahavior of the male hamster. Science 187:961-963.

Richards, D.B. and Stevens, D.A. 1974. Evidence for marking with urine in rats. Behav. Biol. 12:517-523.

Ropartz, P. 1966. Mise en évidence d'une odeur de groupe chez les souris par la mesure de l'activité locomotrice. C. R. Acad. Sci. 262 (D):507-510.

Ropartz, P. 1967. Mise en évidence du rôle de l'olfaction dans l'agressivité de la souris. Rev. Comp. Anim. 2:97-102.

Ropartz, P. 1968a. Le rôle de l'olfaction dans le comportement social des souris males. Rev. Comp. Anim. 2:1-39.

Ropartz, P. 1968b. The relation between olfactory stimulation and aggressive behavior in mice. Anim. Behav. 16:97-100.

Ropartz, P. 1975. Pheromones and aggressive behaviour in female mice. Paper presented at the 7th Annual Meeting of the European Brain and Behaviour Society, Munich.

Ropartz, P. and Haug, M. 1975. Olfaction and aggressive behaviour in female mice. Pages 411-412 in D. A. Denton and J. P. Coughlan (eds.), Olfaction and taste, Vol. V. Academic Press, New York.

Schultz, E.F., and Tapp, J.T. 1973. Olfactory control of behavior in rodents. Psychol. Bull. 79:21-44.

Scott, J.P., and Fredericson, E. 1951. The causes of fighting in mice and rats. Physiol. Zool. 24:273-309.

Sieck, M.H. and Baumbach, H.P. 1974. Differential effects of peripheral and central anosmia producing techniques on spontaneous behavior patterns. Physiol. Behav. 13:407-425.

Singer, A.G., Agosta, W.O. O'Connell, R.J., Pfaffman, C., Bowen, D.V., and Field, F.H. 1976. Dimethyl disulfide: An attractant pheromone in hamster vaginal secretion. Science 191: 948-950.

Stark, B., and Hazlett, B.A. 1972. Effects of olfactory experience on aggression in Mus musculus and Peromyscus maniculatus. Behav. Biol. 7:265-269.

Stevens, D.A. and Gerzog-Thomas, D.A. 1977. Fright reactions in rats to conspecific tissue. Physiol. Behav. (in press).

Stevens, D.A. and Köster, E.P. 1972. Open-field responses of rats to odors from stressed and nonstressed predecessors. Behav. Biol. 7:519-522.

Suchowsky, G.K., Pegrassi, L., and Bonsignori, A. 1969. The effect of steroids on aggressive behavior in isolated male mice. In: S. Garattini and E.B. Sigg (eds.),"Aggressive Behavior." Excerpta Medica Monog., Amsterdam.

Svare, B., and Gandelman, R. 1975. Aggressive behavior of juvenile mice: Influence of androgen and olfactory stimuli. Develpm. Psychobiol. 8:405-415.

Thiessen, D. and Rice, M. 1976. Mammalian scent gland marking and social behavior. Psychol. Bull. 83:505-539.

Thor, D.H., Carty, R.W., and Flannelly, K.J. 1976. Prolonged peripheral anosmia in the rat by multiple intranasal applications of zinc sulphate solution. Bull. Psychon. Soc. 7:41-43.

Tollman, J., and King, J.A. 1956. The effects of testosterone proprionate on aggression in male and female C57 BL/10 mice. Anim. Behav. 4:147-149.

Tucker, D. 1963. Olfactory, vomeronasal and trigeminal receptor responses to odorants. Pages 45-69 in Y. Zotterman (ed.), Olfaction and taste. MacMillan, New York.

Turner, B.H. 1970. Neural structures involved in the rage syndrome in the rat. J. Comp. Physiol. Psychol. 71:103-113.

Uhrich, J. 1938. Social hierarchy in albino mice. J. Comp. Psychol. 25:373-413.

Vandenbergh, J.G. 1971. The effects of gonadal hormones on the aggressive behavior of adult golden hamsters (Mesocricetus auratus). Anim. Behav. 19:589-594.

White, M., Mayo, S., and Edwards, D.A. 1969. Fighting in female mice as a function of the size of the opponent. Psychon. Sci. 16:14-15.

Whittier, J.L., and McReynolds, P. 1965. Persisting odours as a biasing factor in open-field research with mice. Canad. J. Psychol. 19:224-230.

Wilson, E.O., and Bossert, W.H. 1963. Chemical communication among animals. Pages 673-710 in G. Pincus (ed.), Recent progress in hormone research, Vol. 19. Academic Press, New York.

OLFACTION IN RELATION TO REPRODUCTION IN DOMESTIC ANIMALS

R. Mykytowycz

Division of Wildlife Research, CSIRO, P.O. Box 84

Lyneham, A.C.T. 2602, Australia

SUMMARY

In a consideration of the role of odour in reproduction of domestic animals their behaviour generally and the necessity for them to communicate with one another must be taken into account.

Since in mammals, reproduction is under multisensory control, their odour signals, unlike the pheromones of some insects, will rarely be the sole factor regulating breeding. However they may often ensure its efficiency and aid natural selection.

Examples from mammal species which were the subjects of intensive ethological studies show that under natural conditions, smell is involved at different phases of reproduction. Thus in most species the acquisition of breeding space, social status, precopulatory behaviour, coitus, mother-child relationship, and imprinting to own species and social unit, are based on olfactory signals.

Domestication has selected against many forms of behaviour which depend on olfactory communication, but despite all the selection pressures domestic animals retain their abilities to produce and perceive odours and given the opportunity will revert to the natural forms of behaviour in which olfaction plays an important role.

There are different sources of odours in animals - vaginal discharge, urine, saliva, faeces, skin glands, seminal fluid, and embryonic fluid - and all of them have been found to contribute to messages related to reproduction in domestic animals.

There is a tendency to pay attention to the overt responses produced by odours. It is often overlooked that smell - apart from its own specific pathways - also affects the sensitivity of other sensory systems.

There is a need for the systematic investigation of the effects of odours on the endocrine systems in domestic animals.

The manipulation of olfactory stimuli is now successfully applied in some managerial situations as for instance in detecting oestrus in pigs or facilitating the fostering of newborn lambs. There are also other situations in which odour stimulation may prove useful, such as efforts to increase the rate of conception during artificial insemination, the breeding of rare individuals kept in isolation from conspecifics, or stimulation of milk ejection.

Quite apart from any practical gains, a better knowledge of olfaction will permit a more complete understanding of the biology of domestic animals.

INTRODUCTION

It is commonly accepted now that besides exteroceptive stimuli such as light, temperature, humidity and food, social factors also play an important role in reproduction. In the past, lack of information on the behaviour of mammals generally and of domestic species in particular has inhibited the study of the effects of social environment on reproduction. Only recently has the ethology of domestic animals started to attract the serious attention of investigators.

Progress in ethology has brought clearly to our notice that higher animals are not automata programmed to perform certain stereotyped activities but, that they are, like ourselves, members of social entities exposed to various stimuli coming from conspecifics with which they share a common environment. Each species of animal displays a characteristic pattern of behaviour which under free-living conditions provides a framework within which natural selection can operate.

Animals need to communicate. To maintain the characteristic forms of behaviour and consequently the orderly functioning of social units signals have to be exchanged between individuals and groups of animals concerning their behavioural and physiological states as well as about the intentions of the participants. Although in some instances these signals are passed or received unintentionally, in others information is deliberately transmitted and sought.

The great majority of behaviour patterns in animals have become ritualised. Ritualisations lead to more efficient stimulators or releasers of more efficient patterns of action in other individuals; they reduce intraspecific aggression; and they serve as sexual or social bonding mechanisms. Ritualisations can be broadly defined as displays. Displays, although basically visual, may also involve other senses. This paper deals specifically with the sense of olfaction.

Olfactory communication. It has often been suggested that olfactory signals were the first to be utilised for communication by animals. Comparative anatomical studies of the evolution of olfactory systems indicate a regression of the sense of smell from primitive to highly developed vertebrates (Herrick, 1933).

Although all senses may be used to transmit messages related to reproduction, olfaction seems to be the one most frequently employed for this purpose. Indeed it has been demonstrated for many species that the production and perception of odours is strongly related to the reproductive status of the individuals (Mykytowycz, 1970).

Deeper interest in the role of olfaction in reproduction originated with entomologists who introduced the term 'pheromone' to denote an odorous substance or mixture of substances produced by exocrine glands which modify the behaviour and/or development of conspecific individuals. Pheromones are classified as 'releaser pheromones' and 'primer pheromones' according to the type of response they produce. Useful introductions to pheromones in mammalian reproduction, are the reviews by Whitten and Champlin (1973) and Bronson (1974).

Multisensory control of mammalian reproduction. Because of the influence of the entomologists' results with pheromones, there is a tendency to think that olfaction may be of the utmost importance in the reproduction of vertebrates. In fact, while the mating of insects, some fishes and birds depends to a considerable extent upon information transmitted across specific sensory channels, in mammals mating is under multisensory control. Male mammals particularly do not rely on information from a single stimulus for the initiation of copulatory behaviour (Bermant and Sachs, 1973). In many species a series of stimuli including postures, sounds, movements or odours are linked together in a definite order to form a complex display. A potential sexual partner may respond to the early stimuli but may fail to respond to subsequent ones. The sexual activity of rams for example is influenced only slightly by loss of hearing, more by the inability to smell and most by loss of vision. Partner seeking however is affected only in those rams which are unable to smell (Fletcher and Lindsay, 1968).

It is important to recognise that the 'reactivity' of these mechanisms can vary from species to species, strain to strain and even from individual to individual. The 'reactivity' levels are not directly controlled by hormonal levels (Vandeplassche and Spincemaille, 1967).

Due to the complexity of the mammalian nervous system and the importance of early experience, the applicability of the term 'pheromone', which implies a rather specific and unconditional reaction to the stimulus, has often been challenged by various workers who have expressed the opinion that the term probably has been overused. It has been suggested that in the case of mammals it would be more appropriate and correct to talk in terms of olfactory communication, i.e. the sending and perception of olfactory signals (Bronson, 1974).

Smell in the reproduction of free-living and laboratory animals. Reproduction does not begin with mating. Prior to this, members of most species have to acquire a space in which they will feel confident and secure. In gregarious animals, as exemplified by the wild rabbit, this is of particular importance (Mykytowycz, 1974). Not only is space required but also a social position which permits an individual to move freely without restrictions imposed by conspecifics. It has often been reported that in the presence of a dominant animal a subordinate one may become psychologically castrated. Only within a place where it feels confident and unharassed can an individual attain the physiological state necessary for breeding success.

Odour plays an important role in the acquisition and maintenance of breeding space and social position. At the outset of and throughout the breeding season territorial marking and competition for social rank intensify and are evident in all animals (Mykytowycz, 1973).

To ensure successful mating which would lead to fertilisation, pregnancy and later birth and survival of the progeny the animals have to find a fitting partner of the opposite sex. To achieve this, members of most species become engaged in courtship displays which induce various physiological changes and bring both partners to a reproductive pitch. Courtship allows mutual attraction and repellance to operate and the mood of the participants is also of importance in this process.

Although flirting is a usual and important component of human sexual behaviour its importance in domestic animals has been ignored. Yet we know that even lower animals become involved in 'flirtations'. Thus in sticklebacks (Gasterosteus sp.) or fruitflies only a small proportion of courtships lead to fertilisation. In fact the attractiveness of Drosophila females to males is at least partly due to the operation of a pheromone (Cook and Cook, 1975).

During courtship females demonstrate their receptiveness, while males display their readiness to mate by intensive searching for females, by driving and prompting them, and by strengthening their territorial marking.

In many species the formation of pair bonds and the forcible exclusion of competitors is a further step in ensuring the selection of the right breeding partner. In all these activities, which induce not only overt behavioural responses but also physiological changes, olfaction plays a predominant role.

As recent observations show there is also a synchronisation of plasma hormone levels and patterns of behaviour in long-term human sexual partners and it has been suggested that odour plays at least an indirect role in this phenomenon (Henderson, pers. comm.).

In many species individuals display a remarkable degree of synchronisation of oestrus cycles and breeding under natural conditions and in captivity. The synchronisation of breeding under natural conditions assures a wider range of advantages to the individuals and the species as a whole. It can also be of value in the management of domestic animals. In the past, synchronisation has been considered to be essentially controlled by endogenous factors but now the involvement of behavioural factors and particularly of odours in this phenomenon has been demonstrated experimentally. Domestic mice and other species of rodents have been intensively studied in this respect during the past twenty years. Not only stimulation but also inhibition of oestrus cycles can be induced by odour. Thus when female mice are housed together they exhibit fewer periods of oestrus. As studies of pregnancy blocking in mice and other rodents show, odours are also important in the successful maintenance of pregnancy (Bronson, 1974; Whitten, 1975; Whitten and Champlin, 1973). During studies of rabbits (Oryctolagus cuniculus) under natural and quasi-natural conditions the social environment, which in this species is heavily based on olfactory signalling, has been repeatedly found to be responsible for an individual's success or failure in reproduction (Mykytowycz and Fullagar, 1973).

The full maternal cycle from parturition to weaning involves a variety of sensory modalities but the degree of their involvement varies from species to species. Olfactory signalling is of major importance in some species. Thus to secure access to food a newborn individual has to be able to recognise its mother and the right feeding place. Apart from the need to locate the food, infants require a zone of emotional security within which they can learn to cope first with their immediate and then with the more distant physical and social environment. The ability to identify and familiarize with members of their own group and knowledge of the range of their own social units' territory protect immature animals from many dangers. The involvement of olfaction in the acquisition of such protection has been demonstrated experimentally for cat-kittens by Rosenblatt (1972) and for other animals (Mykytowycz, 1974).

For the survival of a given species it is essential that its
members can recognise and direct their reproductive activities
towards the right partner. This is of particular importance in the
case of sympatric species. Early in their life animals become
imprinted to their own species and numerous experiments have demonstrated very clearly that odour is of the utmost importance in this
phenomenon (Mykytowycz, 1970; Müller-Schwarze, 1974).

The sexual development of an individual can also be influenced
by the odours of conspecifics. Thus, the maturation of young female
mice may be accelerated by the smell of adult males and inhibited by
that of females (Vandenbergh, Whitsett and Lombardi, 1975).

Domestication limits olfactory communication. From the very
early stages of domestication there has been an effort to suppress
some forms of behaviour and force the activities of animals into
patterns acceptable to man. Generally the most suitable individuals
and species for domestication were those which could adapt their
pattern of social behaviour to a limited space and crowding without
any dysfunction of the endocrine systems.

Many forms of behaviour such as territoriality, dominance,
aggression and the formation of groups and pairs were very strongly
selected against, and there was therefore very strong selection
against the olfactory signals on which these activities are based.

Despite all these pressures there is ample evidence that in
domestic animals, just as in their wild relatives, the production
and perception of odours fluctuate in relation to changes in
reproductive physiology. Given the opportunity, all domestic species
will revert to natural forms of behaviour in which olfaction plays
an important role in reproductive activities, and natural selection
will operate.

Indeed the earliest experimental reports on the involvement of
olfaction in the reproduction of mammals came from observations on
domestic species. For instance, Stieve (1927) reported that in male
rabbits constantly exposed to the odour of females, the testes remain
active all year round. Kelley (quoted by Whitten, 1975) demonstrated
that the vaginal discharge from oestrous ewes stimulates mating in
rams and Coleman's observations (quoted by Whitten, 1975) suggested
that odours from males influenced the oestrous cycles in ewes.

The existence of sexual dimorphism in the odour of domestic pigs
has probably been known for as long as pork has been consumed and the
characteristics of a boar's odour have apparently often been used by
breeders in selecting stud animals.

Odour of vaginal discharge. The stimulating effect of vaginal
discharge on male sexual activity has been exhaustively studied durin

the past few years by Michael and his associates in primates. They established that in rhesus monkeys mixtures of volatile aliphatic acids (acetic, propanoic, methylpropanoic, butanoic, methylbutanoic and methylpentanoic), which they term 'copulins', possess sex attractant properties. The levels of copulins in women rise during the follicular phase of the menstrual cycle. Their production - both in rhesus females and in women - depends upon the bacterial flora of the vagina. Gonadal steroids exert their action indirectly by determining the availability of substrate for bacterial flora (Michael, Bonsall and Warner, 1975).

In many species the raising and rapid waving of the tail, and exposure of the clitoris by prolonged contraction of the vulva and presentation of hind quarters, encourage nosing which is often followed by massaging and licking of mucus secretion by males. Frequent urination which washes down the vaginal discharge further promotes the spread of the odour.

Kelley (quoted by Whitten, 1975) provided the first experimental evidence on the signalling properties of oestrous vaginal discharge. He smeared the perineal regions of non-oestrous ewes with the vaginal discharge obtained from oestrous ones. The females treated in this way attracted the attention of males and stimulated copulatory approaches. Kelley suggested that failure or delay in the discovery of ewes on heat may be due not only to a subacute sense of smell in the male but also to a low concentration of odoriferous substances in the females. Some ewes are comparatively undemonstrative and oestrus could only be detected by the ram on account of the odour signals.

Odour of urine. Urine is also behaviourally involved at various phases of reproduction and the levels of steroids present in it reflect clearly the reproductive status of an individual. It is known that some of the steroids have characteristic odours detectable even by the human nose (Kloek, 1961).

Tetrapod vertebrates possess two separate olfactory systems, one functioning through the olfactory bulb and the other through the vomeronasal (Jacobson's) organ. Both provide pathways to the hypothalamus but in macrosmatic animals the second pathway which is particularly sensitive to compounds of low-volatility, seems to function as a special receptor for sex hormones.

Flehmen which is displayed by males of many species of domestic animals evidently facilitates entry of urine into Jacobson's organ. Species which can close the external nares without retracting the upper lip can bring urine into the vomeronasal organ without exhibiting flehmen (Estes, 1972). Clear evidence of vomeronasal reception by females is lacking but 'sniff-yawning' has been suggested to be the substitute for flehmen in female bovines. It occurs in the context of special maternal acitivies eg. licking of the offspring.

The regular sampling of the females' urine is a daily routine of males of many species. Through the examination of urine, males can detect oestrus long before the characteristic behaviour or changes in the external genital organs associated with heat become visible (Estes, 1972).

Bulls will run from a distance of 40 metres to cows which urinate or adopt the characteristic urinating posture (Sambraus, 1973). In cases when vision is obscured even the sound of the passing of urine will attract a male. Females when approached by a male usually facilitate examination by producing small quantities of urine - often only a few drops. The ability of bulls to differentiate between oestrous and non-oestrous urine has been demonstrated experimentally (Sambraus and Waring, 1975).

Polyuria and the deliberate use of urine during reproductive activities is a characteristic not only of females but also of males. In males however urine seems to signal not so much the individual's reproductive status but rather his social dominance and control of the space in which the interaction takes place. In fact it has been repeatedly observed that in gregarious species both these attributes are linked together and it is difficult to sort out which one appears first.

Territoriality and territorial marking intensify generally in most species during the breeding season. Odour widens the sphere of influence of an individual and repels conspecific competitors often without the involvement of fighting (Mykytowycz, 1974). Males of different species use different ways to disseminate the odour of their own urine. In some species, including the moose (Alces americanus) the males select a patch of ground on which they stamp, paw and urinate. A muddy excavation results in which the male rolls himself repeatedly. The cow with which the bull is mating also uses this wallow. Females seek and actually fight with others for the right to use an established wallow (Fraser, 1968). The male reindeer (Rangifer tarandus) digs with his forefeet and urinates on the soil and then rubs his nose in urine patches for as long as ten minutes or more. Females also rub their noses against these urine patches.

Billy-goats and males of other species urinate on themselves during the breeding season, directing urine onto their abdomens, briskets and beards.

A male reindeer when sexually stimulated spills small quantities of urine onto his hind legs. Dietland and Christine Müller-Schwarze (1972) suggested that in female pronghorn antelope (Antilocapra americana) this form of behaviour may assist in the dissemination of of the odour of urine. Reindeer also possess tarsal glands and the chemical composition of their secretion has recently been reported (Andersson et al., 1975).

Not only is self-enurination practised during mating activities but males of some species - notably the rabbit - directly enurinate the females.

Extensive experiments with domestic mice and other rodents show that a male pheromone present in urine not only accelerates oestrus and facilitates mating but in some instances may induce the blocking of pregnancy.

As mentioned above, sexual development of young females can be slowed down by exposure to the odour of virgins and of females in late pregnancy and accelerated by the odour of male urine. The pheromone responsible for this last phenomenon is a substance bound to or a portion of a protein (Vandenbergh, Whitsett and Lombardi, 1975). It has also been suggested that the presence of a boar will accelerate puberty in female pigs acting through the mediation of a pheromone (Brooks and Cole, 1970).

Odour of faeces. Although there is ample information on the role of faecal odours in communication in free-living animals no attempts have been made to demonstrate this function in domestic species - particularly not in relation to their reproductive activities.

The odour of faeces may be determined genetically or by environmental factors such as the type of food consumed and the action of bacterial flora. In the first case the admixture of secretions from glands located in the anal region contributes significantly to the individual characteristics of faeces (Mykytowycz, 1970).

It seems that the faeces of both sexes may contain messages related to reproduction. Thus bulls spend more time examining the faeces of oestrous than of non-oestrous cows. Application of faeces from oestrous cows to the tails of non-oestrous ones increases their attractiveness to males (Donovan, 1967).

Bulls of some breeds of cattle frequently pass small amounts of faeces during mating and this is accompanied by 'pumping' of the tail (Fraser, 1968). Tail pumping suggests that massaging of the anal glands may be involved. Camels in rut, besides increasing the secretion of odour from their occipital glands also frequently display diarrhoea.

In rabbits (Mykytowycz, 1974) and rats (Leon, 1974) the odour of the mother's faeces is of particular importance to the newborn because of its ability to regulate their movements. During the period from 14 to 27 days post-partum, lactating female albino rats

produce a special type of faeces - caecotrophe - which contains a
maternal pheromone. This pheromone regulates the movements of the
progeny during the period of their lives when they start to famil-
iarise themselves with the environment. Besides the food base,
bacterial strains contribute to the individually characteristic
odour of the maternal pheromone (Leon, 1975). Bacterial strains
can also be involved in the modification of odours from the skin
glands secretions. For example, behaviourally active carboxylic
acids from the anal pockets of the Indian mongoose (Herpestes
auropunctatus) are the products of bacterial metabolism (Gorman,
Nedwell and Smith, 1974). While it is generally realised that urine
is involved in the communication of oestrus in bitches, the similar
role of the secretion from the anal sacs has been overlooked
(Donovan, 1967).

    Odour of saliva. Although the 'anal pole' (ano-genital region)
is best equipped to generate odour and hence attracts most of the
attention of bulls, the head represents a 'minor pole' and mating is
sometimes directed toward this part of the body (Sambraus, 1973).

    One of the sources of odour in the head region is saliva. In the
domestic pig the main odorous component of saliva is 5α-androst-16-
en-3-one with a urine-like smell which is contained in the submaxillary glands. This steroid can be reduced to 5α-androst-16-en-3α-ol
and 5α-androst-16-en-3β-ol, both with a musk smell.

    The submaxillary glands act only as a reservoir for this subst-
ance which like the androgen is synthesized in the Leydig cells but
does not possess androgen's activity. Sexual dimorphism exists in
the submaxillary glands of pigs. They are more strongly developed
in males than females (Booth, Hay and Dott, 1973).

    The head to head position seems to be the most frequent pattern
of contact when a boar meets an oestrous gilt. Salivation by the
male is characteristic during a meeting. Boars from which the
submaxillary glands were removed surgically were less able to elicit
the 'immobilisation reflex' in sows (Perry, Patterson and Stinson,
1973). Signoret who pioneered the field of reproductive behaviour
in the pig demonstrated that stimulation of oestrus sows by odour
alone elicited the mating stance in 81% of them, and by auditory
stimuli in 71%. Simultaneous presentation of these stimuli increased
the result to 90% (Hafez and Signoret, 1969).

    Aerosols containing steroid compounds found in the saliva of
males are used commercially at present to detect oestrus in sows.
When blown into the face of an oestrous female they release the
mating stance. Signoret and Bariteau (in press) have studied the
effects of five synthetic odorous steroids and found that three of
them released the standing reaction in sows (62.5, 41.4 and 51.1 per
cent of cases respectively) just as frequently as the steroid contai

ed in boar's saliva (52.6 per cent of cases). They also reported that following artificial insemination the rate of farrowing was higher (65.9 per cent) in sows which displayed a positive response after spraying with the odorous steroids in comparison with females which were not exposed to the steroids (46.4 per cent) or which did not display a reaction to them (44.1 per cent). A report that this method of heat detection lowers the rate of pregnancies (Meding, 1972) suggests that the source of odour may be of importance.

Odour-producing skin glands. Saliva is not the only source of odour in the head region. Thus it has been suggested that exhaled air may contain olfactory signals. In most species odour-producing skin glands are situated in different areas on the head (Mykytowycz, 1970). The occurrence, structure and function of skin glands have been reviewed frequently. These apocrine and sebaceous skin glands are specialised organs functioning for the purpose of communication. Although their territorial role is most frequently discussed, their direct relation to reproduction is also known.

There is general agreement that androgens increase the size and functional capacity of sebaceous glands in all species. This uniformity of opinion does not exist however with regard to the effect of oestrogens and progestagens. Whatever the mechanism of action of sex hormones on odour production is, the behavioural stimulation is also of importance. The results of recent studies by Claus (in press) show that the levels of both testosterone and the salivary steroid 5α-androst-16-en-3-one in the blood plasma and fat tissue of boars fluctuate synchronously after exposure to oestrous sows.

There is evidence that the skin can convert steroid hormones to new metabolites. Considering this, one can reason that if the skin secretes pheromones that are steroids they could also be synthesised in the skin. Evidence for such steroid conversions is particularly strong in the case of the androgens (Strauss and Ebling, 1970).

The skin glands are the most important source of odour in animals. Their odours can also be incorporated in the odour of urine, saliva, faeces and body tissue generally. The reduction of meat taint after removal of odoriferous glands has been observed in many animals.

For practical reasons the chemistry of the sex odour in pork has attracted more attention than the odour of any other species but the problem has not been completely solved as yet. Sex-taint is only one of the many odours which are readily detectable in pigs of both sexes (Jonsson and Wismer-Pedersen, 1974). More systematic studies are required to determine the sources of these commercially undesirable smells and explain their relation to behaviour. Of all domestic animals the pig seems to be best equipped with odour-producing, externally-secreting skin glands. It is possible that not only the preputial but also other glands such as the anal, metacarpal and Harderian as well as the gonadal status of an individual affect the tainting of pork.

Generally very little is known about the histology and chemistry of externally secreting skin glands in domestic animals. There is no doubt however that their secretory activities fluctuate in relation to the reproductive state (Fraser, 1968) and social status of an individual similarly as in some laboratory animals which were the subjects of controlled ethological studies (Mykytowycz and Dudziński, 1966; Drickamer, Vandenbergh and Colby, 1973).

In addition to the salivary odour mentioned above the smell of preputial secretion can elicit the mating response in oestrous sows. The fluid from the preputial pouch is not effective however at lower temperatures and it must therefore be warmed to body temperature (Hafez and Signoret, 1969).

Odour of body fluids. Similarly as with preputial secretion, the seminal fluid of a boar when deposited on the snout of a sow which displays incomplete signs of heat will release the response to the haunch pressure stimulus.

The odour of embryonic fluid seems to play a role in the selection of a birth-site in sheep. The ewe giving birth, as well as other ewes, will spend considerable time licking and sniffing the ground where the foetal fluids are spilt at the beginning of the birth process (Smith, 1965).

Earlier information on the involvement of olfaction in reproduction of domestic animals. Although there has always been an awareness of the involvement of odour in the reproduction of domestic animals and the literature is rich in general references to this problem, the information is fragmented and often made in passing hence does not provide us with a complete picture of the role of olfaction.

Apart from the behavioural role of odour in pigs only very few selected topics of practical importance have been intensively investigated, as for instance the role of olfaction in the fostering of new-born sheep (Baldwin and Shillito, 1974), and maternal imprinting in goats (Klopfer and Gamble, 1966).

The lack of a wider specific interest in odours was due mainly to the fact that under conditions of domestication animals can be bred successfully without relying too much on their sense of smell.

Although under natural conditions most species of animals rely heavily on olfactory signals in establishing contacts with conspecifics for purposes of reproduction, in captivity animals rapidly develc

new methods of locating mates or may do without them altogether, as
the practice of artificial insemination illustrates. Even the
leading of a bull toward a familiar mating place without sight of a
female may act as an adequate sex stimulus (Fraser, 1968). Although
response to sex odour may be inborn, as for example the response of a
naive stallion to the oestrous mare's urine (Wierzbowski, 1959),
animals can also be taught to accept various other stimuli which suit
the management. Consequently although imprinting to the odour of its
own species is one of the fundamental phenomena, acceptance of
artificial odours with the same effect has frequently been demonstrated experimentally (Mykytowycz, 1970).

     Need to study priming effects of pheromones. One of the reasons
for the incomplete knowledge of the role of olfaction in the behaviour
of domestic animals is the fact that for the convenience of the
experimenters there is a tendency to pay attention only to immediate
overt responses with obvious involvement of olfactory behaviour. In
this respect it is also easier to observe the behaviour of males.
Less attention has been paid to the fact that the sense of smell, in
addition to its own specific way of action, in cortical associations
serves as a non-specific activator for all activities in the neopall-
ial cortex. The olfactory apparatus, apart from the detection and
analysis of odour, seems to interact in various ways with other
senses. It activates or sensitises the nervous system as a whole and
certain appropriately attuned sensory-motor systems particularly. In
this way it lowers the threshold of excitation for all stimuli and
induces differential reinforcement or inhibition of specific types
of responses (Herrick, 1933). Thus for instance olfaction may
increase the sensitivity and excitation of the visual receptors. The
smells of bergamot oil, geraniol and camphor increase the acuity of
the green sensitive apparatus and lower that of the red sensitive
apparatus of the human eye (Kravkov, 1939). A somewhat wider discussion of the possible interactions between olfaction and the other
senses is given by Diakow (1974) and also by Darby, Devor and Chorover
(1975) in a paper in which they report that the vaginal discharge of
female hamsters is not critical for mating but rather facilitates it.
They speculate that odour enlarges or at least modifies the field of
somato-sensory response similarly as the stimulation of 'attack sites'
in the hypothalamus sensitises the snout in the cat and oestrogen
sensitises the sensory field supplied by the pudendal nerve in the
rat. The sensitivity of the olfactory system fluctuates in relation
to the hormonal status of an animal. Conversely, the removal of the
olfactory bulb in female rats causes a lowering of oestrogen level
and administration of oestrogen to bulbectomised females increases
significantly their readiness to mate (Diakow, 1974).

     The effect of olfaction increases when combined with stimulation
through the other senses. Experiments with mice show that the
pheromone contained in the male's urine and tactile stimuli from a
male act synergistically to accelerate puberty in young females
(Bronson and Maruniak, 1975). In pigs the mating stance response of

oestrous sows increases with the simultaneous application of sound
and olfactory stimuli (Signoret and du Mesnil du Buisson, 1961).

Possible applications of pheromones under conditions of
domestication. Descriptive studies of behaviour are essential and
self-contained, but for a complete understanding of the role of odour
in reproduction parallel enquiries into the pheromone/endocrine
interactions are required.

Some progress has been made on the physiological effect of odours
during studies involving mainly laboratory rodents but work with
domestic animals with the exception of the pig has scarcely begun.
And yet those working in the field of reproduction of domestic animals
are particularly well placed to carry out this type of investigation.

Generally speaking the manipulation of olfactory stimuli may not
induce radical changes in the breeding of domestic animals but it is
possible that a better knowledge of olfaction may assist in some
special situations. For example the application of olfactory stimuli
may be of some assistance in efforts to synchronise breeding, or
stimulate the ejection of milk.

Odour is involved not only in the establishment of a social
relationship between mother and child but also in the stimulation of
the mother's endocrine system. There is experimental evidence that
in rodents olfactory cues derived from nestlings are responsible for
the release of prolactin from the pituitary gland. Other olfactory
stimuli may also affect milk ejection (Grosvenor and Mena, 1974).
Exposure to the olfactory cues from the pups increased the level of
plasma corticosterone in lactating rats. Experimental manipulation
of glucocorticoids improves lactational performance (Zarrow et al.,
1972).

The manipulation of odour may also bring some advantages in
situations where it is desirable to breed from a single, valuable
animal kept in isolation from conspecifics. It may be worthwhile
considering the importance of environmental, group, and species odour
as well as the mutual attraction based on olfaction when confronting
two potential breeding partners. It has been reported that a change
in housing can negatively affect reproduction including delay of
puberty in pigs.

In view of the information on the gonadotrophin-stimulating
activity of male pheromones, it is likely that they could be used to
improve the effectiveness of artificial insemination. The odour of
male urine has been found to affect the enzymatic activity in the
uteri of domestic mice during early oestrus and early pregnancy.
Marchlewska-Koj reported during the VIIIth International Congress on
Animal Reproduction and Artificial Insemination that the males of
different strains stimulated to different degrees the production of

dehydrogenases. Poor stimulation was correlated with a higher rate of pregnancy blocking induced by urine.

CONCLUSION

Independently from possible practical gains, the information on the role of olfaction is needed to complete our knowledge of the biology of domestic animals. Recently individuals and groups in different countries have expressed their concern at the lack of understanding of the effect of modern intensive management on the well-being of animals. The emergence of interest in the ethology of domestic animals gives an assurance that in the future we will be in a better position to judge how to treat the animals in our care.

Now it is generally realised that biostimulation is essential for the elicitation of full reproductive responses in animals and that odour is important in this process (Fraser, 1968). It is not possible to obtain a complete picture without considering the reactions of animals to odour stimuli at the physiological level.

There is also a need to study systematically and quantitatively the overt behaviour of domestic animals in relation to odours. This has been started recently by various groups of research workers in different parts of the world. There is also a need to record the animal management practices handed down from generation to generation. Their review may permit us to share the useful experience of others and may help to revise our present management and suggest areas for future research activities.

REFERENCES

Andersson, G., Andersson, K., Brundin, A., and Rappe, C. 1975. Volatile compounds from the tarsal scent gland of reindeer (Rangifer tarandus). J. Chem. Ecol. 1: 275-281.

Baldwin, B.A., and Shillito, E.E. 1974. The effects of ablation of the olfactory bulbs on parturition and maternal behaviour in Soay sheep. Anim. Behav. 22: 220-223.

Bermant, G., and Sachs, B.D. 1973. Courtship and mating, pp. 194-238, in G. Bermant (ed.). Perspectives on Animal Behavior, Scott, Foresman and Co., Glenview.

Booth, W.D., Hay, M.F., and Dott, H.M. 1973. Sexual dimorphism in the submaxillary gland of the pig. J. Reprod. Fert. 33: 163-166.

Bronson, F.H. 1974. Pheromonal influences on reproductive activities in rodents, pp. 344-365, in M.C. Birch (ed.). Pheromones, North-Holland Publ. Co., Amsterdam.

Bronson, F.H., and Maruniak, J.A. 1975. Male-induced puberty in female mice: evidence for a synergistic action of social cues. Biol. Reprod. 13: 94-98.

Brooks, P.H., and Cole, D.J.A. 1970. The effect of the presence of a boar on the attainment of puberty in gilts. J. Reprod. Fert. 23: 435-440.

Claus, R. (in press). Boar taint: a model for male pheromone physiology, in Proc. VIIIth Int. Congr. Anim. Reprod. A.I., Kraków, 1976.

Cook, R., and Cook, A. 1975. The attractiveness to males of female Drosophila melanogaster: effects of mating, age and diet. Anim. Behav. 23: 521-526.

Darby, E.M., Devor, M., and Chorover, S.L. 1975. A presumptive sex pheromone in the hamster: some behavioral effects. J. Comp. Physiol. Psychol. 88: 496-502.

Diakow, C. 1974. Male-female interactions and the organization of mammalian mating patterns, pp. 227-268, in D.S. Lehrman, J.S. Rosenblatt, R.A. Hinde and E. Shaw (eds.). Advances in the Study of Behaviour, Vol. 5, Academic Press.

Donovan, C.A. 1967. Some clinical observations on sexual attraction and deterrence in dogs and cattle. Vet. Med. Small Anim. Clin. 62: 1047-1051.

Drickamer, L.C., Vandenbergh, J.G., and Colby, D.R. 1973. Predictors of dominance in the male golden hamster (Mesocricetus auratus). Anim. Behav. 21: 557-563.

Estes, R.D. 1972. The role of the vomeronasal organ in mammalian reproduction. Mammalia 36: 315-341

Fletcher, I.C., and Lindsay, D.R. 1968. Sensory involvement in the mating behaviour of domestic sheep. Anim. Behav. 16: 410-414.

Fraser, A.F. 1968. Reproductive Behaviour in Ungulates, Academic Press, London.

Gorman, M.L., Nedwell, D.B., and Smith, R.M. 1974. An analysis of the contents of the anal scent pocket of Herpestes auropunctatus (Carnivora:Viverridae). J. Zool., London. 172: 389-399.

Grosvenor, C.E., and Mena, F. 1974. Neural and hormonal control of milk secretion and milk ejection, Vol. 1, pp. 227-276, in B.L. Larson and V.R. Smith, (eds.). Lactation, Academic Press, New York.

Hafez, E.S.E., and Signoret, J.P. 1969. The behaviour of swine, 2nd Edition, pp. 349-390, in E.S.E. Hafez (ed.). The Behaviour of Domestic Animals, Bailliere, Tindall and Cassell, London.

Herrick, C.J. 1933. The functions of the olfactory parts of the cerebral cortex. Proc. N.A.S. 19: 7-14.

Jonsson, P., and Wismer-Pedersen, J. 1974. Genetics of sex odour in boars. Livestock Prod. Science 1: 53-66.

Kloek, J. 1961. The smell of some steroid sex-hormones and their metabolites. Reflections and experiments concerning the significance of smell for the mutual relation of the sexes. Psychiat. Neurol. Neurochir. 64: 309-344.

Klopfer, P.H., and Gamble, J. 1966. Maternal 'imprinting' in goats: The role of the chemical senses. Z. Tierpsychol 23: 588-592.

Kravkov, S.V. 1939. The influence of odors upon color vision. Acta Ophtal. 17: 426-442.

Leon, M.A. 1974. Maternal pheromone. Physiol. Behav. 13: 441-453.

Meding, J.H. 1972. Anvendelse af syntetisk fremstillet orneduft, 5α-androst-16-ene-3-one, ved brunstkontrol hos søer og gylte. Copenhagen Royal Agr. Vet. Univ. Steril. Res. Instit. Ann. Rep. 1972, pp. 113-121.

Michael, R.P., Bonsall, R.W. and Warner, P. 1975. Primate sexual pheromones, pp. 417-424, in D.A. Denton and J.P. Coghlan (eds.). Olfaction and Taste, Proc. Vth Int. Sympos. Melbourne, 1974, Academic Press.

Müller-Schwarze, D. 1974. Olfactory recognition of species, groups, individuals and physiological states among mammals, pp. 316-326, in M.D. Birch (ed.). Pheromones, North-Holland Publ. Co., Amsterdam.

Müller-Schwarze, D., and Müller-Schwarze, C. 1972. Social scents in hand-reared pronghorn (Antilocapra americana). Zool. Afr. 7: 257-271.

Mykytowycz, R. 1970. The role of skin glands in mammalian communication, pp. 327-360, in J.W. Johnston Jr., D.G. Moulton, A. Turk (eds.). Advances in Chemoreception. Vol. 1. Communication by Chemical Signals, Appleton-Century-Crofts, New York.

Mykytowycz, R. 1973. Reproduction of mammals in relation to environmental odours. J. Reprod. Fert. Suppl. 19: 433-446.

Mykytowycz, R. 1974 Odor in the spacing behavior of mammals, pp. 327-343, in M.C. Birch (ed.). Pheromones, North-Holland Publ. Co., Amsterdam.

Mykytowycz, R. and Dudziński, M.L. 1966. A study of the weight of odoriferous and other glands in relation to social status and degree of sexual activity in the wild rabbit, Oryctolagus cuniculus (L.). CSIRO Wildl. Res. 11: 31-47.

Mykytowycz, R., and Fullagar, P.J. 1973. Effect of social environment on reproduction in the rabbit, Oryctolagus cuniculus (L.). J. Reprod. Fert. Suppl. 19: 503-522.

Perry, G.C., Patterson, R.L.S., and Stinson, G.E. 1973. Submaxillary salivary gland involvement in porcine mating behaviour. In Proc. VIIth Int. Congr. Anim. Reprod. A.I., Munich, 1972.

Rosenblatt, J.S. 1972. Learning in newborn kittens. Sci. Amer. 227: 18-25.

Sambraus, H.H. 1973. Das Sexualverhalten der domestizierten einheimischen Wiederkäuer. Advances in Ethology. Beih 12, Z. Tierpsychol. Paul Parey, Berlin.

Sambraus, H.H., and Waring, G.H. 1975. Der Einfluss des Harns brünstiger Kühe auf die Geschlechtslust von Stieren. Z. Säugetierk. 40: 49-54.

Signoret, J.P., and du Mesnil du Buisson, F. 1961. Etude du comportement de la truie en oestrus. In Proc. IVth Int. Congr. Anim. Reprod. The Hague. 2: 171-175.

Signoret, J.P., and Bariteau, F. (in press). The use of odorous synthetic steroids to facilitate oestrus detection in the sow. Effects on farrowing rates after artificial insemination. In Proc. VIIIth Inter. Congr. Anim. Reprod. and A.I. Kraków, 1976.

Smith, F.V. 1965. Instinct and learning in the attachment of lamb and ewe. Anim. Behav. 13: 84-86.

Stieve, H. 1927. Die Abhängigkeit der Keimdrüsen vom Zustand des Gesamtkörpers und von der Umgebung. Naturwiss. 15: 951-963.

Strauss, J.S., and Ebling, F.J.   1970.   Control and function of skin
    glands in mammals.   Mem. Soc. Endocrinology 18: 341-371.
Vandenbergh, J.G., Whitsett, J.M., and Lombardi, J.R.   1975.   Partial
    isolation of a pheromone accelerating puberty in female mice.
    J. Reprod. Fert. 43: 515-523.
Vandeplassche, M., and Spincemaille, J.   1967.   Some aspects of normal
    and abnormal sexual behaviour in farm animals.   Ann. Endocrinol.
    28: 815-819.
Whitten, W.K.   1975.   Responses to pheromones by mammals, pp. 389-395,
    in D.A. Denton and J.P. Coghlan (eds.).   Olfaction and Taste,
    Proc. Vth Int. Symp., Melbourne, 1974, Academic Press.
Whitten, W.K., and Champlin, A.K.   1973.   The role of olfaction in
    mammalian reproduction, Vol. II, pp. 109-123, in R.O. Greep
    (ed.).   Handbook of Physiology, Waverley Press Inc., Baltimore.
Wierzbowski, S. 1959.   The sexual reflexes of stallions.   Roczn.
    Nauk Roln. 73-B-4, 753-788.
Zarrow, M.X., Schlein, P.A., Denenberg, V.H., and Cohen, H.A.   1972.
    Sustained corticosterone release in lactating rats following
    olfactory stimulation from the pups.   Endocrinology 91: 191-196.

SEX PHEROMONES IN GOLDEN HAMSTERS

Robert E. Johnston

Department of Psychology, Cornell University

Ithaca, NY 14853

In this chapter I will review the role of female odors in the causation of sexual behavior of male hamsters. I will not attempt to review related topics such as the role of male odors in female sexual behavior or the relevant work on neural mechanisms of sexual behavior in hamsters (see Devor, this volume). In addition to being a detailed treatment of one species this chapter will highlight some methodological and theoretical issues that are applicable to mammalian pheromone work in general.

At the start I would like to stress several separate but closely related themes that will recur throughout. Whereas several other authors in this volume have discussed the chemical complexity of mammalian scent secretions, I will deal with behavioral complexity. It is often not appreciated that for each chemical secretion suspected of having communication functions a great deal of behavioral work needs to be done in order to understand these functions. Most mammalian scents probably have multiple functions that depend on the social context, and we need to know more about the total array of functions that each scent signal has. Much of our present knowledge comes from carefully controlled laboratory experiments which are specifically designed to demonstrate a particular function by elimination of other relevant variables. Such methods tell us what effects a signal _can_ have in these prescribed situations, and this information is necessary in the early stages of an investigation, but I suggest that we've gotten to the point where we should begin to ask what _do_ animals do rather than what _can_ they do. In other words, what are the roles of a particular odor signal within the contexts of the species total behavioral repertoire? How do chemical signals

interact with signals in other sensory modes? In what ways are
chemical and other signals specialized for their communication
functions for use within a particular ecological niche, micro
habitat, and social organization?

Unfortunately a great limitation in working with golden
hamsters is that very little is known about their ecology and
social organization. However, a few reasonable inferences can
be made. This species of hamster comes from a semi-arid region
in the vicinity of Aleppo, Syria (Aharoni, 1932; Devor, in
Yerganian, 1972); the individuals are primarily nocturnal and
solitary, each individual living in a burrow in which it hoards
large quanites of food (Vinogradov & Argiropulo, 1941; Murphy,
1971). Adult individuals are aggressive toward all other adults,
except that estrous females do tolerate males for a brief period
of time until they are mated (Johnston, 1975b; Payne, 1974; Wise,
1974; Carter, 1973). It is likely that hamsters defend at least
their home burrow (Dieterlen, 1959; Eibl-Eibesfeldt, 1953;
Johnston, 1975b), but as we know nothing about the spatial distribution of burrows or of the activity patterns of individuals
we cannot say anything definite about the exact nature of hamster
social organization. It seems likely that the system is one of
dispersed burrows and large overlapping home ranges.

POTENTIAL SOURCES OF ODOR SIGNALS IN HAMSTERS

Some of the discrete sources of secretions which may function
as chemical signals are listed below and the likelihood of the
use of each in communication is briefly noted. Dispersed sources,
such as hair follicle sebaceous glands, could also have communicative functions but there is as yet no behavioral evidence im-
plicating such glands.

Urine

Hamsters urinate in one or two locations near the nest in semi-
natural living areas in the laboratory and do not appear to dis-
tribute their urine widely by scent marking; other individuals pay
little attention to this urination area (Johnston, 1975b and unpublished observations). In a two-bottle sniffing test males
sniffed the same amount at a bottle containing female urine and a
clean bottle, suggesting that urine odor is not an attractant for
male hamsters as it may be in other species (Johnston, 1974). The
urine of hamsters is quite concentrated, as it is for most rodents
living in arid regions, and one would expect that hamsters and
similar species would tend not to use urine as a source of communicative odors, since in order to spread it around they would
probably have to increase its dilution, and thereby suffer water loss

Feces

Hamster feces are quite dry, but hamsters do deposit small
piles of feces next to or on top of their food hoards; such piles
could serve some communicative function, but no studies have investigated this possibility. Feces could serve as an efficient
means of marking if those deposited in "marking" piles were scented
with glandular secretions whereas other feces, deposited elsewhere,
were not scented, as is the case for European rabbits (Mykytowycz,
1968).

Harderian Glands

The secretion of the Harderian glands is sexually dimorphic in
hamsters, with that from females containing a pigment not present
in male secretion (Christensen & Dam, 1953). After castration
the secretion of males' harderian glands changes to the female
pattern (Clabough & Norvell, 1973). These secretions are distributed around the face and head when hamsters groom (Thiessen,
personal communication), and although one function may be for
temperature regulation by evaporative cooling, hamsters spend
considerable time sniffing the head region during interactions,
suggesting that the Harderian secretion may also be important in
social communication.

Flank Glands

These glands are clearly defined, bilaterally symmetric regions
of specialized sebaceous glands on the dorsal flanks which are sexually dimorphic in size and degree of pigmentation; this dimorphism
is entirely androgen dependent (Hamilton & Montagna, 1950;
Vandenbergh, 1973). Both sexes have a stereotyped scent marking
behavior which deposits the secretion from this gland and probably
deposits other secretions from the side of the head and body as
well (Dieterlen, 1959; Johnston, 1970, 1975b, 1976). Flank marking
is stimulated by the odors of conspecifics, particularly the odor
of others' flank glands (Johnston, 1975b & 1975c), and high frequencies of flank marking are associated with agonistic motivation
and dominance status (Johnston, 1970, 1975b, 1975c; 1976; Drickamer
& Vandenbergh, 1975; Drickamer, Vandenbergh & Colby, 1975). These
observations suggest functions related to aggressive activities,
but attempts to demonstrate a role for this gland in defense of
a home area or as a threat in agonistic encounters have so far
proved unsuccessful (Alderson & Johnston, 1975, 1976; Johnston,
unpublished observations). However, preliminary experiments in
my laboratory indicate that females are less attracted to the odors
of flank-glandectomized males than to the odors of normal males,
suggesting a role in sexual attraction and/or recognition.

Lipkow (1954) claimed that females would not mate with flank glandectomized males, but this is incorrect (Johnston, unpublished observations).

Preputial Glands

Hamsters do have preputial glands (La Velle, 1951), but the functions of these glands have not been investigated. Although males occasionally drag their ano-genital region across the substrate, it is not known if they are depositing secretions of any kind. The preputial or clitoral gland of females may possibly contribute to the vaginal secretion (see below).

Vaginal Secretion

Dieterlen (1959) first noted the copious vaginal secretion and described a vaginal scent marking behavior for its deposition. A pair of large, lateral pouches that apparently serve as a reservior for the secretion lie at the distal end of the vagina (La Velle, 1951), and these pouches and/or the entire vagina seem to be under muscular control since females deposit small amounts of the secretion when vaginal marking and express a much larger amount just prior to copulation. The secretion undergoes regular changes in volume and consistency correlated with the estrous cycle (Orsini, 1961) with maximal production occurring during the estrous (receptive) period. Large amounts of secretion can be collected at the end of the receptive period and up until twenty-four hours later. Although even ovariectomized females have some secretion (Darby, Devor & Chorover, 1975), the cyclic peak in production during estrus is apparently due to the same hormonal changes that result in behavioral receptivity (Brom & Schwartz, 1968).

The actual site(s) of production of the secretion are unknown. The upper vaginal wall has a specialized epithelium which may be the primary source (Deanesley, 1938), but important components could also come from sloughed cells, from the preputial glands attached to the ventral surface of the vaginal wall, from the urethral glands (La Velle, 1951), or from some other unsuspected source. It is also unknown if the behaviorally active components are produced directly by the hamster or indirectly by bacterial action.

RESPONSES OF MALES TO ODORS OF CONSPECIFICS

Starting at the most general level of analysis we can ask male hamsters if they are preferentially attracted to the odors of females over those of males and if so are they also preferentially

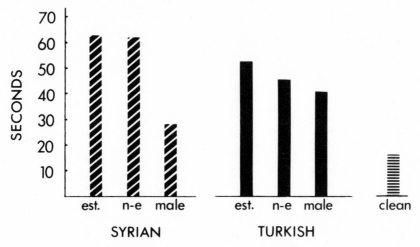

Fig. 1. Mean time spent sniffing by male Syrian hamsters at the
front of bottles containing various conspecific and hetero-
specific stimulus animals—males, estrous females (est.) or
non-estrous females (n-e). Two minute trials; N = 10.

attracted to the odors of estrous females as opposed to non-
estrous females.

Figure 1 shows data for 10 males sniffing at the front of a
bottle containing anesthetized stimulus animals; each male was
tested with all stimuli, one at a time, presented in a counter-
balanced order (Dickie & Johnston, 1976 & in preparation). The
males spend significantly more time sniffing at anesthetized
females than at males (non-estrous female vs male, p < .005;
estrous female vs male, p < .002; t tests), indicating that they
can tell the difference and that they are more attracted to
the odors of females. In this situation the odors of the anesthetized conspecific males were more attractive than the clean bottle
(p < .05). It is interesting to note that although the odors of
male and female Turkish hamsters (M. brandti) were also more
attractive than the clean bottle, the Syrian males did not show
a significant preference for Turkish females over Turkish males.
When M. brandti males were used as test animals a similar pattern
of results was observed.

The second point of interest in Figure 1 is that males were
equally attracted to the odors of estrous and diestrous (day 3)
females. It could be legitimately objected that, since in this
test just one stimulus was presented at a time, the results demonstrate only that the males were highly attracted to the odors
of both estrous and non-estrous females. This experiment does not
demonstrate whether or not males will show a preference when

tested with both females simultaneously, but Landauer has demonstrated that when males are given a simultaneous choice they
still show no preference for the odors of either an estrous or
a diestrous female, and both sexually experienced and inexperienced males are equally indiscriminate (Landauer & Banks, 1973;
Landauer, Banks & Carter, 1976).

Thus male hamsters are preferentially attracted to the odors
of females over those of males but do not prefer odors of
estrous females over those of diestrous females. It is commonly
believed that most male mammals can and do discriminate the
estrous state of females: males of many species "test" females
by sniffing the genital region and experimental results have
confirmed this hypothesis in a number of species such as laboratory rats (Carr, Loeb & Dissinger, 1965; Carr, Loeb & Wylie,
1966; Lydell & Doty, 1972; Pfaff & Pfaffmann, 1969), dogs (Beach
& Gilmore, 1949; Doty & Dunbar, 1974), and sheep (Lindsay, 1965).
I suggest that for hamsters a more important distinction is one
between "soon to be estrus" and "not soon to be estrus", since
during the breeding season female hamsters should be pregnant or
lactating most of the time and only rarely cycling. Thus one
would predict that males would be attracted to the odors of cycling,
non-estrous females, estrous females and late lactation females but
less attracted to the odors of pregnant or early lactation females.
It seems likely that other species might show this same pattern,
particularly those with, (1) a high reproductive potential in
which the females are usually pregnant and/or lactating, and (2)
a solitary and relatively widely dispersed social system, in
which there would be less opportunity for males to obtain information about the females' state at daily intervals than there
would be in more social species. One would expect the females of
solitary species to advertise for several days prior to estrous to
ensure the presence of at least one male, while in more social
species females should provide more precise information about
reproductive condition.

SOURCES OF ATTRACTIVE FEMININE ODORS

What are the sources of the odors which make females so
attractive to males? The vaginal secretion is one obvious possibility
and one approach is to study the behavior of males toward females
lacking the secretion. Marie Kwan, working in my laboratory, has
developed a surgical procedure for removing the vagina and the
glands within and around it (Kwan & Johnston, 1975, and in preparation); in one experiment the attraction of 16 males to normal
and vaginectomized females was studied. During separate two
minute tests the males spent a mean of 62.7 seconds sniffing at
the normal females and significantly less time, a mean of 44.0

seconds, sniffing at the vaginectomized females (p < .002).   In
several other experiments we have shown that areas scent marked by
normal females are more attractive to males than areas marked by
vaginectomized females.   Thus the vagina does seem to be the
source of some of the body odors of females that are attractive
to males.   Although is is unlikely that the vagina is the only
source of such odors, experiments directly comparing the attractiveness of the odors of males and of vaginectomized females
have not yet been completed, so we can present no definite conclusions at this time.

VAGINAL MARKING BY FEMALES

     In order to fully document the existence and functions of a
communication system, one would like to demonstrate specializations
in morphology and behavior of the sender of the signal as well as
effects of the signal on the receiver.   Before proceeding to a
detailed analysis of the effects of the vaginal secretion it is
appropriate to ask in what situations females vaginal mark and
whether or not such marking seems appropriate for the functions
demonstrated by observing responses of males.

     Vaginal marking frequency varies regularly with the estrous
cycle, being greatest the day before estrus, lowest during estrus,
and intermediate on the other two cycle days (Figure 2; Johnston,
1970, 1976).   In Figure 2 the daily marking frequences for three
of five females from one experiment are shown.   Vaginal marking
frequency was significantly higher on day 4 of the cycle than
all other cycle days, and marking on day 1 (estrus) was significantly less than on all other cycle days (Johnston, 1976).
Females marked much more frequently in the vacant home cages of
males (mean for non-estrous days, 10.2 marks per 10 minutes) than
in the cages of females (mean, 2.4 marks; p < .06, non-parametric
Walsh test) or in clean cages (mean, 5.0; p < .06).   The actual
presence of a male also stimulated vaginal marking (Johnston, 1976).
Leonard (1972) has shown that the percentage of trials in which
vaginal marking occurs in the presence of a male is profoundly
affected by pregnancy and lactation--the probability of marking
was quite low during pregnancy and early lactation but increased
gradually throughout lactation.   By the last quarter of lactation
the probability of vaginal marking was as high as when the females
were cycling.   Female hamsters do not have a post-partum estrus,
so they were marking more as their first receptive period approached.   These data alone suggest that vaginal marking serves a sexual
advertisement function.   The reason that females do not mark when
estrous, however, is not entirely clear; possibly it is important
for females to be the primary and most concentrated source of the
odor when they are receptive.

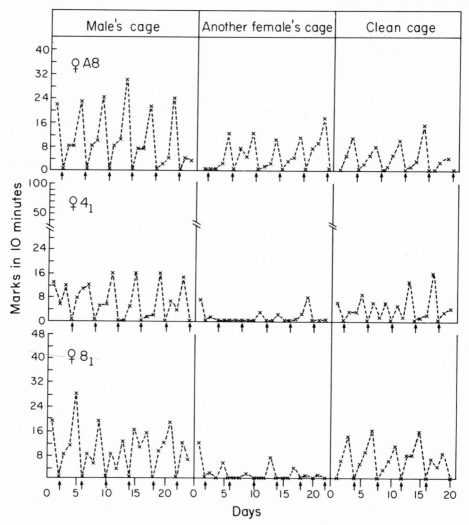

Fig. 2.   Frequency of vaginal marking by females in daily 10 minute
           tests in vacant stimulus cages. (↑) indicates female's
           estrous day.

EFFECTS OF THE VAGINAL SECRETION ON THE BEHAVIOR OF MALES

    Five separate responses of males to the vaginal secretion have
been demonstrated: (1) approach and sniffing at the odor, (2)
licking and consumption of the secretion, (3) increased sexual
arousal and tendencies to engage in copulatory behavior, (4) a
short-latency rise in plasma testosterone, and (5) lowering of
aggressive tendencies. I will present some evidence relevant to
each of these effects and discuss the significance of each.

Attractant Properties of the Vaginal Secretion

The odor of the vaginal secretion is highly attractive to male hamsters in a variety of laboratory situations; in brief 1-4 minute tests males spend 25 to 70 percent of test time sniffing at the odor outlet and often dig and bite at it, as if they were trying to get at the source (Johnston, 1970; 1974; and unpublished observations; Murphy, 1973). The degree of attractiveness is quite strong and seems to be greater than that reported for sexual attractant odors in other species tested in a similar manner--for example sexually experienced male rats spend 10% or less of test time sniffing at the urine of estrous females (Carr, Loeb & Dissinger, 1965; Pfaff & Pfaffmann, 1969). Although male hamsters' responses to the vaginal secretion are generally strong there are large individual differences and the variability in one individual's responses from trial to trial is often quite high. With frequent testing striking habituation effects are observed; for example when males are tested once per day with odor bottles inside their home cages they may stop showing preferential responses to vaginal secretion odor by the third or fourth day. Males' responsiveness is thus labile and subject to alteration by experience.

I have always been somewhat uneasy about using sniffing as a measure of the attractiveness of socially significant odors since animals may sniff for many different reasons and thus the significance of time spent sniffing is difficult to interpret. Recently Mary Dickie and I reported on preliminary experiments to determine, in one standard testing paradigm, how attractive the odor of conspecific vagnial secretion was relative to several other odors, namely vaginal secretion from M. brandti (the Turkish hamster), peanut oil, carrot juice, phenyl-ethyl alcohol, amyl acetate and benzaldehyde (Dickie and Johnston, 1976; and in preparation). In this experiment 10 males were presented with all the stimuli, one at a time in counter-balanced order, inside a bottle in their home cages. The data are shown in Figure 3; it can be seen that the vaginal secretion was sniffed much more than any of the other odors, including freshly made carrot juice (p < .02, non-parametric Walsh test) and peanut oil (p < .01), odors of two highly preferred food items. Time sniffing at phenyl-ethyl alcohol was not different from the time sniffing at a clean bottle whereas times sniffing at both amyl acetate and benzaldehyde were significantly less than at the clean bottle (p < .01), suggesting an aversion to these odors. These data demonstrate that male hamsters show a wide range of sniffing times in this test situation, and that their attraction to vaginal secretion is remarkably strong. We can therefore be more confident about using sniffing time as a measure of attractiveness.

It is interesting to note that the odor of the Turkish hamster vaginal secretion was nearly as attractive as that of the conspecific

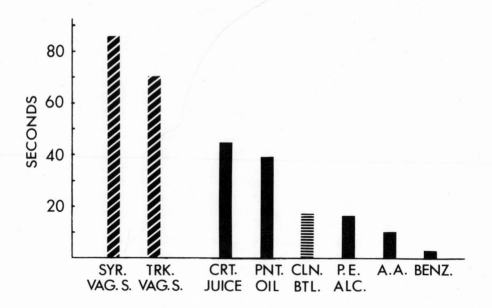

Fig. 3. Mean time that 10 males spent sniffing at the front of a
        bottle containing various odors in two minute tests: Syrian
        hamster vaginal secretion, Turkish hamster vaginal secretion
        carrot juice, peanut oil, clean bottle, phenyl-ethyl alcohol
        amyl acetate, benzaldehyde.

secretion (statistically not different). However, even though both
secretions are highly attractive in this simple approach/no approach test, males of both species prefer the vaginal secretion of
their own species when tested in simultaneous choice tests (Murphy,
personal communication).

      It should be stressed that an actual role for this secretion
in attraction of males from some distance in a naturalistic habitat
has not been demonstrated. We need to know much more about how
this odor functions in the real world--for example what is the
"active space" of the odors emanating directly from a female, what
is the spatial distribution of vaginal marks in the natural habitat,
and can the odor of vaginal marks attract males from a distance
or are they strictly a short-range signal?

Elicitation of Licking

Male hamsters lick and consume the secretion in a variety of naturalistic circumstances, for example when they encounter vaginal marks or when the female extrudes the secretion prior to copulation, as well as when the secretion is presented by an experimenter on glass plates, dummy hamsters, etc. (Johnston, 1970, 1974, 1976; Johnston & Lee, 1976; Murphy, 1973; Murphy & Schneider, 1970; Devor & Murphy, 1973; Darby, Devor & Chorover, 1975). Licking is often elicited by the odor alone; in one experiment males licked 5 seconds or more at the front of a secretion-containing bottle during 69% of trials while they only licked at a simultaneously presented clean bottle during 7% of trials (Johnston, 1974).

The functions of licking and/or ingestion of the secretion are not known. Neither the sense of taste nor ingestion of the secretion is necessary for sexual behavior, since males with their mouths sewn shut still mate (Devor, personal communication), but both taste and ingestion could play some role in normal sexual behavior or in a related physiological processes. An intriguing possibility is that the secretion might have some physiological effects subsequent to ingestion, such as direct action on the brain or neuroendocrine system.

Sexual Excitant Effects

Several laboratories independently demonstrated that the vaginal secretion increases sexual arousal; specifically it has been shown that the presence of the secretion increases the probability and/or frequency of sexual behaviors toward normally inappropriate partners such as docile males, anesthetized males, castrated males, and overiectomized females (Lisk, Zeiss & Ciaccio, 1972; Devor & Murphy, 1972; Murphy, 1973; Johnston, 1972, 1975a; Darby, Devor & Chorover, 1975). The results of one such experiment are shown in Figure 4. Baseline levels of attempted mounting were quite low (one male on one test) and were significantly elevated by addition of vaginal secretion (Johnston, 1975a). There was also a non-significant increase in frequency of attempted mounts on control trials after the males had experience with the vaginal secretion in the test situation, suggesting an effect of test experience on the levels of sexual behavior observed. The role of the vaginal secretion in normal sexual behavior is discussed in more detail in later sections of this chapter.

Effects on Plasma Testosterone

Macrides, et al., (1974) recently demonstrated that when sexually experienced males that had been isolated from the odors of

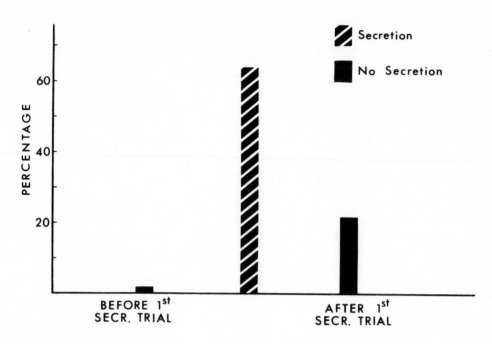

Fig. 4. The percentage of trials in which males attempted to mount
       awake, castrated males scented or not scented with vaginal
       secretion. Control trials are divided into those occurring
       before and after each male's first secretion trial.

conspecifics for three weeks were exposed to the odor of the vaginal
secretion for 30 minutes their plasma testosterone levels increased
twofold. Increases of a similar magnitude were obtained when these
males were given 30 minutes of contact with an estrous female.
Males that were caged in a room with other males, and thus were
not isolated from conspecific odors, showed a significant increase
in plasma testosterone on exposure to a female but the increase
upon exposure to the vaginal secretion alone was not significantly
higher than baseline levels. It is not clear if the secretion
would have effects on testosterone levels in wild living animals
who have relatively frequent experience with conspecifics and
their odors, but the data suggest that the vaginal secretion may
be one of many cues which stimulate testes function.

Reduction of Aggressiveness

Both Murphy (1973) and Johnston (1975) showed that males were
less aggressive when vaginal secretion was added to stimulus
animals, young males or castrated males, respectively. In other

experiments the frequency of flank marking, which is correlated
with dominance and aggressive motivation, was reduced when vaginal
secretion was added to the environment (Johnston, 1970). Although
male hamsters are extremely aggressive toward other males they
are not aggressive toward females on any day of the estrous cycle;
instead they investigate females and may try to mount them
(Johnston, 1970, 1975b; Payne, 1974). The fights that do occur
are nearly always initiated by the females. The vaginal secretion
may be one of the factors that limits aggression between the sexes,
but more work needs to be done to specify the various signals
involved.

THE ROLE OF INDIVIDUAL EXPERIENCE

It should be noted that the influence of individual experience
on males' responses to the vaginal secretion have not been systematically investigated. Sexual experience is not necessary for the
sniffing, licking, sexual arousal or aggression-inhibiting effects
of the secretion (Johnston, 1974, 1975a; Lisk, Zeiss & Ciaccio,
1972) but there may be important quantitative effects of such
experience. It is recognized by many workers in the field that
prior sexual experience greatly influences the amount of sexual
behavior males will show when tested with vaginal secretion on
anesthetized stimulus animals (Johnston, unpublished observations;
M. Murphy, R. O'Connell & F. Macrides, personal communications),
and, as mentioned above, the type of testing regime may influence
the magnitude of sniffing and sexual responses observed to the
secretion. In addition it has been demonstrated that males'
attraction to the secretion, as well as their licking and sexual
responses, can be modified by conditioning procedures: animals
that were made ill after ingesting the secretion showed taste/
odor aversions to the secretion and engaged in less sexual behavior toward estrous females than control animals did (Johnston
& Zahorik, 1975; Zahorik & Johnston, 1976; Johnston, et al., 1976).
It thus seems likely that relevant social experiences of adult
males may affect their responses to this sexual signal. Likewise
social and olfactory experiences of pups during development may
be important for adult responsiveness to olfactory signals, including the vaginal secretion.

NEURAL AND HORMONAL MECHANISMS OF RESPONSE TO THE VAGINAL SECRETION

A number of studies have investigated physiological mechanisms
related to the behavioral effects of the vaginal secretion. In a
study in which males' sniffing responses were observed before and
after castration and then with testosterone replacement therapy,
Gregory, Engle & Pfaff (1975) showed that the attractiveness of
the secretion was testosterone dependent. Recent studies in my

laboratory have also shown that the onset of responses to the vaginal secretion in males correlates nicely with the time of puberty and rising levels of plasma testosterone, as determined by Lach, et al., (1976). In addition we have found that testosterone proprionate injections in pre-pubertal pups results in pre-pubertal responsiveness to the secretion (Johnston & Coplin, unpublished observations). Sexual behavior itself is dependent on gonadal steroids in male hamsters (Beach & Pauker, 1949) but the relative importance of hormone-dependent arousal due to olfaction as opposed to other systems is unknown.

I can only briefly mention recent work on the neural mechanisms of olfactory influences on male sexual behavior. A functional olfactory system is necessary for the initiation, although not the maintenance, of male sexual behavior (Murphy & Schneider, 1970; Devor & Murphy, 1973; Lisk, Zeiss & Ciaccio, 1972). Some progress has been made in analyzing the role of different olfactory systems. Both the vomeronasal organ receptors and the main olfactory receptors seem to be important in mating behavior, whereas only the main olfactory system seems to be involved in sniffing and attraction to the vaginal secretion (Powers & Winans, 1973, 1975). Devor (1973) has also demonstrated a neuroanatomical separation of sniffing and mating: posterior lesions of the lateral olfactory tract resulted in elimination of mating behavior but not sniffing at the vaginal secretion, whereas more rostral lesions eliminated both mating and sniffing. These studies provide exciting new information about the functions of different parts of the olfactory system and about brain mechanisms of male sexual behavior. The increasing sophistication of such studies is due in part to the relative importance of olfaction in sexual behavior of male hamsters and in part to the detailed knowledge of the effects of the vaginal secretion. As more becomes known about the behavioral functions of other odor signals and about the chemistry of socially significant odors it will be possible to make more and more sophisticated neuroanatomical and neurophysiological investigations.

ROLE OF THE VAGINAL SECRETION IN REPRODUCTIVE BEHAVIOR

The experiments cited above have indicated important effects that the vaginal secretion can have in the appropriate experimental circumstances. But what role does the secretion play in normal sexual behavior? How does this chemical cue interact with other signals in the causation of male sexual behavior? These questions are more subtle and difficult to investigate than those raised above, and there is little information that is directly relevant. I will deal briefly with two preliminary and entirely different approaches which attempted to answer the following, more specific questions: (1) is the vaginal secretion either necessary or sufficient for male sexual behavior, and (2) is the fade-out time, a design feature of chemical signals, related to its proposed communicative functions?

Necessity and/or Sufficiency of Vaginal Secretion

First let us ask if the presence of the vaginal secretion is necessary for the occurrence of sexual behavior. Since most aspects of mammalian behavior are under multi-sensory control one would predict that the presence of the secretion would not be required. It has been repeatedly demonstrated, however, that a functional olfactory system is necessary for the initiation of copulation, and such results raise the possibility that a single olfactory cue might be critical. Marie Kwan and I have recently completed an experiment in which the behavior of males toward normal and vaginectomized females was compared (Kwan & Johnston, 1975, and in preparation). The vaginectomized females displayed apparently normal behavior, including regular, 4-day estrous cycles. In brief two-minute tests (which minimized the importance of intromissions or lack thereof) males engaged in equal amounts of sexual behavior toward tube-tied (normal) and vaginectomized females (Figure 5), demonstrating that the vaginal secretion is not necessary. Two other observations support this conclusion. First,

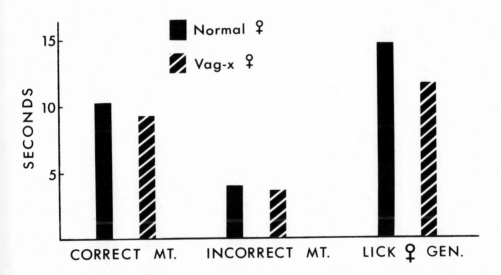

Fig. 5.   Sexual behavior of males toward normal and vaginectomized females during two minute tests. (MT. refers to mount; GEN. to genitals).

Tiefer and Johnson (1975) showed that male hamsters would mount and attempt to copulate with other males that had been treated with estrogen and progesterone and that were exhibiting lordosis. Second, males who have been severely defeated in fights sometimes assume an immobile submissive posture which resembles the lordosis posture of females, and in response the dominant male occasionally mounts the subordinate male (Johnston, 1970 & unpublished observations). Although none of these observations alone provide iron-clad proof that the vaginal secretion is not necessary for male sexual behavior, all three observations together are persuasive and indicate that the control of such behavior is dependent on a variety of redundant sensory cues.

Even though it is not necessary, the vaginal secretion does have powerful effects on the behavior of males. The next question is whether the secretion is sufficient to elicit sexual behavior in the absence of other cues from hamsters. For some species of insects, such as the honey bee, attraction and vigorous copulatory attempts can be obtained from drones when sex pheromones are placed on wooden models (Gary & Marston, 1971). Darby, Devor & Chorover (1975) studied the responses of male hamsters toward a series of models that more and more closely resembled a normal estrous female: a glass bottle, a clay model, a dark furry model, a light furry model, a heated furry model, an anesthetized male, an anesthetized female and an estrous female. The behavior of males toward all of the non-female models was studied both with and without the addition of vaginal secretion. The single most striking aspect of the results was a discontinuity in the amount of sexual behavior shown toward all of the animate stimulus objects and all of the inanimate stimulus objects, with much more sexual behavior shown toward the animate models. For example the two models on either side of this discontinuity were the heated furry model and the anesthetized male; when vaginal secretion was present the males spent a mean of 0.3 seconds mounting the heated furry model versus 33.1 seconds mounting the anesthetized male. It is interesting that even some of the non-sexual behaviors were significantly facilitated by the presence of an "animate" stimulus model; for example, the mean time spent licking the secretion on the heated furry model was 53.8 seconds and on the anesthetized male it was 87.6 seconds (p < .05). These results indicate the importance of other cues (visual, tactile and olfactory) in facilitation of responses to the vaginal secretion.

Thus the vaginal secretion is neither necessary nor sufficient for male sexual behavior, but it is nonetheless one important cue among many that influence males' behavior. The more difficult job lies ahead--determining how various cues interact to influence sexual behavior of both males and females and how these effects are integrated in the nervous system. What other odor cues are involved? How do odor cues interact with the ultra-sonic vocalizations that are produced by both males and females in sexual attraction and arousal (Floody, 1974; Floody & Pfaff, 1976b). Do all these sensory signals act in an additive fashion on a single sexual arousal system or do the various signals have discrete effects on different aspects of sexual behavior?

Persistence of Vaginal Scent Marks

A characteristic unique to chemical signals is their potentially great persistence in time. It is a possibility that this feature, the so called fade-out time, has been selected in the course of evolution so that it would be suited to the communicative function served. Wilson & Bossert (1963) have shown that the fade-out times of some insect signals do indeed vary in predicted directions: ant alarm pheromones are extremely volatile and become inactive 35 seconds after release whereas the effect of ant trail substances lasts about 100 seconds and sexual attractants persist considerably longer. If the function of the vaginal scent marks were primarily for sexual attraction one might expect a persistence of 1-2 days at the most, since females mark most frequently the day before estrous and remain receptive for about 12 hours.

In a series of recent experiments in my lab we have studied the persistence of both vaginal marks and flank marks in an attempt to determine if this feature varied in functionally ap-

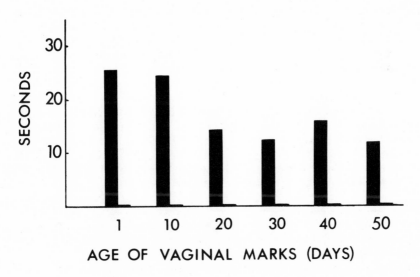

Fig. 6. Mean time spent sniffing and licking by eight males at vaginal marks of various ages and at clean control spots.

propriate ways (Johnston & Lee, 1976; Johnston & Schmidt, unpublished observations). In the vaginal marking experiments females marked glass plates which were then stored (at 70±2°F and 50% relative humidity) for various periods of time. During experimental trials males were given the opportunity to explore a box in which one half of the floor had been marked while the other was clean. The amount of time males spent sniffing and licking the marked places and a matched set of clean places was measured. The results of one such experiment are shown in Figure 6. It can be seen that even vaginal marks that were deposited 50 days prior to the tests were perceived and were sniffed and licked for a considerable amount of time by the males. This strikingly long persistence was totally unexpected, and it indicates that even if the behaviorally active components of the secretion are quite volatile, some of them must be released at an extremely slow rate. Two disclaimers should be made about these data. First, we don't know what the persistence of marks marks might be in nature--exposure to sun, rain, dust, etc., could greatly shorten the persistence of the marks while particularly low humidity (Regnier, this volume) or a more adsorbant substrate might lenghten the persistence. Second, we have only demonstrated the persistence of the marks as measured by sniffing and licking responses. Whether such old scent marks would affect males' sexual behavior, aggressive behavior, or testosterone levels is unknown. Nonetheless these results raise some interesting questions about the function and evolution of this chemical signal. Of what value is it to female hamsters to have such a persistent scent mark? Are there functions that the secretion has for which such long persistence is necessary? Alternatively, is this feature of the vaginal secretion neutral in terms of natural selection? If there is no selective cost of long persistence perhaps hamsters have evolved an extremely efficient means to insure persistence. Although we have no answers yet, this approach has provided some fascinating data and is an approach that should be applied to other mammalian species.

CHEMICAL NATURE OF VAGINAL SECRETION PHEROMONES

It is often assumed in pheromone work that some small subset of the total number of compounds in a secretion constitute "the signal", or at least that such a small subset can produce effects that are similar in kind to the effects that the whole secretion produces. A usually unstated but implied corollary is that the majority of the compounds in such secretions are not important for communication functions. No mammalian system is known in enough detail to evaluate these assumptions. The hamster vaginal secretion contains at least 250-300 compounds and it is difficult to imagine that most of these substances are not involved in any way in communication. Nonetheless analysis of the secretion must begin

by fractionation and attempts to identify important components or
sets of components.

The greatest success in analysis of the vaginal secretion has
been reported by Singer, O'Connell and their colleagues at Rockefeller
University (Singer, et al., 1976). They reported that dimethyl
disulphide can be isolated from the vaginal secretion and that males
are highly attracted to its odor in a sniffing test. Since both
the magnitude of the responses and the percentage of animals
responding to dimethyl disulphide were lower than to the whole
vaginal secretion Singer, et al., concluded that dimethyl disulphide was just one of a number of compounds involved in attraction of males. I have attempted to replicate these findings
using sniffing tests and application of dimethyl disulphide to
anesthetized male hamsters, but in my experiments males did not
demonstrate convincingly strong responses to this substance.
O'Connell, Singer and I have been communicating but at the present
time we do not know the reason for our discrepant results.

Two lines of evidence indicate that relatively non-volatile
components of the vaginal secretion may be involved in its behavioral activity as well as the volatile compounds being investigated at Rockefeller. The experiments mentioned above on
the persistence of the vaginal marks indicate that males detect
and respond by sniffing and licking at vaginal marks that are
50 days old. Secondly, Karl Frey and I have completed a whole series
of experiments with relatively volatile and non-volatile fractions
of the vaginal secretion (Johnston, 1975e; Johnston & Frey, 1976).
We separated these fractions by placing the secretion in a vacuum
at 10^{-6} Torr. for 24 hours and collected the volatiles in a U-
shaped section of the tube surrounded by liquid nitrogen. In all
three separate experiments for sniffing, licking, and sexual behavior, the relatively non-volatile fraction proved to be more
active than the relatively volatile fraction. For example in the
sex behavior tests, in which eight males were tested with anesthetized males as stimulus animals, the mean seconds of sniffing and
licking at the genital region per four minute trial was 1.1 with
unscented stimulus animals, 28.8 with volatile fraction, 87.5
with non-volatile fraction, and 101.6 with whole secretion. The
seconds of sniffing and licking at stimulus animals were significantly greater in the presence of both volatile and non-volatile
fractions than toward unscented controls. However the level of
sniffing and licking was significantly less in the volatile
fraction condition than in either the non-volatile fraction or
whole secretion conditions. The mean amount of mounting and
"crawling on" the stimulus animals was 0.1 seconds with unscented
controls, 3.1 seconds with volatile scented animals, 4.5 with non-
volatile scented animals and 8.7 with whole secretion scented
animals. All three odors caused a significant increase in mounting
attempts over control trials. The level of mounting and crawling

on during the volatile and non-volatile conditions was not significantly different, but the level of mounting during the whole
secretion condition was significantly greater than during the
volatile fraction condition. We conclude from these experiments
that relatively non-volatile components of the secretion must
be involved in some way in the normal functions of this secretion.
They could have direct effects on behavior, facilitate the effects
of more volatile components, or merely serve to bind and slowly
release the volatile components; at present we don't know what
role they have. It is interesting that non-volatile components
of scent secretions have been implicated in sexual recognition
by odor in guinea pigs (Berüter, Beauchamp & Muetterties, 1973)
and in puberty acceleration of female mice (Vandenbergh, Whitsett
& Lombardi, 1975.)

CONCLUDING REMARKS

The vaginal secretion of hamsters has recently become one
of the most thoroughly understood signals among mammalian pheromone
systems. As reviewed in this chapter, a considerable amount of
information has accumulated about the effects of this signal on
male receivers and about the causation of scent marking and the
release of the signal by female senders. Because of this knowledge and because the secretion has relatively powerful effects,
the male hamster has become a useful model for neuroanatomical
and neurophysiological investigations of the role of odors in
male sexual behavior. It is likely that the principles discovered in such investigations will be applicable to many other
mammals, since similar behavioral effects are known or suspected
in a large number of other species.

Despite the relative abundance of information about the
functions of the hamster vaginal secretion, we are a long way from
understanding this communicative system. We need to know much
more about how this signal functions in the real world, how it
interacts with other signals in facilitating sexual interactions,
and what is the nature of the active chemical components. For
example at the present time the details of how a male and a female find each other in order to copulate are totally unknown.
It is estimated that the ultra-sonic calls made by females most
frequently when they are estrous could be perceived by males in
ideal conditions 11-23 meters from the source (Floody, 1974;
Floody & Pfaff, 1976a). Could the vaginal secretion emanating
directly from a female serve as an attractant over greater distances? Alternatively, do the females lay odor trails by vaginal
marking to facilitate attraction of males? To answer such questions
information is needed about both behavior of hamsters in natural
habitats and about the habitats themselves. Would local micro
air currents allow long distance communication by odor? What is

the nature of the vegetation, and would it interfere with chemical or auditory communication? What is the substrate on which hamsters scent mark and what is the persistence of marks on such substrates? Do the animals choose particular substrates on which to mark to influence persistence of the marks? In addition to the need for this type of ecological information we need to know much more about other signals that the hamsters are using for communication. First we must discover what they are (e.g., ultra-sonics, other odors) and then proceed to find out their functions and how the various signals interact. With the little knowledge that we already have it is possible to begin to appreciate our ignorance.

REFERENCES

Aharoni, B. 1932. Die Muriden von Palastina und Syrien. Z. Säugetierk: 7, 166-240.

Alderson, J., & Johnston, R. E. 1975. Responses of male golden hamsters (Mesocricetus auratus) to clean and male scented areas. Behav Biol: 15, 505-510.

Alderson, J. & Johnston, R. E. 1976. The effects of flank gland odor and familiarity of the environment on investigation of scented areas by male hamsters. Submitted for publication.

Beach, F. A., & Gilmore, R. W. 1949. Responses of male dogs to urine from females in heat. J. Mammal: 30, 391-392.

Beach, F. A. & Pauker, R. S. 1949. Effects of castration and subsequent androgen administration upon mating behavior in the male hamster (Cricetus auratus). Endrocrinol: 45, 211-221.

Berüter, J. Beauchamp, G. K., & Muetterties, E. L. 1973. Complexity of chemical communication in mammals: urinary components mediating sex discrimination by male guinea pigs. Biochem. Biophys. Research Commun.: 53, 264-271.

Brom, G. M., & Schwartz, N. B. 1968. Acute changes in the estrous cycle following ovariectomy in the golden hamster. Neuro-endocrinology: 3, 366-377.

Carr, W. J., Loeb, L. S., & Dissinger, M. L. 1965. Responses of rats to sex odors. J. Comp. Physiol. Psychol: 62, 336-338.

Carr, W. J., Loeb, L. S., & Wylie, N. R. 1966. Responses to feminine odors in normal and castrated male rats. J. Comp. Physiol. Psychol., 62, 336-338.

Carter, C. S. 1973. Stimuli contributing to the decrement in sexual receptivity in female golden hamsters (Mesocricetus auratus) Anim. Behav.: 21, 827-834.

Christensen, F., & Dam, H. 1953. A sexual dimorphism of the Harderian glands in hamsters. Acta. Physiol. Scand.: 27, 333-336.

Clabough, J. W., & Norvell, J. E. 1973. Effects of castration, blinding, and the pineal gland on the harderian glands of the male golden hamster. Neuroendocrinology: 12, 344-353.

Darby, E. M., Devor, M., & Chorover, S. L. 1975. A presumptive sex
    pheromone in the hamster: Some behavioral effects. J.
    Comp. Physiol. Psychol.: 88, 496-502.
Deanesly, R. 1938. The reproductive cycle of the golden hamster
    (Cricetus auratus). Proc. Zool. Soc. Lond.: Ser. A, 31-37.
Devor, M. 1973. Components of mating dissociated by lateral
    olfactory tract transection in male hamsters. Brain Research:
    64, 437-441.
Devor, M., & Murphy, M. R. 1972. Social agnosia produced by
    peripheral olfactory blockage in hamsters. Amer. Zool.: 12,
    653.
Devor, M., & Murphy, M. R. 1973. The effect of peripheral olfactory
    blockade on the social behavior of the male golden hamster.
    Behav. Biol.: 9, 31-42.
Dickie, M., & Johnston, R. E. 1976. Olfactory preferences of male
    golden hamsters (Mesocricetus auratus) to natural and synthetic
    substances. Paper presented at Southeast Regional Meetings
    of Animal Behavior Society, Charlottesville, Va.
Dieterlen, F. 1959. Das Verhalten des syrischen Goldhamsters
    (Mesocricetus auratus Waterhouse). Z. Tierpsychol.: 16,
    47-103.
Doty, R. L., & Dunbar, I. 1974. Attraction of beagles to con-
    specific urine, vaginal and anal sac secretion odors. Physiol.
    & Behav.: 12, 825-833.
Drickamer, L. C., & Vandenbergh, J. G. 1973. Predictors of social
    dominance in the adult female golden hamster (Mesocricetus
    auratus) Anim. Behav.: 21, 564-570.
Drickamer, L. C., Vandenbergh, J. G., & Colby, D. R. 1973.
    Predictors of dominance in the male golden hamster (Mesocricetus
    auratus). Anim. Behav.: 21, 557-563.
Eibl-Eibesfeldt, I. 1953. Zur Ethologie des Hamsters (Cricetus
    cricetus L.). Z. Tierpsychol.: 10, 204-254.
Floody, O. R. 1974. Ultrasonic communication and the hormonal
    modulation of aggressive behavior in the female hamster.
    (Doctoral Dissertation, Rockefeller University).
Floody, O. R., & Pfaff, D. W. 1976a. Communication among hamsters
    by high-frequency acoustic signals. I. Physical Characteristics
    of hamster calls. J. Comp. Physiol. Psychol.: in press.
Floody, O. R., & Pfaff, D. W. 1976b. Communication among hamsters
    by high-frequency acoustic signals. II. Determinants of
    calling by females and males. J. Comp. Physiol. Psychol.:
    in press.
Gary, N. E., & Marston, J. 1971. Mating behavior of drone honey
    bees with queen models. (Apis mellifera L.) Anim. Behav.:
    19, 299-304.
Gregory, E., Engel, K., & Pfaff, D. 1974. Male hamster preference
    of female hamster vaginal secretions: Studies of experimental
    and hormonal determinants. J. Comp. Physiol. Psychol.: 89,
    442-446.

Hamilton, J. B., & Montagna, W. 1950. The sebaceous glands of the hamster I. Morphological effects of androgens on integumentary structures. Amer. J. Anat.: 86, 191-233.

Johnston, R. E. 1970. Scent marking, olfactory communication and social behavior in the golden hamster, Mesocricetus auratus. (Doctoral Dissertation, Rockefeller University, 1970). Dissertation Abstracts International, 1972, 32 (University Microfilms No. 72-12,666).

Johnston, R. E. 1972. Sex pheromones of the golden hamster. Amer. Zool.: 12, 662. (Abstract).

Johnston, R. E. 1974. Sexual attraction function of golden hamster vaginal secretion. Behav. Biol.: 12, 111-117.

Johnston, R. E. 1975a. Sexual excitation function of hamster vaginal secretion. Anim. Learn. & Behav.: 3, 161-166.

Johnston, R. E. 1975b. Scent marking by male hamsters I. Effects of odors and social encounters. Z. Tierpsychol.: 37, 75-98.

Johnston, R. E. 1975c. Scent marking by male hamsters II. The role of flank gland odor in the causation of marking. Z. Tierpsychol.: 37, 138-144.

Johnston, R. E. 1975d. Scent marking by male hamsters III. Behavior in a semi-natural environment. Z. Tierpsychol.: 37, 213-221.

Johnston, R. E. 1975e. Behavioral responses of male hamsters to relatively volatile and involatile fractions of vaginal secretion. Paper presented at Eastern Conference on Reproductive Behavior, Nags Head, N.C.

Johnston, E. E. 1976. The causation of two scent marking behaviors in female golden hamsters. Anim. Behav.: in press

Johnston, R. E., & Frey, K. 1976. Responses of male hamsters to relatively volatile and non-volatile components of the vaginal secretion. In preparation.

Johnston, R. E., & Lee, N. A. 1976. Persistence of the odor deposited by two functionally distinct scent marking behaviors of golden hamsters. Behav. Biol.: 16, 199-210.

Johnston, R. E. & Zahorik, D. M. 1975. Taste aversions to sexual attractants. Science: 189, 893-894.

Johnston, R. E., Zahorik, D., Zakon, H., & Immler, K. 1976. Effects of aversions to vaginal secretion on sexual behavior of male hamsters. Paper presented at a meeting of the Animal Behavior Society. Boulder, Colorado

Kwan, M., & Johnston, R. E. 1975. Responses of male hamsters to females in the absence of feminine sex pheromones. Paper presented at meetings of Animal Behavior Society, Wilmington, NC

Lach, L. F., Whitsett, J. M., Vandenbergh, J. G. & Colby, D. R. 1976. Physical and behavioral aspects of sexual maturation in male golden hamsters. J. Comp. Physiol. Psychol.: In press.

Landauer, M. R., & Banks, E. M. 1973. Olfactory preferences of male and female golden hamsters. Bulletin Ecological Society of America: 54, 44. (Abstract)

Landauer, M. R., Banks, E. M., & Carter, C. S. 1976. Sexual and
   olfactory preferences of naive and experienced male hamsters.
   J. Comp. Physiol. Psychol.: In press.
LaVelle, F. W. 1951. A study of hormonal factors in the early
   sex development of the golden hamster. Carnegie Contrib.
   Embryol.: No. 223, Vol. 356 (222-230), Washington, D.C., 19-53.
Leonard, C. M. 1972. Effects of neonatal (day 10) olfactory
   bulb lesions on social behavior of female golden hamsters.
   (Mesocricetus auratus) J. Comp. Physiol. Psychol.: 80, 208-215.
Lindsay, D. R. 1965. The importance of olfactory stimuli in the
   mating behavior of the ram. Anim. Behav.: 13, 75-78.
Lipkow, J. 1954. Über das Seitenorgan des Goldhamsters. Z. Morph.
   u. Ökol. Tiere.: 42, 333-372.
Lisk, R. D., Zeiss, J., & Ciaccio, L. A. 1972. The influence of
   olfaction on sexual behavior in the male golden hamster.
   (Mesocricetus auratus) J. Exp. Zool.: 181, 69-78.
Lydell, K., & Doty, R. L. 1972. Male rat odor preferences for
   female urine as a function of sexual experience, urine age,
   and urine source. Hormones and Behavior: 3, 205-212.
Macrides, F., Bartke, A., Fernandez, F., & D'Angelo, W. 1974.
   Effects of exposure to vaginal odor and receptive females on
   plasma testosterone in the male hamster. Neuroendocrinology:
   15, 355-364.
Murphy, M. R. 1971. Natural history of the Syrian golden hamster-
   a reconnaissance expedition. Amer. Zool.: 11, 632.(Abstract)
Murphy, M. R. 1973. Effects of female hamster vaginal discharge
   on the behavior of male hamsters. Behav. Biol.: 9, 367-375.
Murphy, M. R., & Schneider, G. E. 1970. Olfactory bulb removal
   eliminates mating behavior in the male golden hamster.
   Science: 167, 302-304.
Mykytowycz, R. 1968. Territorial marking by rabbits. Sci. Amer.:
   218, 116-126.
Orsini, M. W. 1961. The external vaginal phenomena characterizing
   the stages of the estrous cycle, pregnancy, pseudopregnancy,
   lactation, and the anestrous hamster (Mesocricetus auratus
   Waterhouse). Proc. Anim. Care Panel: 11, 193-206.
Payne, A. P. 1974. The aggressive response of the male golden
   hamster towards males and females of differing hormonal
   status. Anim. Behav.: 22, 829-835.
Pfaff, D., & Pfaffmann, C. 1969. Behavioral and electrophysiological responses of male rats to female rat urine odors. In
   Olfaction and Taste, Proc. IIIrd Int. Symp., C. Pfaffmann, Ed.
   258-267. Rockefeller University Press, New York.
Powers, J. B., & Winans, S. S. 1973. Sexual behavior in peripherally anosmic male hamsters. Physiol. & Behav.: 10, 361-368.
Powers, J. B., & Winans, S. S. 1975. Vomeronasal organ: critical
   role in mediating sexual behavior of the male hamster. Science
   187, 961-963.

Singer, A. G., Agosta, W. C., O'Connell, R. J., Pfaffmann, C., Bowen, D. V., & Field, F. H. 1976. Dimethyl disulphide: an attractant pheromone in hamster vaginal secretion. Science: 191, 948-950.

Tiefer, L., & Johnson, W. 1975. Neonatal androstenedione and adult sexual behavior in golden hamsters. J. Comp. Physiol. Psychol.: 88, 239-247.

Vandenbergh, J. G. 1973. Effects of gonadal hormones on the flank gland of the golden hamster. Hormone Research: 4, 28-33.

Vandenbergh, J. G., Whitsett, J. M., & Lombardi, J. R. 1975. Partial isolation of a pheromone accelerating puberty in female mice. J. Reprod. Fertil.: 43, 515-523.

Vinogradov, B. S. & Argiropulo, A. I. 1941. Fauna of the U.S.S.R. Mammals. Key to Rodents. (Israel program for scientific translations, 1968).

Wilson, E. O., & Bossert, W. H. 1963. Chemical communication among animals. Rec. Progr. Hormone Res.: 19, 673-710.

Wise, D. A. 1974. Aggression in the female golden hamster: effects of reproductive state and social isolation. Hormones and Behavior: 5, 235-250.

Yerganian, G. 1972. History and Cytogenetics of Hamsters. Prog. Exp. Tumor. Res.: 16, 2-41.

Zahorik, D. M., & Johnston, R. E. 1976. Taste aversions to food flavors and vaginal secretion in golden hamsters. J. Comp Physiol. Psychol.: 90, 57-66.

CHEMICAL SIGNALS AND PRIMATE BEHAVIOR

Richard P. Michael and R.W. Bonsall

Department of Psychiatry, Emory University
Atlanta, Georgia 30322
and Georgia Mental Health Institute
1256 Briarcliff Road, Atlanta, Ga., 30306

INTRODUCTION

People hold strong views about sex and, so it seems, almost equally strong opinions about smell. Put the two together, and there need be no surprise that this area of scientific inquiry is controversial. However, we prefer to avoid the somewhat sterile semantic controversy surrounding the use of the term "pheromone" as it is applied to the reproductive behavior of higher mammals. Whether or not it is useful to borrow a term from insect physiology and apply it to the complexities of primate behavior can be debated far into the night. What concerns us now are the observations that (1) odoriferous substances are produced within the vaginae of rhesus monkeys, (2) their production depends upon a bacterial system that is influenced by the balance of ovarian steroids in the body, (3) these substances escape externally and volatilize, (4) male rhesus monkeys normally make frequent olfactory investigations of the females' perinea, and (5) when the airborne signal is detected olfactorially, there is a change in the male's level of sexual activity. If the foregoing statements can be substantiated, we would then have strong evidence for the existence of a chemical channel of communication between the female and the male in this primate species.

In this chapter we review some of the available evidence. However, it should be understood at the outset that the majority of these experiments was designed to maximize the conditions under which olfactory cues might operate, and to minimize the role of other variables. This was necessarily done in artificial, con-

trolled laboratory conditions, and it is really not possible to
make logical inferences from such results about the role of chemical communication in the biology of the species under natural
conditions: this would be a highly appropriate task for the field
zoologist. There is also the question of the importance of olfaction relative to that of other channels of afferent stimulation.
Few if any mammalian forms rely exclusively on one signal and one
sensory modality for behavior that is crucial to the survival of
the species, and the reproductive behavior of higher primates
depends on a wide variety of signals and inputs, each of which
contributes more or less importantly to the events culminating in
mating. In an earlier experiment (Michael and Keverne, 1968), we
clearly demonstrated that anosmic male rhesus monkeys with prior
sexual experience will copulate quite normally, but this surely
should not be interpreted to mean that they will not use the olfactory channel of communication when it is open to them.

There is also the question of how the olfactory signal exerts
its effect once it has been received. At this point one leaves
the problem of inter-individual communication to consider the
physiological processes and mechanisms within the individual by
which an afferent input is coded and exerts its behavioral effects.
This is a fascinating and important question but one that is pred-
icated upon the obvious fact that a signal must first be received.
It is highly probable that learning plays a major role in these
processes because a prolonged period of maturation and development
is essential for the normal adult pattern of sexual behavior in
male rhesus monkeys. We know from numerous studies that the immature rhesus male needs contact with conspecifics during critical
periods in development, and an opportunity for cognitive rehearsals,
if fully adult levels of sexual performance are to be attained.
The studies of Mason, Harlow and colleagues have shown that when
the male rhesus is deprived of these early experiences, normal
adult patterns of sexual behavior are totally disrupted, and it is
unlikely that the maturation and development of the olfactory mechanisms crucial for normal behavior would be excluded from the
influence of past experiences.

PATTERNS OF BEHAVIOR

In our neuroendocrine studies of rhesus monkey behavior, pairs
of jungle-bred animals of opposite sexes are observed in quiet
observation rooms from behind one-way vision mirrors for 60 min
periods. During this time all the behavioral interactions are
scored and recorded (Michael et al., 1966). Mating in this species
consists of a series of mounts by the male on the female usually
with intromission and pelvic thrusting. The mounting series (about
3-20 mounts) is terminated by a mount in which ejaculation occurs.
After ejaculation, there is a pause of some thirty minutes before

the start of another mounting series. There may be from zero to
as many as seven mounting series, each with an ejaculation, during
a 1 hr test period; one or two ejaculations per test would be
fairly usual. Between each mount, and also during the post-ejacula-
tory pause, animals usually groom each other intensively. To
control for individual differences and partner preferences, an
experimental animal is provided with more than one partner, and
contemporaneous changes in behavior with all partners are looked
for. Also, pairs are studied longitudinally and are tested day-by-
day often for many months, so that each pair can be used as its own
control. Pregnancies are prevented by ligating the Fallopian tubes.
When pairs are tested systematically on a day-by-day basis throughout the menstrual cycle, it is possible to relate the number of
mounts per test to the day of the female's cycle. Figure 1 shows
these data for 12 pairs and 12 cycles, and a picture of some confusion emerges when the group data are plotted. However, when
each pair is considered as a unit, and the twelve cycles are plotted
separately, it can be seen that different patterns characterize
the behavioral interactions of different pairs (Fig. 2). These
individual differences both in the actual levels of behavior and
in the types of patterns observed make for considerable difficulty
in handling these data statistically. Nevertheless, one of the
most consistent features of the interaction between a male and
female rhesus monkey is the decline in mounting activity and in
ejaculatory activity early in the luteal phase of the female's
cycle. This fact emerges from a population study of 32 pairs
observed during 75 menstrual cycles over a 5-year period (Fig.3).

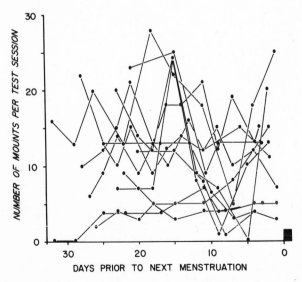

Fig. 1. Relation between mounting activity and stage of the
menstrual cycle in 12 pairs of rhesus monkeys: a picture of
some confusion is seen.

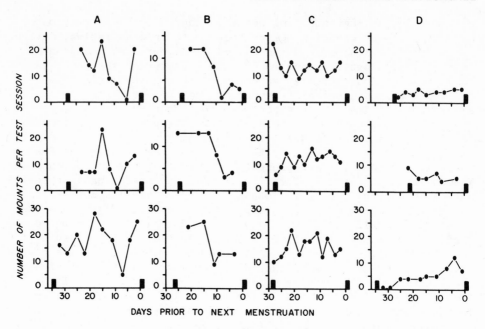

Fig. 2 The same data as in Fig. 1 plotted on an individual
basis: four distinct but different patterns of mounting
activity emerge. (Michael et al., 1972)

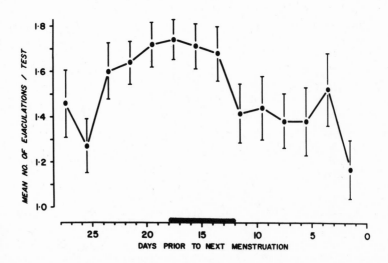

Fig. 3 Decline in ejaculatory activity of male rhesus monkeys
early in the luteal phase of the female cycle (32 pairs, 75
menstrual cycles, 2-day means). Vertical bars give S.E.'s of
means. Horizontal bar gives expected time of ovulation
(Michael et al., 1972)

HORMONAL FACTORS AND FEMALE ATTRACTIVENESS

Sexual behavior in female primates may be conveniently considered under two headings (i) female receptivity: the willingness to accept the male and be mated, and (ii) female attractiveness: the ability to stimulate the male's sexual interest. These variables are not entirely independent and should never be considered as such, but from the practical standpoint of behavioral testing the distinction is a useful one because receptivity can be operationally defined in terms of the female's behavior (numbers of refusals, invitations, etc.), and attractiveness in terms of the male's behavior (numbers of mounting attempts, ejaculations, etc.). However, males differ in the amount of stimulation they require, and a potent male needs fewer mounts to achieve ejaculation than a poorly motivated, less potent male. For this reason, we have to select, and work with, males that are responsive to changes in the females' stimulus properties when conducting experiments on female attractiveness. Nevertheless, difficulties arise because of incompatibilities between the animals being tested and this is part of the phenomenon of partner preferences that is such a striking feature of the sexual interactions of these primates. Attractiveness depends on multi-sensory stimulation of the male by the female, and it is probable that individual males differ in the amount of attention given a particular modality. Differences in the female ages, relative sizes, aggressivity, dominance, and also the frequency and form of her sexual displays enter into the situation. Many of these factors are mediated by visual pathways, but tactile stimulation, both in the form of grooming activity and in the nature of the afferent stimulation provided by the vagina, makes a significant contribution. Finally, there is the afferent stimulation that is mediated by olfactory pathways. To assess the role of the latter, we need to separate its contribution from a matrix of other forms of afferent stimulation. This is not an easy task, and considerable ingenuity must be exercised in the design of experiments to achieve it effectively: if the complexities of the situation are ignored, efforts are likely to be unsuccessful (Goldfoot et al., 1976).

ODORIFEROUS SUBSTANCES IN PRIMATE VAGINAL SECRETIONS

A gas chromatograph fitted with a single column packed with Carbowax-terephthalic acid (5% on Chromosorb W, AWDMCS) was used for separating volatile constituents of vaginal secretions collected from ovariectomized rhesus monkeys receiving 10µg/day estradiol benzoate. An effluent splitter fed both a flame ionization detector and an open port that was used for odor detection. Two people recorded their observations on the strip chart recorder along with the signal from the flame-ionization detector. Secretions were collected by lavage with 1ml distilled water from animals restrained

in a net. Secretions were extracted with 2ml of ethyl ether, and
the extract was concentrated by rapid distillation to about 50μl;
5μl of this was then injected into the gas chromatograph. To
obtain optimal detection and resolution of the various components,
the oven was programmed from 50 - 200°C at 5°/min with a carrier
flow rate of 30ml/min. Under these conditions no odors were
detected (apart from the solvent) until the oven temperature reached
90°. Between 90° and 150°, a series of vapors with various degrees
of pungency and "cheesiness" were noted.

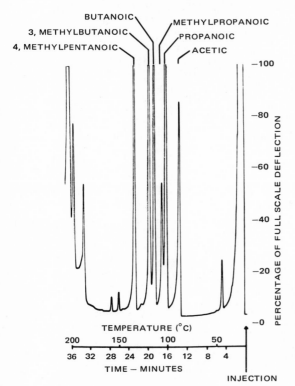

Fig. 4. Gas chromatogram of the
extract of a vaginal secretion
from an estrogen treated rhesus
monkey showing the six major
C_2 - C_6 aliphatic acids. The
small peak at 5 min is n-pentanol,
the concentration marker.

Above 150° the acrid odor of degra-
ded compounds prevented further
observations. A series of six
major peaks was detected by the
gas chromatograph within the
90° - 150° region. These substance
were trapped individually into
melting point tubes and subjected
to mass spectrometry (Curtis et al., 1971). Their mass spectra
and gas chromatographic retention times were shown to be identical
to authentic acetic, propanoic, methylpropanoic, butanoic, 3-methyl-
butanoic and 4-methylpentanoic acids (Fig. 4). Other odoriferous
compounds, separated initially by liquid phase chromatography
(Michael et al., 1971), and then methylated and identified by gas
chromatography and mass spectrometry, were 3-phenylpropanoic (hydro-
cinnamic) acid which has a sweet odor and 3-(4-hydroxyphenyl)propan-
oic (hydrocoumaric) acid which has an odor reminiscent of raspber-
ries. To measure the concentrations of the different volatile
fatty acids in vaginal secretions, the extraction procedure was
modified. First, the secretions were made acid to optimize the
extraction of organic acids, and second, an internal concentration
marker, n-pentanol, was added to the extracting ether. Concentra-

tions of the acids in the sample could now be calculated by relating peak areas to the area of the concentration marker peak and multiplying by calibration constants determined for standard solutions of authentic acids processed in parallel with the samples. Vaginal secretions collected by lavage from eight different primates, all intact and receiving no hormone treatment, have been analyzed by this technique (Michael et al., 1975a). All the species produced secretions containing volatile fatty acids though the relative proportions varied somewhat between species.

Secretions collected by lavage from women in a preliminary study also contained volatile fatty acids (Michael et al., 1972). In a later study involving 47 women vaginal secretions were collected by a new technique. Specially cleaned, small tampons were prepared, and these were placed after use in bottles containing methanol to prevent bacterial action (Michael et al., 1974, 1975b). Analysis of these samples showed that acetic acid predominated in all subjects but the 30% of the women, whom we termed "producers", provided samples containing more than 10μg of acids in addition to acetic: however, only 8.8% of the samples from women contained more than 10μg of the C_3 to C_6 acids.

MECHANISM OF PRODUCTION OF VOLATILE FATTY ACIDS

Bacterial Origin

When vaginal secretions from estrogenized monkeys were incubated at 37° the volatile fatty acid content increased several fold. This effect was prevented by autoclaving or by adding penicillin. Inoculation of sterilized secretions, or any proteinaceous medium, with fresh secretions also caused the acid content to increase during incubation. These observations, together with the characteristic sigmoidal shape of the incubation curve, strongly suggested that the acids were products of the vaginal microflora (Bonsall and Michael, 1971; Michael et al., 1972). Individual bacterial types isolated by subculture from vaginal secretions did not synthesize acids in the proportions found in secretions (Michael et al., 1976). It is tentatively concluded that a natural mixture of micro-organisms is responsible for the normal production of volatile fatty acids. Some bacteria use amino acids as a source of metabolic energy in an anaerobic system known as the Stickland reaction after its discoverer (Stickland, 1934; Nisman, 1954). It involves the coupling of reductive deamination to oxidative deamination and decarboxylation and produces volatile fatty acids. This reaction and other more familiar anaerobic catabolic pathways have been suggested as a biosynthetic route for the production of volatile fatty acids from proteins, carbohydrates and lipids in secretions (Michael et al., 1975b). The hypothetical route predicts that the proportions of the different acids produced will depend on the nature of the nutrients present. This seems to be the case.

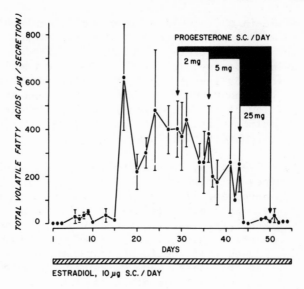

Fig. 5. Stimulation by estradiol and suppression by
progesterone of the acid content of vaginal secretions.
Each point is the mean of samples from 4 ovariectomized
rhesus monkeys. Vertical bars give standard errors of means.

When ejaculate was present as the major substrate in the vaginae
of ovariectomized females not receiving estrogen, the pattern of
acids found in lavages differed markedly from that in estrogen

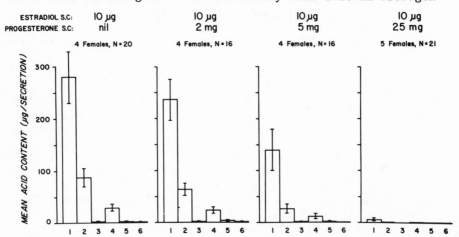

Fig. 6. Graded inhibition by progesterone of the acid content
of estrogen-stimulated secretions. Secretions were collected
after 17 days of estrogen and one day of progesterone treatment.
Acids: 1 = acetic, 2 = propanoic, 3 = methylpropanoic, 4 = butanoic, 5 = methylbutanoic, 6 = methylpentanoic. N = number of
samples. Vertical bars give standard errors.

stimulated secretions when mucus and cell debris were major substrates (Michael et al., 1976).

Effect of Estradiol

Vaginal lavages collected from ovariectomized rhesus monkeys
that were not receiving hormone treatments contained very little
solid matter and had barely detected levels of acids (Michael
et al., 1971, 1976).   Secretions collected by lavage from ovariectomized monkeys given 10 g estradiol benzoate per day contained
mucus, solid debris and elevated levels of acids. The time-course
of this effect is shown in Fig 5.   In addition to the dramatic increase on the 16th day, there was a measurable and significant increase on the 6th day.

Effect of Progesterone

A graded dose-response effect was observed when progesterone
was administered to ovariectomized rhesus monkeys receiving 10 g
estradiol benzoate per day (Fig 6). At 2mg per day, progesterone
was without discernible effect; at 5mg per day, the content of
acids was reduced more or less uniformly by 50%; and at 25mg per
day, levels were reduced virtually to those of ovariectomized,
untreated females. The time-course of the effect of progesterone
on the vaginal acid levels is shown in Fig 5. At 25mg progesterone,
levels dropped almost to zero within 28 hr of treatment.

Variations during the Menstrual Cycle

Vaginal secretions (N=390) and plasma samples (N=492) were
collected from five female rhesus monkeys throughout 31 menstrual
cycles. Collections were made on alternate days except near mid-
cycle when daily blood samples were drawn. Females were given
behavior tests with males throughout the period of study (results
reported elsewhere). Plasma steroids (estradiol, progesterone,
testosterone and dihydrotestosterone) were assayed in duplicate
by a method involving chromatography on Sephadex LH20 (Bonsall and
Michael, unpublished). In Figure 7, three-day means for estradiol
and progesterone are shown with the cycles aligned on the day of
the estradiol peak (Day 0). The acid content of the vaginal secretions was within the range for ovariectomized, estrogen-treated
females but varied considerably both between females and from day-
to-day in each female. There were two peaks in total acids during
the cycle, one in the mid-follicular phase and a second, larger
one, in the mid-luteal phase. The same pattern was shown by the
straight chain acids, but branched chain acids were relatively
higher at the beginning of the cycle. This confirmed our original
findings (Bonsall and Michael, 1971).

In studies on two other primates, changes in the volatile
fatty acids were more clear-cut. Secretions collected during

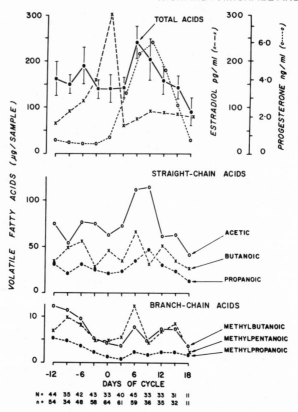

Fig. 7. Simultaneous determinations of volatile fatty acids
in vaginal secretions of rhesus monkeys, and estradiol
and progesterone in plasma samples (5 females, 31 cycles).
Data are aligned on the day of the estradiol peak (Day 0)
and calculated as 3-day means. N = number of vaginal secretions. n = number of plasma samples. Vertical bars
(total acids) give standard errors of means.

the menstrual cycles of baboons (Papio anubis), like those from
rhesus monkeys, showed considerable day-to-day variability. However,
in the periovulatory phase (5 days immediately before sexual skin
deflation) levels were significantly higher than during the first
seven (p<0.05) and last seven (p<0.001) days of the cycle (Michael
et al., 1972). Combined data from 47 women during a total of 86
menstrual cycles also showed higher levels of acids in the second
quarter of the cycle and a progressive decline during the second
half. This variation was statistically significant (Michael et al.,
1975b) and was most pronounced in the more odoriferous C_3 - C_6 acids
(Fig 8). Women "producers" were responsible for most of the variation seen in the combined data, and the fifteen subjects who were
using oral contraceptive preparations during the study had lower
overall levels of acids, and showed no significant changes during
the menstrual cycle (Michael et al., 1974).

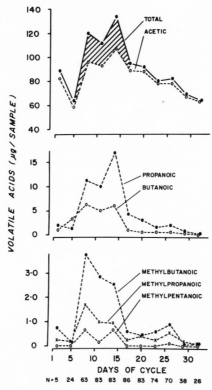

Fig. 8.   Changes in the content of both total and individual
fatty acids in vaginal secretions in successive 3-day periods
of the cycles of 47 women.   Hatched area (top graph) shows
the contribution of acids other than acetic to the increase
near mid-cycle.   N = number of samples (Michael et al., 1975b).

DYNAMICS OF PRODUCTION AND VOLATILIZATION:
THEORETICAL CONSIDERATIONS

     Changes in the volatile fatty acid content of vaginal secre-
tions during the menstrual cycle present an apparent paradox
because the plasma estradiol peak appears to coincide with a
trough between two peaks in vaginal acid content.   This is in
direct contrast with the effect of exogenous hormone treatments
on vaginal acids.   However, a consideration of the dynamics of the
system suggests an explanation.   The concentration of acids depends
on (i) the availability of substrate and (ii) the time for which
the substrate is exposed to the anaerobic action of the bacteria.
In untreated, ovariectomized females, the lack of substrate prevents acid production, and the low level of acids within the
vagina is reflected in the amount externalized and available for
detection by the male.   In estrogen-treated females, however,
substrate is in excess and continued incubation in vitro will
further increase acid levels.   Thus, the period of exposure to

bacterial action becomes the limiting factor. The increased secretion encountered at mid-cycle passes more rapidly through the
vagina and prevents the accumulation of bacterial products, but
at the same time, presents a larger amount externally to the male.
Thus, in the presence of excess substrate, a direct positive relationship between the content of acids in a lavage and the amount
available externally cannot be assumed; indeed an opposite relationship may hold. Other factors are also important in determining
the volatilization of the acids from the perineal region. These
include pH, moisture and surface temperature. The posture of the
female during the male's genital inspection also influences the
vapor concentration for detection by the male. Our preliminary
results indicate that all these factors are optimized by estrogenization.

OLFACTORY CUE-SEEKING BEHAVIOR: GENITAL SNIFFING

In anatomical terms, there is a relative reduction in the mass
of neural structures related to olfactory processes in Old World
primates compared with other primates and lower mammals. Old
World primates also lack large, specialized, odor-producing skin
glands. Nevertheless, the active seeking by males of olfactory
or gustatory cues emanating from the female appears to be as
common in these primates as in other mammalian orders. There are
numerous reports both from the field and from the laboratory of
genital sniffing, and these have been reviewed elsewhere (Michael
et al., 1976). It is quite clear, then, that development of the
neocortex and the dominance of the visual sense in higher primates
is not associated with greatly diminished frequency of genital
sniffing, despite the overall reduction in the role of olfaction.
In male rhesus monkeys, genital sniffing is commonly observed both
in our own and in other laboratories. When the female is receptive, the sniffing generally occurs during sexual presentation.
However, if the female is unreceptive and is not presenting sexually,
the male may push or pull the female into a position that enables
him to sniff the perineum: it seems unlikely that this active
behavior is either totally redundant or vestigial. Indeed, both
chimpanzees (van Lawick-Goodall, 1969) and gorillas (Hess, 1973)
conduct rather similar olfactory, gustatory and tactile inspections
of the female's genitalia.

OLFACTORY SIGNALS AND THE SEXUAL BEHAVIOR OF MALE RHESUS MONKEYS

Our studies on the role of chemical communication in primate
sexual behavior have involved two experimental paradigms. The first
was intended to establish if males used olfactory cues to determine
a female's sexual status. The second was intended to establish a
behavioral bioassay to help identify the substances involved.

Experiments with Reversible Anosmia

When estrogens were given to ovariectomized females in physiological doses both receptivity and attractiveness were increased. The effect on attractiveness was maximized by intra-vaginal rather than subcutaneous administration and, under these conditions, the effect on receptivity was minimized (Michael & Saay-man, 1968). Applying estrogens to the sexual skin increased its redness but produced no changes in the males' behavior (Michael & Keverne, 1968). Local estrogenization of the vagina was, there-fore, used to produce changes in attractiveness independently of changes in the female's behavior.

The behavior tests were performed in large, twin compartment cages equipped with movable partitions. Pressing a lever on one side of the partition a predetermined number of times caused a servomotor to raise the partition and permitted access to the other side. Males were trained to press the lever at least 250 times, first for food rewards and, when they were familiar with the situation, for female rhesus monkeys. Having obtained access, the male was tested with the female for 60 min, and an experienced observer scored all aspects of the pair's behavior. If the male failed to press the lever to criterion within 30 minutes, the pair was tested later the same day in a non-operant cage situation: this preserved the continuity of the behavioral data. Each male was paired with three ovariectomized females and one of these (the control) received 10µg estradiol benzoate subcutaneously throughout. Most males discriminate between estrogenized and non-estrogenized females in performing the operant task, but for these experiments we used males that would not perform at all for ovariectomized, untreated partners. After several pretreatment tests, intravaginal estrogen administration was started and males were made anosmic (Michael & Keverne, 1968) by inserting plugs of gauze impregnated with bismuth-iodoform-paraffin paste above the superior turbinate bones (the nerve supply to the organ of Jacobson was also cut). This left a free nasal airway. Males continued to behave towards the experimental females as if they had not been estrogenized during the period of nasal blocking (Fig. 9, upper and lower left). Males only started performing the operant task for access to the now attractive females after the nose plugs were removed and their olfactory acuity had returned. Males generally worked for access to, and mounted, control females throughout the experiment indicating that the nose blocking procedure did not greatly impair their potency.

A second experiment, similar in general design to that de-scribed, was concerned with the inhibitory effects of progesterone on female attractiveness. Here, females were initially estrogenized and males worked for access to them, copulating in every test.

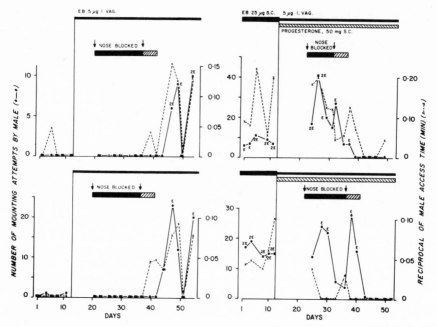

Fig. 9. <u>Upper and lower left</u>: When anosmic, males were
not stimulated by the intravaginal estrogen treatment of
females. <u>Upper and lower right</u>: When anosmic, the same
two males were not inhibited by progesterone treatment of
estrogenized females. Thus, both the stimulatory and
inhibitory effects of hormones were not manifest until the
males' olfactory acuity had returned. The recovery period
is indicated by the hatched area after the removal of
nasal plugs. E = ejaculation.

Females were then treated with large doses of progesterone for the
remainder of the experiment. These doses (50mg per day) would
normally reduce the male's interest and abolish the sexual activity
of the pair (Michael et al., 1976). However, when males were anosmic, they continued to press for access to, and mount, females as
though they were still in the initial estrogenized condition. This
behavior continued until after the nasal plugs had been removed
and their olfactory acuity returned (Fig. 9, upper and lower right).

These two experiments using reversible anosmia demonstrate
quite clearly an effect of olfactory signals on the sexual behavior
of rhesus males. They further demonstrate that olfactory signals
are not necessary for the display of normal sexual behavior which
depends heavily upon the male's prior experience with females.
Rather, we view olfactory signals as being used to detect changes
in the female's sexual status. Thus, in the behavioral bioassay
described below, we used animals that had been allowed to become
thoroughly familiar with each other.

Measurement of Sex-Attractant Properties - The Behavior Bioassay

   The dyadic testing situation in these laboratories has been
described briefly above, and it forms a basis for the bioassay
of sex-attractants. Its most important features are: (1) testing
cages are small enough (1.2 cubic metres) to prevent the female
from easily avoiding the male when he is motivated to make contact;
(2) cages are acoustically isolated and observers are behind one-
way vision screens; (3) males are of known potency and responsive
to changes in the female; (4) females are unreceptive and unattractive after ovariectomy; (5) animals are thoroughly familiar with
each other and aggressive episodes are infrequent and not severe.
The testing sequence consists of: (i) a rather prolonged period
of baseline tests, (ii) 5-10 pretreatment tests in which the female

TABLE 1. SUMMARY OF RESULTS OBTAINED IN THE BIOASSAY OF SEX
          ATTRACTANTS IN VAGINAL SECRETIONS

Test material	Effective	Reference
(1) secretions from:		
rhesus, ovex + EB	Yes	Michael & Keverne, 1970
rhesus, ovex + EB + P	No	Michael et al., 1976
rhesus, intact	Yes	Michael et al., 1976
baboon, intact	Yes	Michael et al., 1972
human, intact	Yes	Michael, 1972
(2) extracts of secretions from estrogenized donors:		
ether extracts	Yes	Keverne & Michael, 1971
acidic re-extract	Yes	Curtis et al., 1971
(3) fractions from ion exchange chromatography:		
void volume	No	Curtis et al., 1971
fatty acids	Yes	Curtis et al., 1971
phenylpropanoic	No	Curtis et al., 1971
hydroxyphenylpropanoic	No	Curtis et al., 1971
(4) gas chromatographic trapping:		
$C_2 - C_5$ acid fraction	Yes	Michael et al., 1971
(5) synthetic mixture:		
$C_2 - C_5$ acids	Yes	Curtis et al., 1971

EB = Estradiol benzoate, P = Progesterone.

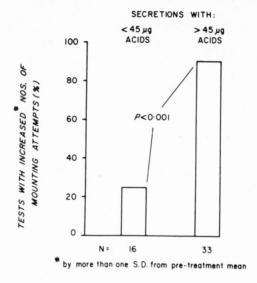

Fig. 10. Increased effectiveness of vaginal secretions in stimulating mounting attempts when the volatile fatty acid content exceeded 45µg. N = number of tests.

is netted and control material is applied to her sexual skin area, (iii) 5–10 treatment tests when an extract or other test substance is similarly applied, and (iv) a withdrawal period when behavior returns to baseline. Problems of shifting sensitivity and non-linear dose-response relationships are inherent in bioassays; these dictate that positive results carry more weight than negative ones.

Results obtained in assays of a variety of test substances are given in Table 1. These results demonstrate quite conclusively that the volatile fatty acids are responsible, at least in part, for the sex-attractant activity of vaginal secretions. Additional evidence for this was obtained in bioassays of secretions from intact females that were also assayed for their volatile fatty acid content. Although only two cycles have been studied to date, neither the sex-attractant activity of the secretions nor their acid content varied consistently with the day of the menstrual cycle of the two donors. However, the proportion of tests showing a positive response was significantly higher ($P<0.001$) when the secretions applied contained more than 45µg of acids than when they contained less than 45µg acids (Fig. 10).

OLFACTORY EXPERIMENTS WITH HUMAN SUBJECTS

We wished to determine if human subjects could detect volatile fatty acids at the concentrations present in natural secretions, and whether they could reliably discriminate between vaginal secretions from rhesus monkeys under different hormonal conditions, and also from human subjects collected at different times
during the menstrual cycle. Ten male and ten female volunteers
were involved in this study. They consisted of university students
and laboratory personnel. The technique was modeled on the forced-
choice method described by Henkin et al. (1962), in which the
subject is presented with a randomized sequence of three vials and
asked to identify (or guess if necessary) which vial smelled differently from the other two. This method was used both to determine detection thresholds, and to test the ability to discriminate
between different samples. For detection thresholds, a set of
tenfold serial dilutions of the test material was used, and this
was presented in both ascending and descending order. Results are
based on a total of thirty presentations of each concentration or
material, six on each of five days. Purely random guessing gives
an expected frequency of ten "correct" identifications in thirty
attempts, 16 correct is significant at P<0.02 and 19 at P<0.001.
A mixture of volatile fatty acids was made up in water to resemble
the proportions found in the secretions of estrogen-treated rhesus
monkeys: acetic 300, methylpropanoic 15, butanoic 100, methylbutanoic 30, methylpentanoic 30 μl per 100ml water. A one thousand
fold dilution of this mixture was detected significantly by the 20
subjects, a ten thousand fold dilution was not. The overall results for the panel of testers show that the detection threshold

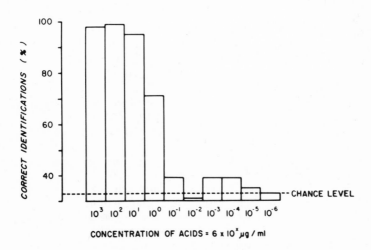

Fig. 11. The detection threshold by a panel of human
testers for a mixture of volatile fatty acids lies between
0.6 and 6.0μg/ml.

lies between 0.6 and 6µg/ml (Fig.11). At a total acid concentration of 6µg/ml only methylbutanoic acid was above the threshold
reported by Amoore et al. (1968). It seems, therefore, that
either the acids in the mixture have an additive effect or methylbutanoic acid is the most important in determining the detection
threshold of the mixture. Subjects were notably successful at
discriminating between different concentrations, and some reliably
identified a solution containing 69.3µg/ml from one containing
60.0µg/ml. On the basis of these results we anticipated that
volunteers would be able to distinguish an aqueous solution of the
ether extractable fraction of estrogen-stimulated secretions from
water, since these normally contain more than 100µg total acids.

Secretions were collected from 4 ovariectomized rhesus monkeys
both before treatment and after three weeks of daily injections
of 10µg estradiol benzoate. The ether extractable fraction of
these secretions was redissolved in water and 8 subjects were
tested for their ability to discriminate between the pretreatment
and treatment secretions, and also between each of the secretions
and the extract of a water blank. With a chance expectation of
33%, subjects discriminated between the estrogen-stimulated secretion and the water blank on 72% of the presentations, between the
non-estrogenized secretion and the water blank on 56.2% of the
presentations, and between the two secretions on 50.8% of the
presentations. All these results were statistically significant
and they demonstrate that the human sense of smell is sufficiently
sensitive to detect the hormonal status of the rhesus monkey.

A panel of five men and three women was tested for their
ability to discriminate between vaginal secretions collected from
women at different stages of the menstrual cycle. Samples were
collected by the tampon method during the first, middle and last
thirds of the menstrual cycles of two women. Samples from each
third were processed by the method of Michael et al. (1974), and
were combined to smooth the day-to-day variation in vaginal acid
production. Similarly processed unexposed tampon samples were
used as blanks. Each test consisted of discriminating between
three vials, two of which contained the same substance. All
possible combinations of the four samples (first third, middle
third, last third, blanks) were tested thirty times for each subject in five testing sessions. Figure 12 shows the results obtained
for secretions from a woman "producer" [mean acid content: acetic
183.1, propanoic 82.9, methylpropanoic 6.1, butanoic 34.1, methylbutanoic 14.1µg] and from a woman "non-producer" [mean acid content:
acetic 80.4, propanoic 1.9, methylpropanoic 0.4, butanoic 0.3,
methylbutanoic 0.2µg]. All members of the panel could reliably
discriminate all the secretion samples from the blanks, and most
could discriminate between the three samples from the producer.
None could reliably discriminate between the samples from the

non-producer, although the overall results were consistently
above chance level. These results together with those of
Doty et al. (1975) show that humans possess a receiving mechanism
sufficiently sensitive to detect the changes in the chemical
composition of human vaginal secretions occurring during the
menstrual cycle. That a proportion of normal women have an
emitting mechanism has already been established (Michael et al.,
1974, 1975b) but we have at present no data for the human species
to indicate that the information, once received, is used and
acted upon.

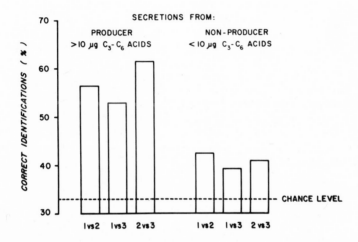

Fig. 12. Discrimination by a panel of human testers between
vaginal secretion samples collected during the first (1),
middle (2) and last (3) thirds of the menstrual cycle from
a woman producer and a non-producer. Those from a producer
were reliably discriminated.

Acknowledgements

This work was supported by U.S.P.H. Service Grant No. MH 19506
from the National Institute of Mental Health. The Grant Foundation
provided essential equipment. The Georgia Department of Human
Resources provided general research support. The help of these
agencies is gratefully acknowledged. Miss Susan Rodman conducted
some of the olfactory tests, and these results formed part of her
Honors thesis in Psychology. We wish to thank all the volunteers
who co-operated in these studies.

REFERENCES

AMOORE, J.E., VENSTROM, D., and DAVIS, A.R. 1968. Measurement of specific anosmia. Perceptual and Motor Skills, 26:143-164.

BONSALL, R.W. and MICHAEL, R.P. 1971. Volatile constituents of primate vaginal secretions J. Reprod. Fertil. 27:478-479.

CURTIS, R.F., BALLANTINE, J.A., KEVERNE, E.B., BONSALL, R.W., and MICHAEL, R.P. 1971. Identification of primate sexual pheromones and the properties of synthetic attractants Nature 232:396-398.

DOTY, R.L., FORD, M., PRETI, G. and HUGGINS, G.R. 1975. Changes in the intensity and pleasantness of human vaginal odors during the menstrual cycle. Science, 190:1316-1318.

GOLDFOOT, D.A., KRAVETZ, M.A., GOY, R.W. and FREEMAN, S.K. 1976. Lack of effect of vaginal lavages and aliphatic acids on ejaculatory responses in rhesus monkeys: behavioral and chemical analyses. Hormones and Behavior 7:1-27.

HENKIN, R.I., GILL, J.R. and BARTTER, F.C. 1962. Effect of adrenal insufficiency and of corticosteroids on smell threshold. Clin. Res. 10:400-411.

HESS, J.P. 1973. Some observations on the sexual behavior of captive lowland gorillas, pp507-581, in R.P. Michael and J.H. Crook (eds.), Comparative Ecology and Behaviour of Primates. Academic Press, London and New York.

KEVERNE, E.B. and MICHAEL, R.P. 1971. Sex attractant properties of ether extracts of vaginal secretions from rhesus monkeys J. Endocr. 51:313-322.

LAWICK-GOODALL, J. VAN 1969. Some aspects of reproductive behaviour in a group of wild chimpanzees, Pan troglodytes Schweinfurthi, at the Gombe Stream Chimpanzee Reserve.J. Reprod.Fert. Sup. 6:353-5.

MICHAEL, R.P. 1972. Determinants of primate reproductive behaviour Acta endocr. Copenh. Suppl. 166:322-361.

MICHAEL, R.P. and KEVERNE, E.B. 1968. Pheromones in the communication of sexual status in primates. Nature 218:746-749.

MICHAEL, R.P. and KEVERNE, E.B. 1970. Primate sex pheromones of vaginal origin. Nature 225:84-85.

MICHAEL, R.P. and SAAYMAN, G.S. 1968. Differential effects on behaviour of the subcutaneous and intravaginal administration of oestrogen in the rhesus monkey (Macaca mulatta). J. Endocr.41:231-246

MICHAEL, R.P., BONSALL, R.W. and WARNER, P. 1974. Human vaginal secretions: volatile fatty acid content. Science 186:1217-1219.

MICHAEL, R.P., BONSALL, R.W. and KUTNER, M. 1975b. Volatile fatty acids, "copulins", in human vaginal secretions. Psychoneuroendocrinology 1:153-163.

MICHAEL, R.P., BONSALL, R.W. and WARNER, P. 1975a. Primate sexual pheromones, pp417-424, in D.A. Denton and J.P. Coghlan (eds.), Proceedings of the Fifth International Symposium on Olfaction and Taste. Academic Press, New York.

MICHAEL, R.P., BONSALL, R.W. and ZUMPE, D. 1976. Evidence for chemical communication in primates. Vitamins and Hormones. in press

, HERBERT, J. and WELEGALLA, J. 1966. Ovarian
grooming behaviour in the rhesus monkey (*Macaca mulat-*
aboratory conditions. *J. Endocr.* 36:263-279.

.P., KEVERNE, E.B. and BONSALL, R.W. 1971. Pheromones:
of male sex attractants from a female primate *Science*
66.

R.P., ZUMPE, D., KEVERNE, E.B. and BONSALL, R.W. 1972.
crine factors in the control of primate behavior.
ogr. *Hormone Res.* 28:665-706.

B. 1954. The Stickland reaction. *Bacteriol. Rev.* 18:16-42.
AND, L.H. 1934. The chemical reactions by which *Cl. sporo-*
nes obtains its energy. *Biochem. J.* 28:1746-1759.

A REVIEW OF RECENT PSYCHOPHYSICAL STUDIES EXAMINING THE POSSIBILITY

OF CHEMICAL COMMUNICATION OF SEX AND REPRODUCTIVE STATE IN HUMANS

Richard L. Doty

Monell Chemical Senses Center* and Department of
Otorhinolaryngology and Human Communication, University
of Pennsylvania Medical School
*3500 Market Street
Philadelphia, Pennsylvania 19104

Numerous anthropological reports suggest a close relationship
of odors to sexual processes in some human groups. For example,
vegetable juices with smells reminiscent of seminal fluid or vaginal
secretions have been used as aphrodisiacs in several primitive cultures (Haire, 1940). In a southwest Pacific society studied by
Davenport (1965), a form of love magic is based upon the similarity
of vaginal odors to those of fish. To attract fish, men of this
community use a red ground cherry attached to the leader of a
trolling line. After a fish has been caught in this manner, the
cherry is believed to have the power to attract women. In some
societies, animal secretions from species with marked physical
stamina and long, violent heat periods are smeared ritually on
sacred objects, as well as on the body. Such practices may form
one basis for the widespread modern usage of animal products such
as musk, civet and ambergris in perfumery.

Although many such reports exist (cf. Bloch, 1934; Brody,
1975; Comfort, 1971; Doty, 1972; Ellis, 1934; Hediger, 1968;
Moncrieff, 1944), few scientific data are available concerning the
types of biological information that can be conveyed between
humans on the basis of body odors, as well as the influences of
such odors on human sexual behaviors. In some cultures, the ordin-
ary salutation between friends is by the smelling of one another,
suggesting at least one context in which natural or artificial
odors may be of considerable social import (Ellis, 1934). Unfortunately, as Sagarin (1964, p. 976) suggests, "Researchers on
olfaction are inclined to exaggerate the importance of odors in
the sexual and amorous life of man, but offer little scientific

evidence to validate their theories."

It is the purpose of the present chapter to review recent experiments which have examined the ability of adult humans to detect or communicate basic biological information by means of conspecific odors. The influences or correlates of gender, endocrine state, and pregnancy upon olfactory sensitivity have been reviewed in detail elsewhere (Doty, 1976b) and will not be examined in this chapter.

Perception of Male and Female Odors

There is little doubt that human beings can detect differences between the sexes on the basis of mutually exclusive odors, e.g., the odors of vaginal secretions, smegma, and seminal fluid. However, it is not clear whether such discriminations can be made on the basis of odors coming from sources more or less similar for each sex (e.g., breath, urine, sweat, sebaceous sebum, etc.). A few recent studies present data pertinent to this topic, although the results are not unequivocal.

To my knowledge, only one study has examined whether humans can determine gender on the basis of conspecific urinary odors.* Beauchamp and Desor (1976) successfully trained three men and three women to distinguish between the odors of urine samples of four men and four women by providing feedback as to the sex of the urine donors. However, when new samples of either male or female urine odors were presented, the subjects were unable to generalize their responses, although they were aware of the uniqueness of the new samples' odors. These findings suggested that humans can be trained to discriminate between urine samples of individual donors, and to assign such samples either a male or female referent, but that a clear-cut ability on their part to detect sex differences, per se, is unlikely. Although the urinary cues responsible for the individual identifications are not known, dietary factors may have played a role and conceivably could have masked cues delineating the sex of the donors, if such cues were indeed present.

Several psychophysical studies of axillary odor have recently appeared. Russell (1976) had 16 male and 13 female college freshmen wear T shirts for 24 hours without bathing or using deodorants. The armpit sections of the shirts were then presented to these

---

* In many nonhuman mammals, urine is perhaps the most ubiquitous medium for the chemical communication of reproductive state and gender (Beauchamp, Doty, Moulton and Mugford, 1976; Doty and Dunbar, 1974).

individuals in a triangle odor test in which (a) the subject's own T shirt, (b) a strange male's T shirt, and (c) a strange female's T shirt served as stimuli. Each subject was first asked to identify his or her own odor, and then to report which of the two remaining odors came from a male. Thirteen of the 16 males and 9 of the 13 females correctly performed both of these tests, suggesting to Russell that "at least the rudimentary communications of sexual discrimination and individual identification can be made on the basis of olfactory cues." He further stated, "During informal questioning after the test the male odours were usually characterized as musky and the female odours as sweet."

While this study provides interesting information, it is likely that the purported sex discrimination was based on the tendency of observers to associate stronger odors with male donors and weaker odors with female donors. Circumstantial support for this hypothesis comes from several quarters, including studies indicating that women have smaller apocrine sweat glands than men (e.g., Hurley and Shelley, 1960) and that apocrine gland size directly correlates with body odor intensity (e.g., Shehadeh and Kligman, 1963). Recent psychophysical experiments by McBurney, Levine and Cavanaugh (1976) and Doty, Kligman, Leyden and Orndorff (1976) also provide evidence for this notion.

McBurney et al. (1976) had 11 male graduate students wear T shirts for two days without showering. In a subsequent test session, these students, along with five male and nine female undergraduates, evaluated the relative unpleasantness of the shirts' odors by a magnitude estimation procedure. In addition, the undergraduates evaluated, from their odor impressions, the likely personality of the unknown donors on seven-point bipolar adjective scales, including masculine-feminine, sociable-unsociable, dirty-clean, etc. The raters were in high agreement with respect to the relative unpleasantness of the various samples, as indicated by Kendall's coefficient of concordance. The correlations between the unpleasantness ratings and a number of the adjective scales denoting seemingly masculine traits proved positive and statistically significant (e.g., masculinity, 0.89; athletic, 0.79; strength, 0.83). Positive correlations were also found between the unpleasantness of the odors and a number of seemingly undesirable traits (e.g., unsociable, 0.96; dirty, 0.92; unintelligent, 0.90; nervous, 0.87; unattractive to the opposite sex, 0.92), although it is not clear to what extent these semantic ratings were intercorrelated with one another. These authors, like Russell (1976), report that some of the observers were able to recognize their own odors from the stimulus array, finding them generally less unpleasant than the others.

Doty, Kligman, Leyden and Orndorff (1976) found a strong and inverse relationship between the pleasantness of axillary odors and

their intensity. The stimuli in this series of experiments were
axillary odors collected on gauze pads that had been taped in the
armpits of men and women for approximately 18 hours. In their
first experiment, 10 male and 10 female observers provided magnitude estimates of both the intensity and pleasantness of the
sweat stimuli taken from 5 men and 5 women. The stimuli were
presented to the observers in sniff bottles (cf. Doty, 1975). The
observers were not informed as to the nature of the odorous
materials. Analyses of variance indicated that the sweat odors
of the donors, as perceived by the observers, differed significantly in both intensity and pleasantness. On the average, the
odors from the females were perceived as less intense and less
unpleasant than those from the males. The correlation between the
intensity of the odors and their pleasantness was −0.96 (Spearman
r). High agreement between the relative ratings of the male and
female observers was also observed (Spearman r = 0.98). The
females assigned larger intensity and unpleasantness magnitude
estimates than did the males, although it is not clear if this
reflected a perceptual phenomenon or a preference for the use of
larger numbers as moduli.

In three subsequent similar experiments, Doty, Kligman, Leyden
and Orndorff told the observers the exact nature of the stimuli.
In the first of these experiments, a set of nine male odors and a
blank control were presented to 10 male and 10 female observers
in separate sniff bottles following these instructions:

> You will be presented with a series of sniff
> bottles containing human sweat odor. We wish
> you to tell us which sex each of the odor samples
> comes from. The set of odors may include samples
> from both men and women, or from only men or from
> only women. Thus, some may be from males, some
> from females, or, alternatively, all may be from
> females or all from males. Therefore, don't allow
> yourself to assume that some predetermined number
> of one or the other sex is represented. If you have
> any questions, please feel free to ask them.

Following this task, the observers gave magnitude estimates of the
relative pleasantness and intensity of the odor samples. In the
second experiment, a set of nine female odors and a blank control
were presented to 10 male and 10 female observers with identical
instructions. In the third, five male and five female odors
were presented.

The results of these three experiments are presented in
Tables 1 - 3. It is apparent from Tables 1 and 2 that, even when
all male or all female stimuli were presented, most of the observers assigned both male and female ratings. A tendency to

assign the stronger odors a male referent and the weaker odors a female referent clearly was present. Thus, Spearman correlations between the median intensity magnitude estimates and the percent of observers assigning the various samples a male referent were, for the three respective experiments, 0.58, 0.90 and 0.98. The blank stimuli were nearly always assigned a female referent (Tables 1 - 3). The data of the third experiment indicate the observers correctly assigned the male referent to the male odors 54% of the time, and the female referent to the female odors 68% of the time. It would appear from these data that the observers were unable to clearly distinguish between the male and female odors and that in their attempts to do so, they relied strongly on the intensity of the stimuli.

It should be noted that the presence of axillary hair greatly increases the substrate for bacterial growth and the surface area for the diffusion of odor. Since many women in our culture shave their axillary region, the generally weaker female odors may reflect this fact to some extent. Figure 1, from Hurley and Shelley (1960, p. 90), shows the marked influence of shaving the armpit area on perceived axillary odor.

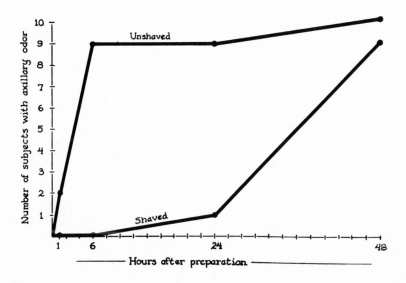

Figure 1. Influence of the shaving of the axillary region upon the presence of noticeable odor. Data based upon the shaven and unshaven axilla of ten healthy adult males. Two observers independently established the ratings in an in vivo test. From Hurley and Shelley (1960, p. 90). Courtesy of Charles C. Thomas, Publisher, Springfield, Illinois.

TABLE 1
SEX ASSIGNMENTS AND MAGNITUDE ESTIMATE RATINGS OF
THE INTENSITY AND PLEASANTNESS OF HUMAN AXILLARY ODOR

SAMPLE #:	ALL MALE ODORS									BLANK
	1	2	3	4	5	6	7	8	9	
% OF MALE OBSERVERS ASSIGNING SAMPLE TO MALE CATEGORY	0	30	50	50	0	50	20	60	90	10
% OF FEMALE OBSERVERS ASSIGNING SAMPLE TO MALE CATEGORY	60	70	30	80	0	20	50	80	50	20
MEAN PERCENT OF ASSIGNMENTS TO MALE CATEGORY	30	50	40	65	0	35	35	70	70	15
MEDIAN INTENSITY MAGNITUDE ESTIMATE	5.5	2.0	4.0	5.5	1.0	1.0	1.5	4.0	8.75	3.25
MEDIAN PLEASANTNESS MAGNITUDE ESTIMATE	-2.0	-.75	-3.0	-2.0	0	-.5	0	-1.75	-5.5	-1.5

$r = .58$

$r = -.89$

TABLE 2

SEX ASSIGNMENTS AND MAGNITUDE ESTIMATE RATINGS OF
THE INTENSITY AND PLEASANTNESS OF HUMAN AXILLARY ODOR

SAMPLE #:	1	2	ALL FEMALE ODORS							BLANK
			3	4	5	6	7	8	9	
% OF MALE OBSERVERS ASSIGNING SAMPLE TO MALE CATEGORY	70	0	80	80	70	50	40	0	0	0
% OF FEMALE OBSERVERS ASSIGNING SAMPLE TO MALE CATEGORY	60	20	90	50	50	30	0	10	50	0
MEAN PERCENT OF ASSIGNMENTS TO MALE CATEGORY	65	10	85	65	60	40	20	5	25	0
MEDIAN INTENSITY MAGNITUDE ESTIMATE	6.0	3.0	9.0	6.0	6.5	4.0	4.5	1.5	2.5	1.0
MEDIAN PLEASANTNESS MAGNITUDE ESTIMATE	-3.5	0	-5.0	-3.0	-3.0	-1.5	-.5	0	0	0

$r = .90$

$r = -.86$

TABLE 3
SEX ASSIGNMENTS AND MAGNITUDE ESTIMATE RATINGS OF
THE INTENSITY AND PLEASANTNESS OF HUMAN AXILLARY ODOR

SAMPLE #:	1	2	3	4	HALF MALE – HALF FEMALE ODORS * 5	6	7	8	9	BLANK
% OF MALE OBSERVERS ASSIGNING SAMPLE TO MALE CATEGORY **	100	50	70	40	10	10	10	50	60	0
% OF FEMALE OBSERVERS ASSIGNING SAMPLE TO MALE CATEGORY	80	50	60	70	20	30	0	20	50	20
MEAN PERCENT OF ASSIGNMENTS TO MALE CATEGORY	90	50	65	55	15	20	5	35	55	10
MEDIAN INTENSITY MAGNITUDE ESTIMATE	8.0	4.5	7.5	4.5	1.0	1.0	.75	4.0	5.5	1.0
MEDIAN PLEASANTNESS MAGNITUDE ESTIMATE	-3.5	0	-5.0	-3.0	-3.0	-1.5	-.5	0	0	0

$r = .98$

$r = -.86$

* SAMPLES 1-5 WERE MALE, AND SAMPLES 6-9 WERE FEMALE
** IN EACH OF THESE EXPERIMENTS THERE WERE 10 MALE AND 10 FEMALE OBSERVERS

In addition to the aforementioned urine and axillary odor studies, several experiments have examined body odors from other sources. For example, Wallace (1976) demonstrated that college students could be trained to discriminate between the hand odors of two men, two women, and a man and a woman. The male hand vs. female hand discrimination was the easiest to make, although significant differences were found for all of the comparisons. Wallace reported that the hand odors of identical twins were discernable, particularly when they were placed on diets differing in the amount of garlic present. Wallace interprets her findings as probable evidence for sexual, genetic and dietary influences on the production of hand odor.

In summary, the aforementioned studies suggest that human beings can detect and reliably respond to individual differences in conspecific urine, sweat, and possibly sebaceous odors. The influences of factors such as emotion, activity, and genetics in producing the discriminable cues is not known. Two of the aforementioned studies suggest that humans can recognize their own axillary odor. However, none of these studies provide strong evidence that sex, per se, is easily discriminable from such odor sources. In general, observers appear to associate intense odors with males and weak odors with females, although the intensity of body odors does not appear to provide an unambiguous basis for the establishment of gender.

Perception of Reproductive Stage by Odors

It is generally believed that axillary odor and possibly vaginal odor becomes noticeable only at the time of puberty. Furthermore, such odors are believed to decline after menopause. Such beliefs are supported by histological findings in the case of the apocrine glands, where prepubertal and postmenopausal glands are typically smaller with fewer active secretory cells (cf. Montagna and Parakkal, 1974). However, no systematic psychophysical studies confirming these beliefs have been made.

Only a few studies are available which have established whether human odors fluctuate during the menstrual cycle. Doty, Ford, Preti, and Huggins (1975) obtained intensity and pleasantness magnitude estimates from male and female observers for the odors of human vaginal secretions sampled from consecutive phases of 15 ovulatory menstrual cycles of four women. On the average, secretions from preovulatory and ovulatory phases were slightly weaker and less unpleasant in odor than those from menstrual, early luteal, and late luteal phases (Figure 2). However, considerable variation in odor patterns was present across cycles from the same donor, as well as across cycles from different donors (Figure 3). These findings suggest that human vaginal

Figure 2 (right). Mean magnitude
estimates of intensity and pleasantness of human vaginal secretion
odors sampled from consecutive
phases of the menstrual cycle. Vertical bars indicate standard errors
of the mean. Each data point represents an average of 256 cases.
Phases are designated as follows:
1, menstrual; 2, preovulatory; 3,
ovulatory; 4, early luteal; 5, late
luteal. Adapted from Doty, Ford,
Preti & Huggins (1975) with permission from the American Association for the Advancement of Science.

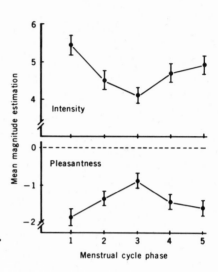

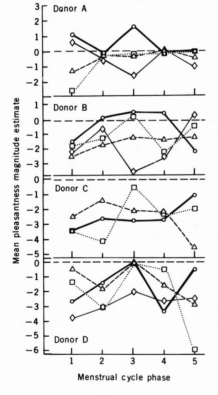

Figure 3 (left). Mean magnitude estimates of the pleasantness of human
vaginal secretion odors sampled from
consecutive phases of 15 menstrual
cycles of four donors. First cycle,
o—o ; Second cycle, △--△ ; Third
cycle, ◻···◻ ; Fourth cycle, ◇—◇ .
Phases are designated as follows:
1, menstrual; 2, preovulatory; 3,
ovulatory; 4, early luteal; 5, late
luteal. Adapted from Doty, Ford,
Preti & Huggins (1975) with permission from the American Association
for the Advancement of Science.
Copyright 1975 by the American
Association for the Advancement of
Science.

odors may change slightly in both pleasantness and intensity during
the menstrual cycle, but do not support the notion that such odors,
in an out-of-context in vitro test situation, are markedly attractive. The perception of such odors in in vivo situations presumably
differ, at least in some cases, from the results of this experiment,
depending upon the partners involved, their ages, sexual proclivities,
histories, and numerous situational variables. It should be emphasized, however, that in our in vitro test, the average of the
pleasantness ratings did not extend far from the neutral zero point
(Figure 2). The marked day-to-day and cycle-to-cycle variability
in the vaginal odors suggests that it is unlikely that humans can
reliably determine the time of ovulation on the basis of vaginal
odor cues alone (Figure 3).

In an analogous experiment using equivalent psychophysical procedures and experimental design, Doty (1976a) examined the relative
intensity and pleasantness of odors of human female urine sampled
from the various stages of the menstrual cycle. Fresh urine was collected in the morning at two or three day intervals across 12 ovulatory cycles of four women and immediately frozen to -60° C. until
psychophysical testing. Before testing, the urine samples of a given
cycle were thawed to room temperature and presented to observers in
100 ml. glass sniff bottles that were covered on the outside by
aluminum foil. Two layers of sterile gauze were taped over the open-
ings of the jars, and the jars were sealed with caps lined with
Teflon when not in use.

The major finding of this experiment is presented in Figure 4.
An overall analysis of variance indicated that significant variations
in the intensity and pleasantness of the urine odors were present
across the cycle phases although, as is apparent in this figure, such
changes were extremely small and accounted for a negligible proportion
of the total variance. The most salient effect appears to be an increase in the intensity of the urine during the preovulatory phase
(Phase 2). Clearly, the urine odors were given larger unpleasantness
magnitude estimates than those given to the vaginal odors (cf. Figure
2). Unlike the case of the vaginal odors, the pleasantness and intensity urine magnitude estimates were not strongly related (Pearson r =
-0.27, n.s.).

Anecdotal and biochemical evidence suggests the possibility that
human breath odors may change during the menstrual cycle. Reports in
the early literature suggested that the odor of menstrual breath has
a peculiar quality, described by some as being similar to chloroform
or violets (cf. Ellis, 1934, p. 64). Although metabolic by-products
of alveolar respiration may be partially responsible for fluctuations
in breath odor during the menstrual cycle, several authors suggest an
important role of more peripheral metabolic factors (cf. Tonzetich
& Kestenbaum, 1969). For example, several reports suggest that

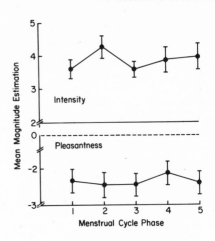

Figure 4. Mean magnitude estimates
of the intensity and pleasantness of
the odors of human urine sampled
from consecutive phases of the menstrual cycle. Vertical bars indicate standard errors of the mean.
Each data point represents an average of 160 cases. Phases are designated as follows: 1, menstrual; 2,
preovulatory; 3, ovulatory; 4, early
luteal; 5, late luteal.

increases in exfoliation of the oral mucosa, gingival inflammation,
and increased numbers of total oral bacteria occur midcycle when the
levels of certain gonadotropins and gonadal steroids are elevated
(Lindhe and Attström, 1967; Loe, 1965; Loe & Silness, 1963; Main and
Richie, 1967; Prout and Hopps, 1970). Such factors may account for
the accelerated production of volatile sulfur compounds (e.g., hydrogen sulfide, methyl mercaptan, dimethyl sulfide) noted by Tonzetich,
Preti and Huggins (1976) to occur midcycle. Unfortunately, no psychophysical studies have established whether or not specific phases of
the menstrual cycle can be reliably determined on the basis of oral
odors.

Conclusions

    Perhaps the major point to be gleaned from this review is that
human beings possess the ability to discriminate between one another
on the basis of bodily odors, regardless of whether they are urinary,
axillary or palmar in origin. This finding supports the numerous
anecdotes on this point (cf. Doty, 1972). Additional research is
needed to establish the relative roles of diet, genetics, activity
patterns, etc. in producing the cues upon which such discriminations
are based.

    The second major point suggested by the present set of studies
is that it is unlikely that adult humans can communicate, in reasonably clear fashion, information pertaining to gender or reproductive
state by means of odors from sources common to both sexes (e.g., urin,
axillary sweat). However, the possiblity of the communication of
gender by means of sexually-exclusive odors (e.g., seminal fluid, vaginal secretions) would appear likely. The variability in the odors o
vaginal secretions and urine sampled across menstrual cycles suggests

that these odors cannot be used in reliable fashion to establish the stage of the reproductive cycle.

References

BEAUCHAMP, G. and DESOR, J.A. 1976. Discrimination of sex and individual identity from human urinary odors. In preparation.

BEAUCHAMP, G., DOTY, R.L., MOULTON, D., and MUGFORD, R. 1976. The pheromone concept in mammals: A critique. In R.L. Doty (Ed.), Mammalian olfaction, reproductive processes, and behavior, Academic Press, New York, in press.

BLOCH, I. 1934. Odoratus sexualis, Panurge Press, New York.

BRODY, B. 1975. The sexual significance of the axillae. Psychiatry 38: 278-289.

COMFORT, A. 1971. Likelihood of human pheromones. Nature 230: 432-434.

DAVENPORT, W. 1965. Sexual patterns and their regulation in a society of the southwest Pacific. In F.A. Beach (Ed.), Sex and behavior, Wiley, New York, 164-207.

DOTY, R.L. 1972. The role of olfaction in man -- sense or nonsense? In S.H. Bartley (Ed.), Perception in everyday life, Harper & Row, New York, 143-157.

DOTY, R.L. 1975. An examination of relationships between the pleasantness, intensity and concentration of 10 odorous stimuli. Percept. Psychophys. 17: 492-496. (a)

DOTY, R.L. 1976. Changes in the intensity and pleasantness of urine odors during the menstrual cycle. Unpublished experiment.

DOTY, R.L. 1976. Reproductive endocrine influences upon human nasal chemoreception: A review. In R.L. Doty (Ed.), Mammalian olfaction, reproductive processes, and behavior, Academic Press, New York, p. 295-321.

DOTY, R.L. and DUNBAR, I.A. 1974. Attraction of Beables to conspecific urine, vaginal secretion, and anal sac odors. Physiol. Behav. 11:825-833.

DOTY, R.L., FORD, M., PRETI, G. and HUGGINS, G.R. 1975. Human vaginal odors change in intensity and pleasantness during the menstrual cycle. Science 190: 1316-1318.

DOTY, R.L., KLIGMAN, A., LEYDEN, J. and ORNDORFF, M. Communication of gender via human axillary odors depends upon intensity-related factors. Submitted to Nature.

ELLIS, H. 1936. Studies in the psychology of sex, Random House, New York.

HAIRE, N. 1940. Encyclopedia of sexual knowledge, Eugenics Publishing Co., New York, 355-386.

HEDIGER, H. 1968. The psychology and behaviour of animals in zoos and circuses, Dover, New York.

HURLEY, H.J. and SHELLEY, W.B. 1960. The human apocrine sweat gland in health and disease, Thomas, Springfield.

LINDHE, J. and ATTSTRÖM, R. 1967. Gingival exudation during the
menstrual cycle. J. Periodont. Res. 2: 194-198.
LOE, H. 1965. Periodontal changes in pregnancy. J. Periodont. 36:
209-217.
LOE, H. and SILNESS, J. 1963. Periodontal disease in pregnancy. I.
Prevalence and severity. Acta Odont. Scand. 21: 532-551.
MAIN, D.M.G. and RICHIE, G.M. 1967. Cyclic changes in oral smears
from young menstruating women. Brit. J. Derm. 79: 20-30.
McBURNEY, D.H., LEVINE, J.M., and CAVANAUGH, P.H. 1976. Psycho-
physical and social ratings of human body odor. Bull. Personal.
Soc. Psychol., 1976, in press.
MONCRIEFF, R.W., 1944. The chemical senses, Leonard Hill, London.
MONTAGNA, W. and PARAKKAL, P.F. 1974. The structure and function
of skin, Academic Press, New York.
PROUT, R.E.S. and HOPPS, R.M. 1970. A relationship between human
oral bacteria and the menstrual cycle. J. Periodont. 41: 98-101.
RUSSELL, M.J. 1976. Human olfactory communication. Nature 260:
520-522.
SAGARIN, E. 1964. The sense of smell and sex. In A. Ellis and A.
Abarbanel (eds.), The encyclopedia of sexual behavior, Hawthorn
Books, New York, 979-986.
SHEHADEH, N. and KLIGMAN, A.M. 1963. The bacteria responsible for
axillary odor. J. Invest. Dermatol. 41: 3.
TONZETICH, J. and KESTENBAUM, R.C. 1969. Odour production by human
salivary fractions and plaque. Archs. Oral. Biol. 14: 815-827.
WALLACE, P. 1976. Individual discrimination of humans by odor.
Physiol. Behav., in press.

PHYSICAL AND COGNITIVE LIMITATIONS ON OLFACTORY PROCESSING IN HUMAN BEINGS

William S. Cain

John B. Pierce Foundation Laboratory and Yale School
of Medicine
New Haven, Connecticut 06519

Chemical communication via natural secretions is thought to play at best a minor role in the lives of human beings, but a major role in the lives of many other species. It is unlikely that the primary reason for this difference lies strictly in the realm of sensory functioning. Even if man is not the most sensitive creature on earth his sensitivity is nonetheless remarkable and undoubtedly rivals that of many other animals. Furthermore, electrophysiological evidence suggests that there is considerable similarity in olfactory reception throughout the vertebrates (Döving, 1966; Döving and Lange, 1967; Köster and MacLeod, 1975). Accordingly, the study of man's sense of smell may teach much about olfaction in other vertebrates.

A pivotal difference between human beings and those species that do communicate via secretions seems to stem from cognitive correlates of olfactory stimulation. As Zwaardemaker (1930) observed: "We men do not think, as do the osmatic animals, in olfactory images. ... Probably in osmatic animals a world of richly varied beauty is awakened which we do not understand, because our brain is not adapted to such a task." It is difficult indeed to imagine what the olfactory world of the so-called macrosmatic animals is like. Man cannot easily conjure up images from any of the lower senses, including olfaction, but when he does, the images are commonly indistinguishable from actual sensory stimulation. That is, they are tantamount to hallucinations. Most other species may be spared this embarrassment. Moreover, they may be able to "operate" on olfactory images in somewhat the same way that man can operate on auditory or visual images. Man can, for instance, conjure up the image of a cube and then cause it to rotate, or he can conjure up the image of another person's voice

saying anything he wishes. He clearly lacks this talent in the
olfactory domain. Nevertheless, there are cognitive aspects of
man's olfactory functioning, and at least one, memory for odors,
can play a role in what may seem superficially to be purely
sensory tasks.

This report deals with two aspects of man's olfactory functioning. The first concerns the simple matter of intensity discrimination. The second concerns identification of odor quality.
Whereas the first involves sensory factors only, the second
involves an interplay between sensory and cognitive processes.

DIFFERENTIAL SENSITIVITY

Psychophysical research in the 19th century established two
enduring views of man's ability to perceive odors. One was that,
for many odoriferous substances, man's absolute sensitivity is
remarkably good. The other was that, for most or all substances,
man's differential sensitivity is notably poor. Gamble (1898)
concluded that in general one concentration would be just noticeably different from another if the two differed by about 25 to
33%. This difference is roughly three times that necessary to
resolve successively presented visual or auditory stimuli.

Subsequent studies of differential sensitivity for smell
agreed quite well with Gamble's estimate. Table 1 shows results
obtained by Hermanides in 1909 (Zwaardemaker, 1925). The average
difference threshold was 37%. As in Gamble's study, variation in
the threshold was not large enough nor reliable enough to prompt
the conclusion that differential sensitivity truly differed from
odorant to odorant. More than a half century later, Stone (1964)

TABLE 1. DIFFERENCE THRESHOLDS MEASURED BY HERMANIDES (1909)

Odorant	Difference Threshold	
	Weak	Strong
Isoamyl acetate	30%	24%
Nitrobenzene	25	26
Terpineol	40	36
Muscone	45	46
Ethyl disulfide	30	36
Guaiacol	35	46
Valeric acid	23	38
Pyridine	30	30
Skatole	60	62

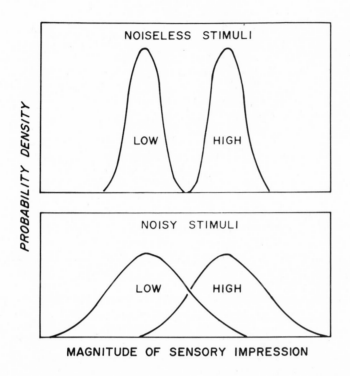

FIG. 1. The impression that stimuli will make on a sensory system
can be represented by probability density functions. Ability to
resolve the difference between pairs of stimuli (high vs low)
will depend both on the difference between the means and on the
variances of the distributions. Hence, it will be easier to resolve the difference between the stimuli that gave rise to the
upper pair of distributions than the stimuli that gave rise
to the lower pair. The difference between the means is the
same in both cases. The upper pair could represent the case
where the stimulus is controlled with precision and, accordingly,
where the variability arises primarily from the organism. The
lower pair could represent the case where the stimulus is controlled only poorly. Variance arising from the stimulus will
add to the variance arising from the organism and will be indistinguishable from it in the human subject's view.

confirmed the early view. The difference threshold for a variety
of odorants was uniformly about 28% (see also Stone and Bosley,
1965).

What factors control capacity to discern small changes in the
magnitude of stimulation? Gustav Fechner, the father of psychophysics, established the view that the difference threshold reflected a relatively stable barrier that had to be surmounted in
order for resolution to be accomplished. Before the end of the
19th century, however, Fullerton and Cattell (1892) registered a
strong objection to Fechner and all others who adopted "...the
curious supposition that stimuli seem exactly alike so long as the
difference is less than a certain amount, whereas, when the difference is made greater than this amount, it becomes suddenly
apparent. This is by no means the case. The clearness with which
a difference is distinguished varies gradually from complete doubt
to complete certainty." In their wise view, the experiment on
differential sensitivity measured the moment-to-moment fluctuations in the impression that pairs of stimuli made on the sensory
system. The difference threshold computed from the data merely
reflected the size of these variations rather than the location
of a barrier. Hence, it could be said that the capacity to discern small changes in stimulation is controlled by "noise"
arising primarily in the sensory system.

Fullerton and Cattell's notions reached full quantitative
development in the modern theory of signal detection. The theory
emphasizes the view that the "noise" that limits resolution can
arise from many sources, not only from the sensory system. Noise
in the stimulus, for example, is just as limiting as biological
noise. Unless the noise in the stimulus is exceedingly low or
unless it can be quantified and subsequently discounted, its
limiting influence will go undetected (see Fig. 1). Has noise in
the stimulus contributed substantially to the size of difference
thresholds measured for smell?

I set out to re-examine the question of differential sensitivity for smell with an eye toward the limitations that fluctuations in the stimulus may impose on psychophysical performance.
Subjects were asked to decide in a forced-choice situation, which
of two similar concentrations was stronger. Each pair of concentrations was presented again and again until two untrained
subjects had made a total of 400 judgments per pair. This procedure was repeated for various pairs of concentrations of three
odorants: n-butyl alcohol, n-amyl alcohol, and ethyl n-butyrate.

The data from this task led to the construction of various
psychometric functions, such as the one shown on the left side of
Fig. 2, where percent correct in the forced-choice task is plotted

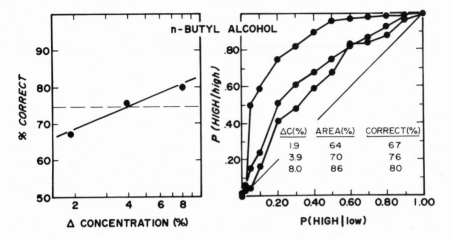

FIG. 2. Left side: Percent correct in a two-alternative forced-choice task versus difference in vapor-phase concentration. Each point is based on 400 observations of stimuli delivered by means of an air-dilution olfactometer. Right side: Receiver operating characteristic (ROC) curves for the pairs of concentrations used in the psychophysical task. The curves were constructed from chromatographic analysis of the stimuli. The inset permits comparison of area under the curves to psychophysical performance.

against difference in concentration. The difference in concentration where the function crosses the dashed line can be taken as a measure, k, of the keenness of discrimination. k is formally equivalent to the difference threshold of classical psychophysics. On the average, k was 19%, lower than previous estimates but still higher than that for most other modalities. There were, however, large and reliable differences from one odorant to another and, for n-butyl alcohol, k was consistently less than 10%. The happy implication is that, for some odorants, intensity resolution in olfaction may rival that in audition and vision.

The psychophysical results comprised only half the data from the experiment. The other half were obtained from analysis of vapor-phase concentration. The stimuli had been presented to the subjects in many small vessels that were designed to permit ready sampling of odorant via the gas sampling valve of a gas chromatograph. Accordingly, it was possible, within certain limits, to treat the chromatograph as an additional subject in the experiment. The device was presented one hundred freshly prepared samples of each pair of the concentrations presented to the human subjects. The areas under the peaks produced by the chromatograph were integrated for a measure of the mass of odorant in the vapor phase.

On the assumption that the higher and lower nominal concentrations of each pair varied independently, the chromatographic data were used to generate receiver operating characteristic (ROC) curves like those shown on the right side of Fig. 2. The points on the curve represent at each of many decision criteria the conditional probability that a vapor-phase concentration of a particular magnitude would be deemed high, given that the sample was actually drawn from a vessel containing the higher nominal concentration, versus the conditional probability that the same vapor-phase concentration would be deemed high, given that the sample was actually drawn from a vessel containing the lower nominal concentration. The important feature is that the curves reflect the limitations on psychophysical performance imposed by the stimulus. According to a theorem derived from signal detection theory, the percentage area under the curves will equal the maximum performance that could be attained in a two-alternative forced-choice task (Egan, 1975).

The ROC curves revealed that variation in vapor-phase concentration often limited strongly the optimal performance achievable psychophysically. In view of the limitations, the subjects performed surprisingly well. Ratio of percent correct measured psychophysically to percentage area under the ROC curves averaged 0.96 for n-butyl alcohol, 0.98 for ethyl n- butyrate, and 0.87 for n-amyl alcohol. Hence, noise in the stimulus seemed to contribute substantially to the size of k. If this is true, then k should decrease if the noise in the stimulus is decreased. This indeed occurred when various concentrations of butyl alcohol were presented by means of an olfactometer, rather than by means of vessels; k was only 4.2%, almost an order of magnitude lower than the commonly accepted value (Fig. 2). Variation of stimuli presented via the olfactometer was only half that of stimuli presented via vessels. Nevertheless, the performance of the subjects still rivaled and even exceeded predictions derived from the ROC curves.

Figure 3 shows the relation between percent correct measured psychophysically and predictions of optimal performance derived chromatographically. In five of the 21 cases shown here, psychophysical performance exceeded predictions of optimal performance. These deviations could represent the operation of chance, or they could reflect some tendency for the chromatographic analysis to overestimate the variability in the stimulus. Neither possibility, however, vitiates the excellent correlation (+ 0.87) between the

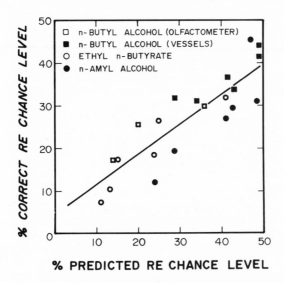

% PREDICTED RE CHANCE LEVEL

FIG. 3. Psychophysical performance (% correct) versus predictions of optimal performance (% predicted) derived chromatographically.

predictions and the psychophysical outcome. In fact, any variability added by the sampling procedure would have depressed the correlation below its true value.

Through the use of a parameter d', an index of sensitivity employed commonly in signal detection theory (Green and Swets, 1966), it was possible to convert the measure <u>percent correct</u>, into a measure of variance. This maneuver permitted calculation of how much variance in the psychophysical data could be accounted for by the variance in the stimulus. Averaged across all experiments, the stimulus accounted for 75% of the variance. Hence, biological noise accounted for only one quarter of the variance in the psychophysical data. When the variance contributed by the stimulus was discounted, the average value of k dropped from 19% to 11%, and ranged from 5% for n-butyl alcohol to 16% for n-amyl alcohol.

In summary, the chromatograph displayed considerable ability to predict how well subjects could resolve differences in odor intensity. The predictive power of the chromatograph derives from the severe limitations that the fluctuations in the stimulus place on the optimal psychophysical performance. When the stimulus-imposed limitations are discounted, the sense of smell, notorious for its inability to register differences in intensity, seems as keen as the higher senses.

IDENTIFICATION OF ODOR QUALITY

Man can perceive an incredibly rich variety of odor qualities.
So vast and complex is the qualitative spectrum that no classification scheme has ever proved satisfactory (Cain, 1976). Furthermore, the minimum number of verbal descriptors necessary to characterize variation in quality is exceedingly large, probably
greater than one hundred (Dravnieks, 1975). These descriptors
are generally derived from the stimulus objects for smell (e.g.,
musky, fruity), or are borrowed from other sense departments
(e.g., sweet). Hence, odors have failed to give rise to their
own vocabulary.

Since the qualitatively rich world of odors defies easy classification, man must frequently deal with an odor as a unique
entity, rather than as a representative of some qualitative class.
To designate quality in terms of stimulus object seems consistent
with the requirement to deal with odors as unique. By the same
token, it seems reasonable to expect that identification of
stimulus objects by smell would be easy.

Most persons believe themselves able to identify many objects
without error by smell alone. If you ask how many, they will say
"hundreds," but until recently the laboratory data said "about
sixteen." Why the enormous discrepancy?

No doubt most persons can discriminate among thousands of
odorants. The task of absolute identification is, however, considerably more demanding than discrimination. Even when made
aware of the easily-missed distinction between discrimination and
identification, the layman will almost surely continue to claim
that his ability to identify clearly outstrips the laboratory estimate of about sixteen. The powerful assistance of contextual
cues, present in abundance in everyday life but absent from the
laboratory, will bring the two discrepant estimates more closely,
but not completely into register. Certain findings now suggest
that the layman has a case. The capacity for identification
actually exceeds sixteen, though it may be smaller than "hundreds."

In a widely-cited experiment, Engen and Pfaffmann (1960)
tested how well subjects could identify odorants presented one at
a time from batteries of odorants that contained from five to 36
members. The subject sought to identify each with a label he had
previously generated during familiarization trials with the 36
possible stimuli in the experiment. Accordingly, each subject
used his own unique set of responses. Throughout testing, he was
reacquainted with the "correct" (i.e., previously generated) label
whenever he failed to identify a stimulus correctly. The number
of correct identifications increased with the number of odorants

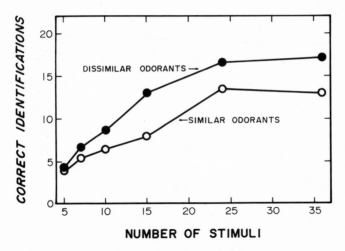

FIG. 4. Average number of correct identifications versus number of odorants in an inspection set. Dissimilar odorants represented a wide range of qualities. Similar odorants smelled sweet and fruity. Data from Engen and Pfaffmann (1960).

in a set, and reached a maximum of 17 for the set of 36 (Fig. 4). It is a fruitful exercise to search for the reasons why performance was not better. The search may uncover some aspects of how man's olfactory perceptions and apperceptions are constructed.

Similarity among the odorants in a set would seem a priori to have a potent influence on accuracy of identification. Nevertheless, when Engen and Pfaffmann presented sets with relatively similar smelling members, performance was almost as good (maximum identification = 13.5) as when they presented sets with different smelling members. Perhaps a very rigorous choice of similar smelling odorants could have caused greater deterioration, but similarity or dissimilarity is apparently not an overwhelming determinant of performance.

The search for the various determinants of stimulus identification leads beyond the realm of sensation into the realm of memory. A prerequisite to correct identification entails recognition that an odorant had in fact been presented previously. Recognition memory for odors received very little attention until the 1970's, and the results of two similar investigations appear in Fig. 5 (Engen and Ross, 1973; Lawless and Cain, 1975). In both, subjects attempted to decide in a two-alternative forced-choice situation which odorants they had smelled during previous, brief (i.e., one-trial) inspection.

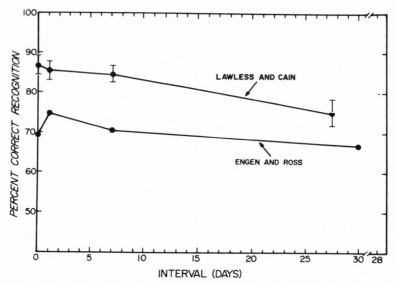

FIG. 5. Percent correct recognition in a two-alternative forced-choice task versus interval after a single exposure to the various test odorants. The points plotted at day <u>zero</u> were obtained from recognition tests performed only a few minutes after initial exposure.

Two characteristics of the forgetting curves in Fig. 5 are noteworthy. One is their incredibly slow decline. Insofar as recognition memory will limit performance in the identification task, it will matter little whether ability to identify is tested 10 min. or 1 mo. after inspection. Long-lasting memory for odors has, incidentally, been noted often by writers, most notably Proust. Most persons have occasionally had the Proustian experience of recognizing a scent they had not smelled for many years. Frequently, the flash of recognition is accompanied by memory of an entire scene, complete with emotions. It is interesting that writers gave little or no attention to such experiences until the 19th century. Shakespeare, for instance, made no use of the reminiscent quality of odors (McKenzie, 1923).

The second noteworthy feature of the forgetting curves is that they start below 100%. Although pictures would be recognized with nearly perfect accuracy a few minutes after inspection (Shepard, 1967), odors are not. Errors made in an immediate test were not restricted to odors that lacked salience along the dimensions of pleasantness, familiarity, etc. Instead, the errors were distributed rather uniformly across all types of odors. The only characteristics that seem to influence initial recognition are the size of the inspection set and the similarity between the old

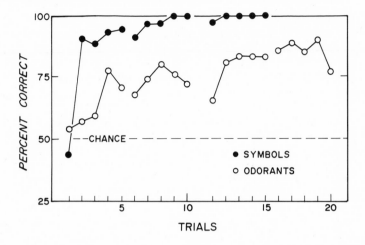

FIG. 6. Percent correct versus number of trials in a paired associate learning task where the stimuli were four schematic symbols or four odorants and the responses were the numerals 1
through 4. A rest interval of 5 min intervened between trials
5 and 6, and between trials 15 and 16. Twenty-four hours intervened between trials 10 and 11. Data from Davis (1975).

and new members of the test pair. If the distractor odorants are
chosen to be particularly similar to the previously inspected
odorants, initial recognition will decrease. Similarly, if the
size of the inspection set is large, recognition will decrease.
This variable is responsible primarily for the difference between
the two forgetting curves in Fig. 5.

The results shown in Fig. 5 and other results where recognition was tested for odorants presented one at a time, rather than
in a forced-choice task, suggest that recognition memory will place
some limitation on ability to identify odorants. An even greater
limitation may arise, however, from failure to associate the
appropriate label with the odorant. When subjects were required
to associate numerical labels with odors in a paired-associate
learning task, learning proceeded very slowly (Fig. 6). After
twenty trials distributed over two days, subjects were still unable to assign the appropriate numerical labels to four test odorants with perfect accuracy (Davis, 1975). A corresponding task
performed with schematic symbols, rather than odorants, yielded
perfect performance after only nine trials. These data cannot be
generalized directly to Engen and Pfaffmann's identification

experiment, where subjects generated their own labels. Nevertheless, it is clear that the associative connection between an odor
and a label will develop only with considerable practice.

Although labels that subjects generate themselves may be
more adequate for subsequent identification of an odor than some
arbitrary label such as a numeral, they may fall far short of the
ideal. The connection between odors and language appears to be
relatively weak. In everyday life, a person will frequently
recognize an odor and will feel that its name is at his disposal,
but try as he may, he cannot retrieve the name from memory (see
Brown and McNeil, 1966). This tip-of-the-tongue phenomenon is
quite frustrating, particularly to experimental subjects who are
asked to label odorants. Failing frequently to come up with the
veridical name, the subjects will use the best available approximation, but they will realize that it is only an approximation.
On a subsequent occasion, they may use another label, discarding
or forgetting their previous attempt. Hence, the association
between the odor and the subject-generated label is weaker than
it may seem. In some respects, the subject must frequently re-
learn his own previous assocation.

Recently, Desor and Beauchamp (1974) set out to re-examine
man's ability to identify odorants. One important feature was
that they chose only very familiar odorants -- odorants of high
"ecological validity." These included popcorn, beer, chocolate,
human urine, cat feces, and many other readily available substances. Engen and Pfaffmann had also used some commonly encountered substances, but on the whole, their odorants were less
familiar than those of Desor and Beauchamp. One other difference
between the investigations was that Desor and Beauchamp required
their subjects to use veridical labels rather than associations.
Whenever a subject used an incorrect label, he was corrected.
Under these circumstances, subjects were able to identify correctly 32 out of 32 odorants with relatively little training.
Subsequently, they managed to identify almost 60 out of a set of
64, again with only moderate training (see Fig. 7). With more
training and larger stimulus sets, the subjects might possibly
have gone on to even higher levels of performance.

Desor and Beauchamp speculated that the superior performance
of their subjects probably arose from the use of stimuli rich in
information. The term information can mean many things and I am
not sure how Desor and Beauchamp meant to use it. It could be
taken to mean mere complexity in terms of the variety of molecules constituting the stimulus. If mere complexity were the key
to easy identification, however, then a mixture of substances
chosen randomly from the chemist's shelf should be as easy to
identify, once it had been given a label, as coffee, for instance.

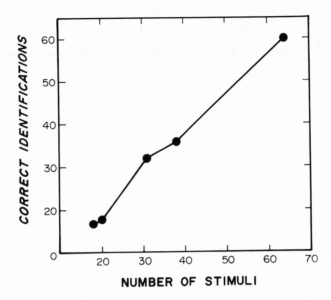

FIG. 7. Average number of correct identifications versus number of odorants in an inspection set. Data from Desor and Beauchamp (1974).

This seems unlikely. A more likely explanation arises from consideration of recognition memory and paired-associate learning, i.e., from cognitive features of the task.

Desor and Beauchamp's subjects needed to achieve perfect recognition in order to achieve perfect identification. The use of ecologically meaningful, familiar stimuli presumably facilitated perfect recognition. Although familiarity did not seem to have an important influence in the recognition memory experiments discussed above, this variable may emerge as important only in situations, such as the identification task, that involve practice with the stimuli. Consistent with this notion was Davis's (1975) finding that subjects learned to associate numerals more rapidly with familiar odors than with relatively unfamiliar odors.

Perhaps the most important advantage that Desor and Beauchamp conferred on their subjects was to combine the use of veridical and hence familiar labels with familiar odors. The bulk of the practice in the paired-associate learning task of connecting the veridical name with the odor had occurred in the lives of the subjects long before the experiment was ever conceived. Use of a correction procedure helped to revive previously established connections and hence prompted the subjects through the mental gap that so frequently and inexplicably separates an odor from its name.

CONCLUSIONS

Re-examination of the tasks of intensity discrimination and odor identification has revealed that man's ability to register and process olfactory information is less limited than hitherto thought. In the task of intensity discrimination, the limitations on performance are largely physical and therefore independent of olfaction. These physical limitations (viz., fluctuations in the stimulus) have undoubtedly inflated the estimates of the difference threshold for other species as well as for human beings. To control precisely the vapor-phase concentration of an odorant introduced into a small animal's test chamber is a formidable task. The difference threshold measured in such a set-up may reflect more the experimenter's success at stimulus control than the animal's capacity to discriminate. This factor seems to deserve particular consideration in light of the report (Davis, 1973) that the difference threshold in the rat, an animal that depends heavily on olfaction, is an inordinate 200%.

The analysis presented here implies that inadequately forged verbal-olfactory links may impose the primary limitation on man's ability to identify odors. It is conceivable that man may be better equipped to fix the identity of objects by smell at a "preverbal" level than at the verbal level tapped by the typical odor identification task. "Preverbal" identification is perhaps most appropriately viewed as a kind of recognition. But it comprises more than mere recognition that an odor has been smelled before. It comprises recognition of the odor for exactly what it is, but without its verbal label. While walking through the street, for instance, a person may pick up the scent of perfume and recognize that it is the perfume her mother always used, but its name may not come to mind. It is perhaps at this level that the dog, for example, may "identify" the scent marks of other dogs in the neighborhood. At this level of "identification" or recognition, human beings may be approximately as capable as most other species. This is not to claim, however, that man will be so confident in his judgment as the dog. Regarding the dog, McKenzie (1923) remarked: "He can recognise his master by sight, no doubt, yet, as we know, he is never perfectly satisfied until he has taken stock also of the scent, the more precisely to do so bringing his snout into actual contact with the person he is examining. It is as if his eyes might deceive him, but never his nose."

Note: Supported by Grant ES00592 from the National Institutes of Health.

REFERENCES

Brown, R., and McNeill, D. 1966. The "tip of the tongue" phenomenon. J. Verb. Learn. Verb. Behav. 5:325-337.

Cain, W. S. 1976. History of research on smell, in E. C. Carterette and M. P. Friedman (eds.), Handbook of Perception, Vol. 6, Tasting, Smelling, Feeling, and Hurting. Academic, New York. In press.

Davis, R. G. 1973. Olfactory psychophysical parameters in man, rat, dog, and pigeon. J. Comp. Physiol. Psychol. 85: 221-232.

Davis, R. G. 1975. Acquisition of verbal associations to olfactory stimuli of varying familiarity and to abstract visual stimuli. J. Exper. Psychol. 104:134-142.

Desor, J. A., and Beauchamp, G. K. 1974. The human capacity to transmit olfactory information. Percept. Psychophys. 16: 551-556.

Döving, K. B. 1966. An electrophysiological study of odour similarities of homologous substances. J. Physiol. (London) 186:97-109.

Döving, K. B., and Lange, A. L. 1967. Comparative studies of sensory relatedness of odours. Scand. J. Psychol. 8:47-51.

Dravnieks, A. 1975. Evaluation of human body odors: methods and interpretations. J. Soc. Cosmet. Chem. 26:551-571.

Egan, J. P. 1975. Signal Detection Theory and ROC Analysis. Academic, New York.

Engen, T., and Pfaffmann, C. 1960. Absolute judgments of odor quality. J. Exper. Psychol. 59:214-219.

Engen, T., and Ross, B. M. 1973. Long-term memory of odors with and without verbal descriptions. J. Exper. Psychol. 100: 221-227.

Fullerton, G. S., and Cattell, J. McK. 1892. On the perception of small differences. Philos. Series, Univ. Pennsylvania 2:10-11.

Gamble, E. A. McC. 1898. The applicability of Weber's law to smell. Amer. J. Psychol. 10:82-142.

Green, D. M., and Swets, J. A. 1966. Signal Detection Theory and Psychophysics. Wiley, New York.

Köster, E. P., and MacLeod, P. 1975. Psychophysical and electrophysiological experiments with binary mixtures of acetophenone and eugenol, pp. 431-444, in D. G. Moulton, A. Turk, and J. W. Johnston, Jr. (eds.), Methods in Olfactory Research. Academic, London.

Lawless, H. T., and Cain, W. S. 1975. Recognition memory for odors. Chem. Senses Flav. 1:331-337.

McKenzie, D. 1923. Aromatics and The Soul: A Study of Smells. Hoeber, New York.

Shepard, R. N. 1967. Recognition memory for words, sentences, and pictures. J. Verb. Learn. Verb. Behav. 6:156-163.

Stone, H. 1964. Behavioral aspects of absolute and differential olfactory sensitivity. Ann. N. Y. Acad. Sci. 116:527-534.

Stone, H. and Bosley, J. J. 1965. Olfactory discrimination and Weber's law. Percept. Mot. Skills 20:657-665.

Zwaardemaker, H. 1925. L'Odorat. Doin, Paris.

Zwaardemaker, H. 1930. An intellectual history of a physiologist with psychological aspirations, pp. 491-516, in C. Murchison (ed.), A History of Psychology in Autobiography. Clark University, Worcester, Mass.

CHEMICAL COMMUNICATION AS ADAPTATION: ALARM SUBSTANCE OF FISH

R. Jan F. Smith

Biology Department, University of Saskatchewan

Saskatoon, Saskatchewan, Canada S7N OWO

ABSTRACT

An alarm pheromone (alarm substance, Schreckstoff) is released when the epidermis of an ostariophysian or gonorhynchiform fish is damaged. Conspecifics respond with a fright reaction. The alarm substance and fright reaction are both absent in some specialized ostariophysians. Others, such as blind cave fish, lose just the fright reaction. Some cyprinids, which have abrasive spawning behavior, lose the alarm substance, but not necessarily the fright reaction, during the breeding season. The development and main-tenance of the alarm substance cells may be an example of animal-altruism maintained by kin selection.

THE FISH ALARM SUBSTANCE-FRIGHT REACTION SYSTEM

When the skin of an ostariophysian or gonorhynchiform fish is damaged (e.g. by a predator), alarm substance cells are broken and release alarm substance (Schreckstoff). Nearby conspecifics smell the alarm substance and show a fright reaction appropriate to their species. Their fright reaction may then be treated as a visual signal by other conspecifics, leading to rapid transmission of the signal through a group of fish. This alarm signal system consists of two main components: a morphological component, the alarm substance cells and a behavioral component, the fright reaction.

Von Frisch (1938) discovered this alarm system. His discovery and the general characteristics of the system have been reviewed by von Frisch (1961) and Pfeiffer (1962, 1963a, 1966b, 1974). In view of the recent and comprehensive review by Pfeiffer (1974) this paper will concentrate on recent developments in the study of fish alarm substance and on the adaptive significance of the alarm substance system. Material covered in earlier reviews will be summarized.

Alarm Substance Cells, the Morphological Component

The alarm substance is contained in large epidermal cells, originally described as "club cells" (Pfeiffer 1960b) although the term club cell has also been applied to other types of non-mucus epidermal secretory cells in fish (Pfeiffer 1960a, b, 1963b, Pfeiffer and Pletcher 1964). Alarm substance cells can be distinguished from mucus cells, the other main fish epidermal secretory cell, by their negative reaction to periodic acid-Schiff's reagent (PAS), round and often centrally located nucleus or nuclei surrounded by a clear halo, and lack of an opening at the skin surface. Mucus cells are typically PAS-positive, have a single flattened basal nucleus and often release their contents at the skin surface through a pore. Alarm substance cells are usually found closer to the basal layer than the mucus cells and are autofluorescent under ultraviolet light (Reutter and Pfeiffer 1973). Pfeiffer (1974) illustrates a variety of alarm substance cells found in ostariophysian and gonorhynchiform fishes. Other morphological and histochemical characteristics of alarm substance cells have been described by Pfeiffer et al. (1971) and Pfeiffer (1963b, 1974).

The identification of ostariophysian "club cells" as the source of alarm substance is based on several lines of evidence. The barbels of the carp, Cyprinus carpio L., and the Asian glass catfish, Kryptopteus bicirrhis Cuv. & Va., lack alarm substance cells which are present in other areas of the skin. Skin extract from barbels does not induce a fright reaction while skin extracts from other areas of the body do (Pfeiffer 1962a). Fathead minnows, Pimephales promelas Rafinesque, lost their alarm substance cells in response to high androgen levels. Skin extract from breeding male fatheads or testosterone treated fish does not induce a fright reaction. Skin extract from females or nonbreeding males, which retain alarm substance cells, does induce a fright reaction (Smith 1973). European minnow, Phoxinus phoxinus L., alarm substance is autofluorescent (Pfeiffer and Lemke 1973) and the contents of the alarm substance cells of Phoxinus show similar autofluorescence (Reutter and Pfeiffer 1973). On a more general level, species which show the fright reaction to skin extract also have alarm substance cells whereas most species which lack a fright reaction to conspecific skin extract also lack alarm substance cells. Blind catfish, Anoptichtyes sp., and piranhas and silv‑

dollars   (Serrasalminae and Mylinae) lack the fright reaction but
have alarm substance cells.   Their skin extract, however, does
elicit fright reactions in other species (Pfeiffer 1974).

Hüttel (1941) suggested that the alarm substance of the European minnow was a purine or pterine-like substance.   But since
Ziegler-Günder (1956) had supposed that purines and pterines were
more concentrated in the dorsal exanthophores, Hüttel's (1941)
suggestion was initially questioned on the basis of von Frisch's
(1941) demonstration that dorsal and ventral skin were equally
effective in eliciting the alarm reaction.   Tucker and Suzuki
(1972) suggested on the basis of recorded electric potentials
from the olfactory lamella of white catfish, Ictalurus catus (L.),
that amino acids or similar compounds mediated the fright reaction
although Tucker (1973) demonstrated a rapid and variable decline
in olfactory and gustatory sensitivity following capture.   Reed
et al. (1972) report that two cyprinid fishes, rosyside dace,
Clinostomus funduloides (Girard) and common shiners, Notropis
cornutus (Mitchill) gave behavioral responses to histamine which
resembled the fright reaction to natural alarm substance.   The
response to histamine was described, however, as weaker than the
response to the natural skin extract.   Spectrophotofluorometric
emission spectra indicated that the natural alarm substance of
rosyside dace and common shiners may be a small ringed or double
ringed compound (Reed et al. 1972).   Pfeiffer and Lemke (1973)
used separation over Diaflo membranes and thin layer chromatography to isolate autofluorescent material from the skin of the
European minnow which elicited a fright reaction in the giant danio,
Danio malabricus (Jerdon).   Similar extracts from the skin of giant
danios and the African long-finned tetra, Alestes longipinnis
(Günther) also elicited the fright reaction of giant danios.
Isoxanthopterin also produced a distinct fright reaction in giant
danios although it differed chromatographically from natural alarm
substance.   Pfeiffer and Lemke (1973) concluded from their findings
and consideration of the previous findings of Hüttel and Sprengling
(1943), Purrmann (1947), Korte and Tschesche (1951) and Ziegler-
Günder (1956) that the alarm substance was most likely a pterin.

Fright Reaction: The Behavioral Component

The European minnow detects alarm substance by smell rather
than taste (von Frisch 1941).   Tucker and Suzuki (1972) reported
an olfactory response to skin extract in white catfish but did not
exclude gustatory responses (Tucker 1973).   Other species have
apparently not been tested experimentally.

The fright reaction varies from species to species.   Schooling
fish such as the European minnow (von Frisch 1941) or fathead
minnow (Smith 1973) often school tightly, swim away from the area

in which the stimulus was perceived then continue to avoid the area for prolonged periods. The seven levels of intensity that were formulated by von Frisch (1941) to describe this typical response by schooling fish have been used extensively to quantify the reaction of fish to skin extracts (Schutz 1956, Pfeiffer 1962, 1963, 1974). Species with more specialized habits often have a fright reaction which differs from this typical response (Pfeiffer 1974). Crucian carp, Carrassius carassius (L.) and tench, Tinca tinca (L.), swim against the bottom and stir up mud. Some bottom species such as the gudgeon, Gobio gobio (L.), loach, Noemacheilus barbatulus (L.) and longnose sucker, Catostomus catostomus (Forster) "freeze" on the bottom taking advantage of their cryptic colouration (Schutz 1956, Pfeiffer 1963a). The hornyhead chub, Nocomis biguttatus (Kirtland), sinks to the bottom and spits gas (Verheijen 1963). Adult creek chub, Semotilus atromaculatus (Mitchill), sink to the bottom but juveniles flee like typical schooling fish (Newsome MS 1975). Other species such as the flying barbs, Esomus sp., flee to the surface and may jump out of the water (Schutz 1956). No doubt investigation will reveal that each species has a fright reaction suited to its own habits and predators.

Although the fright reaction may be initiated by a chemical stimulus it can be transmitted through a school of fish by vision. Blind European minnows react to the alarm substance as individuals while minnows with the olfactory organ destroyed will respond with a fright reaction when intact school-mates react to alarm substance (von Frisch 1941). Schutz (1956) demonstrated that when two aquaria were placed close together with the same number of fish in each tank an alarm substance-induced fright reaction in one tank would trigger a fright reaction in the fish in the adjacent tank although there was no exchange of water or chemical stimuli. Other experiments by Schutz (1956) and Verheijen (1956) confirmed visual transmission of the fright reaction (Pfeiffer 1974).

Taxonomic Distribution of the Alarm Substance System

The alarm substance - fright reaction system described above appears to be restricted to one phylogenetically related group of fish. It was thought that the system was restricted to the Ostariophysi (Pfeiffer 1962, 1963b) until Pfeiffer (1967a,b) reported that a similar system also occurred in the Gonorhynchiformes. Pfeiffer (1969a) has since reported the presence of cells resembling alarm substance cells in galaxiid fish, Galaxias attenuatus (Jenyns). These observations reinforce the suggestions of Greenwood et al. (1966) that the Ostariophysi are closely related to the Gonorhynchiformes and that both groups are derived from the Salmonoids, which include the Gallaxiidae (Pfeiffer 1974).

Within the Ostariophysi and Gonorhynchiformes, species tend
to respond most strongly to alarm substance from their own species
(Schutz 1956). They will also respond to alarm substances from
other species especially those closely related to the receiver
although interfamily responses do occur (von Frisch 1941, Schutz
1956, Pfeiffer 1963c, 1967b, 1974). The observation of Schutz
(1956) and others that species show a lower response threshold to
conspecific alarm substance than to interspecific alarm substance
is in apparent conflict with the conclusion of Pfeiffer and Lemke
(1973) that the alarm substance is identical in European minnows,
giant danios and long finned tetras. In vivo most pterins are
apparently conjugated with protein (Needham 1974), perhaps the
protein component of this association allows species identification.
Other components of crude skin extract may also be species
specific.

Possession of alarm substance and fright reaction is not
universal within the ostariophysian fishes. Several groups lack
one or both components of the alarm system. It is perhaps in
these exceptions that we may best seek evidence of the adaptive
significance of the alarm system.

Very young fish lack the fright reaction for 8-40 days after
the development of alarm substance and schooling behaviour (Table
1). The age at which young fish first produce alarm substance
has apparently not been established although typical alarm
substance cells can be observed in the skin of young fathead
minnows as small as 12 mm. fork length (about 30 days old).

Some species retain alarm substance cells but have lost the
fright reaction. Skin extract of blind cave characins
(Anoptichthyes sp.) is capable of eliciting a fright reaction in
their sighted relatives of the genus Astayanax (Schutz 1956,
Pfeiffer 1963c) but not in the blind characins themselves.
Pfeiffer (1966a) found that all 43 F1 hybrids between two spotted
astayanax, Astayanax bimaculatus (L.), and blind cave characins,
Anoptichthyes antrobius (Alvarez), responded to fright substance
with a fright reaction while 197 of 210 F2 fish responded. The
F1 X Anoptichthyes backcross produced 55 fish of which 47 showed
the fright reaction. Pfeiffer suggested that the response to alarm
substance in these characins is governed by two independently
segregating pairs of alleles with one dominant gene in either pair
being sufficient for a positive response. Thines and LeGrain
(1973) found that both blind characins, Anoptichthyes jordani
(Hubbs and Innes), and the blind cave cyprinid, Caecobarbus geertsi
(Boulenger), respond to skin extract with a feeding response
rather than a fright reaction. In addition to the blind cave fish,
the piranhas and silver dollars (Serrasalminae and Mylinae) also
have alarm substance cells capable of eliciting a reaction in
other fish but lack the fright reaction themselves (Pfeiffer 1963b,

Table 1. Age at which young cyprinids develop fright reaction, alarm substance and schooling. Data from Schutz (1956) and Pfeiffer (1963a, 1966a) modified from table in Pfeiffer (1963b).

Species	earliest fright response	earliest + test for alarm substance[a]	earliest schooling
		Age in Days of:	
zebra danio Brachydanio rerio (Ham.-Buch.)	28	20	20
redside shiner Richardsonius balteatus (Richardson)	42	28	--
European minnow Phoxinus phoxinus (L.)	51	38	26
dace Leuciscus leuciscus (L.)	59	--	31
bitterling Rhodius amarus (Bl.)	40[b]	0[b]	0[b]
Hybrid F1 Astayanax bimaculatus (L.) X Anoptichthyes antrobius (Alvarez)	46-56 (22 mm.)	--	--

a - earliest dates tested, b - days after leaving mussel where eggs develop.

1974). Pfeiffer (1963b) has suggested that loss of the fright reaction in cave fish is related to reduced predation in their specialized habitat. Piranhas may be adequately protected by their teeth, loss of fright reaction in the herbivorous silver dollars is less easy to explain.

Several groups of ostariophysian fishes lack both the alarm substance and fright reaction. These include the electric knife fishes (Gymnotidae), spotted headstander (Chilodus punctatus Müller & Troschel), pencilfish (Nannostominae), armoured catfish (Loricariidae), and banjo catfish (Aspredinidae) (Pfeiffer 1963b, 1974). Pfeiffer (1974) argues, on the basis of phylogeny, that these fishes have secondarily lost the alarm system. He suggests that this loss is related to their specialized habits.

In addition to young fish and species with permanent loss of
one or both components of the alarm system, a fourth category of
fishes show seasonal loss of alarm substance cells during the
breeding season (Smith 1973, 1974, 1976; Smith and Murphy 1974).
This seasonal alarm substance loss was first noted (Smith 1973)
in male fathead minnows, Pimephales promelas (Rafinesque), which
lose their alarm substance cells during the summer breeding season
and regain them in the fall (Fig. 1). Male fatheads clean a
spawning surface on rocks or driftwood by vigorous rubbing with
their dorsal pad and, after spawning, continue to rub the egg mass
with the pad (McMillan and Smith 1974). The dorsal pad consists
of loose connective tissue and thickened epidermis with many mucus
cells (Smith and Murphy 1974). Development of the dorsal pad,
eruption of tubercles and the loss of alarm substance cells all
occur in response to high androgen levels (Smith 1973,1974). Skin
extract from fish with androgen induced loss of alarm substance
cells does not elicit a fright reaction from conspecifics (Smith

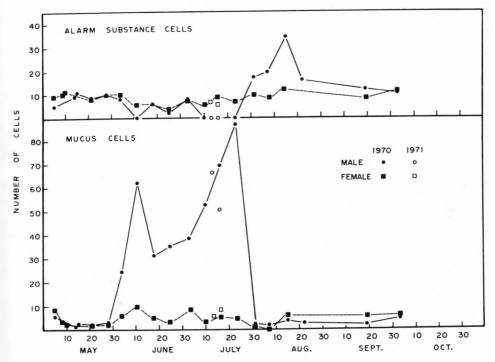

Figure 1. Seasonal variation of alarm substance cells in
the dorsal skin of male and female fathead minnows. Mucus cells
are included as an indicator of the spawning condition of male
fish, which develop a mucus-secreting dorsal pad during the breeding season. The peaks in mucus cell counts and troughs in alarm
substance cell counts correspond to periods of favourable weather.
Cell counts are means for 3 fish of number of cells in an area
0.18 mm X epidermal height. (Original data).

1973), although breeding males retain their fright response to
alarm substance from other individuals (Smith unpublished).
Female fatheads normally retain alarm substance cells through the
year (Fig. 1) but alarm substance cell loss and other male
secondary sexual characteristics can be induced in female fatheads
by treatment with 17 α-methyltestosterone (Smith 1974). The fathead minnow has a long breeding season (May - Aug.) during which
individual fish reach peak spawning condition and lose their
alarm substance cells at different times. Some seasonal decline
in alarm substance cell numbers also occurs in female fatheads
(Fig. 1).

     I have suggested (Smith 1973, Smith and Murphy 1974) that
seasonal alarm substance loss in male fatheads is an adaptation to
their abrasive spawning habits, reducing the chance of the alarm
system "misfiring" and interfering with spawning or parental care.
In an attempt to assess this possibility nine other species of
cyprinids and one catostomid were examined histologically for the
presence of alarm substance cells during the breeding and non-
breeding seasons (Smith 1976). Seasonal alarm substance cell loss
was found in males of seven cyprinid species and in females of
two species (Table 2). Six of these species have abrasive spawning
behavior. The seventh, the pearl dace Semotilus margarita (Cope)
has only been observed spawning over two days in 1928, by Langlois
(1929) who reported no nest digging or other abrasive spawning.
The two cyprinids and the catostomid (the white sucker) which
retained alarm substance cells during spawning are not reported to
dig nests or clean spawning surfaces. Although they lack such
abrasive spawning habits, a statistically significant reduction in
the number of alarm substance cells occurred during the breeding
season in male white suckers. Seasonal alarm substance cell loss
has not been reported in other ostariophysians although data from
female fatheads (Fig. 1) and male white suckers may indicate that
seasonal reductions--rather than complete loss of alarm substance
cells is fairly common.

     The data in Table 2 generally supports the thesis that
seasonal alarm substance cell loss is related to abrasive spawning
behavior. Two problem cases need explanation: pearl dace and female
stonerollers have not been described as digging nests or cleaning
spawning surfaces but they do lose their alarm substance cells.
O.R. Smith (1935) observed both male and female stonerollers with
damaged skins during the breeding season, the result, he felt, of
contact with the sharp tubercles of the males. However Miller
(1962) describing the same species makes no mention of these
phenomena. Both pearl dace and stonerollers are characterized, as
adults, by patches of replaced scales, indicating earlier skin
damage (Langlois 1929, Smith 1935). It appears that these fish
may suffer skin damage during breeding from causes other than nest
building.

Table 2. Relation between abrasive spawning behavior and seasonal loss of alarm substance cells. "Abrasive spawning" indicates nest digging or cleaning of spawning surface. (-) indicates that alarm substance cell loss is nearly complete but not absolute. Based on data from Smith (1973, 1976).

Species	Abrasive Male breeding behavior	Alarm substance cells present in:		
		breeding male	breeding female	non-breeding
Lake chub Couesius plumbeus (Agassiz)	-	+	+	+
Hornyhead chub Nocomis biguttatus (Kirtland)	+	-	+	+
Common shiner Notropis cornutus (Mitchill)	+	-	+	+
Bullhead minnow Pimephales vigilax (Baird & Girard)	+	-	+	+
Bluntnose minnow Pimephales notatus (Rafinesque)	+	-	+	+
Fathead minnow Pimephales promelas (Rafinesque)	+	-	+	+
Stoneroller Campostoma anomalum (Rafinesque)	+	-	(-)	+
Longnose dace Rhinichthyes cataractae (Valenciennes)	-	+	+	
Creek chub Semotilus atromaculatus (Mitchill)	+	(-)	+	+
Pearl dace Semotilus margarita (Cope)	-	-	(-)	+

Table 2. Cont'd.

Species	Abrasive Male breeding behavior	Alarm substance cells present in:		
		breeding male	breeding female	non-breeding
White sucker Catostomus commersoni (Lacepede)	−	+	+	+

These seasonal changes in alarm substance cell number suggest
an explanation for the remarkable radiation of nest-building
cyprinids in North America. Once acquired, seasonal alarm substance loss would open to these cyprinids the sorts of abrasive
digging and surface-cleaning patterns of parental care which are
well exploited by the cichlids and centrarchids but rare in the
ostariophysi (Smith 1976).

THE ADAPTIVE SIGNIFICANCE OF THE FISH ALARM PHEROMONE SYSTEM

The fish alarm substance system seems to belong to the same
class of animal communication systems as the alarm calls of birds,
the visual alarm signals of ungulates and the alarm pheromones of
snails (Snyder 1967), sea anemones (Howe and Sheikh 1975), tadpoles
(Eibl-Eibesfeldt 1949; Hrbacek 1950) and social insects. Some of
these alarm systems pose an evolutionary problem in that they
appear to be examples of altruism: the sender of the alarm signal
may incur increased risk of predation or some other evolutionary
cost in order to increase the survival of other individuals. The
problems posed by such animal altruism have recently been discussed
by Williams (1966), Trivers (1971) and Eberhard (1975). Wilson
(1975, p. 123) summarizes current explanations of the adaptive
value of animal altruism.

1) warning signals function in the breeding season to
protect the brood and mate and are extended into the off season
even though they are no longer adaptive in the absence of strong
kin selection (Williams 1966),

2) warning signals are fixed by interdemic (group) selection
(Wynne-Edwards 1962),

3) warning signals are fixed by kin selection and sustained
outside the breeding season by the probability that close kin are
near enough to be helped (Maynard-Smith 1965). Eberhard (1975)
suggests that aiding distant kin may be sufficient to maintain
"cheap", high-benefit altruism in a population.

4) warning signals evolve by individual selection, contrary
to appearances. Trivers(1971) suggests that the signaler benefits
by reducing the probability that the predator will form a search
image, continue searching the sender's area, or form a preference
for the sender's species.

A fifth explanation, reciprocal altruism (Trivers 1971) does not seem appropriate to non-mammalian organisms.

The adaptive significance of the fish alarm substance system can be examined from the point of view of the sender or of the receiver. The receiver's response, the fright reaction, seems the easier of the two to account for in adaptive terms. Snyder (1967), in his discussion of an alarm reaction in aquatic gastropods, points out that it is probably to the receiver's advantage to respond to any metabolite released from an injured conspecific. He suggests that the first step in the evolution of an alarm pheromone system would be a fright reaction to the odour(s) of damaged conspecific tissue. The ostariophysian fish have progressed beyond this stage, if they ever went through it. This is demonstrated by the absence of a fright reaction to conspecific skin extracts from skin lacking alarm substance cells such as glass catfish barbel skin or breeding male fathead minnow skin. The advantages of restricting the fright reaction to a specific alarm substance may include: reduced interference with chemosensory response to food, increased species specificity, and selection of the molecules with the best information transmission characteristics. Acquisition of the fright reaction would probably be facilitated if the alarm substance were inherently obnoxious.

It is more difficult to explain the adaptive advantage to the sender of maintaining unifunctional alarm substance cells. The evolutionary "cost" to the sender in this alarm system is not death or injury, which would occur anyway, but rather the cost of growing and maintaining the alarm substance cells in the skin.

Williams' (1966) suggestion that warning signals function primarily during the breeding season to warn offspring is not compatible with the details of the fish alarm system. Young fish do not respond to alarm system until after they are free of parental care and many ostariophysian species lack any association between adults and young. In several species which do have parental care the parental males lose their alarm substance cells during the breeding period (Table 2, Smith 1976). Williams himself (1964), in a discussion of fish alarm substance, argues that the release of the olfactory stimulus is "an incidental effect of damage" and that it "gives no indication of being a means of warning other members of the school". He suggests that the so called alarm substance is a repellent with a primary function of stimulating predators to release the prey fish.

Fixing of alarm signals in a population by interdemic selection (Wynne-Edwards 1962) may be possible. Wilson (1975), for example, presents situations which could conceivably lead to effective interdemic selection. These situations seem, however, sufficiently unlikely that it is worthwhile seeking other explanations for the

fixation of the fish alarm substance system and its retention in
so many species.

Kin selection, at first glance, does not appear applicable
to the fish alarm substance system. Maynard-Smith (1965) and
Hamilton (1964a, b) have emphasized the importance of close
genetic relationship in kin selection. Fish schools are not
usually made up of closely related individuals and parent-off-
spring alarm substance interactions are unlikely (see above).
Eberhard (1975), however, argues that in situations where the
benefit to the receiver is high (or many receivers are benefitted)
and where the cost to the sender is low then kin selection can be
effective as long as the receiver(s) are more closely related to
the sender than the average degree of relationship in the
population as a whole. The fish alarm substance system may fit
these requirements. Although the fish in an interaction may not be
closely related their mere physical proximity suggests that they
are more likely related than spatially separated members of the
population (Snyder 1967). The total benefit of the alarm signal
to all receivers in a single episode could conceivably be quite
high. The cost of maintaining alarm substance cells in the skin
is hard to estimate. If we assume alarm substance cells are
unifunctional then their retention in cave fish and piranhas,
which lack the fright reaction, would indicate a very low cost
indeed.

Triver's (1971) suggestion that the signaler, itself, may
benefit from the ultimate effects of the alarm signal would only
apply to the fish alarm system if the victim (sender) were to
escape from a predator after having sustained skin damage. In
most cases in nature the sender will not be in a position to
benefit from long term effects of his transmission.

Fish alarm substance may benefit the sender by reducing
cannibalism on its own, or closely related, young (von Frisch
1941, Schutz 1956, Pfeiffer 1962, 1963b, 1974). Adult fish
preying on conspecific young would release their species' alarm
substance and frighten themselves. European minnows are sometimes
frightened after eating young minnows (Berwein 1941) and young
northern squawfish, Ptychocheilus oregonense (Richardson), show a
marked fright reaction after eating smaller conspecifics (Pfeiffer
1962c). This effect could explain the presence of alarm substance
in the skin of young fish prior to their development of a fright
reaction.

Verheijen (1962, 1963) questioned the anticannibalism effect or
the basis of his observation that freshly collected creek chub
would eat smaller conspecifics. Pfeiffer (1963) countered that
recently disturbed fish often fail to show the fright reaction.
Later Verheijen and Reutter (1969) showed that water from a tank in

which a pike, Esox lucius (L.), had eaten a roach, Rutilus rutilus
(L.), elicited a fright reaction in blind minnows in 6 of 7 tests
while water from a tank in which a large minnow had eaten a roach
only elicited a fright reaction in 2 cases out of 12, indicating
that alarm substance was less likely to be released by the smaller
teeth of the cyprinid predators.

Moshenko and Gee (1973) found that creek chub in the Mink
River, Manitoba ate primarily noncyprinid fishes even though
cyprinids were more abundant than noncyprinids in the river. They
suggested that this was due to the effect of the alarm substance
on the cyprinid predator. Newsome (MS 1975), however, working
on the same population found that creek chub would eat damaged
cyprinids and concluded that cyprinids were inaccessible to creek
chub for other reasons. Large minnows and roach also chased and
ate small roach without showing a fright reaction (Verheijen and
Reutter 1969).

The observations of Berwein (1941), Pfeiffer (1962c) and
Verheijen and Reutter (1969) show that fright reaction can occur
as a result of intraspecific predation, although it is not universal.
One must invoke kin selection or enhanced probability of escape
to account for a selective advantage to the sender. The anticannibalism theory does not account for the retention of alarm
substance cells in large individuals of non-predatory species.

This discussion has been based on the assumption that alarm
substance cells are unifunctional, serving only to warn conspecifics.
Williams' (1964) suggestion that the primary function of alarm
substance cells is to release a repellent compound to discourage
predators would account for the development and maintenance of
special cells by the sender, including species or age groups which
lack the fright reaction, and, in part, for the acquisition of the
fright reaction by receivers. An analogous situation may occur
in bufonid tadpoles which show a fright reaction in the presence
of bufotoxins (Kulzer 1954). The feeding response of cave fish
to skin extract containing alarm substance and the large number
of predators which prey successfully on ostariophysians are
arguments against the repellent hypothesis. On the second point
it should be noted that antipredator adaptations are seldom
completely effective and that coevolution may result in predators
becoming specifically adapted for overcoming a defense mechanism.

The adaptive significance of fish alarm substance remains an
intriguing question. Further research is required, particularly
on Eberhard's (1975) kin selection theory and on Wilson's (1964)
chemical repellent theory.

REFERENCES

Berwein, M. 1941. Beobachtungen und Versuche über das gesellige
    Leben der Ellritze. Z. Vergl. Physiol. 28: 402-420.

Eberhard, M.J.W. 1975. The evolution of social behavior by kin
    selection. Quart. Rev. Biol. 50: 1-34.

Eibl-Eibesfeldt, I. 1949. Über das Vorkommen von Schreckstoffen
    bei Erdkrötenquappen. Experientia 5: 236.

Frisch, K. Von. 1938. Zur Psychologie des Fisch-Schwarmes.
    Naturwissenschaften 26: 601-606.

Frisch, K. Von. 1941. Über einen Schreckstoff der Fischhaut und
    seine biologische Bedeutung. Z. Vergl. Physiol. 29: 46-145.

Frisch, K. Von. 1961. Gerüche als Lock- und Schreckstoffe. In:
    Ausgewählte Vorträge 1911-1969; ed.: K. Von Frisch,
    (BLV München, Basel, Wien, 1970), pp. 203-214.

Greenwood, P.H., D.E. Rosen, S.H. Weitzman and G.S. Myers. 1966.
    Phyletic studies of teleostean fishes, with a provisional
    classification of living forms. Bull. Am. Mus. Nat. Hist.
    131: 339-455.

Hamilton, W.D. 1964a. The genetical theory of social behaviour.
    I. J. Theoret. Biol. 7: 1-16.

Hamilton, W.D. 1964b. The genetical theory of social behaviour.
    II. J. Theoret. Biol. 7: 17-52.

Howe, N.R. and Y.M. Sheikh. 1975. Anthopleurine: a sea anemone
    alarm pheromone. Science, N.Y. 189: 386-388.

Hrbacek, J. 1950. On the flight reaction of tadpoles of the
    common toad caused by chemical substances. Experientia 6:
    100-102.

Hüttel, R. 1941. Die chemische Untersuchung des Schreckstoffes
    aus Elritzenhaut. Naturwissenschaften 29: 333-334.

Hüttel, R. and G. Sprengling. 1943. Über Ichthyopterin, einen
    blaufluorescienenden Stoff aus Fischhaut. Liebigs Ann. 554:
    69-82.

Korte, F. and R. Tschesche. 1951. Über Pteridine. V. Mitt.
    Die Konstitution des Ichthyopterins. Chem. Ber. 84: 801.

Kulzer,E. 1954. Untersuchungen über die Schreckreaktion bei Erdkrötenkaulquappen (Bufo bufo L.). Z. Vergl. Physiol. 36: 443-463.

Langlois, T.H. 1929. Breeding habits of the northern dace. Ecology 10: 161-163.

Maynard-Smith, J. 1965. The evolution of alarm cells. Am. Natur. 99: 59-63.

McMillan, V.E. and R.J.F. Smith. 1974. Agonistic and reproductive behaviour of the fathead minnow (Pimephales promelas Rafinesque). Z. Tierpsychol. 34: 25-58.

Moshenko, R.W. and J.H. Gee. 1973. Diet, time and place of spawning, and environments occupied by the creek chub (Semotilus atromaculatus) in the Mink River, Manitoba. J. Fish. Res. Bd. Canada 30: 357-362.

Needham, A.E. 1974. The Significance of Zoochromes, Zoophysiology and Ecology 3, Springer-Verlag, N.Y. 429 pp.

Newsome, G.E. 1975. A study of prey preference and selection by creek chub, Semotilus atromaculatus, in the Mink River, Manitoba. Unpublished Ph.D. Thesis, Dept. of Zoology, University of Manitoba. 93 pp.

Pfeiffer, W. 1960a. Über die Verbreitung der Schreckreaktion bei Fischen. Naturwissenschaften 47: 23.

Pfeiffer, W. 1960b. Über die Schreckreaktion bei Fischen und die Herkunft des Schreckstoffes. Z. Vergl. Physiol. 43: 578-614.

Pfeiffer, W. 1962a. The fright reaction of Fish. Biol. Rev. Cambridge Phil. Soc. 37: 495-511.

Pfeiffer, W. 1962b. The fright reaction in American fish. Am. Zool. 2: 437.

Pfeiffer, W. 1963a. The fright reaction in North American fish. Can. J. Zool. 41: 69-77.

Pfeiffer, W. 1963b. Alarm substances. Experientia 19: 113-123.

Pfeiffer, W. 1963c. Vergleichende Untersuchungen über die Schreck-reaktion und den Schreckstoff der Ostariophysen. Z. Vergl. Physiol. 47: 111-147.

Pfeiffer, W. 1966a. Über die Vererbung der Schreckreaktion bei Astyanax (Characidae, Pisces). Z. Vererbungslehre 98: 97-105.

Pfeiffer, W.   1966b.   Die Schreckreaktion der Fische und Kaulquappen.
    Naturwissenschaften 53: 565-570.

Pfeiffer, W.   1967a.   Schreckreaktion und Schreckstoffzellen bei
    Kneriidae und Phractolaemidae (Isospondyli, Pisces).   Natur-
    wissenschaften 54: 177.

Pfeiffer, W.   1967b.   Schreckreaktion und Schreckstoffzellen bei
    Ostariophysi und Gonorhynchiformes.   Z. Vergl. Physiol. 56:
    380-396.

Pfeiffer, W.   1969   Kolbenzellen in der Epidermis der Galaxiidae,
    Galaxias attenuatus (Salmoniformes).   Experientia 25: 432.

Pfeiffer, W.   1974.   Pheromones in fish and amphibia.   In:
    Pheromones, Martin C. Birch (ed.), Frontiers of Biology,
    Vol. 32.   North-Holland Pub. Co., Amsterdam.   495 pp.

Pfeiffer, W. and J. Lemke.   1973.   Untersuchungen zur Isolierung
    und Identifizierung des Schreckstoffes aus der Haut der Elritze,
    Phoxinus phoxinus (L.) (Cyprinidae, Ostariophysi, Pisces).
    J. Comp. Physiol. 82: 407-410.

Pfeiffer, W. and T.F. Pletcher.   1964.   Club cells and granular
    cells in the skin of lamprey.   J. Fish. Res. Bd. Canada
    21: 1083-1088.

Pfeiffer, W., D. Sasse and M. Arnold.   1971.   Die Schreckstoffzellen
    von Phoxinus phoxinus und Morulius chrysophakedion (Cyprinidae,
    Ostariophysi, Pisces).   Histochemische und elektronenmikro-
    skopische Untersuchungen.   Z. Zellforsch. 118: 203-213.

Purrmann, R.   1947.   Pterine.   Fiat-Bericht.   Biochemie 39, I.
    Teil, 84.

Reed, J.R., W. Wieland and T.D. Kimbrough.   1972.   A study on the
    biochemistry of alarm substances in fish.   Proc. of the 26th
    Ann. Conference, S.E. Assoc. of Game & Fish Commissioners,
    Knoxville.   pp. 608-610.

Reutter, K. and W. Pfeiffer.   1973.   Fluorescence microscopical
    demonstration of the alarm substance in the alarm substance
    cells of the European minnow Phoxinus phoxinus (L.).   J.
    Comp. Phys. 82(4): 411-418.

Schutz, F.   1956.   Vergleichende Untersuchungen über die Schreck-
    reaktion bei Fischen und deren Verbreitung.   Z. Vergl.
    Physiol. 38: 84-135.

Smith, O.R. 1935. The breeding habits of the stoneroller minnow
    (Campostoma anomalum Rafinesque). Trans. Amer. Fish. Soc.
    65: 148-151.

Smith, R.J.F. 1973. Testosterone eliminates alarm substance in
    male fathead minnows. Can. J. Zool. 51: 875-876.

Smith, R.J.F. 1974. Effects of 17 α-methyltestosterone on the
    dorsal pad and tubercles of fathead minnows (Pimephales
    promelas). Can. J. Zool. 52: 1031-1038.

Smith, R.J.F. 1976. Seasonal loss of alarm substance cells in
    North American cyprinoid fishes and its relation to abrasive
    spawning behaviour. Can. J. Zool. (in press).

Smith, R.J.F. and B.D. Murphy. 1974. Functional morphology of
    the dorsal pad in fathead minnows (Pimephales promelas
    Rafinesque). Trans. Amer. Fish. Soc. 103: 65-72.

Snyder, N.F.R. 1967. An alarm reaction of aquatic gastropods
    to intraspecific extract. Cornell University, Agricultural
    Experiment Station, New York State College of Agriculture,
    Ithaca, New York, Memoir 403. 122 pp.

Snyder, N. and H. Snyder. 1970. Alarm response of Diadema
    antillarum. Science, N.Y., 168: 276-278.

Thinès, G. and J.M. LeGrain. 1973. Effects of alarm substance
    on the behavior of the cave fish Anoptichthys jordani and
    Caecobarbus geertsi. Ann. Spéléol. 28(2): 291-297.

Trivers, R.L. 1971. The evolution of reciprocal altruism.
    Quart. Rev. Biol. 46: 35-57.

Tucker, D. 1973. Rapid decline of olfactory and gustatory
    receptor sensitivities of wild catfish (Ictaluridae) after
    capture. J. Fish. Res. Bd. Can. 30(8): 1243-1245.

Tucker, D. and N. Suzuki. 1972. Olfactory responses to Schreckstoff
    of catfish. In: Olfaction and Taste 4. Proc. 4th Int.
    Symp. held in Starnberg, Germany, August 2-4, 1971; ed.:
    D. Schneider (Wissenschaftl. Verlagsges. mbH, Stuttgart)
    pp. 121-127.

Verheijen, F.J. 1956. Transmission of a flight reaction amongst a
    school of fish and the underlying sensory mechanisms.
    Experientia 12: 202-204.

Verheijen, F.J. 1962. Alarm substance and intra-specific predation
    in Cyprinids. Naturwissenschaften 49: 356.

Verheijen, F.J.   1963.   Alarm substance in some North American
    cyprinid fishes.   Copeia 1963: 174-176.

Verheijen, F.J. and J.H. Reutter.   1969.   The effect of alarm
    substance on predation among cyprinids.   Anim. Behav. 17:
    551-554.

Williams, G.C.   1964.   Measurement of consociation among fishes
    and comments on the evolution of schooling.   Publ. Mus. Mich.
    State Univ. Biol. Ser. 2: 349-384.

Williams, G.C.   1966.   Adaptation and Natural Selection.   Princeton
    Univ. Press, Princeton.   307 pp.

Wilson, E.O.   1975.   Sociobiology.   Harvard Univ. Press, Cambridge,
    Mass.   697 pp.

Wynne-Edwards, V.C.   1962.   Animal Dispersion in Relation to Social
    Behaviour.   Oliver and Boyd.   Edinburgh.

Ziegler-Günder, I.   1956.   Untersuchungen über die Purin- und
    Pterin-pigmente in der Haut und in den Augen der Weissfische.
    Z. Vergl. Physiol. 39: 163-189.

THE STUDY OF CHEMICAL COMMUNICATION IN FREE-RANGING MAMMALS

L. David Mech and Roger P. Peters

U. S. Fish and Wildlife Service, Ncrth Central Forest
Experiment Station, Folwell Ave., St. Paul, MN 55108
Department of Psychology, Fort Lewis College, Durango,
Colorado 81301

Although investigations into chemical communication in
vertebrates have been conducted for a long period, this field of
study actually came of age only a year or so ago, with the founding
of the Journal of Chemical Ecology. The relatively delayed
development of the field resulted from the extreme difficulty
encountered in attempting to cope with its subject matter.

Most researchers interested in chemical communication ideally
would like to understand the entire process as it functions in
nature. Thus investigating the subject in the wild would seem to
be a natural approach. However, in general this is the most
difficult approach, so many workers have turned to studies of
captive animals to help provide insight into the way chemical
communication functions in the wild.

The excellent reviews of chemical communication that have
appeared recently, such as those of Ralls (1971), Cheal and Sprott
(1971), Eisenberg and Kleiman (1972), Johnson (1973), Ewer (1973),
Mykytowycz (1974), and Cheal (1975) demonstrate the role captivity
studies have played in advancing knowledge about this field.

Studying chemical communication in captive situations has
two main advantages: (1) it allows rigorous experimentation under
controlled conditions, providing "hard" quantitative results, (2)
it can provide hypotheses about the natural functioning of chemical
communication and thus stimulate more productive research on free-
ranging animals.

Nevertheless, at some point we must get into the field to learn how chemical communication operates under natural conditions. Thus it is important to study free-living animals whenever and wherever possible. The type of chemical communication most amenable to observation is scent-marking, because characteristic postures are often associated with application of odor and because the marks themselves sometimes have visual properties. For this reason our discussion deals primarily with scent-marking.

The greatest need in scent-marking studies is good descriptive information. How do individuals of any given species scent-mark? How often? At what times and seasons and in what areas? What are the ages and sex of the markers? What is the mode of marking (what gland or organ is involved)? How long does the mark last? How frequently are marks perceived? And the key question-- what is the effect(s) on the perceiving individual? Many similar questions can be asked about the other forms of chemical communication. By gaining enough information on each of these aspects of olfactory signalling in any one species, and integrating that information, we eventually will understand the chemical communication system for that species. At the same time we will no doubt gain a much greater appreciation about the degree of social organization shown by populations of that species.

However, this is all much easier said than done. There are many major difficulties in trying to study chemical communication in free-ranging mammals: (1) many species are nocturnal or highly secretive and shy, so are hard to observe, (2) often the age and/or sex of individuals cannot be identified, (3) many move around so quickly or travel through such thick vegetation that they are difficult to keep track of, (4) the scent-marks of most mammals cannot be seen, (5) scent-marks may have long-term, and far-ranging, consequences to the receiver, and these effects may be impossible for an investigator to determine, even if he/she could observe the receiver encountering the marks, which in itself usually presents great difficulty, and (6) scent released by one individual when interacting with another cannot usually be perceived by an observer.

Because of all these problems, the study of chemical communication in free-ranging mammals requires considerable ingenuity. Several main approaches have been used successfully, but the choice of species is critical. Given present technology, it would be folly to try to learn much about olfactory signalling in some small subterranean-dwelling mammal, for example. On the other hand, a species such as the barren-ground caribou (<u>Rangifer tarandus</u>), which is large, social, and easily observable, would seem to be a much more suitable species, although it appears that no one has attempted such work with this animal. Large diurnal species that inhabit the African plains would also be excellent subjects to study.

When attempting to deal with nocturnal mammals, one ought to choose species native to temperate and frigid zones. There, at least during winter, an investigator can take advantage of snow-tracking from which to infer considerable information about certain types of scent-marking, especially urine-marking in canids. With care and persistence, one can obtain a precise picture about many aspects of an animal's behavior from its record in the snow.

An extremely valuable supplementary technique of chemical communication studies is to live-trap the subject animals first. Through examining an individual, considerable information can be obtained about its age, sex, and general condition. Blood sampling and analysis can provide excellent insight into the nutritional and physiological state of the subject (Seal et al., 1975), and glandular secretions can be sampled and analyzed from the same individual (Peters and Mech, 1972). In addition, the subject can then be marked in any of several ways so as to be identifiable in the field. This then allows correlation of the background infor-mation about an individual with the observations of its general behavior (e.g. its travels, activity, and usual companions) and with data on its communication.

The most refined type of marking of individuals is the radio-marking technique (Cochran and Lord, 1963), which is of particular value with secretive, wilderness-dwelling species (Mech, 1974). This technique not only permits the identification of an individual (by frequency of radio signal), but also allows an investigator to keep close track of the movements and behavior of that individual and eventually to build a long history about it. Such a history is invaluable in trying to interpret data about that individual's olfactory signalling. Furthermore, the accumulation of histories of several members of the same popula-tion provide additional background data, forming an excellent context within which to interpret the observations of scent-marking and other communicatory behavior.

Another technique that could prove very helpful in studying certain aspects of chemical communication in the wild is the use of a dog (Canis familiaris) specially trained to detect a particular type of scent-mark (Moore and Marchinton, 1974). We have found that even without special training, a dog can detect wolf (Canis lupus) urine marks and distinguish them from those of foxes (Vulpes fulva), so presumably even finer distinctions could be expected with formal training for the purpose.

In studying chemical communication in free-ranging mammals, as much information as possible about the study species should be collected, for it is impossible to say just what facts may be most important. For this reason, it is helpful to have a hypothesis from captivity studies to test in the wild, which then allows a finer focus for the research. Conversely, general work on olfactory signalling in a natural population may provide several hypotheses that can then best be tested with captive animals. Either way, studies in the wild are imperative.

The ways in which the above procedures have been put to use in studies of free-ranging mammals have provided a variety of interesting information. We even believe that it is such studies that have made those of us who are interested in mammalian chemical communication believe that we are finally beginning to understand the process. Although it is impossible in this short review to discuss all the field studies of olfactory communication in mammals, those discussed below illustrate how various techniques have been used and how much information can result from them.

At times a field worker just happens to be in the right place at the right time to observe some undescribed behavior pattern; in such cases, the importance of taking detailed notes and publishing the information cannot be overemphasized. Such an observation was reported by a field biologist who watched a beaver (Castor canadensis) respond to an artificial scent post he had made to entice beavers to a live-trap placed at one of their scent mounds. The scent was made from the castoreum of a freshly killed beaver, mixed with glycerin and anise oil. In the author's words (Aleksiuk, 1968:760): "When I had completed the set and moved about 50 ft. away, the /large male/ beaver swam directly to the trap, climbed out of the water beside it and smelled the scent I had deposited. The animal hissed loudly all the while, which I believe indicates aggression in beaver. It then deposited some of its own castoreum on the site. The sound produced by the emmission of the castoreum could easily be heard from 50 ft. away. The beaver returned to the water and swam about, hissing loudly. It followed me closely, repeating at four more trap sites the performance witnessed at the first site."

A more detailed description of the frequency of scent-marking in wolverines (Gulo gulo) was recently provided by Pulliainen and Ovaskainen(1975). They recorded 26 scent-marks along some 7.5 km of tracks, and mapped the locations of these marks along the route.

Although these are isolated, casual observations, they certainly add more insight into olfactory communication in these two species than was available previously, and suggest approaches that could produce considerably more information.

A more deliberate study of scent-marking was conducted in white-tailed deer (Odocoileus virginianus), which suddenly made

sense out of those many "antler rubs" and "ground scrapes" so
familiar but mysterious to deer hunter and deer biologist alike.
The technique used was to observe individual deer preconditioned
to reduced flight distance (Marchinton, 1969), outfitted with
radio-collars, and released to the wild. The authors (Moore and
Marchinton, 1974:447) were able to describe four types of scent-
marking and conclude the following: "Rubs and scrapes were marked
with visual and olfactory signals. Rubs apparently expressed
dominance during the breeding season whereas scrapes facilitated
communication with does, although both types of signposts probably
also had other functions. Urination on tarsal glands and resulting
scent trails may have functioned to attract does and intimidate
bucks. Our observations suggest that in some situations, bucks
defend areas and utilize signpost marking and displays to delineate
these areas."

Similar investigations, but on black-tailed deer (Odocoileus
hemionus columbianus), were taken several steps further by
Müller-Schwarze (1971, 1972). He observed deer in the wild, but
also combined these studies with intensive observation and
experimentation on animals in an enclosure, and with chemical
analyses of the secretions involved in the olfactory signalling.
Histological studies of the scent glands (Quay and Müller-Schwarze,
1970) helped round out this investigation. Through these
techniques, Müller-Schwarze (1971:141), was able to determine the
following: "For short-range sex, age and individual recognition,
the tarsal scent is most important. Gas Chromatograms show
differences related to sex, age and individuals. Metatarsal scent
is discharged in 'fear-inducing' situations. Female urine attracts
males. Rub-urinating serves as a distress signal in fawns, but as
a threat in adult males and females. The home range of deer is
marked by rubbing the forehead against dry branch tips." He also
was able to develop bioassays for testing fractions of tarsal and
metatarsal scents and to start defining the active components in
both scents.

The technique of supplementing studies of chemical communication in free-ranging mammals with supporting methods from the lab
and from captive situations is no doubt the most fruitful of all
approaches in trying to understand olfactory signalling. One of
the most intensive and successful investigations using this
approach is that involving the wild rabbit (Oryctolagus cuniculus)
in Australia.

Mykytowycz (1974) summarized the results of this long-term
study, based on many of his own publications and those of Myers
and Poole (1959), Myers et al. (1971), and others of his
associates. They learned that the wild rabbit lives in colonies,
with subgroups being territorial and containing both males and

females in separate dominance hierarchies. In addition, each
individual occupies its own space, which may overlap with those
of others. The space of the dominant male in each subgroup
extends throughout the territory of the group. Extra individuals,
mostly males, live outside the subgroup territory.

Scent-marking plays an important role in this spatial
organization, and three glands are involved, each with its
distinct chemical secretion. The submandibular, or chin, gland
and the anal glands function in territoriality, with the chin
gland secretion being applied to prominent objects in the
environment, and the anal gland secretion coating the rabbits'
fecal pellets. The pellets are deposited at definite locations
called "dung hills", which are usually concentrated at points in
the territory most exposed to contact from neighboring groups.
The inguinal gland functions in other behavior, particularly in
helping young rabbits learn the odor of their mother.

The dominant male rabbit possesses the largest glands, and
those with the greatest secretory activity. Although all members
of a group add to the general odor of the territory, it appears
that the dominant male may make the greatest contribution to the
individual odor of the area. Since individual rabbits seek to
live near others, apparently they find colonies and remain there
through detecting the general "rabbitty" odor of each colony.
When a strange rabbit enters an established subgroup, it
cautiously explores the various scent posts, suspends its own
marking behavior, and displays submissively to all permanent
members of the group.

Even within a subgroup territory there are smaller areas
defended by breeding females with kittens, and marked by those
females both with pellets and with urine. Mothers also leave a
few pellets in the nest, and the kittens learn to recognize the
anal odor of the mother and are more attracted to it than to any
other natural rabbit odor. The kittens apparently learn the
general odor of the subgroup as well. This knowledge serves them
well, for experiments have shown that if they were to encounter
females from strange subgroup territories, they would be more
severely attacked than by members of their own territory.

It should be clear from the rabbit studies that such
detailed information about chemical communication automatically
lends a great deal of insight into the social system of the species.
In fact, it appears that one could hardly claim to understand a
mammal's social system until that person gains detailed knowledge
about the species' olfactory signalling.

It was this consideration that led us into scent—marking studies of wolves (<u>Canis lupus</u>). Mech had been investigating the social spacing of a population of wolves in northeastern Minnesota for a number of years (Mech and Frenzel, 1971; Mech 1972, 1973, 1974, 1975). He had learned that the population was organized primarily into territorial packs that were family groups. However, lone wolves also dispersed from the packs and circulated within the population, trying to avoid the packs. If they could locate vacant areas suitable for settling down in with members of the opposite sex, mating, and producing new packs, they did so. As this information was developed, several new questions presented themselves however: Since there was little evidence of direct encounters between packs, how are territories maintained? How do dispersed wolves avoid pack territories? How do males and females find each other? How do they locate areas vacant of other wolves? Obviously scent—marking became the prime suspect in our search for answers to these questions (Peters and Mech, 1975).

Scent—marking has long been thought to function in the spacing of wolves, but it was apparent that getting from suspicion to hard data would be a long jump. The first thing we realized was that although wolves clearly had a well developed scent—marking system, very little was known about it. Only isolated data on types of marks and marking postures were available from captivity studies. The most elementary aspects of the marking system were yet to be described. No one knew how often wolves marked, or under what conditions; what stimulated them to mark; where they marked; or what their response was to previous marks. To compound the problem, wolves live in low densities in inaccessible wilderness areas, and are shy and secretive.

Thus to try to shed some light on the wolf's scent—marking system, we tried to use as many approaches as we could. In doing so we had the benefit of having live-trapped and examined a large number of wolves from our study population, and we had blood-samples, anal gland samples, pack histories, and movement data from them as a background. We could identify each of our study packs and had excellent knowledge about their age and sex composition. We decided that snow-tracking of known packs would be our basic technique, supplemented by aerial observation, captivity studies, and laboratory analyses.

Over a period of three winters, we tracked wolves in the snow for a total of 240 km and recorded 584 raised-leg urinations, which we designated "RLUs" (Peters and Mech, 1975). It became apparent that these RLUs were the most important scent-marks, although scats, scratching, and squat urinations (SQUs) no doubt also functioned as marks at times. From captivity studies we learned that RLUs were performed primarily by the top-ranking

animals, usually the alpha males, but sometimes by alpha females.
In the field we discovered that the RLU rate increased from
January through the breeding season in late February and then
dropped off to about one-third the rate in late March. Thus it
appeared that RLU was at least partly a function of dominance and
breeding. This was confirmed by the fact that the curve for the
rate of squat urination, often performed by the alpha female,
paralleled that of RLU, and frequently SQU and RLU were found
together.

RLUs were most often made on well established roads and
trails, particularly near junctions with other travel routes. It
was not unusual for a pack of wolves to cut cross-country with a
very low rate of marking but then suddenly mark just as they
reached a road or trail. The RLU rate when the wolves traveled
off trails was only 1.7 per km compared to 3.4 when on trails.
At the wolves' usual rate of travel, 8 km per hour, this amounted
to a RLU rate of about one per 3 minutes. If all the olfactory
sign produced by wolves is included, this rate increased to one
potential mark each 2 minutes.

Because the primary stimulus for making a scent-mark is the
presence of an existing mark --familiar or strange, this means
that generally wolves are detecting and making an average of
about one mark every 2 minutes, or 240 meters of travel. We also
had indications that the fresher a mark, the higher the rate of
remarking, even if the original and new marks were both by the
same wolf. At about one week of age, old marks stimulated re-mark-
ing at about the original rate, and by about 3 to 4 weeks they
began to lose their stimulus value.

The rate of RLU marking by a resident pack on each trip was
about 2.7 per km along the pack territory edges, whereas in the
centers it averaged about 1.3. This fact, coupled with observations we had of tracks in the snow showing where packs had reached
the edge of adjacent territories, encountered scent-marks, marked
them heavily themselves, and then turned back into their own
territories, convinced us that RLU scent-marking functioned in
reinforcing wolf pack territories. We could also see how lone
wolves could detect territory edges easily and tend to avoid the
packs, as well as how they could find areas vacant of wolf packs.

Certainly many more details of wolf olfactory signalling
need to be worked out. We are currently trying to obtain more
precise information about the chemical composition of wolf anal
gland secretions, and how they function in the wolf's social
system. Russell Rothman, one of Mech's students, is applying the
snow-tracking technique to radioed lone wolves of both sexes, and
to newly-formed pairs to discern just when scent-marking begins

and how it may be used in pair bonding and territory formation. Nevertheless, already we feel that we have a much better understanding of the wolf's basic social organization than we could ever have had without investigating its scent-marking system.

One further step in the study of chemical communication in free-ranging mammals has now been taken --that of experimentation in the wild. We have placed various kinds of scents in the field and recorded the responses of wild wolves to the scents, and have even dropped wolf scats, frozen wolf urine, and wolf urine-filled balloons out of aircraft in front of wolves to watch their reaction. Although our cursory and preliminary efforts could hardly be labeled "formal experimentation," we did begin to appreciate the possibilities in this approach.

This makes us very pleased to see a recent experimental study of scent-marking in wild red foxes. Henry (in press) tested the hypothesis that urine-marking may serve as a "bookkeeping system" in the fox's scavenging activity. That is, urine marks on small inedible remnants of food such as bone, hide, etc. indicates "no food present" and saves foxes scavenging and digging time, according to the hypothesis.

The basic technique used was to simulate fox caches using varying amounts of food remnants, and odors, in the field, and then to observe free-living, but human-tolerant, foxes as they reacted to the caches. When the foxes urine-marked the caches, these were then regarded as another type of experimental cache, and the later behavior of foxes to the marked caches was observed. In three rigorous and well replicated experiments of this type, Henry established that (1) foxes do urine-mark food remnants, (2) they investigate places having both food odor and urine odor for significantly less time than those with just food odor, and (3) they spend more time investigating places where both urine and a significant amount of food exist than where urine and a small amount of food exist. These results are all consistent with Henry's hypothesis.

Regardless of the results, however, it is the technique that is of interest. Surely many more scent-marking experiments can be conducted in the wild by manipulating a mammal's environment and then observing the creature responding to the known manipulation. When this technique is combined with live-capture and marking techniques so that the identity of, and background information about, the responding individual is known, perhaps we will have reached the epitome of studying olfactory signalling.

Additional studies of chemical communication similar to those examples chosen above for illustrative purposes have been conducted with several other species of mammals in the wild. In each case, a new dimension was added to the general concept of chemical communication as well as to the specific role of olfactory signalling in a particular mammal's social organization. With the increasing interest in sociobiology (Wilson, 1975), all aspects of the study of chemical communication have taken on a new importance. The dovetailing of the new emphasis by theoretical biologists on sociobiology and by field workers, as well as other researchers, on olfactory signalling promises an exciting future in elaborating the intricacies of animal organization.

REFERENCES

ALEKSIUK, M. 1968. Scent-mound communication, territoriality, and population regulation in beaver (Castor canadensis Kuhl). J. Mammal. 49:759-762.

CHEAL, M. 1975. Social olfaction: a review of the ontogeny of olfactory influences on vertebrate behavior. Behavioral Biol. 15:1-25.

CHEAL, M., and SPROTT, R. L. 1971. Social olfaction: a review of the role of olfaction in a variety of animal behaviors. Psychological Rept. 29:195-243.

COCHRAN, W. W., and LORD, R. D., JR. 1963. A radio-tracking system for wild animals. J. Wildl. Mgmt. 27:9-24.

EISENBERG, J. F., and KLEIMAN, D. G. 1972. Olfactory communication in mammals. Pages 1-32 in Johnston, R. F., Frank, P. W., and Michener, C. D. (eds.) Ann. Rev. of Ecol. and Syst. 3.

EWER, R. F. 1973. The carnivores. Cornell Univ. Press, Ithaca, N.Y.

HENRY, J. D. In Press. The use of urine marking in the scavenging behavior of red fox (Vulpes vulpes). Behaviour.

JOHNSON, R. P. 1973. Scent marking in mammals. Anim. Behav. 21:521-535.

MARCHINTON, R. L. 1969. Portable radios in determination of ecological parameters of large vertebrates with reference to deer. Pages 148-163 in Johnson, P. L. (ed.) Remote sensing in ecology. Univ. Georgia Press, Athens.

MECH, L. D. 1972. Spacing and possible mechanisms of population regulation in wolves. Am. Zool. 12(4):9 (abstract).

MECH, L. D.   1973.   Wolf numbers in the Superior National Forest
    of Minnesota.   USDA For. Serv. Res. Pap. NC-97.   North
    Central For. Expt. Sta., St. Paul, MN.   10 pp.

MECH, L. D.   1974.   Current techniques in the study of elusive
    wilderness carnivores.   Proc. 11th Int. Congr. Game
    Biologists: 315-322.   Nat. Swed. Environ. Protection Board,
    Stockholm.

MECH, L. D.   1975.   Disproportionate sex ratios in wolf pups.   J.
    Wildl. Mgmt.   39:737-740.

MECH, L. D., and FRENZEL, L. D.   1971.   Ecological Studies of the
    timber wolf in Northeastern Minnesota.   USDA For. Serv. Res.
    Pap. NC-52.   North Central For. Expt. Sta., St. Paul, MN.
    62 pp.

MOORE, W. G., and MARCHINTON, R. L.   1974.   Marking behavior and
    its social function in white-tailed deer.   Papers Int. Symp.
    on Behaviour of Ungulates: 447-456.   IUCNNC, Morges,
    Switzerland.

MÜLLER-SCHWARZE, D.   1971.   Pheromones in black-tailed deer
    (Odocoileus hemionus columbianus).   Anim. Behav. 19:141-152.

MÜLLER-SCHWARZE, D.   1972.   Social significance of forehead
    rubbing in blacktailed deer (Odocoileus hemionus columbianus).
    Anim. Behav.   20:788-797.

MYERS, K., and POOLE, W. E.   1961.   A study of the biology of the
    wild rabbit, Oryctolagus cuniculus (L.), in confined
    populations.   II.   The effects of season and population
    increase on behavior.   CSIRO Wildl. Res.   6:1-41.

MYERS, K., HALE, C. S., MYKYTOWYCZ, R., and HUGHES, R. L.   1971.
    The effects of varying density and space on sociality and
    health in animals.   Pages 148-187 in Esser, A.H. (ed.)
    Behavior and environment.   Plenum Press, N.Y., London.

MYKYTOWYCZ, R.   Odour in the spacing behaviour of mammals.   Pages
    327-343 in Birch, M.C. (ed.).   Pheromones.   North-Holland,
    Amsterdam.

PETERS, R. P. and MECH, L. D.   1972.   Scent-marking and territory
    maintenance in wolves.   Bul. Ecol. Soc. Am.   53(2):19
    (abstract).

PETERS, R. P., and MECH, L. D.   1975.   Scent-marking in wolves.
    Amer. Sci.   63:628-637.

PULLIAINEN, E. and OVASKAINEN, P. 1975. Territory marking by a wolverine (Gulo gulo) in northeastern Lapland. Ann. Zool. Fennici: 12:268-270.

QUAY, W. B., and MÜLLER-SCHWARZE, D. 1970. Functional histology of integumentary glandular regions in black-tailed deer (Odocoileus hemionus columbianus). J. Mammal. 51:675-694.

RALLS, K. 1971. Mammalian scent marking. Science 171:443-449.

SEAL, U. S., MECH, L. D., and VAN BALLENBERGHE, V. 1975. Blood analyses of wolf pups and their ecological and metabolic interpretation. J. Mammal. 56:64-75.

WILSON, E. O. 1975. Sociobiology. Belknap, Harvard Univ. Press, Cambridge, Mass.

TWO HYPOTHESES SUPPORTING THE SOCIAL FUNCTION OF ODOROUS

SECRETIONS OF SOME OLD WORLD RODENTS

D. Michael Stoddart

University of London King's College, Department of

Zoology, Strand, London, W.C.2

INTRODUCTION

The functions of odorous secretions produced by mammals are largely unknown though their involvement in social signalling is often regarded as a foregone conclusion. In only a few instances have their effects on reproduction, territoriality and social life been unequivocally demonstrated; in even fewer have their chemical compositions been investigated (Mykytowycz 1972; Ralls 1971; Stoddart 1976). As far as the rodents are concerned most research into the role of odorous secretions has been directed towards *Mus musculus L.* with *Rattus norvegicus* (Berkenhout), *Peromyscus maniculatus* (Wagner), *Mesocricetus auratus* (Waterhouse) and *Meriones maniculatus* (Milne-Edwards) receiving less, but still significant attention (Beach and Jaynes 1956; Bowers and Alexander 1967; Bronson 1971; Lisk, Zeiss and Ciaccio 1972 and Thiessen 1968). Almost all studies have been carried out under laboratory conditions. Most rodent species have escaped attention possibly because laboratory stocks are not available. Yet it is from naturally structured and freeliving populations that fundamental observations can be made on secretion complexity and quality in relation to sex, sexual condition, age, season of the year, etc., from which reasoned interpretation of the influences of such factors are possible. Observations of this nature are free from the constraints imposed by laboratory conditions. This study was conducted on seven species in three genera of Old World rodents to examine two hypotheses of fundamental importance to the formulation of future controlled condition experimental investigations.

If odorous secretions of wild rodents do, indeed, have a

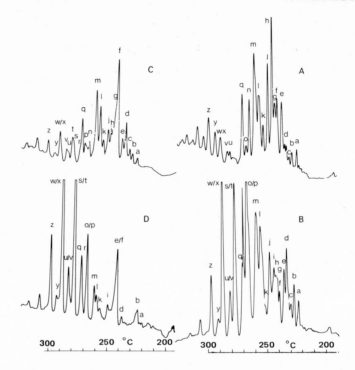

Figure I A. *Apodemus flavicollis* ♂ adult. Brno, Czechoslovakia
Aug. 5 1972. Sample size: 3μℓ Attenuation x 500
B. *Apodemus microps* ♂ adult. Lednice, Czechoslovakia
Aug. 9 1972. Sample size: 3μℓ Attenuation x 1000
C. *Apodemus sylvaticus* ♂ adult. Sussex, England April 18
1975. Sample size 3μℓ Attenuation x 500
D. *Apodemus gurkha* ♂ adult. Dhonpatan, Nepal April 1
1974. Sample size 4μℓ Attenuation x 500

biological significance as social signallers and are not simply
excretory products, logical reasoning would suggest that their
compositions ought not to be primarily under the control of
fluctuating environmental factors. A simple hypothesis would expect greater similarity of structure between species within a
genus than between genera, and substantial similarity between individuals of a species occupying a wide distribution range. By
analogy with visual social integrating signalling devices found in
animals, the hypothesis would expect the basic secretion to be
under genetic control. If, however, the secretion plays a role
in the maintenance of social organisation of a population a second
hypothesis would predict some observable qualitative and/or quantitative change in response to changing ecological and associated
demographic conditions by the same analogy. This paper reports
on an examination of the evidence supporting these two hypotheses.

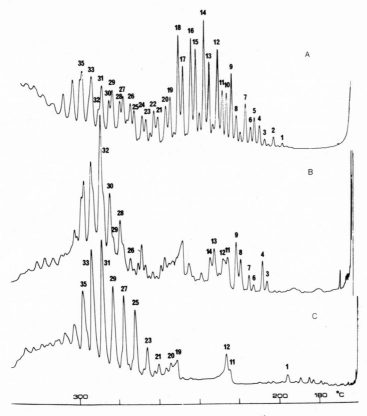

Figure II A.   *Arvicola terrestris amphibius* ♂adult.   Sussex,England
June 8 1973.   Sample size: 3μℓ Attenuation x 1000

B.   *Arvicola terrestris scherman* ♂adult.   Bulle, Switzer
-land   Aug. 21 1972.   Sample size: 3μℓ Attenuation
x 1000

C.   *Arvicola sapidus* ♂adult.   Grenoble   France   Aug.18
1973.   Sample size: 3μℓ Attenuation x 1000

MATERIAL USED

All the animal secretions used in this study were taken from
freshly killed specimens and quickly transferred to ½ dram vials
containing about 50μℓ of 'Aristar' xylene which acted both as
preservative and solvent for the secretion.   The locations and
dates of capture of all material are shown in Table I.   A minimum
of six specimens of each species was used in this study.

Table 1. Material used in this study

Genus and Species	Location	No. of specimens used in the study	Date of Capture
Arvicola terrestris amphibius(L.)	Warnford,Hants, U.K.	30	Throughout 1973 and 1974
	Washington,Sussex, U.K.	10	October 1972
	Woodbastwick Fen, Norfolk, U.K.	12	February 1974
A.t. scherman Miller	Bulle,Fribourg, Switzerland	7	August 1972 and in captivity
A.t. terrestris L.	Udoli,Moravia,ČSSR	8	July 26-Sept 6 1970
A. sapidus Miller	Grenoble , France	11	Aug 26-Sept 8 1972
Apodemus flavicollis (Melch.)	Brno and Lednice, Moravia, ČSSR	20	Aug 2-Aug 11 1972
	Rogate, Sussex,U.K.	6	Sept 29 1972
A. sylvaticus (L)	Elsted and Rogate, Sussex,U.K.	30	Throughout 1975
	Mukut, Nepal	6	May 1 1974
A.microps Kratochvil & Rosicky	Lednice,Moravia ČSSR	9	Aug 8 1972
A.gurkha Thomas	Dhonpatan, Nepal	6	Mar 24-April 1 1974
Gerbillus campestris Levaillant	Al Hoceima, Morocco	7	August 1974

The secretions from *Arvicola* used were those produced by the paired lateral flank organs; from *Apodemus* those produced by the caudal organ, and from *Gerbillus* those produced by the ventral organ. For methods of collection see Stoddart, et.al., 1975 and Stoddart 1973.

METHODS OF STUDY

The principal technique upon which this study is based is that of gas-liquid chromatography. The analytical instrument used was a Pye-Unicam series 104 fitted with dual heated flame ionisation detectors.   The columns were glass, 1.5m long, 4mm. i.d. and packed with 3% dimethyl silicone gum on diatomite CQ support material of mesh size 100-120. The carrier gas was nitrogen which,

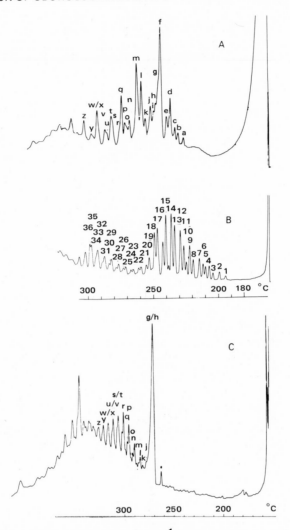

Figure III   A. *Apodemus sylvaticus* ♂adult Sussex, England April
                18 1975. Sample size: 3μℓ Attenuation x 500
             B. *Arvicola terrestris* ♂adult Sussex, England July
                14 1975. Sample size: 3μℓ Attenuation x 2000
             C. *Gerbillus campestris*♂adult Al Hoceima, Morocco Aug.
                20 1974. Sample size: 3μℓ Attenuation x 500 (N.B.
                Temperature programme: 156°C for 5 minutes then
                2.0°C min. $^{-1}$).

unless otherwise stated, flowed at a rate of 60 ml. min $^{-1}$. A
temperature programme of 170°C for 5 minutes rising thereafter by
1.5°C min $^{-1}$ to 310°C was used throughout, unless otherwise

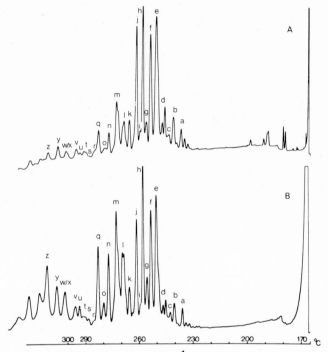

Figure IV A. *Apodemus flavicollis* ♂adult. Sussex, England Sept.
               12 1972.   Sample size 1µℓ Attenuation x 1000
            B. *Apodemus flavicollis* ♂adult.   Brno,Czechoslovakia
               Aug. 4 1972. Sample size: 1.2µℓ Attenuation x 1000
               (N.B. temperature programme for both samples 170°C
               for 5 minutes then 1°C min $^{-1}$ carrier flow rate
               54 ml. min $^{-1}$).

indicated.

        Analysis of each chromatogram involved the measurement of the
height of each peak of interest and the expression of that peak as
a proportion of the total peak height. This technique of relative
peak importance is considered valid since no check could be made
on concentration of secretion extracts (Stoddart, et.al., 1975).
About half the chromatograms were measured manually; the remainder by computing integrator.

        Many comparisons are best made by visual inspection of
chromatograms.   Where differences between seasons, populations,
families and years within any one species are concerned, either
correlation or factorial analysis has been used.

        Two standard peak enumeration series are used in this study:

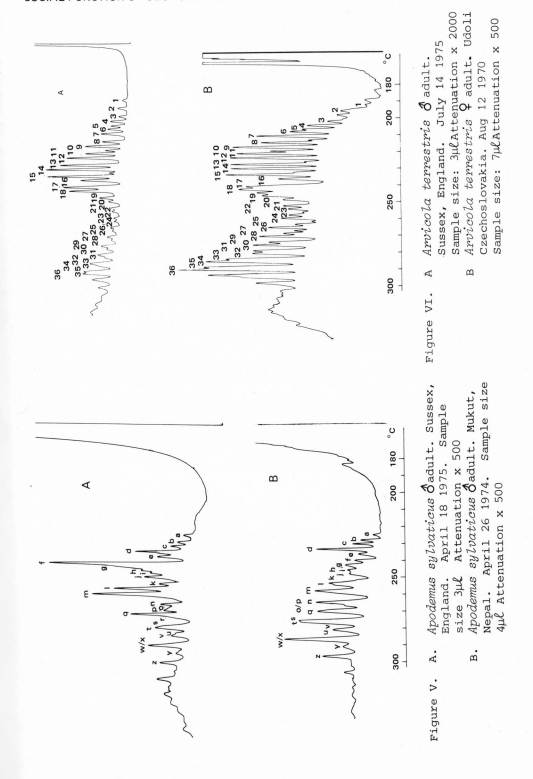

Figure V.   A.   *Apodemus sylvaticus* ♂ adult. Sussex, England.   April 18 1975.   Sample size 3μℓ   Attenuation x 500

B.   *Apodemus sylvaticus* ♂ adult. Mukut, Nepal.   April 26 1974.   Sample size 4μℓ Attenuation x 500

Figure VI.   A   *Arvicola terrestris* ♂ adult. Sussex, England.   July 14 1975   Sample size: 3μℓ Attenuation x 2000

B   *Arvicola terrestris* ♀ adult. Úĺdoli Czechoslovakia. Aug 12 1970   Sample size: 7μℓ Attenuation x 500

one, a litteral one, for secretions from *Arvicola spp.* and the
other, a numerical one, for *Apodemus spp.* and *Gerbillus*. These
series were used initially on standard stocks of *Arvicola terrestris
amphibius* and *Apodemus sylvaticus* respectively. Correspondence of
peaks between species and genera was verified by reinforcement with
the standard stock of *Arvicola* or *Apodemus*. It was further aided
by the use of traced overlays of the initial samples of these genera.
These two techniques were found to be highly reliable.

All chromatograms illustrated are wholly typical of the type
of specimen which they represent.

RESULTS AND DISCUSSION

In interpreting the chromatograms illustrating this paper for
similarity and difference, three features should be looked for.
One is the range of elution temperatures of the components or component complexes, another is the presence or absence of a particular
component and the third is the relative importance of a particular
component as a fraction of the whole mixture. Comparisons of
actual peak heights between chromatograms are of no value since
injected samples were of unknown and, very likely, unequal concentrations.

Examination of Figs. I, II and III indicates that the first
hypothesis cannot effectively be destroyed. In Fig. I the differences between the species are slight, for example the apparent
loss of '*p*' in *A.flavicollis* and the distinctly double pronged
appearance of '*d*' in the same species; the relative diminution of
'*l*' and '*m*' in *A. gurkha*; the relative magnification of '*s/t*' and
'*w/x*' in *A. microps* and *A. gurkha* etc. With the exception of a
few low boiling point peaks in *A. flavicollis* and *A. gurkha*, which
may be contaminants, there are no peaks eluting at temperatures
below about 220°C. Between genera, however, the differences are
more marked (Fig. III). The most immediate difference is the
profusion of low boiling point components in *Arvicola*. The first
important component in *Apodemus* '*a*' is identical with *Arvicola*
'11'. *Apodemus* '*c*' equates with *Arvicola* '13', '*e*' with '15',
'*g*' with '17', '*h*' with '18' and '*j*' with '19'. With the exception of phenylacetic acid which elutes close to the solvent peak,
Gerbillus has no peak of importance eluted at lower boiling point
to that of *Apodemus* '*e*'. This genus is characterised by large
amounts of components of very high boiling point and only slight
appearance of components with a boiling point of less than about
260°C (Cholesterol is present in the fused '*g/h*' peak).

The similarities within the genus *Arvicola* (Figs. II and VI)
are less obvious than within *Apodemus* (Fig. I), but are nevertheless present. All three subspecies of *Arvicola terrestris* have

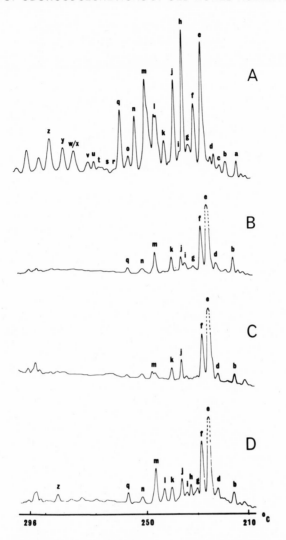

Figure VIIa *Apodemus flavicollis*. Brno, Czechoslovakia Aug. 1972
    A. ♂adult   Sample size: 1.5μℓ Attenuation x 1000
    B. ♀adult   Sample size: 2μℓ Attenuation x 1000
    C. ♂juvenile Sample size: 3μℓ Attenuation x 1000
    D. ♂adolescent Sample size 1.5μℓ Attenuation x 1000

the same components present, although their relative importances
differ. *A. sapidus* shows a strong development of high boiling
point components with an interesting dearth of components eluting
between 210°C and 240°C.

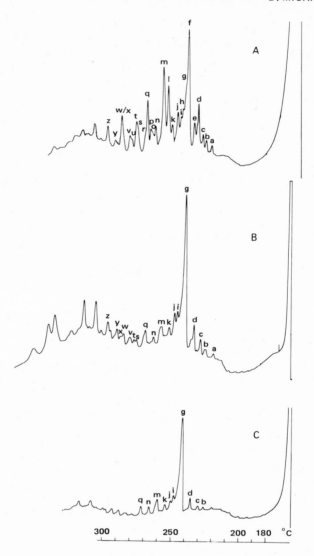

Figure VIIb *Apodemus sylvaticus* Sussex, England Feb. and April
            1975.
            A. ♂adult Sample size: 3µℓ Attenuation x 500
            B. ♀adult Sample size: 3µℓ Attenuation x 500
            C. ♂juvenile Sample size: 3µℓ Attenuation x 1000

Fig. IV shows two typical male *Apodemus flavicollis*, one from
Sussex, England and the other from the environs of Brno, Czechoslovakia. The differences between them are very slight and are restricted to the relatively greater importance of very high temperature eluted components in the central European specimens.

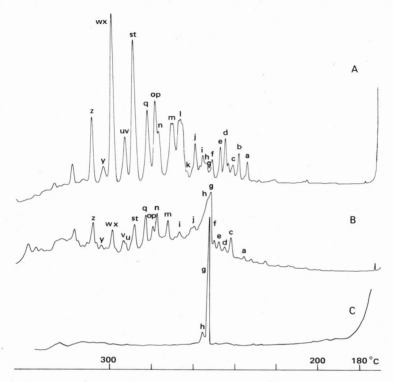

Figure VIIc *Apodemus microps* Lednice, Czechoslovakia August 1972
   A. ♂adult   Sample size: 3μℓ Attenuation x 1000
   B. ♂juvenile Sample size: 3μℓ Attenuation x 1000
   c. ♀adult   Sample size: 3μℓ Attenuation x 500

Fig. V shows two typical male *Apodemus sylvaticus*, one from Sussex, England and the other from Mukut, Nepal.   With the exception of a strong development of peak '∫' in English specimens, the compositions of the two secretions from the two regions are very similar.   The Nepalese specimens have relatively higher concentrations of peaks '*n*', '*∫/t*' and '*w/x*' than the English, but contain neither more nor less components.

Fig VI shows two typical *Arvicola terrestris;* one a male (subspecies *amphibius*) from England, the other a female (subspecies *terrestris*) from northern Moravia, Czechoslovakia.   Although there are some differences in relative concentrations of components it should be noted that the same range of boiling point substances are present in each and the overall pattern of relative abundance of components is similar.

   Further evidence suggesting that environmental factors play

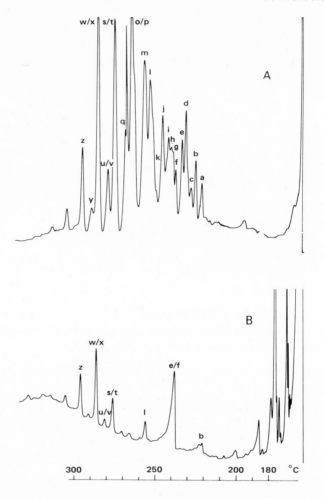

Figure VIId *Apodemus gurkha* Dhonpatan, Nepal March 1974
         A. ♂ adult   Sample size: 4μℓ Attenuation x 500
         B. ♀ adult   Sample size: 5μℓ Attenuation x 1000

little part in controlling secretion quality has been provided by
Stoddart et al. (1975).   They took flank organ secretions from
8-12 week old *Arvicola terrestris* that were bred in captivity and
housed and fed under identical conditions.   When subjected to a
two-factor factorial analysis the average peak height data showed
there to be a high degree of similarity within families and little
between families (F(57,160) = 6.329, P < 0.001).   Analysis of
difference between three widely dispersed populations showed that
location was a major factor influencing peak pattern.   This
occurred even though the age and sex composition of the samples

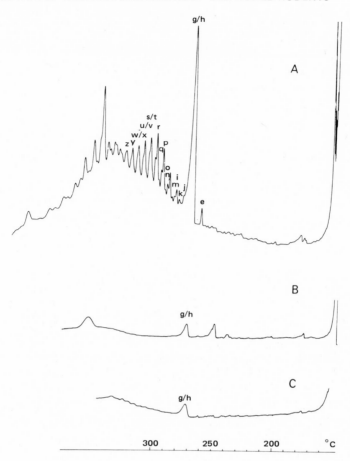

Figure VIII   *Gerbillus campestris* Al Hoceima, Morocco Aug. 1974
             A. ♂ adult   Sample size: 3μℓ Attenuation x 500
             B. ♀ adult   Sample size: 2μℓ Attenuation x 500
             C. ♂ juvenile Sample size: 2μℓ Attenuation x 500

were not identical (F(38,300) = 8.812, P < 0.001).   Other characteristics indicate that populations are reproductively isolated
(Stoddart 1970), and in view of all these data the first hypothesis
cannot be shown to be incorrect.

    The most fundamental physiological and sociological difference
between members of a population is their sex.   The second hypothesis might expect a sexual difference in secretion quality between
the sexes.   For the four species of *Apodemus* and for *Gerbillus*
this is the case, but no obvious sex differences can be detected
in either species of *Arvicola*.   The sex differences in *Apodemus*
spp. are illustrated in Fig. VII a,b,c, and d.   Assuming both

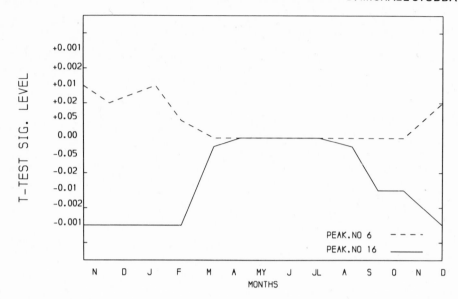

Figure IX Level of significant monthly change in the proportion
of two arbitrarily selected component complexes from *Arvicola
terrestris amphibius* flank organ secretion as displayed by gas
chromatography and measured against a May/June standard.

sexes of each species eat similar diets, and there is no evidence
that this is not so, the existence of marked differences suggests
that the secretions are not excretory products containing unwanted
ingested substances.

    The large number of specimens of *A. sylvaticus* allowed an
intensive examination of intrapopulation characteristics.   Using
the adult male shown in Fig. VIIb as the standard,females showed
an absence of peaks '*c*', '*e*', '*l*', '*o*', '*p*', '*s*' and '*t*'.   Other
components, on a relative concentration basis, were much reduced.
A Spearman rank correlation test was conducted on the patterns of
relative concentrations of constituents in a sample of seven males
and six adult females from the same population captured between

18th and 20th April 1975.   All were in reproductive condition.
The test indicated a very high level of dissimilarity (r=-0.052).
It is felt that similar differences would be revealed by other
species in this genus, but additional tests were not possible.   In
Gerbillus sp. the difference between males and femals is even more
sharply marked (Fig. VIII).

When reproductive condition is considered, marked differences
between adults and juveniles are apparent in *Apodemus flavicollis,*
sylvaticus and *microps,* and *Arvicola t. amphibius* - the small
numbers of the other species did not permit an examination of this
characteristic.   The greatest differences between adult and
juvenile male *Arvicola terrestris* were seen in components 12, 18
and 20 and between adult and juvenile females in components 12 and
17 (Stoddart et al. 1975).   Figs. VII a-c show the pattern for
juvenile males in three species of *Apodemus.*   Scarcity of juvenile
females in all species makes no comparison between them and adult
females possible.   It was fortunate that a few specimens of what
can best be described as 'adolescent' male *A. flavicollis* were
obtained.   The structure of their secretions are shown in Fig.
VIIa.   They reveal an intermediate pattern of complexity between
juvenile and adult.   The similarity between adult female and
juvenile male in all species of *Apodemus* is clearly obvious, and
for a sample of four females and four juvenile male *A. sylvaticus*
the Spearman rank correlation test showed close similarity
(r = 0.728).   From an olfactory point of view young males resemble
their mothers much more than they do their fathers.   It can be
postulated that the transition to adulthood occurs quickly with
adolescent intermediates being rare in any one trapped sample.

The *A. sylvaticus* population was sampled at various times of
the year which allowed an analysis to be made of changes that
occur as the breeding season progresses.   A sample of nine adult
males caught at the start of the breeding season (mid April in
1975) was compared with four adult males caught in the middle of
the breeding season (early August 1975).   The Spearman rank
correlation test indicated that the similarity was little
(r = 0.478).   That this was not due to any possible masking effect
of individual difference is suggested by the fact that a sample of
seven adult males from the population in April 1974 were very
similar in pattern to nine adult males from the same population one
year later, (r = 0.630).   It thus seems as if the time of year
is an important factor in influencing pattern characteristics.   It
was observed further that mid-summer males possessed three extra
peaks corresponding exactly to *Arvicola* peaks No. 3, 4 and 6
($C_{24} H_{48} O_2$, $C_{24} H_{48} O_2$, $C_{26} H_{50} O_2$).   These extra peaks were not
seen in Spring (April) or Winter (December).   Significant differences
between *Arvicola terrestris* caught in June and September are also
seen (Stoddart et al. 1975).   These times are approximately early

and late in the breeding season.   Much greater differences are
seen in females than in males, involving ten of the twenty numbered
component complexes.   The peak containing Cholest-5-en-3β ol (16)
differs more than any other, being half as important in September
as in June.

    Although clearly marked differences do no exist between adult
and juvenile *Arvicola terrestris*, a far greater amount of variation
is seen in chromatograms prepared from extracts from sexually mature
individuals than in extracts prepared from immature individuals.
In order to overcome any individual differences inherent in secretions taken from natural populations, two litters each containing
two young males were reared in captivity and maintained under
identical lighting, heating and feeding conditions.   Their flank
organs were sampled regularly each fortnight for about eight months.
The first samples were taken when they were about 140 days old;
this was the first occasion upon which sufficient secretion could
be obtained for analysis.   Two of the voles, one from each litter,
produced less secretion than the minimum required for satisfactory
analysis (ca. 0.5 mg) on several occasions.   They were excluded
from the investigation.   The reproductive condition of the other
two was carefully noted at each sampling time and the resultant
data grouped according to whether the testes were abdominal or
scrotal.   Occasions upon which their position was judged intermediate were omitted as were the two occasions immediately adjacent
to a change in testis position.   Such precautionary adjustments
left 12 suitable occasions for analysis of   ♂ 023 (6 immature and
6 mature) and 8 for ♂ 027 (4 immature and 4 mature).   The mean
peak height proportions for these occasions, and the variances about
them, are shown in Table 2.   Two points are immediately obvious.
The first is that the total amount of variation (Σs^2) within the
samples taken during the time of testicular inactivity is substantially lower than that occurring during the time of their activity.
The second point is that the same trend is apparent in both specimens.

    The high value for ♂ 023 (s) can be explained in part by point
-ing out the inclusion of two sets of observations obtained when
the vole had scrotal testes prior to the start of the period of
winter inactivity.   If these data are ignored, the total variation
observed during the period of post-winter testicular activity falls
to 12.01.   These preliminary data, although based on only two
specimens and hence to be interpreted with great caution, suggest
that youth exerts a marked influence over variability of secretion
concentration.   It is by no means clear why such variation should
occur within the secretory output of a single individual during the
time it is in reproductive condition.

    Preliminary conclusions of investigations currently in progress

Table 2. Mean peak height proportions, and their variances, for two captive *Arvicola terrestris*. Data are grouped according to reproductive condition.

Component Number	Composition ($RCO_2\bar{R}$)	023 A n=6 $\bar{X}$	023 A n=6 s^2	023 S n=6 $\bar{X}$	023 S n=6 s^2	027 A n=4 $\bar{X}$	027 A n=4 s^2	027 S n=4 $\bar{X}$	027 S n=4 s^2
1	$C_{22}H_{44}O_2$	0.45	0.151	0.73	0.109	0.76	0.065	0.58	0.003
2	$C_{23}H_{46}O_2$	1.42	0.140	1.33	0.324	1.15	0.022	0.91	0.004
3	$C_{24}H_{48}O_2$	1.16	0.042	0.88	0.110	0.92	0.016	0.86	0.520
4	$C_{24}H_{48}O_2$	2.63	0.133	2.63	0.128	1.76	0.109	1.91	0.102
5	$C_{24}H_{48}O_2$	3.22	0.602	4.12	0.207	1.80	0.219	2.47	0.327
6	$C_{25}H_{50}O_2$	2.90	0.070	2.09	0.413	2.46	0.041	2.02	0.127
7	$C_{25}H_{50}O_2$	3.75	0.060	3.83	0.220	2.58	0.050	2.52	0.006
8	$C_{26}H_{52}O_2$	3.13	0.048	2.32	0.275	2.64	0.066	2.66	0.256
9	$C_{26}H_{52}O_2$	7.20	0.206	7.37	1.466	5.87	0.790	6.45	1.16
10	$C_{26}H_{52}O_2$	4.72	0.078	6.10	1.336	3.35	0.252	4.17	0.186
11	$C_{27}H_{54}O_2$	6.64	0.423	4.52	1.046	7.21	0.043	6.23	1.02
12	$C_{27}H_{54}O_2$	9.06	0.313	8.81	1.321	9.03	0.009	8.15	0.314
13	$C_{28}H_{56}O_2$	8.84	0.340	6.85	0.877	9.15	0.304	8.54	1.63

Table 2. continued....

Component Number	Composition (RCO_2R^-)	023 A n=6 $\bar{x}$	s^2	023 S n=6 $\bar{x}$	s^2	027 A n=4 $\bar{x}$	s^2	027 S n=4 $\bar{x}$	s^2
14	$C_{28}H_{56}O_2$	13.60	0.309	14.52	2.352	14.55	0.989	14.99	0.539
15	$C_{29}H_{58}O_2$	10.47	0.485	8.41	0.844	12.33	1.29	10.06	2.50
16	$C_{29}H_{58}O_2$	5.10	0.116	6.04	0.467	5.00	0.123	4.94	0.258
17	$C_{30}H_{60}O_2$	5.39	0.202	6.55	7.260	6.89	0.484	6.78	0.980
18	$C_{30}H_{60}O_2$	6.98	0.280	8.68	2.360	8.35	0.120	7.59	0.309
19	$C_{30}H_{50}O_2/C_{31}H_{62}O_2$	2.03	0.028	2.38	0.213	3.38	0.273	4.75	2.318
20	$C_{30}H_{60}O_2/C_{31}H_{62}O_2$ $C_{32}H_{64}O_2$	1.14	0.007	1.80	0.262	1.79	0.067	4.09	1.927
		$\Sigma s^2=4.033$		$\Sigma s^2=21.59$		$\Sigma s^2=5.332$		$\Sigma s^2=14.018$	

seem to indicate that similar changes occur in natural populations.
A cycle of concentration change in most components can be observed
annually (Fig. IX).   This indicates that subjects in captivity
react much as do others in the wild.   Data are not yet available
on the behavioural significances of these changes.

CONCLUSIONS

There is sufficient evidence derived from laboratory studies
on rodent social behaviour to indicate that odorous substances
play an important role.   Although precautions are usually taken
to ensure that the investigator is dealing with odour cues only,
the ecological interpretation of even the best documented phenomena
is, to say the least, difficult (Stoddart 1974).   The studies
described in this paper were so designed to overcome the limitations
imposed by laboratory studies such that a fresh look could be taken
at the functional significance of rodent sebaceous odours.   The
evidence presented and examined does not destroy either of the two
fundamental hypothesis presented in the introduction.   The next
step in the approach herein adopted is to compare the attributes of
a well documented example of a mammalian visual signalling system
of known social significance with the observed attributes of rodent
odours, and Table 3 sets out a comparison between the antlers of
red deer *(Cervus elaphus L)* and rodent odour characteristics as
revealed by this study.   Also included are references to other
mammalian studies.   Although the rodent model is far from complete,
all known odour characteristics are directly comparable with visual
characteristics of antlers.   It is of the greatest interpretational
significance that there is none which lacks a visual counterpart.

It is vital that future studies be directed towards a field
demonstration of the effect, at population level, of artificial
modification of the odorous environment.   Jones and Nowell (1973),
Müller-Velten(1966) and others have demonstrated at the individual
level the effect of artificial manipulation of odours in the laboratory but have not extended their studies into the field.   Without a population field test, which would complement the elegant
experimental work of Lincoln et al. (1971) confirming the social
role of antlers to red deer, the model must remain hypothetical.

This is a new and exciting field which may be expected to bear
important fruits for students of population control, ecology,
behaviour and social biology.   As in many others, it is easy to
overlook basic principles and to jump to conclusions; thus it
would seem advisable to proceed thoroughly and in a stepwise
manner.   It is hoped that the small step of this contribution to
the debate will add to the growing momentum of the long march
ahead.

Table 3. A comparison of a) the main components of the red deer antler model with b) odorous secretions from rodents (this study) and c) other small mammals.

	a) Features of antlers of red deer (*Cervus elaphus* Lincoln et.al.1971 Fraser Darling 1936)	b) Features of rodent odorous secretions (this study)	c) Corroborative evidence from other small mammalian species
1.	Present in males, absent in females	Greater complexity in males; poorly developed in females	Rabbits Mykytowycz 1972 Sugar-gliders Schultze-Westrum 1968 Possums Thompson and Pears 1962
2.	Present in juvenile and adolescent males in reduced form	Poorly developed in juvenile males: adolescents intermediate	
3.	Cycle of signal strength related to breeding season	Cycle of component concentration shows maximum amount of change at onset and cessation of breeding	
4.	Quality of signal related to blood level of testosterone	?	
5.	Complexity of signal influenced by age	Age exerts a marked influence on complexity	Rabbits Mykytowycz 1972
6.	Strength of signal related to carrier's level of dominance in the social hierarchy	?	Rabbits Mykytowycz 1972. Sugar-gliders Schultze-Westrum 1968 Rats Bowers and Alexander 1967
7.	Configuration of visual signal influences reaction by conspecifics	?	Mice Jones and Nowell 1973 Mice Müller-Velten 1966 Sugar-gliders Schultze-Westrum 1968

Table 3 continued...

a) Features of antlers of red deer (*Cervus elaphus* Lincoln et.al.1971 Fraser Darling 1936)	b) Features of rodent odorous secretions (this study)	c) Corroborative evidence from other small mammalian species
8) Overall size of signal influences by quality and quantity of diet	?	
9) Genetic differentiation of antler type in different populations	Greater similarity within sub-species, populations and families than between these groups	Bank voles Godfrey 1958

Acknowledgements

I am grateful to Dr. G.B.Corbet for the supply of *Apodemus sylvaticus* and *gurkha* from Nepal; to M. Rawlings and University College London for *Arvicola sapidus;* to D.A.P. Bundy for *Gerbillus* and to G. Tomkins for allowing me to use his unpublished data in Fig IX.   I am further grateful to Academician J. Kratochvil, Czechoslovak Academy of Science, Institute of Vertebrate Zoology, Brno, for providing facilities for field work in Czechoslovakia in 1970 and 1972, and to Dr. A Meylan, Service de Zoologie des Vertébrés, Nyon, Switzerland for providing both specimens and facilities for work on *Arvicola terrestris scherman*.   Grateful acknowledgement is made to the Science Research Council and the University of London Central Research Fund for financial support.

REFERENCES

Beach, F.A. and J. Jaynes 1965.   Studies on maternal retrieving in rats.   III Sensory cues involved in the lactating female's response to her young.   Behaviour 10, 104-125.

Bowers, J.M. and B.K. Alexander 1967.   Mice: Individual recognition by olfactory cues.   Science 158, 1208-1210.

Bronson, F.H. 1971.   Rodent pheromones.   Biol. of Reprod. 4, 344-357.

Fraser Darling, F. 1937. "A herd of red deer.   A study in Animal Behaviour." O.U.P. 215 + X.

Godfrey, J. 1958.   The origin of sexual isolation between bank voles.   Proc. Roy. Phys. Soc. Edinburgh, 27, 47-55.

Jones, R.B. and N.W. Nowell 1973.   The effect of urine on the investigatory behaviour of male albino mice.   Physiol. and Behav. 11, 35-38.

Lincoln, G.A. and R.W. Younger and R.V.Short 1970.   The social and sexual behaviour of the red deer stag.   J. Reprod. Fert., Suppl. 11, 71-103.

Lisk, R.D., J. Zeiss and L.A. Ciaccio 1972.   The influence of olfaction on sexual behaviour in the golden hamster *(Mesocricetus auratus)*. J. Exptl. Zool. 181, 69-78.

Müller-Velten, H. 1966.   Über den Angstgeruch bei der Hausmaus *(Mus musculus L)* Z. Vergl. Physiol. 52, 401-429

Mykytowycz, R. 1972.   The behavioural role of the mammalian skin
    glands. Naturwissenschaften 59, 133-139.

Ralls, K. 1971.   Mammalian scent marking. Science 171, 443-449.

Schultze-Westrum, T. 1968.   Innerartliche Verständigung durch
    Düfte beim Gleitbeutler *Petaurus breviceps papuanus* Thomas
    (Marsupialia, Phalangeridae) Z. Vergl. Physiol. 50, 151-220.

Stoddart, D.M. 1970.   Tail tip and other albinisms in voles of the
    genus *Arvicola* Lacepede 1799. in Variation in Mammalian
    Populations." ed. R.J.Berry and H.N.Southern. Symp. Zool.
    Soc. Lond. No. 26, 271-282.

Stoddart, D.M. 1973.   Preliminary characterisation of the caudal
    organ secretion of *Apodemus flavicollis*. Nature, Lond. 246,
    501-503.

Stoddart, D.M. 1974.   The role of odor in the social biology of
    small mammals in "Pheromones", ed. M.C.Birch. Elsevier
    North Holland Publ. Comp. 297-315.

Stoddart, D.M. 1976. "Mammalian odours and pheromones." Studies in
    Biology. Edward Arnold, London (in press).

Stoddart, D.M. R.T.Aplin and M.J.Wood, 1975. Evidence for social
    difference in the flank organ secretion of *Arvicola*
    terrestris (Rodentia : Microtinae) J. Zool., Lond. 177,
    529-540.

Thiessen, D.D. 1968. The roots of territorial marking in the
    Mongolian gerbil: A problem of the species-common topo-
    graphy. Behav. Res. Meth. Instr. 1, 70-76.

Thomson, J.A. and F.N. Pears 1962. The functions of the anal glands
    of the brushtail possum. Vict. Nat. 78, 306-308.

THE SEARCH FOR APPLICATIONS OF CHEMICAL SIGNALS

IN WILDLIFE MANAGEMENT

Stephen A. Shumake

United States Fish and Wildlife Service
Wildlife Research Center
Denver, Colorado 80225

IMPORTANCE OF CHEMICAL SIGNALS TO WILDLIFE

Chemical signals play a vital role in the lives of many wildlife species. Several review articles (Bronson, 1971; Chael and
Sprott, 1971; Ralls, 1971; Eisenberg and Kleiman, 1972; Estes, 1972;
Mykytowycz, 1972) have described the diversity of chemical signals,
their glandular or excretory origins, and their effects on behavior
and reproduction in a variety of wild animals. Most often, chemical
signals (pheromones) from conspecifics have been investigated. However, interspecific signals or allomones (Eisenberg and Kleiman,
1972) as well as odors and flavors of certain food items (Hansson,
1973; Rice and Church, 1974) and prey odors (Burghardt, 1970) have
been studied and have been found to have survival and chemical signal
value to many species.

Successful use of many identified and synthesized insect
pheromones for managing populations and controlling damage to crops
and forests suggests a large potential for use of chemical signals
to managing wildlife. Unlike insects, however, the higher vertebrates
do not automatically respond every time the appropriate chemical signal is presented. Factors such as previous experience (Stern, 1969,
1970; Carr et al., 1970; Lydell and Doty, 1972), level of specific
circulating hormones (Carr et al., 1966; Doty, 1974; Price, 1975),
and social dominance level (Krames et al., 1969; Jones and Nowell,
1973) will often determine whether or not the appropriate response
is made. The young, inexperienced animal has the genetic capability
(Mayr, 1974) of using chemical signals, but its proficiency in using
and responding to them is often shaped through subsequent social
experience (Lott and Hopwood, 1972).

Even the social insects are capable of habituation to sex pheromones if the stimulus is presented too frequently (Shorey, 1974). In studying the responses of higher vertebrates to chemical signals, the problem of habituation can be compounded by a variety of other mechanisms. For example, in tests of food odor attractants with rodents, the animal's ingestion of a food will control its olfactory response to that specific food (LeMagnen, 1963; Larue, 1975).

This complexity of response is among the reasons why chemical signals have not been widely used in wildlife management. Almost no work has been done on the practical application of chemical signals for managing birds, and work with mammals has been limited to attempts at directly or indirectly reducing damage to man's livestock, crops, and other products. For a number of years, the U.S. Fish and Wildlife Service has been involved in research to develop and improve methods for safely and effectively reducing such damage by wildlife. The rest of this discussion will focus on the three major damage situations where chemical signals have been used for this purpose: coyote predation on livestock, deer and elk damage to trees, and rat damage to crops.

In these wildlife damage situations, chemical signals have been used or proposed for five different kinds of behavior modification: (1) to attract the animal to an area, place, or object (area attractant or lure); (2) to repel or frighten it from an area (area repellent); (3) to encourage it to eat a food (food attractant or bait enhancer); (4) to repel it from eating a food (food repellent); and (5) to disrupt its normal reproductive patterns. The types of chemical signals investigated for these purposes have been:

(1) signalling pheromones such as sex attractants; alarm, fear, or stress odors; and territorial marking pheromones;
(2) primer pheromones such as those associated with estrus cycling or pregnancy block effects;
(3) attractant food flavors;
(4) repellent or aversive food flavors; and
(5) allomones from attractive prey, avoided predators, or competitive sympatric species.

APPLICATIONS TO PREVENTING COYOTE PREDATION

Coyote Damage

Each year, millions of dollars worth of sheep and other livestock, are lost to coyotes throughout the western United States (Balser, 1974a, 1974b). In spite of man's many efforts to control sheep predation by reducing coyote numbers, the population continues to expand and is increasing in some areas. The species is extremely adaptable and will probably continue to inflict severe economic losses unless more effective management tools are developed through research. One of the present research priorities is developing better area attractants. A plausible subject for future work is developing effective area repellents.

Area Attractants

Effective coyote attractants are needed for two purposes: monitoring relative abundance and increasing the number of animals reached by control measures. These are also the two main practical applications of insect pheromones (Tette, 1974).

Since 1972 a survey has been conducted each fall in an attempt to monitor the relative abundance of predators in the western states. Scent capsules are placed in circles of sifted dirt at intervals on transects; each such scent station is examined daily for tracks, and the number visited by coyotes is used as an index to their abundance in the area (Linhart, 1973). The lure that has been used in the scent capsules is air-inoculated fermented egg product, commercially manufactured mainly as an eye-gnat attractant, but found several years ago to attract coyotes to live traps. Since the natural fermented egg product is subject to batch-to-batch variations in odor quality and since its commercial formulation may be changed in the near future, we are currently developing a synthetic fermented egg (SFE) formulation to replace it. By using a combination of gas chromatography and mass spectrometry, we have so far identified over 70 compounds in the natural product. With the help of a human odor panel (Guadagni, 1968), whose members were selected by the criteria of Wittes and Turk (1968), specific concentrations of these compounds have been incorporated into a synthetic material that the panel could discriminate from the natural product less than 60% of the time. We chose the human odor panel technique because of inherent difficulties involved in evaluating many fermented egg fractions with coyotes even in pens. This human odor panel technique has been used to assess rabbit scent gland odor intensity (Hesterman and Mykytowycz, 1968) but this technique does have some limitations as pointed out by Dryden and Conaway (1967). For example, shrews respond to side gland odors that are undetectable by humans. No discrimination test or a test for similarity judgments have been made with the coyote since our main priority was to develop an SFE formulation equal to the natural fermented egg as a coyote attractant. Thus far, pen tests and preliminary field tests near Zapata, Texas, have indicated that the SFE is at least as attractive as the natural product.

The development of other good lures that are selective for coyotes should permit more reliable censusing (for example, to correlate local livestock losses with coyote activity), more effective
use of control devices such as traps and the M-44 cyanide cartridge,
and more effective delivery of control chemicals such as toxicants,
chemosterilants, or aversive agents. Over the years, trappers and
others engaged in operational field control have used a vast assortment of organic scents and homemade formulations to attract coyotes,
but little has been done to test the effectiveness of these materials. We are currently planning field tests with some of the commonly used ingredients, including synthetic musks, essential oils,
decomposing animal products, urine, and glandular secretions, to
evaluate the behavioral responses these materials elicit.

One group of compounds that may warrant further investigation
for formulating selective attractants for coyotes, and possibly for
other mammals, is the volatile short-chain fatty acids. These compounds make up about 75% of the natural and synthetic fermented egg
attractants. Coyotes are also attracted to a synthetic monkey pheromone preparation (Keverne and Michael, 1971; Curtis et al., 1971)
containing a specific ratio of five of these acids (acetic, iso-
butyric, isovaleric, isocaproic, propionic). In the Texas field
test, this preparation elicited somewhat fewer visits than the two
fermented egg products but more pronounced behavioral responses
(such as rolling and digging). Exactly why coyotes are attracted
to different ratios of volatile short-chain fatty acids is unclear
at present. However, these compounds apparently occur widely in
mammalian signalling systems. For example, they have been identified
in the anal glands of coyotes and dogs (Preti et al., 1976), red
foxes (Albone and Fox, 1971), weasels and Indian mongooses (Gorman,
1976), and guinea pigs (Beruter et al., 1974), and in vaginal secretions of Rhesus monkeys (Curtis et al., 1971) and man (Michael et
al., 1974). There is evidence that mongooses use individual differences in volatile fatty acid ratios as a code to identify each other
(Gorman, 1976). Thus, coyotes may be attracted to certain fatty
acid ratios out of curiosity and to investigate other, unidentified
coyotes, or because these ratios mimic odor cues from competing predators or prey. Volatile fatty acids are also known to be the products of protein decomposition. Because coyotes frequently scavenge
for food, some animals could be attracted to the short-chain fatty
acid odor due to a resemblance to carrion odor. The question of why
coyotes are attracted to different volatile fatty acid ratios needs
to be examined closely to define future applications for this group
of compounds.

Area Repellents

Although area attractants are currently the most important
research priority, area repellents may have some potential for

controlling coyote predation on sheep. We know, for instance, from the work of Peters and Mech (1975) that dominant male wolves have a marked influence on the territorial boundaries of their packs. It would be somewhat surprising if coyotes did not show a similar respect for the territorial scent marks of dominant males. It is highly unlikely that the scent of dominant male coyote urine placed near a flock of sheep would lead to avoidance of this easily accessible and abundant prey item for any sustained period. However, on open range where the sheep are moved every few days to new grazing pastures pre-marking the acreage with active components in dominant male urine could perhaps have a temporary repellent influence, and could be used to protect sheep especially as the herd is moved into new grazing pasture across coyote trails and established territories. An important question in this context is whether there are certain chemical compounds only produced by dominant animals or whether the response of subordinate animals is conditioned through physical agonistic encounters.

Other odors of natural origin might also be considered as area repellents. Most odors that have been tested with coyotes have been selected from various classes of chemical stimuli or lachrymators used as dog repellents (Lehner et al., 1976). But dog repellent chemicals are often selected on the basis of human responses to irritating or objectionable odors. The canid nose, of course, may tolerate these and respond more strongly to odors of which we are unaware. For example, it is often said that wild mammals can detect human odor, and Taylor et al. (1974) found that captive Norway rats responded to it, but the active components in human scent were not identified. Natural prey species such as certain lizards, snakes, moles, shrews, skunks, and toads may have effective chemical predatory defense mechanisms (Bolles, 1970) that allow them to travel unmolested through the territories of predators. As far as I know, no one has ever attempted to isolate repellent aversive compounds from these species for use in predator control. Working against this concept, however, is the fact that coyotes are extremely adaptable and will accept a wide variety of food items (Sperry, 1941) including prey usually considered malodorous or distasteful.

Most species seem to have mechanisms to ensure species isolation and prevent hybridization. Moore (1965) and Doty (1973a, 1973b), for example, found that deer mice were tuned to the odors of their own species or strain and rejected others. Coyotes apparently will socially accept and readily interbreed with red wolves and domestic dogs (Gibson, 1974; Riley and McBride, 1975). However, they may be rejected or be repelled by odors from other canids such as grey wolves or foxes. Bears and large felids, such as the mountain lion, may also produce territorial marks that coyotes avoid. As with pheromones, however, the coyote's response to allomones from competitive predators or avoided prey has never been adequately explored.

APPLICATIONS TO PREVENTING DEER AND ELK BROWSING

Deer and Elk Damage

As with many forms of wildlife, deer and elk can provide recreational sport hunting, and aesthetic resources to the public. At the same time, however, deer and elk may do appreciable damage to Douglas-fir seedlings (Kverno, 1964) in the Pacific Northwest. When browsing becomes too severe, the seedlings grow too thin and mature much later. The Denver Wildlife Research Center and our Forest Animal Damage Field Station at Olympia, Washington, have been involved in the development of deer repellents for a number of years (Campbell and Bullard, 1972).

Food Repellents

Of 255 chemicals that have been screened for repellent activity with black-tailed deer, only 3 - a putrified fish preparation, fermented egg product, and an animal rendering material, Wilson blood meal - have shown reliable repellent effects. Their repellency probably relates to the natural tendency of ungulates to avoid decomposing plant and animal material as food. The putrified fish and natural fermented egg products have been used to repel black-tailed deer from developing Douglas-fir seedlings. Concentrations as low as 10 ppm of certain components within putrified fish have been shown to repel captive deer in pens (Campbell and Bullard, 1972). The synthetic fermented egg (SFE) product previously described, proved to be equally repellent as the natural product. Later in development, to improve the SFE repellent activity, we again used human odor panel testing techniques to develop seven enhanced SFE preparations. Classes of compounds found in the natural product such as esters, amines, aldehydes, short chain fatty acids, sulfur compounds, C_7-C_9 fatty acids, and fecal odors (skatol and indole) were added to SFE until the odor panel could barely detect the odor change. These same materials were then tested on pelletized Douglas-fir feed with captive deer and elk in Olympia, Washington. The results indicated that the sulfurous compounds added to SFE provided more repellency than the untreated SFE for both deer and elk. The enhanced SFE work should allow us to develop a more simplified formulation for controlled release of volatiles after the active fractions have been identified.

Another natural repellent product found in the Pacific Northwest is wild ginger (Asarum caudatum). This plant material is rarely, if ever, browsed by deer and elk. Campbell and Bullard (1972) report that ground ginger shows a high repellency when added in ground form to feed. Field tests have shown high repellency to deer when this material is applied to Douglas-fir seedlings (Personal communication, D. L. Campbell). Whether the effect is from odorous oils or bitter-tasting compounds is still, however, unknown.

Black-tailed deer are repelled from food by application of conspecific metatarsal scent placed underneath their feeding bowls (Muller-Schwarze, 1971). Only 42.3% of the food from the treated bowl was eaten compared with 97.3% from the untreated bowl. The responses of deer to this scent are somewhat complicated inasmuch as visual stimuli such as erected tail and auditory stimuli such as hissing or foot stomping also accompany release of the metatarsal scent when deer are stressed. Rochelle et al. (1974) also evaluated black-tail deer metatarsal scent placed on Douglas-fir branches but found the repellency to be short term, and they subsequently discontinued their research on deer-alarm pheromones.

Muller-Schwarze (1972) has also explored the concept of feeding suppression with allomones from predators. Six hand-reared black-tailed deer would not feed from bowls scented with feces from coyotes or mountain lions. Fecal scents from other non-sympatric predators produced less suppression. It should be noted, however, that fecal odors in general (Campbell and Bullard, 1972) are repellent to deer. Whole body odors, or odors from various glands of predators have not yet been tested for repellency with deer or elk.

APPLICATION TO PREVENTING AGRICULTURAL RODENT DAMAGE

Agricultural Rodent Damage

If either food sources, water, or cover are removed from vacant buildings, there is generally no infestation problem (Chitty, 1954; Barnett, 1963) with commensal rodents. However, the rodent problems often found in agricultural regions, especially where there is damage to the growing crop before harvest, cannot be controlled so easily. Fortunately, in the mainland U. S., rodent populations seldom, if ever, reach levels where entire crops are destroyed. Isolated crop areas are affected by such species as the pine and meadow voles in the Northeast in fruit orchards (Byers, 1975), deer mice in the Northwest in newly seeded Douglas-fir plantations (Kverno, 1964), and the Polynesian rat (Rattus exulans) in Hawaiian sugar-cane plantations (Hood et al., 1970). In other parts of the world, especially true tropical regions, rats cause extensive damage to grain and vegetable crops. The traditional approach to management of agricultural rodents has been population reduction by means of toxic baits and trapping.

Sex Attractants

The application of toxic baits for rodent control will probably be the main control tool for the next decade. A great deal of effort has been expended on improving bait palatability, trying to overcome bait shyness and toxiphobia, and developing bait formulations for weatherability.

The application of sex attractant pheromones to the toxic baiting method is somewhat questionable. For example, male hamsters show both licking of vaginal secretion and olfactory attraction to this secretion (Murphy, 1973). It has recently been demonstrated (Zahorik and Johnston, 1976) that male hamsters can also be conditioned to avoid water flavored with vaginal secretion after injection of a sublethal toxicant. Olfaction is so important to this species that olfactory bulb ablation in males will eliminate mating (Murphy and Schneider, 1970). Even though olfaction plays such an important role in the reproduction of this species, vaginal secretion as a sex attractant did not retard development of toxiphobia or sub-lethal aversion, but instead, served as an avoidance signal. Thus, sex pheromones that have biological survival value can be used as avoidance cues when they are associated with sub-lethal toxicosis. Nevertheless, such a material could work better than expected if the rate of feeding is increased to the point where the rodent ingests a lethal dose before the sub-lethal aversion occurs.

In an attempt to evaluate certain sources of sex attractant pheromones in ricefield rats (R. r. mindanensis), we have evaluated the response of sexually experienced males to female R. r. mindanensis odors using an in-cage preference apparatus (Shumake and Bullard, unpublished). The odor source materials were preputial gland exudate, fresh urine, and vaginal lavage material. Each material was collected from female conspecifics at the four estrus stages as determined by the vaginal smear method. As shown in Table 1, only proestrus urine produced a reliable increase in time spent by male rats in the presence of odor as compared with deionized water. Essentially the same experiment was replicated with female rats, but this time hair samples from the base of the males' tails were used in place of the vaginal wash. The females were tested this time as they cycled through various estrus stages; no statistically significant preferences were evident.

The proestrus female urine odor was the only material showing a reliable effect in the male rats, therefore we attempted to determine its effect on a normally poorly accepted bait. Using quinine levels (Shumake, Thompson, and Caudill, 1971) known to be rejected by wild Norway rats, we found that the proestrus urine produced no more effectiveness than tap water.

TABLE 1. ELAPSED TIME RESPONSE (SECONDS) OF
MALE RICEFIELD RATS EXPOSED TO FEMALE ODORS
(MEAN ± SE)

Stage of estrus of female sample	n	Odor source			
		Preputial gland	Urine	Vaginal wash	Water (control)
Diestrus	20	19.8 ± 6.5	10.6 ± 4.7	9.2 ± 3.0	23.7 ± 8.1
Proestrus	20	19.3 ± 8.4	100.5 ± 80.9*	49.5 ± 43.4	13.2 ± 8.1
Estrus	20	12.2 ± 3.2	59.5 ± 44.3	24.5 ± 10.1	29.9 ± 13.7
Metestrus	20	30.7 ± 11.8	7.6 ± 2.3	18.2 ± 5.9	7.5 ± 3.1

* P (test sample > water control) = 0.05.

Attractant Signals from Familiar and Preferred Food

Lacking information regarding other potential sources of pheromones in ricefield rats, we turned our efforts toward developing a highly palatable food flavor for rice bait for control of this species. Since rice, when it is available, is the most common food item found in ricefield rat stomachs (Tigner, 1972), we reasoned that this highly preferred and familiar food could be made more attractive to the rats in the field by the addition of certain naturally occurring rice flavor components.

A preliminary test with eight materials representing different components of rice grain (Bullard and Shumake, 1976) was run. Rice bran oil, ether extract of rice, endosperm, rice polish, bran, rice bran volatiles, and rice volatiles were all evaluated on groups of rats with a brief exposure taste preference device previously described (Thompson and Grant, 1971; Thompson et al., 1972). Only the rice volatiles material proved to be superior to untreated rice in many replications of the original preference test and in overnight individual animal preference tests. Results of a semi-natural test, shown in Table 2, confirmed the extent of this preference when whole grain rice, granulated rice, rice treated with 1% soybean oil, and rice treated with rice volatiles were used as bait bases.

As a final test of the toxic bait enhancing properties of the rice volatiles material (Bullard and Shumake, 1976), two groups of

TABLE 2.   GRAMS OF RICE BAIT CONSUMED BY FIVE RICEFIELD RATS
EACH IN FOUR ENCLOSURES
(MEAN ± SE)

	Bait			
Day	(A) Whole grain rice	(B) Granulated whole grain rice	(C) Granulated whole grain rice treated with 1.0% soybean oil	(D) Granulated whole grain rice treated with trapped rice volatiles
1	8.1 ± 1.7	12.9 ± 2.7	22.8 ± 2.9	35.1 ± 3.2**
2	6.6 ± 2.1	8.9 ± 1.6	21.9 ± 2.3	35.1 ± 3.2*
3	5.3 ± 2.0	15.5 ± 2.7	25.9 ± 2.5	35.9 ± 3.3**
4	6.4 ± 2.4	14.6 ± 2.9	16.4 ± 3.0	34.4 ± 2.8**
5	5.8 ± 2.0	12.0 ± 2.3	19.8 ± 3.0	34.6 ± 2.6**
6	4.5 ± 1.7	15.6 ± 2.6	21.6 ± 2.6	36.8 ± 3.1***
7	5.5 ± 2.6	18.9 ± 2.4	15.0 ± 2.2	35.8 ± 2.7**
8	7.0 ± 2.8	19.1 ± 3.3	21.9 ± 3.0	32.4 ± 1.7**

  * $P(D>A)<0.10$
 ** $P(D>A)<0.05$
*** $P(D>A)<0.01$

ricefield rats were offered two food cups for 7 hours per day for
3 days; one contained only the control formulation of untreated
ground rice and the other cup contained this bait treated with 0.2%
zinc phosphide, a commonly used acute rodenticide. The bait carriers were the control formulation for one group of rats and the
rice volatiles formulation for the other group. There was much
greater zinc phosphide dosage in the rice volatiles group (mean 33.9
mg/kg/rat/day versus 19.8 mg/kg/rat/day) and significantly greater
mortality in the rice volatiles bait carrier group (14 dead versus
8 dead) after a 14 day mortality follow-up.

The advantage to working with flavors (both odor and taste stimuli) associated with a preferred food base is that rats eat the food base more rapidly when the flavors are intensified. Since the rice flavor is highly familiar to ricefield rats, toxiphobia to the bait base per se is reduced.

To improve the rice flavor additive quality to an even greater degree, we have also evaluated rice variety preference in R. r. mindanensis in order to extract volatiles from the most preferred varieties. The rats show a consistent preference (P < .01) for one particular variety, FK-178A, that has been commonly grown in the Philippines. IR-20 and C-4 were less preferred (P < .05) to FK-178A and Milagrosa. All four varieties, however, were preferred (P < .01) to the domestic California Brown Pearl (C.B.P.) rice. These same preference patterns were shown in both live-trapped adult animals and in laboratory born and reared ricefield rats. The preference pattern for certain varieties as shown by the live-trapped rats was much more distinctive, although similar, to the pattern shown by laboratory-reared rats.

Rice volatiles materials from FK-178A rice were compared with the volatiles from our standard California Brown Pearl rice in a later test to verify that rice flavor was the critical factor controlling the preference responses. Mean preference for FK-178A flavored rice was 64.3% compared with mean preference of 58.5% for C.B.P. flavored rice. This difference, although small, was reliable indicating that the rice variety preference data reflect primarily a flavor variable rather than texture, grain size, or nutritional level. Thus, volatile rice flavor materials collected from FK-178A should produce an even greater effect on zinc phosphide bait consumption than was shown with the C.B.P. rice flavor.

Signals to Disrupt Reproductive Patterns

The mode of action of certain primer pheromones is at present still somewhat debatable. The early work into this area (Bruce, 1959, 1960, 1961; Parkes and Bruce, 1961) implied that the odor of strange male mouse urine had the direct effect of blocking pregnancy. The evidence for direct pharmacological action of this foreign male odor is not supported by the fact that direct injections of strange male urine into pregnant female mice does not produce this effect (Mykytowycz, 1970). An alternate explanation may lie in the signalling function of the strange male odor. That is, the pregnant female is stressed by the strange male odor, and perhaps, innately recognizes the danger of carrying to term. Parkes and Bruce (1961) imply such a mechanism in their work. That is, if the stud male is of her own strain, then males of another strain are alien and block her pregnancy. However, if she first mates with a male of a foreign strain, then males of her own strain are "alien" and pregnancy is again blocked.

In attempting to apply this pregnancy-blocking effect to population reduction of rodent species, we must deal with a complexity of problems. First, pregnancy is only blocked in 70-80% of the females. Kennelly et al. (1972) found that sterilizing 80% of male Norway rats had a negligible effect on a population over a 3-month period. One should expect birth control measures applied to female rodents to have a much greater effect on the population than male sterilization. However, their polygamous nature and capability of producing up to 8 - 10 litters per year can also work against female chemosterilant concepts. Second, the matter is somewhat further complicated by data (Lott and Hopwood, 1972) that indicate an experience factor in the olfactory pregnancy-block effect. Female mice (Mus musculus) are apparently sensitized to the alien male odor only if given sufficent sexual experience with a given male. It would thus appear that the pregnancy-block effect is extremely difficult to apply as a practical reproduction suppressor in the control of rodents.

Application of other primer pheromone effects such as the suppression of estrus cycling in mice (Whitten, 1959; Chipman and Fox, 1966) in the absence of male urine odor or the development of pseudo-pregnancies (Lee and Boot, 1956) are also difficult in operational terms. Such an application might involve the development of interfering odor materials that would not permit the females to perceive male urine odors existing in the ecosystem. Large-scale field application of such a material, however, would probably present many logistical and economical problems.

The feasibility of controlling rodents by confusing males with area-wide perfusion of female sex odors also appears to be extremely remote. Rodents do not rely exclusively on olfactory cues for mate seeking, courtship, or copulatory behavior. In general, olfactory bulb ablation has only minor influence (Johnston 1970, 1974; Bermant and Taylor, 1969) on the sexual behavior of male laboratory rats.

Area Repellents

Perhaps the greatest potential use of rodent pheromones lies in alarm, fear, or aggression-producing odors. When the predator population is high, perhaps the natural response of rodents to predator odors induces stress and produces a natural rodent population decline. Alarm pheromones from rodents or allomones from predators could have a dispersal effect on rodent populations. Stevens and Saplikoski (1973) have reported evidence of an alarm substance that is present in the blood and muscle tissue from conspecific rats. Hornbuckle

and Beall (1974) have confirmed these results. Fulk (1972) reported
that voles tend to avoid places frequented by short-tailed shrews,
an occasional predator. The odor source was apparently from shrew
droppings.

To systematically assess the potential of alarm or stressor
produced pheromones in rodents will probably require the comparative approach. The Polynesian rat (Rattus exulans) would be an
excellent candidate species for evaluation of the dominant male
stressor pheromone concept. Wirtz (1973) reported that the Polynesian rat has never been studied in its social context to any
extent. Their midventral sebaceous glands as described by Quay and
Tomich (1963) appear to be homologous to those found in gerbils
(Thiessen et al., 1971). The report of Thiessen and Dawber (1972)
indicated that Mongolian gerbils, a desert species, will attempt to
cross water and many will drown rather than encroach upon territory
marked out by dominant males. A behavioral method for bioassaying
ventral scent gland secretion of the gerbil has also been published
(Thiessen et al., 1974) and the active fraction was identified as
phenylacetic acid. If the same dominant male scent gland marks
exert as much influence on polynesian rats as in the gerbil or
hamster (Johnston, 1975), it may be possible to protect Hawaiian
sugarcane plantations with a synthesized dominant male odor. The
concept would be even more promising if, for example, habituation
effects could be delineated and defined in gerbils. The question
here again is, will dominant male odor per se act as a durable barrier over a period of months or years? There are risks in expecting
a repellent to continue working in the absence of further reinforcement, but other authors (Bolles, 1970; Hinde and Stevenson-Hinde,
1973; Muller-Schwarze, 1974) indicate that innate responses to
naturally dangerous olfactory signals should not habituate, nor
should they require learning experiences. Our work on odor attractants (Bullard and Shumake, 1976) indicates that odors and taste
stimuli that occur naturally in the ecosystem have much more durable
effects than those selected from non-biological origins.

SUMMARY OF CHEMICAL SIGNAL APPLICATIONS TO WILDLIFE MANAGEMENT

In the course of attempting to apply chemical signals to the
management of certain wildlife species, we have tentatively developed
the following guidelines:
(1) attractive odors and tastes have to originate from or be
related to products that occur naturally in the ecosystem of the
species to achieve success;
(2) odors that have specific communication content in one
context (i.e., genital fluid ingestion for sex pheromones) do not
necessarily lead to applications in another context (i.e., toxic
bait consumption enhancement);

(3) repellent or stress provoking odors (signalling pheromones) may have as much or more application to wildlife management
than the attractant pheromones;

(4) certain short chain fatty acids (i.e., the end-products of
the action of bacteria upon protein breakdown products) are found
in various ratios across diversified species and may form a chemical coding system for inter- and intra-specific signals; and

(5) other long chain fatty acids (e.g., oleic, linoleic, and
stearic) are found in both plant and animal tissue and are probably the end-products of fat breakdown. These materials also
appear to have attractant function for several species when added
to a food source (Hansson, 1973).

Chemical signals form an integral part of the life of many
wildlife species. The sources of these signals are not only from
conspecifics but are also related to food items (prey and plant
material), and possibly to the avoidance of competitive (Doty,
1973a) or predatory species (Müller-Schwarze, 1972). The stress
provoking and dispersal capabilities of alarm or fear pheromones
(Stevens and Saplikoski, 1973; Müller-Schwarze, 1974) in rodents
and predators should be evaluated. Various means of delivering
controlled dosages of chemical control agents (chemosterilants,
toxicants, and aversive agents) under natural field conditions
should be further explored in order to attempt application with
attractant pheromones. The primer pheromone application to wildlife management presents extreme difficulty. Most mammalian species do not depend upon odor alone for mate detection and selection. Finally, although we are beginning to understand the importance of chemical signals as they affect the behavioral ecology of
many wildlife species, a great many basic and comparative studies
must be undertaken before field applications can be attempted.

ACKNOWLEDGMENTS

Thanks are due to Mrs. A. H. Bean, Mr. R. W. Bullard, Mr. D. L.
Campbell and Dr. R. T. Sterner for critically reading the manuscript.

This work was supported, in part, with funds provided to the
U.S. Fish and Wildlife Service by the U.S. Agency for International
Development under the project "Control of Vertebrate Pests: Rats,
Bats, and Noxious Birds," PASA RA(ID) 1-67.

REFERENCES

Albone, E. S., and Fox, M. W. 1971. Anal gland of the red fox. Nature 233:569-570.

Balser, D. S. 1974a. A review of coyote control research. Proc. Sixth Vert. Pest. Conf. 6:171-177.

Balser, D. S. 1974b. An overview of predator-livestock problems with emphasis on livestock losses. Trans. North Am. Wildl. Nat. Resour. Conf. 39:292-300.

Barnett, S. A. 1963. The Rat: A Study in Behavior. Aldine, Chicago.

Bermant, G., and Taylor, L. 1969. Interactive effects of experience and olfactory bulb lesions in male rat copulation. Physiol. Behav. 4:13-17.

Berüter, J., Beauchamp, G. K., and Muetterties, E. L. 1974. Mammalian chemical communication: perineal gland secretion of the guinea pig. Physiol. Zool. 47:130-136.

Bolles, R. C. 1970. Species specific defense reactions and avoidance learning. Psychol. Rev. 77:32-48.

Bronson, F. H. 1970. General discussion (chapters 8-12), pp. 385-404, in J. W. Johnston, Jr., D. G. Moulton, and A. Turk (eds.), Communication by Chemical Signals, Vol. I. Appleton-Century-Crofts, New York.

Bronson, F. H. 1971. Rodent pheromones. Biol. Reprod. 4:344-357.

Bruce, H. M. 1959. An exteroceptive block to pregnancy in the mouse. Nature 184:105.

Bruce, H. M. 1960. A block to pregnancy in the mouse caused by proximity to strange males. J. Reprod. Fert. 2:138-142.

Bruce, H. M. 1961. Time relations in the pregnancy in the mouse caused by proximity to strange males. J. Reprod. Fert. 1:96-103.

Bullard, R. W., and Shumake, S. A. 1976. Food base flavor additive improves bait acceptance by ricefield rats. J. Wildl. Manage. (in press).

Burghardt, G. M. 1970. Chemical perception in reptiles, pp. 241-308, in J. W. Johnston, Jr., D. G. Moulton, and A. Turk (eds.), Communication by Chemical Signals, Vol. I. Appleton-Century-Crofts, New York.

Byers, R. E. 1976. New compound (RH787) for use in control of orchard voles. J. Wildl. Manage. 40:169-171.

Campbell, D. L., and Bullard, R. W. 1972. A preference testing system for evaluating repellents for black-tailed deer. Proc. Fifth Vert. Pest. Conf. 5:56-63.

Carr, W. J., Loeb, L. S., and Wylie, N. R. 1966. Response to feminine odors in normal and castrated male rats. J. Comp. Physiol. Psychol. 62:336-338.

Carr, W. J., Krames, L., and Castanzo, D. J. 1970. Previous
     sexual experience and olfactory preference for novel versus
     original sex partners in rats. J. Comp. Physiol. Psychol.
     71:216-222.
Chael, M. L., and Sprott, R. L. 1971. Social olfaction: a review
     of the role of olfaction in a variety of animal behaviors.
     Psychol. Rep. 29:195-243.
Chipman, R. K., and Fox, K. A. 1966. Oestrus synchronization and
     pregnancy blocking in wild house mice (Mus musculus). J.
     Reprod. Fert. 12:233-236.
Chitty, D. 1954. The study of the brown rat and its control by
     poison. pp. 160-305, in D. Chitty (ed.), Control of Rats and
     Mice. Oxford, London.
Curtis, R. F., Ballentine, J. A., Keverne, E. B., Bonsall, R. W.,
     and Michael, R. P. 1971. Identification of primate sexual
     pheromones and the properties of synthetic attractants.
     Nature 232:396-398.
Doty, R. L. 1973a. Odor preferences of female Peromyscus manicula-
     tus bairdi for male mouse odors of P m. bairdi and P. leucopus
     noveboracensis as a function of estrus state. J. Comp. Physiol.
     Psychol. 81:191-197.
Doty, R. L. 1973b. Reactions of deer mice (Peromyscus maniculatus)
     and white-footed mice (Peromyscus leucopus) to homospecific
     and heterospecific urine odors. J. Comp. Physiol. Psychol.
     84:296-303.
Dryden, G. L., and Conaway, C. H. 1967. The origin and hormonal
     control of scent production in Suncus murinus. J. Mammal.
     48:420-428.
Eisenberg, J. F., and Kleiman, D. G. 1972. Olfactory communica-
     tion in mammals. Annu. Rev. Ecol. System. 3:1-32.
Estes, R. D. 1972. The role of the vomeronasal organ in mammalian
     reproduction. Ext. Mammal. 36:315-341.
Fulk, G. W. 1972. The effect of shrews on the space utilization
     of voles. J. Mammal. 53:461-478.
Gibson, P. S. 1974. Coyotes and related Canis in Arkansas. Coyote
     Res. News 3:4.
Gorman, M. L. 1976. A mechanism for individual recognition by
     odour in Herpestes auropunctatus (Carnivora: Viverridae).
     Anim. Behav. 24:141-145.
Guadagni, D. G. 1968. Requirements for coordination of instrumen-
     tal and sensory techniques, pp. 36-48, in American Society for
     Testing and Materials (ed.), Correlation of Subjective-
     Objective Methods in the Study of Odors and Taste. ASTM STP
     440, Philadelphia, Pa.
Hansson, L. 1973. Fatty substances as attractants for Microtus
     agrestis and other small rodents. OIKOS. 24:417-421.
Hesterman, E. R., and Mykytowycz, R. 1968. Some observations on
     the intensities of odors of anal gland secretions from the
     rabbit Oryctolagus cuniculus (L.). CSIRO Wildl. Res. 13:71-81.

Hinde, R. A., and Stevenson-Hinde, J. 1973. Constraints on Learning. Academic, London.

Hood, G. A., Nass, R. D., and Lindsey, G. D. 1970. The rat in Hawaiian sugarcane. Proc. Fourth Vert. Pest. Conf. 4:34-37.

Hornbuckle, P. A., and Beall, T. 1974. Escape reactions to the blood of selected mammals by rats. Behav. Biol. 12:573-576.

Johnston, R. E. 1970. Scent marking, olfactory communication and social behavior in the golden hamster, Mesocricetus auratus. Ph. D. dissertation. The Rockefeller University, New York, N. Y. 145pp.

Johnston, R. E. 1974. Sexual attraction function of golden hamster vaginal secretion. Behav. Biol. 12:111-117.

Johnston, R. E. 1975. Scent marking by male golden hamsters (Mesocricetus auratus). III. Behavior in a seminatural environment. Z. Tierpsychol. 37:213-221.

Jones, R. B., and Nowell, N. W. 1973. Aversive and aggression-promoting properties of urine from dominant and subordinate male mice. Anim. Learn. Behav. 1:207-210.

Kennelly, J. J., Johns, B. E., and Garrison, M. V. 1972. Influence of sterile males on female fecundity of a rat colony. J. Wildl. Manage. 36:161-165.

Keverne, E. G., and Michael, R. P. 1971. Sex-attractant properties of ether extracts of vaginal secretions from rhesus monkeys. J. Endocr. 51:313-322.

Krames, L., Carr, W. J., and Bergman, B. 1969. A pheromone associated with social dominance among male rats. Psychon. Sci. 16: 11-12.

Kverno, N. B. 1964. Forest animal damage research. Proc. Second Vert. Pest. Conf. 2:81-89.

Larue, C. J. 1975. Comparison of the effects of anosmia induced by either peripheral lesion or bulbectomy upon the feeding pattern of the rat. J. Physiol. 70:299-306.

Lee, S. van der, and Boot, L. M. 1956. Spontaneous pseudopregnancy in mice II. Acta Physiol. Pharmacol. Neerl. 5:213-215.

Lehner, P. N., Krumm, R., and Cringan, A. T. 1976. Tests for olfactory repellents for coyotes and dogs. J. Wildl. Manage. 40:145-150.

LeMagnen, J. 1963. Sensorial control in the regulation of food intake. Prob. Actuels Endocr. Nutr. 7:147-171.

Linhart, S. B. 1973. Predator Survey-Western U. S. United States Fish and Wildlife Service. Wildlife Research Center, Denver, Colorado. 54pp.

Lott, D. F., and Hopwood, J. H. 1972. Olfactory pregnancy-block in mice (Mus musculus): an unusual response acquisition paradigm. Anim. Behav. 20:263-267.

Lydell, K. and Doty, R. L. 1972. Male rat odor preferences for female urine as a function of sexual experience, urine age, and urine source. Horm. Behav. 3:205-212.

Mayr, E. 1974. Behavior programs and evolutionary strategies. Amer. Sci. 62:650-659.

Michael, R. D., Bonsall, R. W., and Warner, D. 1974. Human vaginal secretions: Volatile fatty acid content. Science 186: 1217-1219.

Moore, R. E. 1965. Olfactory discrimination as an isolating mechanism between Peromyscus maniculatus and Peromyscus polionatus. Am. Midl. Nat. 73:85-100.

Müller-Schwarze, D. 1971. Pheromones in black-tailed deer (Odocoileus hemiones columbianus). Anim. Behav. 19:141-152.

Müller-Schwarze, D. 1972. Response of young black-tailed deer to predator odors. J. Mammal. 53:393-394.

Müller-Schwarze, D. 1974. Olfactory recognition of species, groups, individuals and physiological states among mammals, pp. 316-326, in M. C. Birch (ed.), Pheromones, North Holland, Amsterdam.

Murphy, M. R., and Schneider, G. E. 1970. Olfactory bulb removal eliminates mating behavior in the male golden hamster. Science 167:302-304.

Murphy, M. R. 1973. Effects of female hamster vaginal discharge on the behavior of male hamsters. Behav. Biol. 9:367-375.

Mykytowycz, R. 1970. The role of the skin glands in mammalian communication, pp. 327-360, in J. W. Johnston, D. G. Moulton, and A. Turk (eds.), Communication by Chemical Signals, Vol. I. Appleton-Century-Crofts, New York.

Mykytowycz, R. 1972. The behavioral role of the mammalian skin glands. Naturwissenschaften. 59:133-139.

Parkes, A. S., and Bruce, H. M. 1961. Olfactory stimuli in mammalian reproduction. Science 134:1049-1054.

Peters, R. P., and Mech, L. D. 1975. Scent-marking in wolves. Am. Sci. 63:628-637.

Preti, G., Muetterties, E. L., Furman, J. M., Kennelly, J. J., and Johns, B. E. 1976. Volatile constituents of dog (Canis familiaris) and coyote (Canis latrans) anal sacs. J. Chem. Ecol. 2:177-186.

Price, E. O. 1975. Hormonal control of urine-marking in wild and domestic Norway rats. Horm. Behav. 6:393-397.

Quay, W. B., and Tomich, P. Q. 1963. A specialized midventral sebaceous glandular area in Rattus exulans. J. Mammal. 44: 537-542.

Ralls, K. 1971. Mammalian scent marking. Science 171:443-449.

Rice, P. R., and Church, D. C. 1974. Taste responses of deer to browse extracts, organic acids, and odors. J. Wildl. Manage. 38:830-836.

Riley, G. A., and McBride, R. T. 1975. A survey of the red wolf (Canis rufus), pp. 263-277, in M. W. Fox (ed.), The Wild Canids, Van Norstrand Reinhold, New York.

Rochelle, J. A., Gaudity, I., Oita, K., and Oh, J. 1974. New
    developments in big game repellents, pp. 103-112, in Proc.
    Symp. on Wildlife and Forest Management in the Pacific North-
    west, Oregon State Univ., Corvallis.
Shorey, H. H. 1974. Environmental and physiological control of
    sex pheromone behavior, pp. 62-95, in M. C. Birch (ed.),
    Pheromones, North Holland, Amsterdam.
Shumake, S. A., Thompson, R. D., and Caudill, C. J. 1971. Taste
    preference behavior of laboratory versus wild Norway rats.
    J. Comp. Physiol. Psychol. 77:489-494.
Sperry, C. C. 1941. Food habits of the coyote. U.S. Fish and
    Wildlife Service Research Bulletin, 4, 70pp.
Stern, J. J. 1969. Copulatory experience and sex odor preferences
    of male rats. Proc. 77th Annu. Convent., Amer. Psychol. Assn.,
    Washington, D. C. pp. 229-230.
Stern, J. J. 1970. Responses of male rats to sex odors. Physiol.
    Behav. 5:519-524.
Stevens, D. A., and Saplikowski, N. 1973. Rats' reaction to con-
    specific muscle and blood: evidence for an alarm substance.
    Behav. Biol. 8:83-92.
Taylor, K. D., Hammond, L. E., and Quy, R. J. 1974. The response
    of captive wild rats (Rattus norvegicus) to human odour and
    to the odour of other rats. Ext. Mammal. 38:581-590.
Tette, J. P. 1974. Pheromones in insect population management,
    pp. 399-410, in M. C. Birch (ed.), Pheromones, North Holland,
    Amsterdam.
Thiessen, D. D., Owens, K., and Lindzey, G. 1971. Mechanism of
    territorial marking in the male and female Mongolian gerbil
    (Meriones unguiculatus). J. Comp. Physiol. Psychol. 77:38-47.
Thiessen, D. D., and Dawber, M. 1962. Territorial exclusion and
    reproductive isolation. Psychon. Sci. 28:159-160.
Thiessen, D. D., Regnier, F. E., Rice, M., Goodwin, M., Isaaks, N.,
    and Lawson, N. 1974. Identification of a ventral scent mark-
    ing pheromone in the male Mongolian gerbil (Meriones unguicu-
    latus). Science 184:83-85.
Thompson, R. D., and Grant, C. V. 1971. Automated preference test-
    ing apparatus for rating palatability of food. J. Exp. Anal.
    Behav. 15:215-220.
Thompson, R. D., Shumake, S. A., and Bullard, R. W. 1972. Method-
    ology for measuring taste and odor preference of rodents.
    Proc. Fifth Vert. Pest. Conf. 5:36-42.
Tigner, J. R. 1972. Seasonal food habits of Rattus rattus mindan-
    ensis in Central Luzon. Ph. D. dissertation. University of
    Colorado, Boulder, Colorado. 66pp.
Whitten, W. K. 1959. Occurrence of anoestrus in mice caged in
    groups. J. Endocrin. 18:102-107.

Wittes, J., and Turk, A. 1968. The selection of judges for odor
    discrimination panels, pp. 49-70, in American Society for
    Testing and Materials (ed.), Correlation of Subjective-
    Objective Methods in the Study of Odors and Taste. ASTM STP
    440, Philadelphia, Pa.
Wirtz, W. O. 1973. Growth and development of Rattus exulans. J.
    Mammal. 54:189-202.
Zahorik, D. M., and Johnston, R. E. 1976. Taste aversion to food
    flavors and vaginal secretion in golden hamsters. J. Comp.
    Physiol. Psychol. 90:57-66.

FROM INSECT TO MAMMAL: COMPLICATIONS OF THE BIOASSAY

Robert J. O'Connell

The Rockefeller University

New York, New York   10021

INTRODUCTION

The bioassay is a test procedure which endeavors to use the occurrence of a unique behavioral or physiological response to evaluate the various steps involved in the chemical fractionation, isolation, and identification of the active compounds which occur in an animal's chemical communication system. It is usually assumed that the animal which normally perceives the chemical signal under investigation is best suited to act as the biological detector for the presence of active compounds. Therefore, the central element in the bioassay is the experimental animal with its range of sensory capabilities and its behavioral and physiological repetoire. Accordingly, the typical bioassay represents a coalition between natural product chemistry, sensory physiology, and ethology. Each of these disciplines has its own historical perspective, theoretical expectation, armament of technique and experimental strategy. Consequently, it is easy to appreciate how it is that complications should arise from time to time in the design and execution of a suitable bioassay. Since the intricacies of natural product chemistry have already been thoroughly evaluated, I shall concentrate on the complications which arise in the interaction between sensory physiology and ethology.

Nearly all of these complications represent failures to appreciate the complex web of inputs and outputs which center on the experimental animal. For instance, there is the rather common shortcoming of many bioassays in which the nature of the chemical signals involved are ignored both as mutable chemical entities and as potent mixtures of sensory stimuli. Further, there is the consistent desire on the part of many investigators to use behavioral or physiological measures which have tenuous or poorly understood relationships with the normal responses elicited by the chemical signals being evaluated.

Because of the ubiquitous nature of chemical communication systems, and the consequent bewildering variety of experimental designs encountered in their study, I shall restrict this discussion to an evaluation of assays for olfactory signals in only a few species of insect and mammal. Because there have been a relatively large number of successful bioassays for insect pheromones, I shall rely on the knowledge gathered in this area to illuminate some of the dangers and pitfalls which can be expected in the design and implementation of bioassays for mammalian pheromones.

GENERAL CONCEPTS

It is axiomatic that there should first be some certitude about the fact that chemical signals are actually involved in modulating an animal's behavior. Further, the chemical modality within which these signals are perceived should also be known. Without these assurances, little importance can be attached to the outcome of the "bioassay." Once it has been firmly established that there are in fact olfactory pheromones operating in a particular animal species, it becomes necessary to determine which behavioral responses are normally elicited by these compounds. This determination is best made in the field with feral animals, where the full expression of the chemical and behavioral attributes of the chemical communication system can be expected to operate. Short of this ideal, it is necessary to insure that the experimental setting in which behavior is to be observed is not so constrained as to preclude expression of the usually obtained responses or so unnatural that behaviors other than those normally elicited are produced. Further, it should be recognized that the behavioral responses elicited by olfactory pheromones are often strongly influenced by the context within which they are elicited (Johnson, 1973; Mykytowycz, et al., 1976; Thiessen, 1973). For example, Thiessen and his colleagues have made an extensive investigation of scent marking behavior in the Mongolian gerbil (Thiessen, 1973; Thiessen, et al., 1970; 1971; 1972; Wallace, et al., 1973). They demonstrated that this response is elicited in experimental settings which are associated with territoriality, exploration, social dominance, aggression, and in the female, maternal responsibility. It is reasonable to assume that other behavioral responses to the chemical signals present in the scent mark, or other sources of odors, will also be dependent upon the context within which they are perceived. Certainly it would be an error to assume that a particular behavioral or physiological response must invariably follow the presentation of a particular chemical signal.

To further complicate the situation, attention must be paid to the fact that the behavioral responses obtained with olfactory pheromones are easily modulated by a variety of other environmental factors (Shorey, 1974). For instance, Hindenlang and his colleagues (1975) have shown that male potato tuberworm moths are strongly attracted to extracts of the female pheromone gland just before dawn. They are relatively insensitive to it six hours later. The absolute level of lighting, the ambient temperature, the age of the animals, in addition to the time of day are known to modulate the magnitude of this attraction. We have observed that the attraction of male hamsters to the odor of estrous female vaginal discharge follows a yearly cycle. Approach, sniffing and digging patterns are virtually eliminated during the late winter and early spring in spite of the fact that other aspects of the animal's normal sexual behavior are unimpaired during this time. Similar environmental variables are known to influence the production and release of pheromone compounds in females (Cardé et al., 1974; Sower et al., 1972). These few brief examples should serve to illustrate the fact that without full and complete knowledge of the behavioral responses of an animal to its olfactory pheromone, it is impossible to design an adequate and valid bioassay.

SPECIFIC EXAMPLES

There are a large number of different behavioral and physiological responses which could be used to follow the various chemical steps in the fractionation, isolation and identification of chemicals signals. These include human psychophysical judgments (Thiessen et al., 1974; Whitten, 1969), instrumental conditioning tasks (Thiessen et al., 1974), changes in heart rate (Hesterman et al., 1976), respiratory rate, or sniffing frequency (Wenzel, 1971) and various measures of electrical activity in the olfactory system (Baker & Roelofs, 1976; Roelofs & Comeau, 1971b). The utility of these response measures depends to a large degree upon their ability to impart specificity to the determination of chemical structure. The more generalized the response in terms of the environmental events which may produce it, the less likely it is to identify the particular chemical species under investigation. Furthermore, unless the relationship between the response actually measured in a bioassay and the behavioral response normally elicited by a chemical signal is already known, it is very difficult to interpret the experimental outcome. For example, there is no a priori reason why a bioassay, which attempts to use

human judgments of the characteristic odor of a particular scent
gland as an end point, should be successful; primarily because
there is a very tenuous relationship between human perceptions
and those of an experimental animal. Dimethyl disulfide was
recently identified in estrous hamster vaginal discharge and was
shown to be a powerful attractant for male hamsters in a laboratory bioassay (Singer et al., 1976). The range of stimulus intensities within which hamster behavior is modulated is well below the human detection threshold for this compound. This is a
common finding for the odors of many of the insect sex pheromones,
and has also been noted for the odor of the rabbit chin gland
(Mykytowycz, 1975). Even when amounts above the human threshold
are used, there is no necessary relationship between the perceived odor quality of the identified active principle and that
of the natural glandular product. The odor of dimethyl disulfide
is very different from the odor of hamster vaginal discharge.
Consequently, human judgments concerning the various steps in the
fractionation, isolation, and identification of the putative
chemical signals of other species are, in large degree, suspect.

Instrumental conditioning techniques are faulted because there
is no assurance that the instrumental responses evoked by chemical
compounds correspond in any way with the responses normally elicited
by the same stimuli. The fact that an animal can learn to associate two odorous stimuli does not guarantee that they are chemically identical. Moreover, when a number of different comparisons are to be made, as they are in a fractionation procedure,
there is no way of insuring that the decision rules employed by
the animal for one of the comparisons between odors are the same
as those employed for a second. All that can be said for such procedures is that the chemicals finally identified may have some odor
similarity with compounds actually present in the secretion without specifying if they have any relationship to the biologically
active portions. Each of the physiological response measures
which have been used as end points have the common drawback that
they do occur in many experimental situations. For example,
there are a great number of individual stimuli operating across
several sensory modalities which can change the respiratory rate
of an animal. Only a small fraction of these stimuli are likely
to be involved in the animal's chemical communication system. An
even smaller number of those will be involved in eliciting a particular behavioral response.

Related difficulties are encountered when the electrical
activity elicited in a portion of the olfactory system is used to
evaluate compounds. The utility of this measure is directly
proportioned to the specificity of the responses elicited by the
active principles of the secretion. If both pheromone and non-
pheromone compounds elicit similar levels of electrical activity,

then this measure cannot be used to identify behaviorally relevant compounds. The basic generality of these and other physiological measures usually precludes their use in a bioassay,
especially when there is a basic lack of information about the
relationship between the natural response and the measure used as
the end point. However, useful information can be obtained with
these measures if the range of sensory stimuli impinging on the
animal is severely restricted. For instance, Roelofs has used
the summated electrical activity of olfactory receptor neurons
located in insect antenna (the electroantennogram) to identify a
large number of different insect sex attractant pheromones. A
degree of specificity was imparted to the electroantennogram by
restricting the compounds used as stimuli to those actually present in the secretion of the pheromone gland. On a probabilistic
base, those compounds which elicit the largest responses are
usually involved in the animal chemical communication system.
On the other hand, smaller electroantennograms are routinely obtained with many other compounds found in the secretion, only
some of which have proved to be behaviorally important (Baker
and Roelofs, 1976). Discrimination between these compounds must
be obtained with other, more specific means (Baker et al., 1976).

     After an appropriate measure of animal behavior has been
selected, attention should be shifted to the control and specification of the olfactory stimulus which is used to elicit the behavioral response. Failure to do so can lead to serious complications (Bedoukian, 1970). The secretory products of an animal's
pheromone gland are usually a complex mixture of different chemical substances, only a portion of which are active in producing
behavioral responses (Singer et al., 1976). It can easily be
appreciated that these mixtures do not remain invariant over
time. They are subject to continual changes in their composition
due to differences in the relative volatility of their components.
These differences result in a progressive decrease in the concentration of the more volatile compounds and a consequent selective
enrichment of the least volatile components. In addition, the
metabolic action of bacterial contaminates cannot be overlooked.
They will selectively remove certain chemicals while adding others
as by-products of their own respiratory processes. Furthermore,
there are a number of direct chemical changes which can be expected to occur in these complex mixtures. For instance, many
Lepidopteran sex attractants are ten to sixteen carbon mono- and
di-unsaturated acetates. Hydrolysis of these compounds to their
respective alcohols is quite common. Since olfactory pheromones
are often active at extreme dilutions, small and seemingly subtle
changes in their chemical composition can induce relatively
large changes in behavior.

Further complications in the specification of the chemical
stimulus can occur whenever an animal produces several different
pheromones, each from a different glandular source. It is clearly
important that they are not mixed inadvertently. Even when there
is only a single source of chemical signals, they should be kept
separate from the remainder of the animal's body surface to prevent admixture of waxes or oils from the integument which can
change the composition of the pheromone secretion and its rate of
volatilization. Once collected, the pheromone should be kept in
a closed, odor free container under storage conditions which will
insure the constancy of its composition. The bioassay apparatus
itself should be kept free from odorous contamination both from
previous exposure to test animals and from odors imparted by contact with the experimentor. A similar range of precautions should
be taken with the individual fractions of the pheromone as they
are obtained during the course of the chemical identification.

Errors in the specification of the effective olfactory stimulus can occur during the bioassay in a variety of different ways.
For instance, the animal being tested should not be allowed to
come into physical contact with the olfactory pheromone being
evaluated. Such contact may add to the secretion odors unique
to the test animal which will then alter the subsequent responses
of other test animals. For example, Black-Cleworth and Verberne
(1975) have demonstrated that there are relatively large changes
in a number of different behavioral responses when male rabbits
are exposed to an environment which had previously been scented
by another male rabbit. Many of these behaviors occurred more
often in response to odors produced by the activity of other
animals than in response to those odors produced by the test animal himself. A second way in which the specifications of the
effective behavioral stimuli can be confounded, occurs when test
animals are evaluated while they are inadvertently exposed to
both the suspected active fraction, and a second animal which is
capable of producing an effective pheromone on its own. For
example, in one series of experiments, Gentry and his colleagues
(1975) evaluated the capability of the oriental fruit moth's sex
pheromone, cis-8-dodecenyl acetate to increase the number of pecan bud moths caught in black-light traps. Traps of this sort are
intrinsically attractive to many insect species. They capture
both male and female pecan bud moths along with a sampling of any
other species which happens to be in the area over the duration
of the experiment. Each of the animals caught in the trap serves
as a potential source of different odorous compounds some of
which could interact in an unspecified way with the synthetic sex
pheromone being evaluated. Therefore, each trap is actually
baited with a complex mixture of compounds. These include cis-8-
dodecenyl acetate, the pheromones released by both male and female
pecan bud moths and those of any other species caught in the trap.

The fact that an unspecified mixture of chemical compounds is being evaluated effectively precludes interpretation of the resulting data other than to demonstrate that some interaction phenomenon between odorous compounds does exist. An analogous sort of difficulty is encountered when bioassays are carried out in areas with high levels of infestation using traps having large exposed trapping surfaces. In this situation unbaited traps will capture both male and female insects of a particular species solely because of their large size. As a consequence, evaluation of single synthetic compounds is impossible. For example, Hendry and his colleagues (1974) have recently employed this bioassay to evaluate the ability of several suspected synthetic pheromones to trap male oak leaf roller moths. They concluded that cis-10-tetradecenyl acetate was a major component of this insect's sex attractant pheromone. However, since all of the compounds were examined in the presence of substantial numbers of both male and female oak leaf roller moths, each of which could contribute its own pheromone , an alternate explanation of the data is possible. It could be that cis-10-tetradecenyl acetate modulates in some manner the innate attractiveness of the trapped females toward additional males and is not, in itself, attractive. A subsequent chemical and behavioral reevaluation of this chemical communication system by Miller and his colleages (1976) failed to detect cis-10-tetradecenyl acetate in female oak leaf roller pheromone glands. Further testing of this compound with small traps appropriately randomized and replicated failed to uncover any attraction of the male moth to this compound. They found instead that a 67 to 33 percent ratio of cis- and trans-11-tetradecenyl acetates is present in females and very effective in the field trapping experiments. These two examples demonstrate that the presence of other animals during a bioassay, each with their own potential for pheromone release, can have profound effects on the outcome of a seemingly simple test procedure.

An important aspect of the bioassay is the manner in which odorous stimuli are presented to the test animal. For example, a serious problem arises when one evaluates very large amounts of putative pheromone in very small chambers. The dangers here are exemplified by a study of Keville and Kannowski (1975) who attempted to bioassay sex excitant pheromones in the confused flour beetle. In these tests, 2 to 2½ milligrams of putative pheromone was placed in a covered petri dish along with 100 male and female beetles. Given that an individual flour beetle weighs from 2 to 5 milligrams, and that the usual amount of pheromone in a single insect is in the nanogram range, it seems reasonable to assume that several thousand bug equivalents of synthetic compound were being evaluated in the single petri dish. Little wonder then that the male beetles became confused and clambered onto the back of any suitable object, male or female, in an attempt to get away

from this overwhelming stimulus onslaught. A further complication
in this particular assay is the fact that several of the identified
compounds are known to act as defensive secretions. Consequently,
one should question the wisdom of forcing the animals to stand in
these compounds during the course of the bioassay. The conclusion
that the tested compounds caused sexual excitation and increased
the number of copulatory attempts in male flour beetles should be
accepted with caution.

Although not specifically an issue in this later example,
one should appreciate that the ability of an animal to accurately
orient to, and approach the source of an olfactory pheromone is
severely compromised in any experimental setting where the entire
apparatus is saturated, either overtly or covertly, with the compounds being tested. This becomes particularly acute in those
instances where the design of the bioassay is such that the animal
must approach the pheromone source to within some arbitrary distance in order to register a positive response. Hindenlang and
his colleagues (1975) observed that crude extracts of female
potato tuberworm moth pheromone gland were attractive to male
moths when assayed in small glass tubes (3.5 diam. x 45 cm long).
A positive response was scored whenever a male was found within
4 cm of the pheromone source during the first 30 seconds of the
test. As stimulus intensity was increased from 1×10^{-6} to
1×10^{-2} female equivalents, the number of males attracted increased from approximately 23% to a maximum of 69% of the exposed
animals. Further increases in intensity caused the percent attraction to decrease until it was back at the 23% level at an
intensity of one female equivalent. The authors concluded that
males could not locate the source of the attractant over a distance
at the highest stimulus intensities, implying that the gradient
along which they were orienting had been eliminated, at least
within the first four cm of the tube. An alternate explanation
for this data involves the authors' observation that the males
made increased numbers of copulatory attempts at the higher stimulus intensities. Perhaps the occurrence of such copulatory
attempts is incompatible with the motor acts required for attraction. It should be noted that only crude extracts have been
assayed as yet; therefore, it is also possible that the increase
in copulatory attempts and the resulting decrease in attractancy
is a direct consequence of the presence of two active compounds
in the crude extract -- one which drives copulatory attempts and
a second which drives attraction. Such an interlocked series of
stimulus-response sequences has been observed in the red-banded
leafroller moth (Baker et al., 1976a) and the oriental fruit
moth (Cardé et al, 1975). These two studies serve to exemplify
one of the cardinal rules in the design of an appropriate bioassay. The assay must be built on the known behavior of the test
animal rather than on the expectations of the experimentor. At

first glance, baiting a trap with presumed attractant pheromone
and then counting the number of animals caught, would seem to be
a relatively straightforward way of measuring attraction. However, its ultimate usefulness relies on the expectation that the
relationship between placing an attractant pheromone in the trap
and then inducing a test animal to enter, is a simple unitary
stimulus-response sequence. As long as there is no solid observational base for this expectation, the data obtained with
trapping bioassays has the potential of being misinterpreted. In
a series of classical studies on the chemical communication system
of the red-banded leafroller moth, Roelofs and his colleagues
(1968; 1971a; 1975) isolated and identified cis-11-tetradecenyl
acetate as the principal component of the female pheromone gland.
In the course of several large scale field tests using standard
trapping techniques, a number of related compounds were identified
which modulated the number of males caught in traps baited with
cis-11-tetradecenyl acetate. For instance, the trans compound when
present in a one to one ratio with the cis compound, decreases the
number of males caught. However, with a cis to trans ratio of 94
to 6, the number of males trapped increases. Further increases in
trapping efficiency were obtained with the addition of dodecyl
acetate to the mixture. A total of nine compounds were characterized as inhibitors and 16 compounds as synergists of the
principal component of the pheromone gland. A subsequent re-
examination of the contents of the female pheromone gland revealed that both of these synergistic compounds (i.e., trans-11-
tetradecenyl acetate and dodecyl acetate) were present in sufficent amounts to account for the observed behavioral synergy.
The underlying physiological mechanism which results in increased
numbers of males caught in synergized traps was assumed to be a
simple increase in the trapping range of the pheromone gland.
However, recent ethological analysis of the behavioral responses
of males in the presence of these compounds, suggests that the observed attraction to a baited trap is not a unitary behavioral
event.

   Cis-11-tetradecenyl acetate alone activates approximately 27%
of the exposed males. Activation includes both wing flapping and
flying. The addition of 8% trans increases the level of activation to 76% of the exposed animals. Once aloft, these males fly
upwind towards the source of pheromone. Of those males arriving
within 1/2 meter of the trap, approximately 69% will turn away
and fly rapidly downwind. However, if dodecyl acetate is added
to the pheromone blend, 100% of those animals within a 1/2 meter
will land and subsequently be caught. These results indicate that
the trans compound synergizes trap catches by modulating upwind
anemotaxis, whereas dodecyl acetate synergizes trapping by increasing the propensity of males to land. Furthermore, the

effectiveness of dodecyl acetate in modulating landing frequency
is related to the physical dimensions of the trap being employed
in the bioassay. Large traps (60 cm radius) show a two-fold increase in trapping efficiency whereas small traps (8 cm radius)
show a twelvefold increase. This difference is presumably due to
the fact that landing is both chemically and visually guided.
Therefore, what seemed at the outset to be a simple stimulus-
response sequence which resulted in a single behavioral event is
in reality a complex series of stimulus-response elements, each
of which is driven by a particular chemical compound, and highly
dependent upon the physical attributes of the bioassay. Without
a thorough ethological analysis of the actual behaviors involved
in this sequence, there would remain many opportunities for error.
For instance, it is relatively unlikely that a compound which
modulates landing behavior would have ever been identified in the
female pheromone gland extract, in the absence of a bioassay procedure which specifically examined this element of the overall behavioral sequence.

                              CONCLUSION

     There are several basic dictums which if followed should
avoid most of the complications which arise in bioassays. 1) The
chemical signals being evaluated must be regarded as olfactory
stimuli with all the attendant precautions relative to maintaining them in their pristine form. 2) The delivery of these signals
to the experimental animal should mimic as closely as possible
the natural route of application. 3) The most specific
biological response appropriate to the kind of chemical
signal being evaluated should be employed as the endpoint in the
chemical fractionation procedure.

     Acknowledgement. I thank Drs. R. Costanzo, M. S. Mayer, and
W. L. Roelofs for their helpful suggestions; S. Laden, M. Lipton,
and J. Clement for typing.   NIH Grant NS08902 and NSF grant
BNS 07707 provided financial support.

REFERENCES

BAKER, T. C., CARDÉ, R. T., and ROELOFS, W. L. 1976. Behavioral
    responses of male Argyrotaenia velutinana (Lepidoptera:
    Tortricidae) to components of its sex pheromone. J. Chem.
    Ecol., In press.
BAKER, T. C., and ROELOFS, W. L. 1976. Electroantennogram
    responses of male Argyrotaenia velutinana (Lepidoptera:
    Tortricidae) to mixtures of its sex pheromone components.
    J. Insect. Physiol., In press.
BEDOUKIAN, P. Z. 1970. Purity, identity and quantification of
    pheromones. Pages 19-34 in J. W. Johnston, Jr., D. G.
    Moulton, and A. Turk (eds.), Communication by Chemical
    Signals. Appleton-Century-Crofts, New York.
BLACK-CLEWORTH, P., and VERBERNE, G. 1975. Scent-marking, domi-
    nance and flehmen behavior in domestic rabbits in an arti-
    ficial laboratory territory. Chemical Senses & Flavor 1:
    465-494.
CARDÉ, R. T., BAKER, T. C., and ROELOFS, W. L. 1975. Ethological
    function of components of a sex attractant system for
    oriental fruit moth males, Grapholitha molesta (Lepidoptera:
    Torticidae). J. Chem. Ecol. 1: 475-493.
CARDÉ, R. T., DOANE, C. C., and ROELOFS, W. L. 1974. Diel
    periodicity of male sex pheromone response and female
    attractiveness in the gypsy moth (Lepidoptera: Lymantriidae).
    Can. Ent. 106: 479-484.
GENTRY, C. R., BEROZA, M., and BLYTHE, J. L. 1975. Pecan bud
    moth: Captures in Georgia in traps baited with the pheromone
    of the oriental fruit moth. Environ. Entomol. 4: 227-228.
HENDRY, L. B., JUGOVICH, J., ROMAN, L., ANDERSON, M. E., and
    MUMMA, R. O. 1974. Cis-10-tetradecenyl acetate, an attrac-
    tant component in the sex pheromone of the oak leaf roller
    moth (Archips semiferanus Walker). Experientia 30: 886-887.
HESTERMAN, E. R., GOODRICH, B. S., and MYKYTOWYCZ, R. 1976. Be-
    havioral and cardiac responses of the rabbit, Oryctolagus
    cuniculus, to chemical fractions from anal gland. J. Chem.
    Ecol. 2: 25-37.
HINDENLANG, D. M., MCLAUGHLIN, R. R., GUILIANO, R. M., and HENDRY,
    L. B. 1975. A sex pheromone in the potato tuberworm moth,
    Phthorimaea operculella (Zeller): Biological assay and pre-
    liminary chemical investigation. J. Chem. Ecol. 1: 465-475.
JOHNSON, R. P. 1973. Scent marking in mammals. Anim. Behav. 21:
    521-535.
KEVILLE, R., and KANNOWSKI, P. B. 1975. Sexual excitation by
    pheromones of the confused flour beetle. J. Insect Physiol.
    21: 81-84.

MILLER, J. R., BAKER, T. C., CARDÉ, R. T., ROELOFS, W. L. 1976. Reinvestigation of oak leaf roller sex pheromone components and the hypothesis that they vary with diet. Science. 192: 140-143.

MYKYTOWYCZ, R. 1975. Activation of territorial behaviour in the rabbit, Oryctolagus cuniculus, by stimulation with its own chin gland secretion. Pages 425-432 in D. A. Denton and J. P. Coghlan (eds.), Olfaction and Taste V. Academic Press, Inc., New York.

MYKYTOWYCZ, R., HESTERMAN, E. R., GAMBALE, S., and DUDZINSKI, M. L. 1976. A comparison of the effectiveness of the odors of rabbits, Oryctolagus cuniculus, in enhancing territorial confidence. J. Chem. Ecol. 2: 13-24.

ROELOFS, W. L., and ARN, H. 1968. Sex attractant of the red-banded leaf roller moth. Nature 219: 513.

ROELOFS, W. L. and COMEAU, A. 1971a. Sex pheromone perception: Synergists and inhibitors for the red-banded leaf roller attractant. J. Insect Physiol. 17: 435-448.

ROELOFS, W. L. and COMEAU, A. 1971b. Sex pheromone perception: Electroantennogram responses of the red-banded leaf roller moth. J. Insect Physiol. 17: 1969-1982.

ROELOFS, W., HILL, A., and CARDÉ, R. 1975. Sex pheromone components of the redbanded leafroller, Argyrotaenia velutinana (Lepidoptera: Tortricidae). J. Chem. Ecol. 1: 83-91.

SHOREY, H. H. 1974. Environmental and physiological control of insect sex pheromone behavior. Pages 62-80 in M. C. Birch (ed.), Pheromones. North-Holland, Amsterdam.

SINGER, A. G., AGOSTA, W. C., O'CONNELL, R. J., PFAFFMANN, C., BOWEN, D. V., and FIELD, F. H. 1976. Dimethyl disulfide: An attractant pheromone in hamster vaginal secretion. Science. 191: 948-950.

SOWER, L. L., SHOREY, H. H., and GASTON, L. K. 1972. Sex pheromones of Lepidoptera. XXVIII. Factors modifying the release rate and extractable quantity of pheromone from females of Trichoplusia ni (Noctuidae). Ann. Entomol. Soc. Am. 65: 954-957.

THIESSEN, D. D. 1973. Footholds for survival. Am. Sci. 61: 346-351.

THIESSEN, D. D. and DAWBER, M. 1972. Territorial exclusion and reproductive isolation. Psychon. Sci. 28: 159-160.

THIESSEN, D. D., LINDZEY, G., BLUM, S. L., and WALLACE, P. 1970. Social interactions and scent marking in the mongolian gerbil (Meriones unguiculatus). Anim. Behav. 19: 505-513.

THIESSEN, D. D., OWEN, K., and LINDZEY, G. 1971. Mechanisms of territorial marking in the male and female mongolian gerbil (Meriones unguiculatus). J. Comp. and Physiol. Psych. 77: 38-47.

THIESSEN, D. D., REGNIER, F. E, RICE, M., GOODWIN, M., ISAACKS, N., and LAWSON, N. 1974. Identification of a ventral scent marking pheromone in the male mongolian gerbil (Meriones unguiculatus). Science. 184: 83-85.

WALLACE, R., OWEN, K., and THIESSEN, D. D. 1973. The control and function of maternal scent marking in the mongolian gerbil. Physiol. Behav. 10: 463-466.

WENZEL, B. M. 1971. Olfaction in birds. Pages 432-448 in L. M. Beidler (ed.), Handbook of Sensory Physiology Vol. IV, Olfaction. Springer-Verlag, Berlin.

WHITTEN, W. K. 1969. Mammalian pheromones. Pages 252-257 in C. Pfaffmann (ed.), Olfaction and Taste III. The Rockefeller University Press, New York.

METHODOLOGY AND STRATEGIES IN THE LABORATORY

D. D. Thiessen [1]

Department of Psychology

University of Texas at Austin, Austin, Texas 78712

ABSTRACT

The investigation of chemocommunication in mammals demands
a flexible methodology and, more importantly, attention to
relevant biological problems, such as the nature of chemosignals,
reactions to signals, biological attributes of the organism,
physiological processes, ecological influences, and the relationship of chemocommunication to genetic fitness. Laboratory
techniques can isolate and identify parameters of the signaling
system, but the full understanding of chemocommunication will
only follow combined laboratory and field investigations. Special
emphasis should be given to the integration of various signaling
systems within the ecological and genetic constraints of the
organism.

Chemocommunication, perhaps more than any other area of
sensory-behavioral investigation, requires detailed attention to
methodology. Unlike related areas of vision and audition where
stimulus configurations can be physically defined and characterized at the receptor, chemosignals are fundamentally invisible,
variable in physical properties and difficult to detect at the
receptor. The problems are manifestly compounded as we search
for neural transducers, central pathways and behavioral responses.
Again, unlike the search for basic attributes of vision and
audition which can be investigated in detached neurophysiological
preparations, chemocommunication must be studied within an integrated biological system. Not only is it imperative to specify
the signal and its transduction into neural energy, but it is also

essential to characterize the organism--its physiology, ecology,
genetics and evolutionary history. The investigation of
vertebrate chemocommunication requires a complex methodology
aimed at understanding the entire organism.

This is not to assert that laboratory methodology in chemocommunication is the sine qua non of a vigorous discipline, or
that our methodology should evolve toward standardization. Quite
the contrary. Ours is a primitive discipline without historical
benchmarks. The problems are not even clearly defined, let alone
the technology. As such, we should not prematurely specify the
tools of the trade nor attempt to solidify methodology in the
laboratory. Rather, our attention should be directed at defining
problems that require solution, and only secondarily at the
articulation of these problems within a methodological context.

> To paraphrase an old saying.
> Beware of the man of one method
> or one instrument, either exper-
> imental or theoretical. He tends to
> become method-oriented rather
> than problem-oriented. The method
> oriented man is shackled; the
> problem-oriented man is at least
> reaching freely toward what is
> most important. Strong inference
> redirects a man to problem-
> orientation, but it requires him
> to be willing repeatedly to
> put aside his last methods
> and to teach himself new ones.
>   (Platt, 1964, p 351)

Paradoxically, my task is to discuss laboratory methodology
while at the same time emphasizing that strategies are more
important than tactics. Specifically, I will outline some
relevant problem areas in chemocommunication and suggest where
laboratory approaches can be applied to the solution of a fraction
of these problems. My primary focus is on mammalian chemocommunication, with illustrations drawn mainly from our own
efforts with the Mongolian gerbil, Meriones unguiculatus. I will
avoid a catalog approach to laboratory methodology, referring
to an excellent discussion of methodologies in olfactory research
in a recent volume edited by Moulton, Turk and Johnston (1975).
Instead, I will present a brief, almost philosophical and certainly
speculative look at problems and problem solving strategies.

Problems Worth Solution

In a developing discipline we can expect surprises and occasional reversals of strategies. For the most part, however, the current problems in chemocommunication (that is, relevant questions) are encompassed in Table 1. They resolve down to questions dealing with (1) the nature of the signal, (2) the primary and secondary reactions to a chemosignal, (3) general attributes of the organism, (4) physiological processes intervening between the signal and the response, (5) ecological features specifying the communication system, and (6) the relationship between chemosignaling and genetic fitness.

Not all of these problems can be attacked in the laboratory. Laboratory procedures can be used to isolate and identify a pheromone system and pick apart its complexities. These same procedures can be used to amplify the intensity of a signal and specify some of its basic physiological and neurophysiological attributes. They can even suggest the functions of a pheromone system. Laboratory procedures cannot, however, provide an integrated picture of chemocommunication within an ecologically valid community of interacting animals. This can only be done in conjunction with field investigations where laboratory hypotheses can be tested against the realities of natural selection. The final criterion of a pheromone system is a chemical information exchange process that directly or indirectly affects reproductive fitness---individual or inclusive fitness (Hamilton, 1964). Field verification of laboratory findings is essential to establish this final criterion.

Laboratory Trends in Signal Identification

Within the framework of problems identified in Table 1, laboratory methodology is demonstrating the power to isolate and identify communication chemicals in mammals. Thus far six putative pheromones have been identified (Table 2), but using numerous behavioral and chemical techniques now available, other active molecules will surely be found. Already the evidence suggests that some of these signals are short range or contact pheromones, whereas others are long distance airborne signals. No generalizations are yet possible about the molecular nature of mammalian signals or their temporal effectiveness. It is informative to note that the greatest success at biochemical identification has occurred with species where the signal originates from discrete glands or tissues that may be specialized for the synthesis of signals. General body exudates such as urine, feces or sweat, which can also function as signals, are proving more difficult to analyze because of their chemical complexity and environmental dependent variability.

Table 1
Classification of Chemosignal Variables According
to Investigative Concerns

A View of Signal Characteristics

Origin of body chemosignals
        Specific source (glandular)
        Nonspecific source (metabolic exudate)
Biochemical identification of chemosignals
Effective range of communication
        Short-range (contact signal)
        Projected (directed signal)
Temporal effectiveness of chemosignals
        Short-term signal
        Long-term signal
Cross-modality signaling (communication syntax)

Detecting Primary and Secondary Effects

Direct and immediate reaction to chemosignals (releaser)
        Orientation and habituation
        Motor adjustments and locomotion
Indirect but immediate reaction to chemosignals (signal)
        Stimulation and arousal of responses
        Facilitation and inhibition of responses
        Interpretation of signal (symbol recognition)
Delayed long term reactions (primer)
        Biasing of physiological and behavioral actions
        Synchronization of physiological and social activities

Important Organismic Parameters

Circadian and circannual rhythms
Age relationships
Sexual dimorphism
Social status variables
Individual, population and species differences
Receptor characteristics
Associative qualities
        Ontogenetic imprinting
        Learning ability
Responses contingent on environmental context

Table 1 (Con't)

Classification of Chemosignal Variables According
to Investigative Concerns

The Search for Physiological Correlates

Metabolic pathways of chemosignal synthesis
Genetic regulatory mechanisms
Hormonal influences
Sensory modulation of signal
  Receptor sensitivity
  Transduction of signal into neural code
Central nervous system regulation
  Control of synthesis, emission and recognition of
   signal
General physiological preparedness
  Motivational set

Notable Ecological Restraints

Geographical distribution of species
Climatic influences
  Short and long term effects
Habitat preference of species
  Energy requirements
  Reactions to prey or predators
Acquisition of environmental chemosignals
  Ingestion of plant and animal chemicals
  Use of symbiotic organisms to produce chemosignals

Attributes of Genetic Fitness

Relationship of pheromone utilization to gene transmission
  Nature of social status and reproductive capacity
  Dispersal, territorial and home range characteristics
Demographic nature of population
  Size, density and flexibility
  Mating strategy, mate preference and family structure
Natural selection of coordinated signaling devices
  Selection pressure toward optimal signal to noise ratio
  Evolution of communication specialists and generalists
  Development of multiple signaling systems

TABLE 2

Putative Mammalian Pheromones

Species	Glandular or Tissue Source	Chemosignal	Nature of Signal	Reference
Boar (Sus scrofa)	Male submaxillary gland	5α-androst-16-en-3-one	Airborne (sex attractant)	Patterson (1968)
Black-tailed deer (Odocoileus hemionus)	Tarsal gland	cis-4-hydroxy-dodec-6-enoic acid lactone	Airborne and close range(identification signal)	Brownlee, Silverstein, Müller-Schwarze & Singer(1969)
Pronghorn antelope (Antilocapra americana)	Subauricular gland	Isovaleric acid	Object marking, airborne (territorial signal)	Müller-Schwarze, Müller-Schwarze, Singer, Silverstein (1974)
Mongolian gerbil (Meriones unguiculatus)	Ventral scent gland	Phenylacetic acid	Object marking, airborne (territorial signal)	Thiessen, Regnier, Rice, Goodwin, Isaacks, Lawson (1974)
Hamster (Mesocricetus auratus)	Vaginal tissue	Dimethyl di-sulfide	Airborne and close range (sex attractant)	Singer, Agosta, O'Connell, Pfaffman, Bowen & Field (1976)
Monkey (Macaca mulatta)	Vaginal tissue	Acetic, propionic, iso-butyric, n-butyric, iso-valeric, & iso-caproic acids	Airborne (sex attractant)	Michael, Keverne & Bonsall (1971)

The microstructure and the kind of innervation of pheromone glands can add to our knowledge of their chemical nature as well as their behavioral functions. Many types of exocrine glands exist, some that suggest direct and simple biosynthetic mechanisms, and some that suggest storage of secretory material and complex synthetic mechanisms (Figure 1). Neural innervation of a pheromone gland, as with the Harderian gland in the orbit of vertebrate eyes (Figure 2), which secretes an attractant pheromone out the nares during a facial groom (Thiessen, Clancy & Goodwin, 1976), may indicate a radically different physiological and behavioral function from that of sebaceous glands which are primarily testosterone dependent (Figure 3). The former is part of a quick acting, short term signaling system, relatively independent of gonadal control (at least in the gerbil); the latter is related to a long acting signaling system more directly linked to reproductive functions. In glands with mixed cellular elements, such as in the mouth gland of the Columbian ground squirrel (Kivett, 1975), where sebaceous and apocrine tissues interlace, differential functions may be involved (Figure 4). As early as 1940 Schaffer postulated that sebaceous glands affect sexual behaviors, whereas apocrine glands convey species-specific odors. According to this speculation, a mixture of the two glands would be evident in reproductive processes related to individual recognition, aggregation and courtship behavior. There is no direct evidence in support of this postulation, but attention to this possibility could be revealing. In general, mammalian laboratory research on chemosignals can be advanced by detailed attention to the structure, type, storage and release capacity and innervation of exocrine glands. Surely, many of these restrictions on chemosignaling will be found at this level of analysis.

The Flexibility of Signaling Systems

The methodology associated with the investigation of mammalian pheromones is formulated along the lines established by entomologists. As such there has been an overemphasis on discovering the source of chemosignals and chemically identifying them, and an underemphasis on transmission characteristics of signals and their biological significance. Without a doubt, knowing the molecular nature of a signal provides an entering wedge into the neuroendocrinological mysteries of an organism, but that knowledge may not provide answers to the seminal questions of evolution and genetic fitness. Reductionism can in some cases even delay our understanding of communication systems.

If secretory portion is:

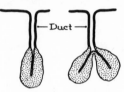

Duct →

1
tubular,

2
flasklike,

3
both,

it is a tubular
exocrine gland.

it is an alveolar or
acinous gland.

it is a tubulo-
alveolar gland.

If duct doesn't branch:

←Duct→

it is a simple gland.

If duct branches:

Duct→

it is a compound gland.

Fig 1.   Diagram showing the different kinds of secretory units of exo?
         glands and the difference between simple and compound glands.
         (From Ham and Leeson, 1961).

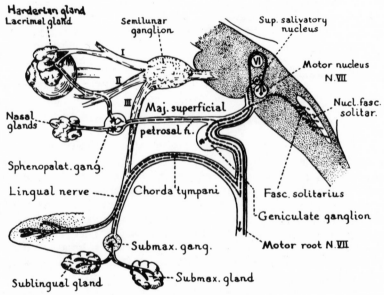

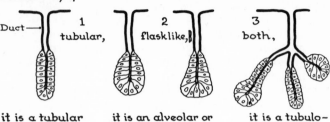

Fig 2.   Neural innervation of eye glands and salivary glands.
         Parasympathetic and sympathetic fibers control glandular
         secretion (Modified from Grossman, 1967).

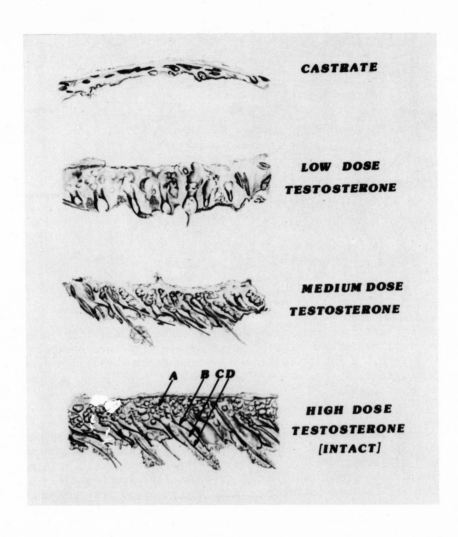

3. Histological changes in the ventral scent gland of the gerbil after castration and after castration followed by low (1 injection), medium (2 injections) and high (3 injections) doses of testosterone propionate (640 µg). Injections one week apart. A = secretory cells, B = lumen, C = holocrine secretion, D = hair.

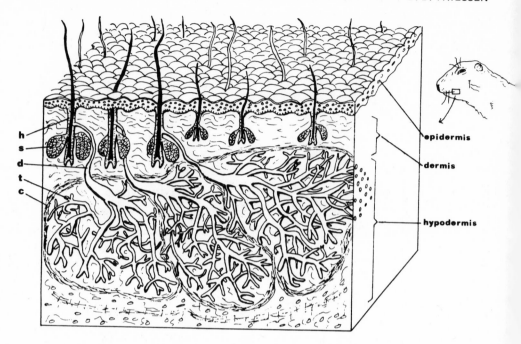

Fig 4.   Section of skin from the oral angle region of the Columbian
ground squirrel showing the branched tubule structure of apocrine
gland tissue.   h = hair follicle, s = sebaceous gland, d = duct from
apocrine gland, t = apocrine gland tubule, c = connective tissue.
(From Kivett, 1975)

        There is growing evidence that many pheromones may not have
unique biosynthetic pathways and species-specificity, as generally
assumed (Thiessen, 1977).   Rather, a sizable number of chemosignals
may be pirated from plants or other ecological sources and
sequestered for use in intraspecific and interspecific communication.
Others may be formed through symbiotic interactions with microorganisms, such as with vaginal and skin bacteria.   The list is
growing for invertebrates but is still short for mammals (Table 3).
In our own laboratory we have evidence that the odor from ventral
scent gland sebum differs in quality depending on the composition
of the diet (Figure 5).   Pups tested in a choice maze prefer the
odors from individuals fed on the same diet as their mothers.
Apparently pups imprint on the chemosignals associated with maternal
care.   It is likely that our earlier identification of phenylacetic
acid as a primary pheromone of the ventral scent gland is rather
specific to Purina Laboratory Chow.   If this proves to be representative of a general case, pheromone identification will be more valuab
in regard to ecological questions than as a means of specifying uniq

TABLE 3

Environmentally Determined Chemosignals

Chemical Source	Chemical Characteristic	Environ-mental Source	Behavioral Effect	Reference
Rat Milk	Unidentified	Diet	Dietary choice	Galef & Clarke 1972; Galef & Henderson 1972
Rat Fecal Material	"Caecotrophe"	Diet	Pup attractant	Leon (1974, 1975)
Mongoose Anal Sac	Short-chain fatty acids	Anal sack bacteria	Individual recognition	Gorman (1976)
Gerbil ventral scent gland	Unidentified	Diet	Pup attractant	Skeen & Thiessen (1976)
Rhesus Vaginal Secretions	Short-chain fatty acids	Vaginal bacteria	Sex attractant	Michael, Keverne & Bonsall (1971)
Human palm	Unidentified	Diet	Individual recognition	Wallace (1976)

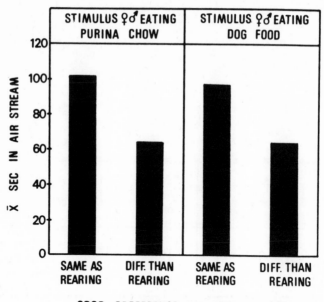

Fig 5. Odor preference of 12-20 day old pups for ventral sebum of conspecific animals after rearing by mothers eating Purina Chow or Pooch dog food.

and invariant signals. Current inconsistencies in the literature regarding primate sex attractants may be resolved at this level (Goldfoot, Kravetz, Goy & Freeman, 1976).

There is some question whether environmentally obtained signals or microfloral stimulated signals should be called pheromones, but as long as they satisfy the usual criteria for information molecules, it seems immaterial what we call them. Certainly at this stage of our ignorance a non-restricted defin- ition of a pheromone is warranted (Table 4). The important point is that understanding the biological significance of chemo- communication may not be advanced by attempts to specify unique molecules and biosynthetic pathways within a restricted definition of pheromones.

Again let me emphasize that the major problem in chemo- communication is to describe the entire information processing system, not merely the molecular attributes of the signal. Obviousl in order to do this we must look at the interaction of all signal

Table 4

A Non-Restrictive Definition of Pheromone

1. Pheromones are biological chemicals derived from cellular
    metabolism or acquired from the environment to integrate
    social and physiological activities.

2. Pheromones must reflect significant events in the sender, meet
    the physical criteria for transmission and influence genetic
    fitness of the sender and receiver.

3. Pheromones are open to environmental and physiological
    perturbations, can be derived from the environment or produced
    by symbiotic organisms.

4. The significance of a chemosignal may be imprinted or learned.
    Receptor elements may be induced by exposure to chemicals,
    mediated, perhaps, by genetic regulatory systems.

emitting and signal receiving systems, and not just the olfactory
system (Figure 6). Very little has been done in this critical
area. As a promissory note we have found that ventral gland marking
in the gerbil correlates significantly with the frequency of
ultrasonic emissions (r = .60) and that ultrasonic calls occur
when animals are exposed to the general body odors of conspecifics.
As with scent marking, ultrasonic calls to odors are sexually

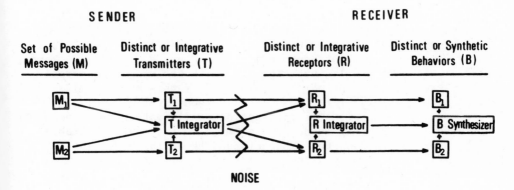

Fig 6. Communication channel illustrating discrete and integrative
properties of signal transmission.

dimorphic, dependent on gonadal secretions and contingent on
social status. While we have only scanty data at this time,
we believe that animals interchange and integrate a number of
messages during confrontations. They appear to use signals
redundantly, in variable amounts depending on the situation, and
in syntactic arrangements. Our task, it seems to me, is to
discover how animals sequence and weave together modes of communication and responses. Information exchanges between animals
are certainly much richer and much more complex than laboratory
investigations have hitherto suggested, a notion which does not
argue against laboratory work. Laboratory methodology is imperative
in this quest, as it is only under controlled conditions that the
invisible network of messages can be made visible and the necessary
manipulations accomplished.

Physiological Concerns

Rapid progress is being made in elucidating the physiological
processes underlying chemocommunication, especially in regard to
reproductive hormones. Laboratory work is indispensible here, but
seasonal and shorter term reproductive changes seen in the field
can also suggest the type of hormone system operating.

Looking just at scent marking species, where data exist at all,
it is apparent that males of most species depend on androgens for
hormonal control of marking, or the species possess other attributes
that point to gonadal control (Table 5, Thiessen & Rice, 1976).
Table 5 incidentally points out serious gaps in our knowledge
about these widely known species.

Hormone studies have been especially rewarding, but we must
admit that they have been somewhat superficial, in that most involve
simple endocrine gland extirpation and hormone replacement.
Only a few investigators have traced ontogenetic processes or
seasonal variations (e.g., the Mongolian gerbil, Rice, 1975;
the pronghorn antelope, O'Gara, Moy & Bear, 1971) and even fewer
have involved a look at neuroendocrine processes (Thiessen & Yahr,
1977). Pauline Yahr's investigations with the Mongolian gerbil
is an exemplary exception (see Chapter 8 of this symposium).
Implanting steroids into the central nervous system, analyzing
brain-pituitary-pheromonal interactions and relating brain hormones
to genetic regulatory mechanisms seem to me the most profitable
laboratory approaches to the understanding of the molecular
regulation of chemosignaling.

TABLE 5

General Characteristics of Scent Gland Marking
in Various Mammalian Species

Characteristics Of Marking	Ground Squirrel	Mongolian Gerbil	Golden Hamster	Guinea Pig	N. American Pika	Sugar Glider	European Rabbit	Pronghorn Antelope	Black-tail deer	Maxwell Duiker	Lemur	Marmoset Monkey
Sexually Dimorphic	+++	+++	+++	+++	+++	+++	+++	+++	+++	+++	+++	0
Age-Dependent	+++	+++	++	+++	0	++	++	0	0	0	+++	++
Seasonal development	+++	++	0	0	+++	0	+++	+++	+++	0	+++	0
Circadian	0	++	0	0	0	0	0	+++	0	0	0	0
Dominance-related	+++	+++	+++	+++	++	+++	+++	+++	+++	+++	+++	+++
Stimulated by Aggression	+++	+++	+++	+++	0	++	+++	+++	+++	+++	+++	+++
Stimulated by sex	0	++	+++	0	0	0	++	+++	0	+++	+++	+++
Androgen dependent in male	++	+++	+++	++	0	0	+++	++	0	0	+++	0

KEY: +++ = Yes   ++ = Probable   0 = Untested

Physiological work is badly needed along other dimensions as well, especially at the level of the receptor. Much of this will have to wait until more is known about odoriferous molecules and their interaction with epithelial receptors. Still, progress is being made (Moulton, Turk & Johnson, 1975). Even simple peripheral reactions to the presence of a pheromone, such as changes in respiration and heart rate, can tell us something about the interface between the environment and deep physiological changes (Figure 7). There are endless questions of this sort that can be and should be addressed in the laboratory.

Genetic and Ecological Attributes of Laboratory Species

The search for pheromones among mammals began with the mouse and is being extended to a number of species; including man. For the most part, however, there has been a general neglect of species characteristics that could facilitate our efforts in the laboratory and the field. We have, for example, concentrated on common laboratory organisms and pheromone systems that are readily manipulated. More attention should be paid to species of sociobiological interest (Table 6) and those that possess attributes that will allow combined laboratory and field investigations. No one species is ideal on all dimensions, but some have outstanding qualities that should hasten our approach toward understanding (Thiessen & Rice, 1976). Ungulates, in particular, with their diverse social systems and ecologies are outstanding candidates (Geist, 1974). The wolf is another exceptional possibility (Peters & Mech, 1975), and among the lagomorphs are the European rabbit (Mykytowycz, 1974), and the American Pika (Barash, 1973; Markham & Whicker, 1973; Sharp, 1973). A number of diurnal squirrels, such as the Columbian ground squirrel (Steiner, 1974) and primitive primates, such as the common marmoset (Epple, 1974) and lemurs (Jolly, 1966) offer exciting primate material. Most of these species can be studied in their normal ecologies as well as in the laboratory or in the zoo; most are diurnal, have specialized scent marking behaviors, are seasonal breeders and are highly social. In some cases the ecological conditions can be successfully mimicked in the laboratory, as with the gerbil in a semi-natural environment (Thiessen & Yahr, 197 7).

Most importantly the pheromone systems studied in the laboratory should ideally have their obvious references in more natural situations. The so-called Lee-Boot, Whitten and Bruce reproductive pheromones are interesting from a physiological perspective, but may be nothing more than laboratory artifacts (Bronson, 1971). Population structures necessary to elicit these responses are rarely encountered in the field. More to the point are pheromone processes that affect development, that function

Table 6

Criteria for Selection of Species
for Laboratory Study

Important Attributes

Sociobiological interest

Known phylogenetic status
        Relatedness among species

Stable population parameters
        Demographic and social characteristics
        Energy requirements and sources
        Prey-predator status

Known physiological, neurological and behavioral traits

Laboratory and field convenience
        Small size
        Discrete glandular sources of chemicals
        Specified distribution and home range
        Diurnal and surface dweller
        Circannual rhythm of reproduction (non-hibernating)

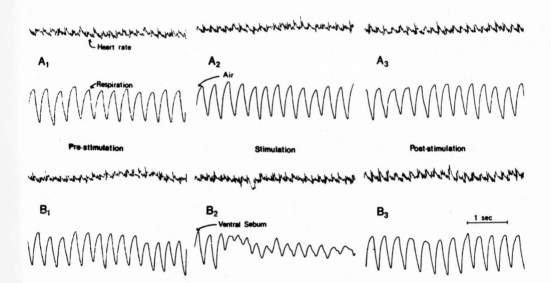

Fig 7.   Heart rate and respiration frequency in restrained animals
         subjected to an air puff to the nostrils (top panel) or
         volatiles from the ventral marking gland (bottom panel).

Table 7

Conditions of Pheromonal Variation

Conditions	Degree of Pheromonal Flexibility		Degree of Pheromonal Modulation	
	Pheromonal Plasticity	Pheromonal Rigidity	Regulated Systems	On-Off Systems
Species Characteristics	Complex organisms with wide tolerance for change	Simple organisms with narrow tolerance for change	Complex, endothermic, K - selected organisms	Simple, ecto-thermic, r - selected organisms
Physiological and Ecological Attributes	Large body size, demographic instability, meterological uncertainty	Small body size, demographic stability, meterological certainty	Slow growth, self-regulated population growth, interdependent homeostatic processes	Rapid growth, ext-ernally regulated population growth, wide fluctuations in metabolism
Behavioral Traits	Wide ranging and dispersing, variable social structure and flexible learning ability	Short ranging and dispersing, invariant social structure and limited learning ability	Complex social organization	Simple or no social organization

in parental activities, that increase or decrease intraspecific aggression, and that act as sex attractants and territorial markers. Of course in our primitive stage of scientific development we should not overlook any and all possibilities, but where choices are available our attention should be drawn toward sociobiological criteria.

One can expect that the social structure of a species will dictate the types of pheromones used and hence what we detect in the laboratory (Table 7). Depending on ecological conditions, species variations and behavior one can expect wide differences in the plasticity of signaling systems and in the nature of their regulation. By knowing the natural history of species the value of our laboratory investigations can be considerably extended. It is only in this way that we can confidently proceed with a reductionistic strategy.

A Summary Comment

Mammalian pheromone research in the laboratory has unique problems, not the least of which is methodological. Strategies should be derived from problem analysis rather than technical capabilities. Laboratory investigators can capitalize on methodological strengths by a judicious selection of species for study and by assuring that their findings are compatible with field observations. Species are opportunistic in mixing signaling devices together, and varying olfactory signals according to ecological conditions. A reductionistic search for molecular characteristics of chemosignaling systems is a powerful strategy but only within a broader biological framework, which entails reference to ecological and evolutionary movers of communication systems, and which has a constant regard for features of communication which have an impact on genetic fitness.

Footnotes

1. Preparation of this chapter was facilitated by NIMH Grant MH 14076-09. Special thanks are due to many colleagues and students, who have contributed to the thinking or investigations referred to in this chapter, especially Jan Bruell, Michael Graham, Michael Goodwin, Andrew Clancy, James Skeen, Gene Ondrusek, Maureen Rice and Pauline Yahr.

BIBLIOGRAPHY

Barash, D. P. 1973. Territorial and foraging behavior of Pika
      (Ochotona princeps) in Montana. Amer. Midl. Nat., 89:
      202–207.

Bronson, F. H. 1971. Rodent pheromones. Biol. Reprod., 4(3):
      344–357.

Brownlee, R. G., Silverstein, R. M., Müller-Schwarze, D. & Singer,
      A. G. 1969. Isolation, identification and function of
      the chief component of the male tarsal scent in black-tailed
      deer. Nature, 221:284–285.

Epple, G. 1974. Olfactory communication in South American
      primates. Ann. N.Y. Acad. Sci., 237:261–278.

Galef, B. G. & Clark, M. M. 1972. Mothers milk and adult
      presence: two factors determining initial dietary selection
      by weaning rats. J. Comp. & Physiol. Psych., 78(2):220–
      225.

Galef, B. G. & Henderson, P. W. 1972. Mother's milk: A deter-
      minant of the feeding preference of weanling rat pups.
      J. Comp. & Physiol. Psych., 78(2):213–219.

Geist, V. 1974. On the relationship of social evolution and
      ecology in ungulates. Amer. Zool., 14:205–220.

Goldfoot, D. A., Kravetz, M. A., Goy, R. W. & Freeman, S. K.
      1976. Lack of effect of vaginal lavages and aliphatic
      acids on ejaculatory responses in rhesus monkeys: Behavior-
      al and chemical analyses. Horm. & Behav., 7(1):1–29.

Gorman, M. L. 1976. A mechanism for individual recognition by odour
      in Herpestes auropandatus (Carnivora: Viverridae). Anim.
      Behav., 24:141–145.

Grossman, S. P. 1967. A Textbook of Physiological Psychology.
      New York: John Wiley & Sons, Inc.

Ham, A. W. & Leeson, T. S. 1961. Histology. Philadelphia:
      J. B. Lippincott Co.

Hamilton, W.D. 1964. The genetic evolution of social behavior.
      J. Theoret. Biol., 7:1–52.

Jolly, A. 1966. Lemur Behavior: A Madagascar Field Study. Chicago: University of Chicago Press.

Kivett, V. K. 1975. Variations in Integumentary Gland Activity and Scent Marking in Columbian Ground Squirrels (Spermophilus C. Columbianus). Ph. D. Dissertation. Dept. Zoology, University of Alberta, Edmonton, Alberta.

Leon, M. 1974. Maternal pheromone. Physiol. & Behav., 13: 441-453.

Leon, M. 1975. Dietary control of maternal pheromone in the lactating rat. Physiol. & Behav., 14:311-321.

Markham, O. D. & Whicker, F. W. 1973. Notes on the behavior of the Pika (Ochotona princeps) in captivity. Amer. Midl. Nat., 89:192-199.

Michael, R. P., Keverne, E. B. & Bonsall, R. W. 1971. Pheromones: Isolation of male sex attractants from a female primate. Science, 172:964.

Moulton, D. G., Turk, A. & Johnston, J. W. Jr. 1975. Methods in Olfactory Research. London: Academic Press.

Müller-Schwarze, D., Müller-Schwarze, C., Singer, A. G. & Silverstein, R. M. 1974. Mammalian pheromones: Identification of active component of the subauricular scent of the male pronghorn. Science, 183:860.

Mykytowycz, R. 1974. Odor in the spacing behavior of mammals. In: M. C. Birch (Ed.), Pheromones. Amsterdam: North Holland Publ. Co., pp 327-343.

O'Gara, B. W., Moy, R. F. & Bear, G. D. 1971. The annual testicular cycle and horn casting in the pronghorn (Antilocapra americana). J. Mammal., 52:537-544.

Patterson, R. L. S. 1968. Identification of 3 α-hydroxy-5 α-androst-6-ene as the musk odour component of boar submaxillary salivary gland and its relationship to the sex odour taint in pork meat. J. Sci. Fd. Agric., 19:434.

Peters, R. P. & Mech, L. D. 1975. Scent marking in wolves. Amer. Sci., 63(6):628-637.

Platt, J. R. 1964. Strong inference: Certain systematic methods of scientific thinking may produce much more rapid progress than others. Science, 146(3642):347-353.

Rice, M. 1975. Relationships among sex behavior, scent marking, aggression and plasma testosterone in the male Mongolian gerbil (Meriones unguiculatus). Unpublished Ph.D. Dissertation, Dept Psychology, University of Texas at Austin, Austin, Tx.

Schaffer, J. 1940. Die Hautdrusenorgane der Saugetiere. Berlin: Wien.

Sharp, P. L. 1973. Behavior of the Pika (Ochotona princeps) in the Kananaskis Region of Alberta. Thesis in the Dept of Zoology, University of Alberta, Edmonton, Alberta, Canada.

Singer, A. G., Agosta, W. C., O'Connell, R. J., Pfaffman, C., Bowen, D. V. & Field, F. H. 1976. Dimethyl disulfide: An attractant pheromone in hamster vaginal secretion. Science, 191:948-960.

Skeen, J. & Thiessen, D. D. 1976. The scent of gerbil cuisine. Sixth Annual Behavior Genetics Association Meeting, June 17-19, Boulder, Colorado.

Steiner, A. L. 1974. Body-rubbing, marking and other scent-related behavior in some ground squirrel (Sciuridae), a descriptive study. Canad. J. Zool., 52(7):889-906.

Thiessen, D. D. 1977. Thermoenergetics and the evolution of pheromone communication. In: E. M. Sprague and A. N. Epstein (Eds.), Progress in Psychobiology and Physiological Psychology. New York: Academic Press, in press

Thiessen, D. D., Clancy, A. & Goodwin, M. 1976. Harderian pheromone in the Mongolian gerbil (Meriones unguiculatus). J. Chem. Ecol., 2(2):231-238.

Thiessen, D. D. & Rice, M. 1976. Mammalian scent gland marking and social behavior. Psych. Bull., 83(4):505-539.

Thiessen, D. D., Regnier, F. E., Rice, M., Goodwin, M., Isaacks, N. & Lawson, N. 1974. Identification of a ventral scent marking pheromone in the male Mongolian gerbil (Meriones unguiculatus). Science, 184:83.

Thiessen, D. D. & Yahr, P. 1977. The Gerbil in Behavioral Investigations: Mechanisms of Territoriality. Austin: University of Texas Press.

Wallace, P. Individual discrimination of humans by odor. unpublished manuscript, 1976.

COMPLEX MAMMALIAN BEHAVIOR AND PHEROMONE BIOASSAY IN THE FIELD

Dietland Müller-Schwarze

State University of New York College of Environmental
Science and Forestry
Syracuse, New York 13210

Abstract: True field bioassays of mammalian pheromones, i.e., experiments with free-living populations, are highly desirable from a
biological point of view but are practical only in rare cases. The
best generally practical approach, i.e., experiments with captive
animals in outdoor settings, has yielded biologically meaningful results despite many environmental factors. Eleven aspects of such
bioassays are discussed. The term "Informer Pheromone" is proposed
in point 9. Point 10 discusses the pathway of the tarsal odor and
the supporting specialized structures. Recent results of experiments
with black-tailed deer and reindeer are given in point 11. These
results deal with geometric isomers, early experience with maternal
odors, taxonomic specificity of an alarm pheromone, the role of antorbital gland secretion in age class recognition in deer, and social
odors in reindeer.

A variety of different bioassays have led to the identification
of six mammalian pheromones (Table 1). The following eleven points
discuss some complications of such bioassays with mammals.

1) The optimal goal: Bioassay in the field. Ideally, the bioassay should take place in the physical setting and social context
typical for normal behavior. In most cases this is not possible due
to the distances, terrain or vegetation involved. In addition, there
are the animals' predictable movements and behavior not to mention
problems in field logistics and weather. In situ experiments have
been accomplished in some cases, such as von Frisch's (1941) study of
the fright substance in minnows (Phoxinus laevis), or scent marking
in highly territorial mammals such as male Thomson gazelles, Gazella
thomsoni (Hummel, 1968).

413

Table 1. Bioassays Used with Mammals

Species	Behavior	Source	Compound(s)	Bioassay	Author
Swine	Courtship	Submaxillary Gland	Steroid	Human Panel	Patterson, 1968
Black-tailed Deer	Attraction Recognition	From Urine to Tarsal Organ	Lactone	Attraction of deer to fractions	Brownlee et al., 1969
Rhesus Monkey	Mating	Vaginal Secretion	5 fatty acids	Recombined fatty acids on : mounting	Curtis et al., 1971
Pronghorn	Marking	Subauricular Gland	Iso-valeric acid	Recombined constituents: Marking	Müller-Schwarze et al., 1974
Mongol. Gerbil	Marking	Ventral Gland	Phenylacetic acid	a) Conditioning b) Approach & Investigation	Thiessen et al., 1974
Golden Hamster	Sexual Attraction	Vaginal Secretion	Dimethyl-disulf.	Approach, Sniff, Dig, (latency & Duration)	Singer et al., 1976

Domestic animals can also be bioassayed in their "field". The dog's (Canis familiaris) response to scent marks can be studied better on routine walks than in the confinement of cages or laboratories (v. Uexkull and Sarris, 1931; Graf and Meyer-Holzapfel, 1974).

Where true field bioassays are impossible or impractical, a combined field and laboratory study is an excellent substitute and has certain advantages over pure field studies. This approach is particularly feasible with species indigenous to the investigator's geographical area. For example, in a study of scent marking in pronghorn (Antilocapra americana), a number of animals were reared, observed and tested in captivity (Müller-Schwarze and Müller-Schwarze, 1972; Müller-Schwarze et al., 1973, 1974) while quantitative field observations were carried out simultaneously in Yellowstone Park and the National Bison Range (Gilbert, 1973). This approach permits the interpretation of the responses of captive animals in terms of the conditions of free-living populations and facilitates the recognition of behavior artifacts that might occur.

2) Semi-field conditions are a good substitute for the true "field". For most mammals, a true bioassay is impossible in the wild. The logistics are often formidable in bringing an odor sample to a point where the species in question lives and where normally it would scent mark or otherwise use social odors. The approach and presence of the investigator would most likely interfere with the animal's behavior. In addition, many mammals are crepuscular or nocturnal and cannot easily be observed in action. Furthermore, the varying movements of wild animals do not provide a "standard situation".

Therefore, semi-field conditions, such as large outdoor pens, are the next best thing to the true field. The experimental animals are thus exposed to weather, natural terrain, vegetation and soil. In addition, they can arrange themselves in social groups typical for the open range. For example, a group of nine captive reindeer, which we studied in a 100 x 100 in pen, segregated themselves during the summer into one of the following subgroups: females, including one cow's male yearling; a "bachelor group" consisting of two males; a solitary mature bull; and a mother with her new calf. In fall, a "harem" was formed consisting of one mature male, and all females with their respective calves or yearlings. Not tolerated near the haren were two males who formed a close social unit (Müller-Schwarze et al., in prep.). These social groupings were the same as those occurring in the wild (Espmark, pers. comm.).

Indirect methods, such as using radiotelemetry to monitor the responses of free ranging test animals to previously placed odor samples have two major drawbacks. One drawback is the varying and often unknown time interval between the placing of the sample and its encounter by the animal. This in turn would affect the evaporating of the sample. The other drawback is the loss of detailed information on the animals' behavior. The first problem can be overcome by proper "formulation" which though advanced in entomology, is still in its infancy when applied to mammals (Silverstein, this volume). In our own experiments, using Trioctanoin as an evaporation inhibitor for isovaleric acid, male pronghorn antelopes (Antilocapra americana) responded with the same intensity to samples with and without Trioctanoin (Müller-Schwarze, unpublished). The main effect of an evaporation inhibitor is prolongation of the activity of the sample rather than an immediate reduction of activity.

3) Detailed study of intricate social behavior in a natrualistic setting is prerequisite for bioassay. Mammals live in a most complex physical and social setting. Before a sensible bioassay can be developed, there must be detailed, systematic observation. The behavior in question must be understood as fully as possible, with a contextual analysis performed. The difficulties of many pheromonal studies result from the expectation that all adult individuals of a given sex and species should respond to a stimulus in the same way. Quite the contrary, differing social status, differences in upbringing, special individual experiences, genetic differences, behavioral polymorphism with regard to sensory modalities, and special relationships between odor donors and recipients may all result in drastic individual differences in response to "standard stimuli". Moreover, often the "response" does not consist of a single, consistent or reproducible motor pattern that the animal would exhibit in different settings including the laboratory. In short, our first concern should be how biologically sound a bioassay will be, and not how practical. Methods should be subservient to the behavioral principle in question, however complicated it may be, and not vice versa. Some phenomena observed in laboratories, such as effects of odors on maturation and reproduction, may not occur in the wild because odor concentrations would be lower (Rogers and Beauchamp, 1976) or because dispersal of individuals would reduce the odor effects (Krebs et al., 1973).

4) The initial bioassay lacks quantitative rigor. After obser vations have led to a good understanding of the behavior involved, experimentation can commence. First, glandular secretions or urine are used as stimuli, then the solvent extracts of these. Next used are fractions and finally single compounds alone and in combinations In the first step, the bioassay with whole secretions (or excretions the stimulus can be quantified by a) the numbers of glands extracted ("gland equivalents"), b) by standardized collection techniques, such as rubbing a cotton swab or teflon stick a given number of

times over a gland of a live animal (Müller-Schwarze, et al., 1974),
c) by assigning points to the effort needed to collect a specified
amount of material from a gland. (For instance, the interdigital
gland of the hindfoot in reindeer (Rangifer tarandus) which may
yield a "visible and strong smelling" secretion at the first insertion of a teflon rod into the gland pouch, may at the other extreme, yield a smell only after prolonged poking (Müller-Schwarze
et al., in prep.)), or d) by the duration of the collection or
"animal hours" when a gland, body part, whole animal, or group of
animals is aerated in order to collect the volatiles on an absorbant,
such as Porapak or Tenax.

Since the constituents of the secretions in question have not
been identified chemically at that point and it is not yet known
which constituent(s) is (are) biologically active, it is not possible
to quantify the stimulus in terms of amounts of specific compounds.
Although large pooled samples can be divided into equal parts to
guarantee uniform amounts of material for each test, different batches
will inevitably differ in their composition.

5) The bioassay yields incomplete responses. It is often
difficult or impossible to release a complete response by chemical
stimuli alone because a complex pattern of stimuli involving different sensory modalities controls the behavior in question.

In a typical sequence of behavior patterns of a mammal, several
sensory modalities play a role. This is more obvious in large animals that are, at least partially, active in daylight. More than
one sensory modality may be involved simultaneously, successively or
vicariously. An example of simultaneous action of several modalities
is the threat behavior of male black-tailed deer: a male faces
another laterally, ruffles his fur over the entire body, raises his
tail, spreads the hair of the tarsal tuft on the inside of the hocks,
lays his ears back, opens the antorbital pouch, throws his head up
quickly while uttering a short snort. This is followed by lowering
the head and blowing out slowly with a hissing sound while the muzzle
is pointing down vertically. Here, olfactory signals from tarsal and
antorbital organs, stationary visual signals such as the raised hair,
motion signals such as the up and down movement of the head, and
auditory signals such as the snort and hiss, are all incorporated
into a complex display. In the Golden hamster (Mesocricetus auratus)
vaginal secretion applied to an anesthesized, castrated hamster will
release sexual behavior in males, whereas the same secretion on a
clay model of a hamster does not (Johnston, 1970, 1974).

Successive activation of different sensory systems occurs in
deer, as for instance, during the rut. The sexes are brought together most likely by means of forehead marks (Müller-Schwarze,
1972) and urine on the ground, but also possibly by the clapping

sounds of antlers thrashing at trees and actual antler fights. A
buck will pursue a doe in a moderately fast trot, with his head
stretched forward and held low while hissing 2-5 times in rapid
succession. The doe will urinate and move on. The male will stop,
sniff and lick the urine on the ground, lift his head and show the
Flehmen (lip curl) response. Then he will resume his pursuit. The
distance between male and female will gradually shorten, and the buck
will smell the doe directly at the anogenital area, tail and tarsal
organ. Finally, mechanical communication occurs: the male places
his chin on the rump of the female, which in turn will stand if in
estrus, and move on if not. The "standing" is the mating stance
and will permit mounting and copulation to occur immediately. The
preliminaries are necessary neither for bringing a doe into estrus,
nor for successful copulation. A doe coming into heat while
separated from males, will be bred immediately when brought together
with a buck. Another example of successive effects in different
modalities is the response to scent marks on visually conspicuous
landmarks by canids such as coyotes, wolves, or arctic foxes (Peters
and Mech, 1975).

A case of vicarious involvment of sensory channels is the
flight reaction in fish. Minnows (Phoxinus phoxinus) may escape
from injured conspecifics in response to the released alarm substance, and in the absence of alarm substance to the visual stimulus of fleeing group members (Schutz, 1956; Pfeiffer, 1975). In
contrast to many highly specific, complex sexual and social behaviors
alarm responses usually can be released fully by alternative stimuli
in different modalities. Such redundancy is to be expected in a
context as vital as predator alarm.

6) The behavioral effect may be a general state rather than a
fixed (motor) response. The best possible outcome of a bioassay is
the elicitation of a specific behavior pattern by a defined chemical
stimulus. However, mammals--in contrast to insect--often do not
respond with a standard response to a standard stimulus. Instead,
the behavior elicited will depend on the situation in which the stimulus occurred. A given stimulus may lead to a typical end situation,
and not to a standard response. For example, a predator stimulus
may release varying responses in a mammal or bird that all have in
common the final situation of being in cover. The animal will flee,
if it is in the open, but will remain motionless if it already is
in cover.

Terms used to describe psycho-physiological states of animals
include "confidence" (Mykytowycz, et al., 1976), "familiarity"
(Kleiman, 1966) and "preference" (Doty, 1975). Other concepts frequently used in mammalian behavior studies are "alarm, "dominance",
"recognition", or "curiosity". Different meanings of each of these
poorly defined terms will inevitably arise as different indicative

behavior patterns are used by different investigators, a situation
which is often mandatory when different species are studied. Most
of the identified pheromone components of mammals do not reliably
release a specific response although all increase the probability
of a given behavior (Table 1).

7) Experiences of the test animals have profound effects on
the outcome of the bioassay. Because it is impossible to carry out
a true bioassay in a true field setting, we experiment with animals
kept and raised in captivity. This treatment in itself as well as
variations in maintenance and rearing procedures may account for
variations in the observed behavior and responses of the bioassay.
Especially critical are the effects on anti-predator behavior. A
few examples from our experience are as follows:

In black-tailed deer, hand-raising has profound effects on
agonistic behavior and alarm behavior, but not on sexual behavior.
For the purpose of this paper, we single out the differences in
alarm behavior between "tame","shy" and "intermediate" individuals.
"Tame" are animals raised individually in close contact with people,
from the first day of life on. "Shy" animals, at the other extreme,
are raised by their own mothers, and "intermediate" animals are hand-
raised by taken from their natural mothers at 3-6 days of age. It
would seem normal for a deer to avoid an area where another deer is
being held or has been caught, since a large amount of alarm odor
from the metatarsal glands would be present in the air. However,
"tame" deer do not show this avoidance response. Instead they
approach such a scene of capture and even solicit social responses
from the experimenters. Secondly, when "tame" and "intermediate"
males were bioassayed for avoidance of metatarsal secretion, the
"intermediate" males avoided feeding near the metatarsal odor
whereas the "tame" deer did not show this avoidance. Thirdly,
"tame" individuals tolerate handling, crating and extended stays in
transport crates very well whereas "shy" ones are extremely stressed
under such conditions, even to the point of death. In short, human
contact early in life has profound and general effects on the entire
"alarm" or anti-predator syndrome. The important aspect in the
context of this paper is the altered response to the chemical alarm
stimuli.

Pooling data from large numbers of test animals is often necessary, but it may obscure significant behavioral differences between
individuals of different social experience and status. An example
is the scent marking by forehead rubbing in black-tailed deer. In
both yearlings and fawns, a formerly dominant male will mark more
often than one that has had experience in being subordinate (Volkman, et al., in press).

8) The bioassay has to take into account experimentally caused
changes in responsiveness. In pheromone experiments the most common change in responsiveness is habituation, i.e., decreasing re-

sponsiveness with repeated stimulation. This occurs in many animals: in insects, such as moths (Shorey and Gaston, 1964); in fish
(Pfeiffer, 1974); and is well known for mammals. In black-tailed
deer depressed responsiveness due to prior stimulation with tarsal
scent can be observed after intervals between exposures of up to
three weeks (Müller-Schwarze, 1971). Mammals typically do not respond in pars pro toto fashion. The odor is indicative of another
individual, but does not by itself constitute the stimulus situation
necessary for response; hence, habituation occurs.

Habituation has been used successfully by Schultze-Westrum
(1965) to demonstrate the discrimination of individual odors in
flying phalangers (Petaurus breviceps): after the response to repeated presentations of an individual's odor had dropped to a certain
level, the odor of a different individual was presented. The response was observed to increase again. This same technique was used
with Lemur fulvus by Harrington (1976).

Habituation has also been used to demonstrate species specificity of a mammalian alarm pheromone. Black-tailed deer habituate
faster to metatarsal odor of white-tailed deer than to their own
(Müller-Schwarze and Volkman, unpublished).

A second, often encountered change in responsiveness is conditioning. The animals may associate the test situation with a
particular experience and increasingly either avoid or seek it.
For instance, black-tailed deer, when repeatedly sprayed with odor
samples, increased their distances to the experimenter from a few
centimeters to several meters. They ultimately had to be tricked
into a position near a hidden person who could then apply the sample
(Müller-Schwarze et al., 1976).

9) "Informer pheromones" and the need for more complex bioassays. Conventional bioassays are not sensitive to many pheromone
effects in mammals. Mammals are notorious for their constant sniffing and frequent licking. Chemical information concerning their
environment, their foods, their conspecifics and their predators is
under continuous processing by their central nervous systems. Only
a very small portion of that continuous processing of information
results in overt behavior unequivocal enough to indicate the specific
motivation induced or maintained. The classification of pheromones
has for the past 15 years consisted of a dichotomy separating primer
and releaser pheromones. Primer pheromones effect slow and longer
term responses of a physiological or developmental nature. Well-
known examples of such effects in laboratory mice are: estrus-
suppression (Lee-Boot Effect), estrus-acceleration (Whitten Effect),
pregnancy-block (Bruce Effect), and stress-syndrome (Ropartz Effect)
Releaser pheromones bring about immediate responses. Examples are
alarm pheromones which cause escape, defense, and general excitement in conspecifics, or scent marks (such as in pronghorn, deer, or
dogs) which, in turn, release marking by another individual.

These two categories overlap. "Immediate" and "long-term" are not clearly delineated, and one pheromone may have priming and releasing effects at the same time, at different times, or in different individuals. The latter is especially true in opposite sexes. More important, these categories do not cover all phenomena that can be the result of a chemical signal by a conspecific. Consequently, a third category should be established.

The traditional dichotomy between primer and releaser pheromones was sufficient for insects, but recent detailed work with mammals proved this classification inadequate. There are chemical signals exchanged between conspecifics that neither release an immediate response nor trigger (or maintain) a longer term physiological process. Rather, these stimuli provide information that is stored in the memory and can be recalled later in a variety of contexts. Thus, any delayed response to the signal is not necessarily specific (as are by definition the responses to releaser and primer pheromones.

Such chemical signals inform (or apprise) an animal. Following the choice of words for the other categories, we propose the term "informer pheromones" for these chemical signals. The main differences between the three types of pheromones are listed in Table 2. Wherever mere discrimination of odors of conspecifics occurs, we are dealing with "informer pheromones" and the vast majority of mammalian odors may fall into this category.

Table 2. Differences between primer, releaser and informer pheromones.

	Primer	Releaser	Informer
Main effect	physiological change	behavioral response	provide information (stored in CNS); discrimination
Mode of action	influences hormonal system (via sense organ, CNS. hormonal system)	release behavior (via sense organ, CNS and efferent NS and effectors)	guide behavior; influence choices of action (via sense organs and CNS)
Time of effect	long term	immediate	delayed
Specificity	high	high	low

Often bioassays with mammalian pheromones encounter difficulties such as inconsistent responses by the animals or mere sniffing and licking instead of any expected specific responses. This can occur even though the chemical stimulus used in the experiment is one that is eagerly sought by the receiving animal and is produced in the emitting animal by means of a highly developed scent gland and/or an elaborate behavior pattern. All this indicates the biological importance attached to the exchange of these chemical signals. An analogy can be drawn to visually-cued human behavior. That is, although the reading of mail is done with great motivation, an ethologist would be hard put to develop a bioassay to determine the responses released by certain parts of its contents. The only effect common to all readers of letters is that information is received and stored; there is no standard response to any standard message. The social context will determine the response(s). In summary, it is suggested that either the term "pheromone" be abandoned altogether for mammals or that the category of "informer pheromones" be introduced.

Recording techniques far more sophisticated than today's bioassay procedures will be required if we are to demonstrate an effect of a chemical signal that is delayed and non-specific, i.e., one that gives direction to later behavior and may be dependent on earlier experience together with other concomitant or subsequent stimuli. It will be necessary to keep a complete record of all behaviors and social interactions occurring after stimulation and to carry out intricate sequential analyses. This is true also for other sensory modalities.

An example may illustrate the point made. Consider a male pronghorn (Antilocapra americana) which encounters a dung pile of a conspecific. If it is the dung pile of another male (either a member of the same bachelor group or a territorial neighbor), he will sniff it, obliterate it by pawing and, in a highly stereotyped fashion, urinate and defecate on the same spot. For this situation we would use the term "releaser pheromone". But if the dung pile happens to be his own, the male will merely sniff it and then continue with his ongoing behavior such as grazing. The sniffing cannot be distinguished from other sniffing, e.g., the sniffing of vegetation. No specific response is exhibited and the designation "releaser pheromone" does not apply. Yet, we would be wrong in dismissing the episode as one with no effect on the sniffing animal. It is safe to assume that information is collected and that there are later behavioral consequences of that experience. "No response" does not mean "without significance" or, in the context of a bioassay, "no biological activity". The "biological activity" takes place in the central nervous system and is not expressed as overt behavior.

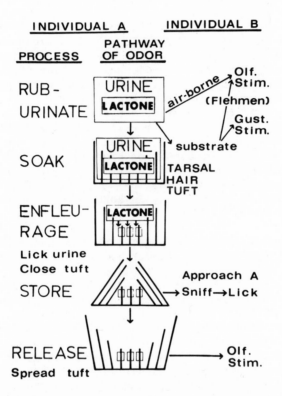

Fig. 1   Function of tarsal hair tuft in black-tailed deer (see text).

10) The bioassay has to consider complicated odor pathways.
Odors may be transferred from one part of the body to another, between individuals, and from the body of an animal to features of
the environment such as in marking and in the use of excrements for
communication. From the environment, they can be picked up again
by other individuals. Thus far, bioassays have used only fairly
simple pathways. Secretions or excretions may serve different
functions, and conspecifics may be exposed to it in different ways.
The exposure can differ according to the broadcasting distance, the
emitting body part, or the numbers of compounds involved.

As an example the pathways of volatile compounds in the tarsal
scent of black-tailed deer (Odocoileus hemionus columbianus) are
shown in Figure 1. An important pheromone component in the tarsal
scent is cis-4-hydroxydodec-6-enoic acid lactone (Brownlee et al.,
1969). This lactone has recently been found in the urine of black-
tailed deer (Claesson and Silverstein, this volume). Urine is
applied to the hair tuft covering the tarsal glands by "rub-
urination". The hocks are rhythmically rubbed together while the
deer slowly urinates. Urine is sprayed about and can serve at least
two functions in this forst mode of broadcasting. In adult males
and females, rub-urination is a threat and prevents conspecifics
from approaching closer tha about 3-5 m. Fawns, however, attract
their mothers by rub-urination. In both cases, communication occurs
over distances of up to about 20 m.

After soaking the tarsal tuft with urine, lactone is extracted
from the urine into the lipid layer on the short, stiff hair in the
center of the tarsal tuft. The lipid stems from the sebaceous glands
of the tarsal organ. No lactone was found in fresh secretion
(Claesson and Silverstein, this symposium). The extraction of volatiles from urine into the lipid layer is essentially the same as
"enfleurage", the chromatographic process used to extract volatiles
from flowers by layers of oil placed on glass plates. In two cases
we succeeded in collecting urine from a deer while it rub-urinated,
both before the urine hit the tarsal tuft and after it had passed.
The urine was extracted with petroleum ether. The solvent was
evaporated and the sample subjected to gas liquid chromatography.
In both cases, there was less material in the sample after urine had
passed the tarsal tuft, than before (Figure 2). This shows that
urinary constituents are collected on the tarsal tuft.

The short, stiff hair in the center of the tarsal tuft is
specialized for holding large amounts of lipids. The Scanning
Electron microscope reveals that its cuticular scales form chambers
that can hold secretion. In addition, the scales carry ridges that
enlarge the surface and serve to hold the lipids in place, much like
projections on roofs prevent the snow from sliding (Müller-Schwarze,
et al., in press).

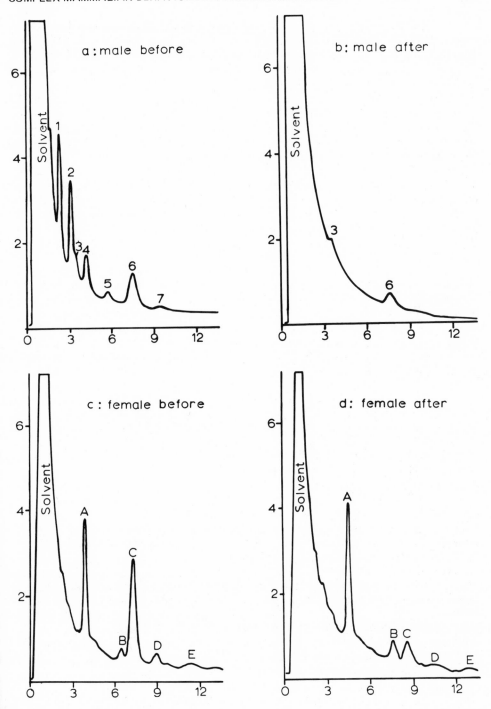

Fig. 2   Changes in volatile constituents of urine during rub-urinating (see text).

The urine is then licked off, and the tuft is closed. The outer hair of the tarsal tuft is longer and not specialized in the way described. It serves as cover for the stored compounds which are less volatile in the lipid than they would be in urine. Although the skin temperature at the hocks is usually lower than either the core temperature or the temperature of urine, it still has, surprisingly, the highest temperature of all body parts when ambient temperatures are low, even al low as -20°C (Stevens, 1972). The stored odoriferous materials can serve communications over very short distances, i.e., in the millimeter and centimeter range. This is the second mode of broadcasting. Therefore, we frequently see approaching, sniffing and licking among conspecifics. A third mode of broadcasting is the spreading of the tarsal tuft which exposes the stored volatiles maximally. This occurs during threat, general excitement, and alarm.

In this example, (i.e., the pathway of the tarsal scent), there are three different phases at which the material involved can be "tapped" by conspecifics for information. They differ in the distances over which information is broadcast, and, therefore, in the behavior required to bring the receptors to the optimal "active space" (Wilson and Bossert, 1963). Thus they differ in the behavioral context in which the odors play a role. For these reasons, a different bioassay is indicated for each of these different response patterns, although they all may be dealing with the same array of chemical compounds.

The existence of specialized scent hairs requires appropriate caution in those experiments in which odor samples are applied to an animal's body. The hair or skin surfaces of other body parts may effect an odor pattern differently from that typical for the specialized scent hair (see also Regnier and Goodwin, this volume).

11) Field bioassays yield reliable results even under poorly controlled conditions. Despite the environmental perturbations of its setting and the varying backgrounds of the experimental animals, a field bioassay can be highly reliable. Five examples from our recent work illustrate this statement:

I. Geometric isomers: Experiments were carried out during the reproductive season of black-tailed deer from November to March, including the most inclement and variable weather periods of the year. The cis-lactone and its trans-isomer were sprayed on the hindlegs of group members. Both adults and fawns responded more to the pheromone (cis-lactone) than to its trans-isomer (Müller-Schwarze et al., 1976).

II. Early experience of maternal odors: Four fawns of black-tailed deer were raised in isolation in outdoor pens. The only "companion" was a mother surrogate consisting of a wooden rack with a built-in bottle, a buzzer, and an "artificial scent gland" made fro

filter paper scented with extracts from glands. The different
extracts used for different animals were from ischiadic glands of
pronghorn, and from tarsal organs of mule and black-tailed deer.

Table 3. Preferences of "maternal" Odors by Black-tailed Deer Fawns.

Age:	2 Wks.		1 Mo.		2 Wks.		1 Mo.	
Individual:	♂P$_1$	♂P$_2$	♂P$_1$	♂P$_2$	♀B	♀M	♀B	♀M
Time at Rear	P B	P B	P B	P B	-	-	-	-
of Dummy	P M	-	-	P M	-	-	-	-
Total Time	-	P B	-	-	P B	-	-	-
at Dummy	-	-	-	P M	-	-	-	-

P$_1$ and P$_2$: fawns raised with pronghorn ischiadic scent; B: fawn
raised with black-tailed deer tarsal scent; M: fawn raised with
mule deer tarsal scent. Preferences are shown as time spent at
dummy. P B means significantly more time was spent at dummy
scented with pronghorn ischiadic scent in choice experiment. Only
significant differences of a level of at least p = 0.05 are included.

The fawns were tested for odor preferences at age 2 weeks and
1 month. In these choices males showed preferences whereas females
did not (Table 3). At the age of 2 1/2 months the fawns were
individually placed together with live female black-tailed deer,
mule deer, and pronghorn. Again, the females showed a preference
for the conditioned odor in only one of four comparisons. The
single male in this test preferred the pronghorn to whose odor he
had been conditioned. In choices including a dummy along with live
animals, the females always stayed closer to it than to any of the
live animals (Table 4). The responses of the fawns in the choice
situations varied with the time of day; most approaches to the two
dummies occurred after 21:00 hrs. The point we wish to make is that
in an outdoor setting the animals' differential behavior can be
measured despite those response differences which are most likely
due to changes in light intensity and temperature.

III. The role of antorbital secretion: The genus Odocoileus
marks woody parts of the vegetation with forehead rubbing and antler
thrashing. Yearling and adult males open the antorbital gland and
touch the substrate with it during rubbing (DeVos, 1967; Müller-
Schwarze, 1971, 1972; Moore and Marchinton, 1974). In a recent

Table 4.   Shortest Distances to Live Animals Maintained by Three
           Fawns of Black-tailed Deer Aged 2.5 Months

Fawn:	♀B	♀M	♂P
Comparison		Preferred Animal	
M – P	P	P	P
B – P	–	B	(P)
B – M	B	B	B
D – M	D	D	M
D – B	D	–	(D)
D – P	D	D	P

D: dummy. For other symbols see Table 3.

experiment (Volkman, et al., in press), we investigated the possible
role of the antorbital secretion in this marking behavior. Yearlings and fawns of both sexes were exposed to teflon sticks treated
with forehead and antorbital secretions either separately or in
combination. The samples were taken from live deer housed separately
from the responding animals. Our main finding was that both yearlings and fawns were able to discriminate the age classes of the
odor donors when the forehead and antorbital secretions were combined. They were unable to do so on the basis of either secretion
alone.

      It is of interest to note that in this study performed in New
York State, the males marked 7 times as often as females, whereas in
an earlier observational study carried out in Utah (Müller-Schwarze,
1972), the ratio was 8 to 1. This shows a remarkable consistency
despite different climates, vegetation and individuals,even though
the more recent study used offspring of the animals used in Utah.

      IV. Species specificity of alarm pheromone: In a study of
alarm odor specificity we exposed black-tailed and white-tailed deer
to the alarm odor of the metatarsal (MT) gland of both species.
They were also exposed to one MT of mule deer. The response we
measured was the inhibition of feeding. Black-tailed deer responded
to their own MT secretion and not to the others, whereas white-tailed
deer responded little or not at all to any of the MT secretions. This
indicated that this secretion may be of minor importance as an alarm
pheromone in white-tailed deer. However, as indicated in point 7,

the background of the particular captive animals may have influenced the response level, and therefore further studies are needed.

The alarm odor was also tested in a series of repeated stimulations of black-tailed deer with MT secretions from white-and black-tailed deer. Rapid habituation to a lower level occurred in response to MT odor of white-tailed deer, whereas the response to MT secretions of black-tailed deer remained high. The response to the control (pure solvent), however, was still lower than that to white-tailed deer MT odor. The strong negative response to odor from black-tailed deer is remarkable, considering the novelty of the odor of white-tailed deer and the thorough familiarity of the deer with their own species' MT odor.

V. Pheromones in reindeer: Like the genus Odocoileus, reindeer (Rangifer tarandus) belong to the Telemetacarpalia branch of the Cervidae, and possess antorbital tarsal, interdigital, and caudal glands, but no metatarsal glands. Our observations (Müller-Schwarze et al., a, in prep.) showed that in social behavior the interdigital and caudal glands play an important role, and the tarsal gland is of minor importance. Accordingly, the tarsal hair tuft is less developed than in Odocoileus, and in rub-urination the interdigital glands are soaked instead of the tarsal tufts. In the absence of a metatarsal gland the caudal gland is a candidate for the source of an alarm pheromone. The caudal gland is located on the ventral side of the tail (Müller-Schwarze et al., b, in prep). It is exposed during flight and alarm when the tail is raised and the long white hair is spread.

Conclusion: There is no single procedure for bioassays of mammalian pheromones, as pointed out by the complications discussed in this paper. It is hoped that our experiences reported here will help us to remain circumspect when interpreting experimental results or when designing new experiments. And, we should not lose sight of the fact that successful experimentation must be preceded by careful observation.

Acknowledgment: Our studies were supported by several grants from the National Science Foundation.

REFERENCES

BROWNLEE, R.G., R.M. SILVERSTEIN, D. MÜLLER-SCHWARZE, and A.G.
    SINGER. 1969. Isolation, identification and function of
    the chief component of the male tarsal scent in black-tailed
    deer. Nature 221, 284-285.

CURTIS, R.T., BALLANTINE, J.A., KEVERNE, E.B., BONSALL, R.W.,
    and MICHAEL, R.P. 1971. Identification of primate sexual
    pheromones and properties of synthetic attractants. Nature
    232, 396-398.

DARBY, E.M., M. DEVOR, and S.L. CHOROVER. 1975. A presumptive
    sex pheromone in the hamster: some behavioral effects.
    J. Comp. Physiol. Psychol. 88 496-502.

DE VOS, A. 1967. Rubbing of conifers by white-tailed deer in
    successive years. J. Mammal. 48, 146-147.

DOTY, R.L. 1975. Determination of odour preferences in rodents:
    a methodological review. Pp. 395-406 in: Methods in
    olfactory research, D.G. Moulton, A. Turk, J.W. Johnson, Jr.,
    eds. Acad. Press.

VON FRISCH, K. 1941. Über einen Schreckstoff der Fischhaut und
    seine biologische Bedeutung. Z. vergl. Physiol. 29, 46-145.

GILBERT, B.K. 1973. Scent marking and territoriality in prong-
    horn (Antilocapra americana) in Yellowstone Park. Mamalia
    37, 25-33.

GRAF, R. and M. MEYER-HOLZAPFEL. 1974. Die Wirkung von Harn-
    marken auf Artgenossen beim Haushund. Z. Tierpsychol. 35
    320.

HARRINGTON, J.E. 1976. Discrimination between individuals by
    scent in Lemur fulvus. Anim. Behav. 24, 207-212.

HUMMEL, H. 1968. Beiträge zur Kenntnis der Pheromone. Ph.D.
    Dissertation Marburg University, 122, pp.

JOHNSTON, R.E. 1970. Scent marking olfactory communication and
    social behavior of the golden hamster, Mesocricetus auratus.
    Ph.D. dissertation. Rockefeller University, New York, N.Y.
    (Copyright, 1971).

JOHNSTON, R.E. 1974. Sexual attraction function of golden hamster vaginal secretion. Behav. Biol. 12, 111-117.

KLEIMAN, D.G. 1966. Scent marking in the Canidae. Symp. Zool. Soc. Lond. 18, 167-177.

KREBS, D.J., M.S. GAINES, B.L. KELLER, J.H. MEYERS, and R.H. TARMARIN. 1973. Population cycles in small rodents. Science 179, 35-41.

MERTL, A.S. 1975. Discrimination of individuals by scent in a primate. Behav. Biol. 14, 505-509.

MOORE, W.G. and L. MARCHINTON. 1974. Marking behavior and its function in white-tailed deer. Pp. 447-457. In: the behavior of ungulates and its relation to management. (V. Geist and F. Walther, eds.) IUCN, Morges, Switzerland.

MÜLLER-SCHWARZE, D. 1971. Pheromones in black-tailed deer. Anim. Behav. 19, 141-152.

MÜLLER-SCHWARZE, D. 1972. Social significance of forehead rubbing in black-tailed deer. Anim. Behav. 20, 788-797.

MÜLLER-SCHWARZE, D. and C. MÜLLER-SCHWARZE. 1972. Social scents in hand-reared pronghorn (Antilocapra americana). Zoologica Africana 7, 251-271.

MÜLLER-SCHWARZE, D., C. MÜLLER-SCHWARZE, and W.L. FRANKLIN. 1973. Factors influencing scent marking in pronghorn (Antilocapra americana). Verh. Ot. Zool. Gesellsch. 66, 146-150.

MÜLLER-SCHWARZE, D., C. MÜLLER-SCHWARZE, A.G. SINGER, and R.M. SILVERSTEIN. 1974. Mammalian pheromones: identification of active component in subauricular scent of pronghorn (Antilocapra americana). Science 183, 860-862.

MÜLLER-SCHWARZE, D., R.M. SILVERSTEIN, C. MÜLLER-SCHWARZE, A.G. SINGER, and N.J. VOLKMAN. 1976. Responses to a mammalian pheromone and to its geometric isomer. J. Chem. Eco. 2, 389-398.

MÜLLER-SCHWARZE, D., N.J. VOLKMAN and K.F. ZEMANEK. In Press. Osmetrichia: Specialized scent hairs in black-tailed deer. J. Ultrastructure Research.

MÜLLER-SCHWARZE, D., L. KÄLLQUIST, and T. MOSSING. In Prep.
    Social behavior and olfactory communication in captive rein-
    deer.

MÜLLER-SCHWARZE, D. and W.B. QUAY. In Prep. The caudal gland in
    reindeer (Rangifer tarandus).

MYKYTOWYCZ, R., E.R. HESTERMAN, S. GAMBALE, and M.L. DUDZINSKI.
    1976. A comparison of the effectiveness of the odors of
    rabbits, Oryctalogus cuniculus, in enhancing territorial
    confidence. J. Chem. Ecol. 2, 13-24.

PATTERSON, R.L.S. 1968. Identification of 3 -Hydroxy-5
    -androst-16-ene as the musk odour component of boar sub-
    maxillary salivary gland and its relationship to the sex
    odour taint in pork meat. J. Sci. Fd. Agric. 19, 434-438.

PETERS, R.P. and MECH, L.D. 1975. Scent marking in wolves. Amer.
    Sci. 63, 628-637.

PFEIFFER, W. 1974. Pheromones in fish and amphibia. Pp. 269-637.
    In: "Pheromones", M.C. Birch, ed. North Holland American
    Elsevier.

ROGERS, J.G. and G. BEAUCHAMP. 1976. Influence of stimuli from
    populations of Peromyscus leucopus on maturation of young.
    J. Mammal. 57, 320-330.

SCHULTZE-WESTRUM, T. 1965. Innerartliche Verständigung durch
    Düfte beim Gleitbeutler Petaurus breviceps papuanus Thomas
    (Marsupialia, Phalangeridae). Z. vergl. Physiol. 50,
    151-220.

SCHUTZ, J. 1956. Vergleichende Untersuchungen über die Schreck-
    reaktion bei Fischen und deren Verbreitung. Z. vergl.
    Physiol. 38, 84-135.

SHOREY, H.H. 1974. Environmental and physiological control of
    insect sex pheromone behavior. Pp. 62-80. In: "Pheromones",
    M.C. Birch, ed. North Holland American Elsevier.

SHOREY, H.H. and L.K. GASTON. 1964. Sex pheromones of noctuid
    moths. III. Inhibition of male responses to the sex
    pheromone in Trichoplusia ni (Lepidoptera:Noctuidae). Ann.
    Entomiol. Soc. Am. 57, 775-779.

SINGER, A.G., W.C. AGOSTA, R.J. O'CONNEL., C. PFAFFMAN, D.V.
    BOWEN, and F.H. FIELD. 1976. D. Methyl Disulfide: An
    attractant pheromone in hamster vaginal secretion. Science
    191, 948-950.

STEVENS, D.S. 1972. Thermal energy exchange and the maintenance of homeothermy in white-tailed deer. Ph.D. dissertation, Cornell University. 231 pp.

THIESSEN, D.D., F.E. REGNIER, M. RICE, M. GOODWIN, N. ISAACKS, and N. LAWSON. 1974. Identification of a ventral scent marking pheromone in the male Mongolian gerbil (*Meriones unguiculatus*). Science *184*, 83-85.

UEXKÜLL, J. VON, and E.G. SARRIS. 1931. Das Duftfeld des Hundes. Z. Hundeforschung *1*, 55-68.

VOLKMAN, N.J., K.F. ZEMANEK, and D. MÜLLER-SCHWARZE. In press. Antorbital and forehead secretions of black-tailed deer (*Odocoileus hemionus columbianus*): their role in age class recognition. *Anim. Behav.*

WILSON, E.O. and W.H. BOSSERT. 1963. Chemical communication among animals. Rec. Progr. Horm. Res. *19*, 673-716.

FUNCTIONAL ANATOMY OF THE MAMMALIAN

CHEMORECEPTOR SYSTEM

P. P. C. Graziadei

Department of Biological Science

Florida State University, Tallahassee, Fla. 32306

INTRODUCTION

The sensation of smell, which is elicited when odorous substances contact the receptor membranes, is perceived by the animal
as a unitary experience affecting his behavior in one of two ways:
attraction or withdrawal. However, if we study the anatomical
substrate to this unitary experience which gives rise to specific
behavioral responses, we find that more than one channel of sensory input is involved. The olfactory nerve itself is but one
of these several channels. Each of the systems involved in odor
perception is morphologically distinct having both its own peripheral distribution and discrete central projection.

Our appreciation for the morphology of what is functionally
considered the receptor organ for odorants has developed over
many years, and we now recognize that it actually consists of
five major components. While the study of these five components
would appear essential for an understanding of animal behavior
to odorants, it is still true that the fine organization and
functional interdependence of these components is as yet incompletely known. Nonetheless, it is the goal of this presentation
to outline and analyze the current knowledge concerning the anatomical structures presumed to be involved with the reception of
odorants and to indicate the more pertinent bibliographical
sources on the subject.

We presently know that in the mammalian nose there are five
neural components:
                1)   the olfactory neuroepithelium proper
                2)   the vomeronasal or Jacobson's organ

3)   the septal organ of Masera
4)   the terminal endings of the nervus
     terminalis
5)   the terminal endings of the
     trigeminal nerve (V cranial nerve).

HISTORICAL OVERVIEW

The olfactory nerve was first described by Massa in 1536
and the olfactory bulb was noted by Vesalius in his well known
EPITOME of 1543. The connections between the two structures
escaped the attention of Vesalius who recognized the bulbs as
the organs of smell but did not relate them to the nerves in the
nose. The first complete report on the course of the olfactory
nerves in the nasal mucosa and their connections to the olfactory
bulbs was provided by Scarpa in 1789 in a thesis "De Auditu et
Olfacto". In 1811 Jacobson described in the septum of the nose
in mammals, a tubular structure that was later commonly called
the Jacobson's organ or the vomeronasal organ. Although the
function of this organ was not determined by Jacobson, its close
association with the olfactory apparatus soon became clear since
its efferent nerves projected to a portion of the olfactory bulb.
In the first half of the last century it was also shown that the
trigeminal nerve distributed fine branches to the olfactory mucosa
as well as the respiratory mucous membranes of the nose (Milne
Edwards 1844, Koelliker 1843, 1856). The nervus terminalis was
first illustrated by Fritsch in 1878 for Selachians and was
subsequently confirmed by a series of authors for a variety of
vertebrates including man (Pinkus, 1894-1895; Wijha, 1994; Locy,
1905; Brookover, 1908, 1910, 1914a,b, 1917; Doellken, 1909;
Herrick, 1909; McKibben, 1911, 1914; Brookover and Jacobson, 1911;
McCotter, 1913; Huber and Guild, 1913; Johnston, 1913, 1914; Holmgren
1918-1920; Larsell, 1919, 1918/19, 1950; Pearson, 1941). The last
of the five structures to be discovered in the nose was the septal
organ, first described by Masera in 1943.

OLFACTORY MUCOSA

The olfactory mucosa proper was first described from an
histological viewpoint by Ecker (1855), Eckhard (1855) and
Schultze (1856, 1862). These authors observed both supporting and
receptor cells. It was not clear however if the receptors were
epithelial secondary cells or primary neurons. Schultze (1862)
was of the latter opinion but was not able at that time to provide
convincing data (Sappey, 1876; Gegenbaur, 1889). Using gold chloride
methods, Babuchin (1872) demonstrated that the olfactory receptors
were indeed provided with an axon and consequently were to be
considered true primary neurons. The observations of Babuchin
(1872) have been subsequently confirmed by many authors (Ehrlich,
1886; Arnstein, 1887; Dogiel, 1887; Grassi and Castronuovo, 1889;

Van Gehutchen, 1890; Retzius, 1892a,b, 1894; Disse, 1892; Peter, 1901; Bowden, 1901; Jagodowski, 1901; Kamon, 1904; Ballowitz, 1904; Kallius, 1905; Read, 1908; van der Stricht, 1909; Alcock, 1910), and the relation of the axons of these primary receptor neurons to the glomeruli of the olfactory bulb was clearly established by Cajal (1911). Comparative observations showed that the olfactory mucosa has a particular basic histological plan: in all vertebrates the neuroepithelium is made up of primary receptor neurons, basal cells and supporting cells. Receptors are consistently provided with cilia at their dendritic bare end. A complete bibliographical survey of the olfactory mucosa from an histological point of view was provided by Kolmer in 1927.

Our understanding of the olfactory mucosa's organization which was derived from the light microscope has been complemented and extended substantially by several recent ultrastructural studies (Yasutade, 1951; DeLorenzo, 1957, 1960, 1963; Trujillo-Cenoz, 1961; Yamamoto, Tonosake and Kurosawa, 1965; Graziadei, 1965b, 1966, 1971a,b; Okano, 1965; Reese, 1965; Bronstein and Ivanov, 1965; Bannister, 1965; Balboni, 1965; Seifert and Ule, 1965; Andres, 1966; Brown and Beidler, 1966; Frisch, 1967; Graziadei and Bannister, 1967; Okano, Weber, and Frommes, 1967; Seifert, 1968, 1970; Thornhill, 1970; Graziadei and Monti Graziadei, 1976a).

There have been few quantitative estimates of the number of receptors in any given animal species (Negus 1958). Allison and Warwick (1949) estimated that 50 million receptors are present on each side of the rabbit olfactory mucosa. When this number is compared with the number of glomeruli (1,900) and mitral cells (45,000) it is clear that the system is capable of considerable convergence of channels between the first and the second neuron of the chain.

The most interesting aspect of the study of the olfactory primary neurons is related to their recently proven ability to periodically replace themselves and to regenerate after injury (Graziadei and Metcalf, 1971; Graziadei, 1973; Graziadei and DeHan, 1973; Moulton, 1975; Graziadei and Monti Graziadei, 1977). Whereas other neurons of the mammalian nervous system do not undergo a continuous turnover, and are not replaced when destroyed, the olfactory neurons have long been suspected of having such a capacity. The surveys of Takagi (1971) and Graziadei and Monti Graziadei (1977) provide a complete analysis of the literature concerning this phenomenon up to the present time. In our own laboratory we have recently shown by means of autoradiographic, ultrastructural and other experimental methods that the olfactory neurons of mammals and other vertebrates do undergo a continuous turnover and that they can be replaced after experimental degeneration by the basal cells of the neuroepithelium which undergo mitotic division and act as true neuroblasts (Figs. 1, 2).

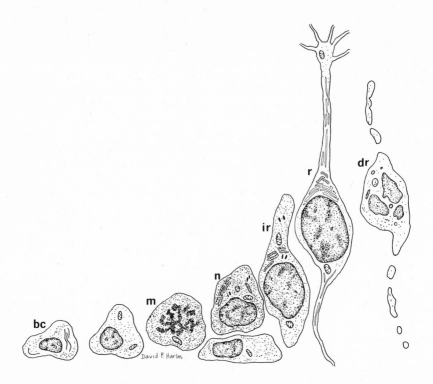

Fig. 1. Some of the most significant stages of the process of
neuron differentiation, maturation and senescence are schemati-
cally illustrated. The diagrams have been derived from our
previous autoradiographic and ultrastructural studies. The
staminal or immature cells (bc) are close to the basal lamina (b),
and correspond to the so called basal cells. They show little
difference from cells at the base of other epithelia. The first
sign of differentiation is an increase in cell volume, appearance
of a clear cytoplasmic matrix and numerous free ribosomes. These
cells are often seen undergoing a mitotic process (n and m).
Immature receptors (ir) maintain the characteristics of (n) cells
but a stumpy dendritic process appears where many basal bodies
are localized. The mature neurons (r) have the characteristic
bipolar appearance with a ciliated dendrite and a fine unmyelinated
axon. Senescence figures are also seen in the epithelium and
appear as morphologically deteriorating receptors (dr) which are
eventually phagocyted by macrophages penetrated into the epithelium. The total life span of each cell has been estimated by
autoradiographic methods in the range of 4 to 5 weeks (Graziadei
and Monti Graziadei, 1977; Moulton, 1975).

Due to the continuous maturation process of the olfactory neurons
we can find in the olfactory neuroepithelium of adult animals all
the transitional stages from the basal neuroblast (true staminal
cell for the olfactory neurons) to the mature and senescent
olfactory neuron.

We have observed that this process of neurogenesis is not
uniformly distributed over the entire sensory surface. We have
shown both autoradiographically and morphologically that some
zones are active in that the immature elements are more numerous
than the mature neurons. On the other hand, other zones are
quiescent in that there are few immature neurons (Graziadei and
Monti Graziadei, 1977) (Fig. 2).

VOMERONASAL ORGAN

The vomeronasal organ was described from an histological
point of view by Balogh (1860), Klein (1881a, b, 1882), and
Piana (1891). These authors observed that the epithelium covering
the walls of the organ was similar to the olfactory neuroepithelium
proper and that it contained primary sensory cells. Later von
Brunn (1892) and Lenhosseck (1892) showed that the vomeronasal
receptors were provided with axons forming the vomeronasal nerve.
The connection of this nerve with the accessory olfactory bulb
was described by a number of authors (Balogh, 1860; DeVries, 1905;
McCotter, 1912, 1919; Bruner, 1914; Herrick, 1921). A controversy
arose in regard to the presence or absence of cilia on the free
endings of the vomeronasal dendrites. The interest in the controversy was motivated by the assumption that it is the cilia which
contain the receptor sites for the odorant molecules. vonBrunn
(1882), Read (1908), Clark (1957) described the vomeronasal
receptors with cilia whereas Allison (1953) did not. Recent
evidence from ultrastructural studies with both the transmission
and the scanning electron microscopes clearly show that the vomeronasal receptors do not have cilia but only microvilli (Graziadei
and Tucker, 1970). It is interesting to note in this context that
the receptors of the vomeronasal organ have responded to the
same odorants as the olfactory receptors proper, although at
different thresholds (Tucker, 1963; Moulton and Tucker, 1964;
Altner and Mueller, 1968). Because of the many structural and
functional similarities between olfactory and vomeronasal receptors it appeared possible that the vomeronasal receptors could
also be regenerated by staminal cells after injury and that
they could even be replaced as a normal life process. In a series
of yet unpublished experiments carried on in our laboratory we
observed that in turtles (Terrapene carolina) (where the vomeronasal nerve is well arranged in discrete bundles of fibers some
4 to 6 mm in length) a 1 to 2 mm plug removed from the nerve was

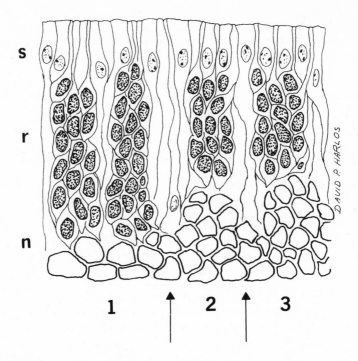

Fig. 2. Diagramatic representation of different regions of the
olfactory epithelium as observed in mouse. The proportion
between immature neuroblasts (n) and mature neurons (r) varies
from region 1 to 2 and 3. Neuroblasts have been represented as
empty profiles. The supporting cells (s) are nucleated but otherwise empty silouettes. Regions such as 2 and 3 have been called
active because of the high number of mitoses observed in morphological preparations and the high concentration of labeled elements
in autoradiographs. Zone 1 has been called quiescent because most
elements are mature and only a few cells at the base are in the
process of differentiation. The diagram also illustrates the
columnar arrangement of the mature neurons and the role of the
supporting cells in partially delimiting these columnar units.

later bridged by regenerating vomeronasal axons. In more
recent experiments in mice we have also observed that labeled
mature receptors can be observed from 20 days after ^{3}H-thymidine
administration. Unlike the olfactory neuroepithelium, there are
no basal cells in the vomeronasal organ. As described in Fig.
3, staminal cells for the vomeronasal receptor neurons are dis-
placed from the two edges of the sensory area where the sensory
epithelium interfaces with the simple columnar ciliated epithelium.

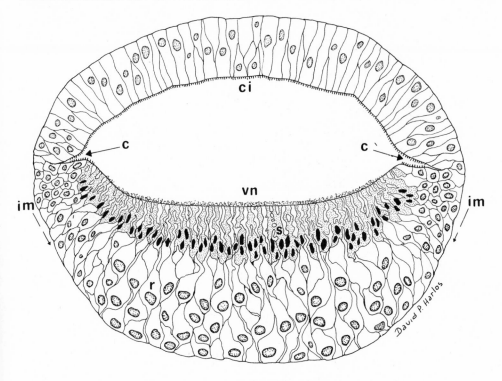

Fig. 3. Cross section of the vomeronasal organ to show the common ciliated (ci) and vomeronasal (vn) epithelia which are in contact in the region indicated at c. Our autoradiographic observations show that the immature elements incorporate ^{3}H-thymidine in the region indicated at (im), subsequently mature and travel downwards in the direction of the little arrows to become mature neurons (r). Supporting cells are indicated at (s).

Our experiments are preliminary but they indicate that the vomero-nasal receptors have indeed the same capacity to regenerate after experimental degeneration and to continuously differentiate even in adult animals.

SEPTAL ORGAN OF MASERA

The septal organ was described relatively recently by Masera in 1943 and at light microscope level it resembles the olfactory epithelium proper. However, macroscopic and microscopic obser-vations indicate that the organ is not simply an ectopic area of olfactory mucosa (Fig. 4). In a recent study on the innervation of the nose in rats Bojsen Moller (1975) described two discrete nerves consistently innervating the sensory organ. These nerves are directed to the olfactory bulb; however, they never fuse or interchange fibers with either the olfactory or the vomeronasal nerve bundles.

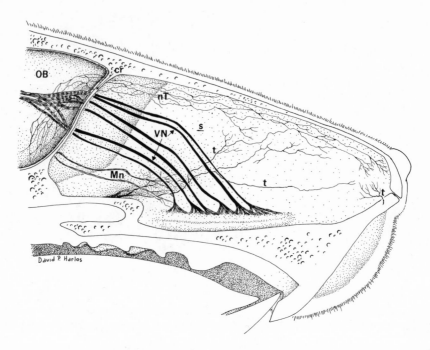

David P. Harlos

Fig. 4. Sagittal section through the nose of a rat at the level
of the septum (s), to show the five nerve components discussed in
the text. OB., olfactory bulb; nT., nervus terminalis; VN.,
vomeronasal nerve; t., branches of the trigeminal nerve; Mn.,
nerves originated from the septal organ of Masera. The region
of the septum close to the olfactory bulb and lamina cribrosa (cr)
which is densely shaded represents the olfactory region proper.

Complementary ultrastructural observations that we are doing in
collaboration with Bojsen Moller indicate that the receptors are
indeed bipolar neurons, but their ultrastructural details are
different from both the olfactory and the vomeronasal receptors.
The differences (Fig. 5) refer mostly to the arrangement of their
organelles and especially to the very unusual arrangement of the
smooth endoplasmic reticulum which occupies a large portion of the
cytoplasm. It would be interesting to investigate the electrophysiological and behavioral characteristics of these receptors
and to establish whether their projections to the olfactory bulb
are restricted to a specific area.

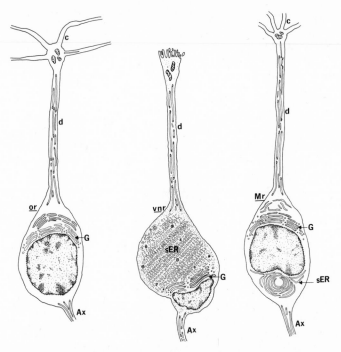

Fig. 5. The three primary receptor neurons of the olfactory,
vomeronasal and Masera organ are diagramatically illustrated.
The olfactory receptor (or) is characterized by cilia (c) and the
long dendrite (d), common to all these primary receptors. The
main difference lies in the displacement and emphasis of organelles in each receptor. The smooth endoplasmic reticulum (sER)
is characteristically arranged in the vomeronasal receptor (vnr)
where it fills most of the perikarion. In the Masera organ receptor (mr) the same organelle often has an "onion rings" appearance
and the Golgi is preponderant.

NERVUS TERMINALIS

      The <u>nervus terminalis</u> was described in Selachians by Fritsch
in 1878 and later in a variety of vertebrates including man. It
derives its name from the close association with the lamina
terminalis (Locy, 1905). It is macroscopically visible only in
dogfish (<u>Squalus Achantia</u>). The difficulty in macroscopically
visualizing the nervus terminalis in mammals is one of the
reasons why we still lack knowledge of its precise central projections and of its electrophysiological and behavioral characteristics. In mammals the nervus terminalis is a plexus of fine
nerve bundles (Fig. 4) on the medial aspect of the bulb directed
towards and branching in the nasal septum (Bojsen Moller 1975).
Although this peripheral distribution in the mucosa of the septum

has been studied with the light microscope, the characteristics
of terminations of these fibers have never been observed at the
ultrastructural level.

In spite of the limited amount of knowledge some facts about
this consistently present structure in vertebrates are firmly
established. Its neurons originate from the primitive olfactory
placode (Simonetta, 1932), which is the source of both olfactory
and veromonasal receptors (Bedford, 1904; Dieulafe, 1906; Brookover,
1910; Brookover and Jackson, 1911; Pearson, 1941; Breipohl, Mestres
and Meller, 1973; Cuschiere and Bannister, 1975a, b). The neurons
migrate centrally along the plexiform extension of the nerve
without collecting in a true unique ganglion like other sensory
neurons do in both cranial and spinal sensory nerves. The cells
are scattered in small clusters on the medial aspect of the olfactory bulb, and peripherally they run along with the branches of
the nervus terminalis in the nasal septum. The widespread distribution of these neurons and the relative limitations of the
silver methods so far employed have left some doubt regarding the
functional nature of these nerve cells. Brookover (1914a, b) and
McKibben (1914) have described multipolar nerve cells and have
suggested on this basis their autonomic nature. Other authors
believed the terminalis to be a sensory nerve because of the close
association with the olfactory nerve and because, according to
their observations, the nerve cells appeared to be bipolar
(Holmgren, 1918-1920; Larsell, 1950). Some preliminary observations of the neurons of the nervus terminalis in fish have
been conducted at ultrastructural level in our laboratory. We
have found that these primary neurons do not have presynaptic
terminals impinging upon them and it is unlikely that they belong
to the autonomic system. It is more probable that they are
sensory neurons.

TRIGEMINAL NERVE

The peripheral portion of the nervus trigeminus is well
known in its peripheral distribution and in its sensory and motor
functions, which have been established and studied for a long
time (Gray 1975). The association of the trigeminus with the
olfactory organ was already indicated by the early observations
of Milne Edwards (1844), Koelliker (1856), and Sappey (1876), who
described the ethmoidal branches of the ophthalmic division and
a portion of the superior maxillary branch supplying fibers to
both the respiratory and the olfactory sensory area of the nose.
Later, Grassi and Castronuovo (1889) and vonBrunn (1892) described
myelinated fibers with fine branching terminals at the boundaries
and within the olfactory sensory epithelium. These fibers are also
commonly known as the vonBrunn's fibers (Cajal 1911).

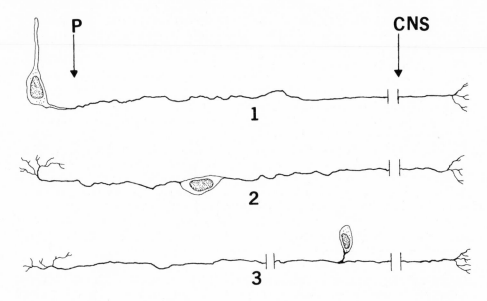

Fig. 6. The diagram shows the peripheral (P) arrangement of the
perikaria (1) of the receptors of the olfactory, vomeronasal and
Masera organs as opposed to the perikaria of the n. terminalis (2)
and trigeminus (3). The neurons of the n. terminalis are scattered along the course of the nerve and some are observed along the
peripheral branches of the nerve in the septum itself, although
not in the epithelium. On the other hand, the perikaria of the
trigeminal fibers instead are located in a typical sensory
ganglion along the nerve and close to its entrance into the CNS.

There is no further report on the structure of these fibers with
the exception of the observations of Graziadei and Gagne (1973)
who reported their ultrastructural characteristics within the
olfactory epithelium. In spite of little anatomical study there
is no doubt that the trigeminus can respond to odors (Tucker, 1963;
Stone, Carregal and Williams, 1966; Stone, Williams and Carregal,
1968; Stone and Robert, 1970; Tucker, 1971). Therefore form an
olfactory point of view, further study of this system would seem
desirable.

CONCLUSIONS

An analysis of the five putative olfactory sensory systems
of the nose shows either an apparent redundancy in distance

chemoreception or a finely tuned apparatus whose components can detect a variety of subtly different chemical signals. In either case it indicates the importance of the information to be collected for the survival of the animal.

There is a common structural basis for the olfactory, septal and vomeronasal receptors (Fig. 6): all have their perikarion located at the periphery in a sensory neuroepithelium. This primitive condition is exceptional among vertebrates where the primary neurons usually have the cell body located in a ganglion. The neuroepithelial location of the perikarion is however commonly observed in invertebrates (Bullock and Horridge 1965, Graziadei 1965). The common peripheral location of the perikaria in these three systems is contrasted by some basic differences in the cytological make up of the receptors (Fig. 5), and in their discrete central projections. The presence of cilia in the olfactory receptors has long been speculated as an essential prerequisite for the reception of odorants (Parker 1922). The lack of these organelles in the vomeronasal receptors seems to contradict this contention as both olfactory and vomeronasal receptors respond to similar stimuli. On the other hand, cilia may be devices which simply increase the free sensory surface area and for some reason the vomeronasal surface need not be increased by such mechanisms over that already afforded by the microvilli. In agreement with this hypothesis is the fact that although the olfactory receptors are freely exposed along the free surface of the nasal cavities, the vomeronasal receptors are confined in a long tubular structure which has a narrow lumen filled with dense mucus. Because the tube is closed at one end, the mucus streams towards the only aperture in a direction which is opposite to the direction that the odor molecules must travel to reach the receptor surface. It is not known why the vomeronasal receptors which respond to odors have morphological arrangements which are in sharp contrast with those believed to favor the reception of airborne chemical stimuli.

We are inclined to believe that the organ of Masera represents a discrete apparatus not only because of the peculiar ultrastructure of its neurons but also for the discreteness of its nervous supply. However, electrophysiological and behavioral studies are needed to define its role in olfaction.

Common to the three systems of primary receptor neurons is the exceptional characteristic of a replacement turnover all through adult life. This phenomenon differs in the olfactory and septal epithelia from that in the vomeronasal organ. In the former as demonstrated by our ultrastructural, autoradiographic and experimental observations, the staminal cells are located at the base of the epithelium from where they gradually differentiate and mature towards the epithelial surface (Graziadei, 1976a,b; Graziadei and Monti Graziadei, 1977). In their process of differentiation and maturation the primary neurons (Fig. 1) dispose themselves into orderly columnar arrangements as shown in Fig. 2 (Graziadei, 1975a,b). We do not know at present if this columnar grouping of neurons (Fig. 2) has functional significance, but its consistent occurrence in the olfactory mucosa suggests at least an orderly development in the process of neuron differentiation. On the other hand, the vomeronasal organ has the staminal cells located on the edges of the neuroepithelial groove where there is a sharp transition between the sensory epithelium and columnar ciliated epithelium. Thus in spite of the common embryological origin of the two receptor systems their neurogenetic processes differ.

We can only speculate as to the reasons why these primary receptors are replaced after birth while all the other neurons of the nervous system are not. The hypothesis that they have adapted to the damaging effects of the environment by replacing themselves does not apply to the vomeronasal neurons which are well protected in a tubular structure. Moreover, it is clear that both the olfactory and vomeronasal neurons undergo a complete life cycle from differentiation to maturity to senescence which is comparable to the life cycle of CNS neurons. The significant difference is that the CNS neurons senesce during the life time of the animal while the chemosensory neurons mature and die in a period of only a few weeks. The turnover of the primary receptors affects by necessity their central connections in the olfactory bulb glomeruli. As a consequence there is in the synaptic zone of the glomeruli a turnover of synapses which is parallel to the peripheral turnover of neurons. The olfactory system is then unique for this continuous morphological change that does not seem to affect the performance of the organ over time. It would also be very interesting to study in this system how the reconnections of the new axons occur both in the normal animal and under a variety of experimental conditions.

There seems to be five morphologically distinct organs which could play a role in the reception of odorous stimuli. Each relates to the central nervous system through different pathways and establish connections with different brain centers. We do not know much about the central pathways of each of these five sensory organs but what little is known suggests that they differ considerably. Where at the perceptual level the information from these several systems is again united to obtain one behavioral response is still unknown.

REFERENCES

Alcock, N. (1910). The histology of the nasal mucous membrane of the pig. Anat. Rec. 4:123-136.

Allison, A.C. (1953). The morphology of the olfactory system in the vertebrates. Biol. Rev. 28:195-244.

Allison, A.C. and Warwick, R.T. Turner (1949). Quantitative observations of the olfactory system of the rabbit. Brain 72:186-197.

Altner, H. and Müller, W. (1968). Eletrophysiologische und elektronenmikroskopische Untersuchungen an der Riechschleimhaut des Jacobonsches Organs von Eidechsen (Lacerta). Z. vergl. Physiol. 60:151-155.

Andres, K.H. (1966). Der Feinbau der regio olfactoria von Makrosmatikern. Z. Zellforsch. mikrosk. Anat. 69:140-54.

Arnstein, C. (1887). Die Methylenblaufärbung als histologische Methode. Anat. Anz. 2:125-135.

Babuchin, A. (1872). Das Geruchsorgan. Sticker, Handb. Lehre den Geweben. 2:964-976.

Balboni, G.C. (1965). Sull' ultrastrutura della mucosa olfattiva in condizioni fisiologiche e sperimentali. Studi Sassaresi 1: 5-30.

Ballowitz, E. (1904). Die Riechzellen des Flussneunauges. Arch. mik. Anat. 65:78-95.

Balogh, (1860). Das Jakobsonsche Organ des Schafes. Sitzungsber. d. Akad. Wien, Mathem.-naturw. 42:449-532.

Bannister, L.H. (1965). The fine structure of the olfactory surface of teleostean fishes. Quart. J. Micr. Sci. 106:333-342.

Bedford, E.A. (1904). The early history of the olfactory nerve in swine. J. Comp. Neurol. 14:390-410.

Bowden, H.H. (1901). A bibliography of the literature on the organ and sense of smell. J. Comp. Neurol. 11:1-11.

Bojsen-Moller, F. (1975). Demonstration of terminalis, olfactory, trigeminal and perivascular nerves in the rat nasal septum. J. Comp. Neurol. 159:245-256.

Breipohl. W., Mestres, P., and Meller, K. (1973). Licht und Elektronmikroskopische Befunde zur Differenzierung des Riechepithels der weissen Maus. Verh. Anat. Ges. 67:443-449.

Brookover, C. (1908). Pinkus' nerve in Amia and Lepidosteus. Science 27:193-194.

Brookover, C. (1910). The olfactory nerve, the nervus terminalis and the pre-optic sympathetic system in Amia calva L. J. Comp. Neurol. 29:49-118.

Brookover, C. (1914a). The development of the olfactory nerve and its associated ganglion in Lepidosteus. J. Comp. Neurol. 24: 113-130.

Brookover, C. (1914b). The nervus terminalis in adult man. J. Comp. Neurol. 24:131-135.

Brookover, C. (1917). The peripheral distribution of the nervus terminalis in an infant. J. Comp. Neurol. 28:349–360.

Brookover, C. and Jackson, T. (1911). The olfactory nerve and the nervus terminalis of ameiurus. J. Comp. Neurol. 21:237–259.

Bronstein, A.A. and Ivanov, V.P. (1965). Electron microscopic investigation of the olfactory organ of Lamprey. J. Evol. Biochem. and Physiol. 1:251–261.

Brown, H.E. and Beidler, L.M. (1966). The fine structure of the olfactory tissue in the black vulture. Federation Proc. 25:329.

Bruner, H.L. (1914). Jacobson's Organ and the respiratory mechanism of amphibians. Gegenbarus Morphologisches Jahrbuch. 48:. 157–165.

Brunn, A. von (1892). Beiträge zur mikroskopischen Anatomie der menschlichen Nasenhöhle. Arch. Mikr. Anat. 39:632–651.

Bullock, T.H. and Horridge, G.A. (1965). Structure and Function in the Nervous System of Invertebrates. Vol II. W.H. Freeman & Co., San Francisco and London.

Cajal, S.R. (1911). Histologie du Systeme Nerveux de l'Homme et des Vertebres. Vol. II. Maloine, Paris.

Clark, W.E. LeGros (1957). Inquiries into the antomical basis of olfactory discrimination. Proc. R. Soc., B 146:299–319.

Cuschieri, A. and Bannister, L.H. (1975). The development of the olfactory mucosa in the mouse: electron microscopy. J. Anat. 119:471–498.

Cuschieri, A. and Bannister, L.H. (1975). The development of the olfactory mucosa in the mouse: light microscopy. J. Anat. 119:277–286.

DeLorenzo, A.J. (1957). Electron microscopic observations of the olfactory mucosa and olfactory nerve. J. Biophys. Biochem. Cytol. 3:839–850.

DeLorenzo, A.J. (1960). Electron microscopy of the olfactory and gustatory pathways. Ann. Otol. Rhinol. Laryngol. 69: 410–420.

DeLorenzo, A.J. (1963). Studies on the ultrastructure and histophysiology of cell membranes, nerve fibers and synaptic junctions in chemoreceptors. In: Olfaction and Taste (Ed. Y. Zotterman), pp. 5–17. Pergamon Press, Oxford.

DeVries, E. (1905). Anatomy. Note on the Ganglion vomeronasale. Proc. K. Akad. W. Amsterdam. 7:704–708.

Dieulafe, L. (1906). The morphology and embryology of the nasal fossae of vertebrates. Annal of Otology, Rhinology and Laryngology 15:512–559.

Disse, J. (1897). Die erste Entwickelung des Riechnerven. Anat. Hefte Abt. 1, 255–300.

Doellken, A. (1909). Ursprung und Zentren des Nervus terminalis. Monatsschr. f. Psych. u. Neurol. 26:10–33.

Dogiel, A.S. (1887). Ueber den Bau des Geruchsorganes bei Ganoiden, Knochenfischen und Amphibien. Arch. mikrosk. Anat. 29: 74–139.

Ecker, A. (1855). Ueber das Epithelium der Riechschleimhaut

und die wahrscheinliche Endigung des Geruchnerven. <u>Zeit</u>.
  <u>Wiss</u>. <u>Zool</u>. 8:303-306.

Eckhard, C. (1855). Ueber die Endigungsweise des Geruchnerven. <u>Beiträge Anat</u>. <u>Physiol</u>. 1:77-84.

Ehrlich, P. (1886). Ueber die Methylenblaureaction der
  lebenden Nervensubstanz. <u>Deutsche med</u>. <u>Wochenschr</u>.
  12:49-52.

Frisch, D. (1967). Ultrastructure of the mouse olfactory
  mucosa. <u>Am</u>. <u>J</u>. <u>Anat</u>. 121:87-120.

Fritsch, G. (1878). <u>Untersuchungen ueber den feineren</u>
  <u>Bau des Fischgehirns</u>. Berlin. (As quoted by W.A.
  Locy, 1905).

Gegenbaur, C. (1889). <u>Traite d'Anatomie Humaine</u>. C. Reinwald,
  Libraire-Editeu, Paris.

Grassi, V. and Castronuovo, A. (1889). Beitrag zur Kenntnis
  des Geruchsorgans des Hundes. <u>Arch mikrosk</u>. <u>Anat</u>.
  34:385-390.

Gray, H. (1975). <u>Gray's Anatomy</u>, 35th British Ed., Saunders
  Philadelphia, Pa.

Graziadei, P.P.C. (1965a). Sensory receptor cells and related neurons in Cephalopods. <u>Cold Spring Harb</u>. <u>Symp</u>. <u>quant</u>. <u>Biol</u>.
  30:45-57.

Graziadei, P.P.C. (1965b). Electron microscopic observations on
  the olfactory mucosa of the cat. <u>Experientia</u> 21:274.

Graziadei, P.P.C. (1966). Electron microscope observation of
  the olfactory mucosa of vertebrates. In: <u>Sixth International</u>
  <u>Congress for Electron Microscopy</u>, Kyoto, p. 487.

Graziadei, P.P.C. (1971a). Topological relations between
  olfactory neurons. <u>Z</u>. <u>Zellforsch</u>. 118:449-466.

Graziadei, P.P.C. (1971b). The olfactory mucosa of vertebrates.
  In: <u>Handbook of Sensory Physiology</u>, Vol. IV. (Ed. L.M.
  Beidler). Springer-Verlag, New York, pp. 27-58.

Graziadei, P.P.C. (1973). Cell dynamics in the olfactory mucosa
  <u>Tissue and Cell</u> 5:113-131.

Graziadei, P.P.C. (1975a). Structural evidence of functional
  units in the olfactory mucosa of mice. In: <u>Neuroscience</u>
  <u>Abstracts Society for Neuroscience</u>, 5th Annual Meeting
  New York City.

Graziadei, P.P.C. (1975b). Neuronal Plasticity in the Vertebrate
  Olfactory Receptor Organ. In: <u>10th Int</u>. <u>Cong</u>. <u>Anat</u>., Tokyo.
  p. 45.

Graziadei, P.P.C. and Bannister, L.H. (1967). Some observations
  on the fine structure of the olfactory epithelium in
  the domestic duck. <u>Z</u>. <u>Zellforsch</u>. 80:220 -234.

Graziadei, P.P.C. and Tucker, D. (1970). Vomeronasal receptors
  in turtles. <u>Z</u>. <u>Zellforsch</u>. <u>mikrosk</u>. <u>Anat</u>. 105:498-514.

Graziadei, P.P.C. and Metcalf, J.F. (1971). Autoradiographic
  and ultrastructural observations of the frog's olfactory
  mucosa. <u>Z</u>. <u>Zellforsch</u>. <u>mikrosk</u>. <u>Anat</u>. 116:305-318.

Graziadei, P.P.C. and DeHan, R.S. (1973). Neuronal regeneration in frog olfactory sytem. J. Cell Biol. 59:525-530.

Graziadei, P.P.C. and Gagne, H.T. (1973). Extrinsic innervation of olfactory epithelium. Z. Zellforsch. mikrosk. Anat. 138:315-326.

Graziadei, P.P.C. and Graziadei, G.A. Monti. (1976a). Olfactory epithelium of Necturus maculosus and Ambystoma tigrinum. J. Neurocytology 5:11-32.

Graziadei, P.P.C. and Graziadei, G.A. Monti. (1976b). The primary olfactory neuron in the vertebrate nervous system. In: Neuroscience Abstracts, Vol. II, Part 1 & Part 2.

Graziadei, P.P.C. and Graziadei, G.A. Monti. (In Press-1977). Continuous nerve cell renewal in the olfactory system. Handbook of Sensory Physiology, Vol. IX (Ed. M. Jacobson). Springer Verlag, New York.

Herrick, C.J. (1909). The nervus terminalis (Nerve of Pinkus) in the frog. J. Comp. Neurol. 19:175-190.

Herrick, C.J. (1921). The connections of the vomeronasal nerve, accessory olfactory bulb and amygdala in amphibia. J. Comp. Neurol. 33:213-280.

Holmgren, N. (1918-1920). Zur Kenntnis des Nervus terminalis bei Teleostiern. Folia Neruobiol. 11:16-36.

Huber, G.C. and Guild, S.R. (1913). Observations on the peripheral distribution of the nervus terminalis in mammalia. Anat. Rec. 7:253-272.

Jagodowski, K.P. (1901). Zur Frage nach der Endigung der Geruchsnerven bei den Knochenfischen. Anat. Anz. 19:257-267.

Jacobson, M. (1811). Description anat. d'un organe observe dans les Mammiferes. Ann. du mus. d'histoire nat. Paris 18:412-431.

Johnston, J.B. (1913). Nervus terminalis in reptiles and mammals. J. Comp. Neurol. 23:97-120.

Johnson, J.B. (1914). The nervus terminalis in man and mammals. Anat. Rec. 8:185-198.

Kallius, E. (1905). Geruchsorgan. Bardeleben, Handb. Anat. Menschen. 5:115-242.

Kamon, K. (1904). Ueber die "Geruchsknospen," Arch. mikrosk. Anat. 64:653-664.

Klein, E. (1881a). A further contribution to the minute anatomy of the organ of Jacobson in the Guinea pig. Quart. J. Mic. Sc., 21:103-118.

Klein, E. (1881b). The organ of Jacobson in the rabbit. Quart. J. Mic. Sc., 21:549-578.

Klein, E. (1882). The organ of Jacobson in the dog. Quart. J. Mic. Sc., 22:229-242.

Koelliker, A. (1843). Über das Geruchsorgan von Amphioxus. Arch. Anat. u. Physiol wiss. Med. 1:33-35.

Koelliker, A. (1856). Elements d'Histologie Humaine. Librairie de Victor Masson, Paris.

Kolmer, W. (1927). Geruchsorgan. In: Handbuch der mikro-
skopische Anatomie des Menschen, 3:192-249. Springer Berlin.

Larsell, O. (1918). Studies on the nervus terminalis: Mammals
J. Comp. Neurol. 30:3-68.

Larsell, O. (1918-19). Studies on the nervus terminalis: Turtle. J. Comp. Neurol. 30:423-443.

Larsell, O. (1950). The nervus terminalis. Ann. Otol.
Laringol. 59: 414-438.

Lenhossek, M. von, (1892). Die Nervensträn und Endigungen
im Jacobson'schen Organ des Kaninchens. Anat. Anz. 7:628.

Locy, W. A. (1905). On a newly recognized nerve connected
with the forebrain of Selachians. Anat. Anz. 26:33-63.

Masera, R. (1943). Sul'esistenza di un particolare organo olfattivo nel setto nasale della cavia e di altri roditori. Arch.
It. Anat. Embriol. 48:157-212.

Massa. (1536). (quoted by Sappey 1877).

McCotter, R. E. (1912). The connection of the vomero -
nasal nerves with the accessory olfactory bulb in the
opossum and other mammals. Anat. Rec. 6:299-318.

McCotter, R. E. (1913). The nervus terminalis in the adult
dog and cat. J. Comp. Neurol. 23:145-152.

McCotter, R. E. (1919). The vomero-nasal apparatus in Chrysemys punctat and Rana catesbiana. Anat. Rec. 13:51-67.

McKibben, P. S. (1911). The nervus terminalis in urodele
amphibia. J. Comp. Neurol. 21:261-309.

McKibben, P. S. (1914). Ganglion cells of the nervus terminalis
in the dogfish (Mustelus canis). J. Comp. Neurol. 24:431-443.

Milne-Edwards, P. (1844). Histoire Naturelle., I. Partie,
Anatomie et Physiologie. Librairie de Victor Masson, Paris.

Moulton, D.G. and Tucker, D. (1964). Electrophysiology of the
olfactory system. Ann. New York Acad. Sci. 116:380-428.

Moulton, D.G. (1974). Dynamics of cell populations in the
olfactory epithelium. Ann. New York Acad. Sci. 237:52-61.

Negus, V. (1958). The Comparative Anatomy and Physiology of
the Nose and Paranasal Sinuses. Livingstone, Edinburgh.

Okano, M. (1965). Fine structure of the canine olfactory hair-
lets. Arch. Histol. Japan. 26:169-185.

Okano, M., Weber, A.F., and Frommes, S.P. (1967). Electron
microscopic studies of the distal border of the canine
olfactory epithelium. J. Ultrastruct. Res. 17:487-502.

Parker, G.H. (1922). Smell, Taste and Allied Senses in the
Vertebrates (J.B. Lippincott, Philadelphia and London).

Pearson, A.A. (1941). The development of the Olfactory nerve,
the nervus terminalis, and the Vomeronasal nerve in man.
Ann. Otol. Rhinol. and Laryngol. 51:317-332.

Peter, K. (1901). Die Entwickelung des Geruchsorgans und
Jacobsonschen Organs in der Reihe der Wirbeltiere.
Hertwig. Handb. Entwick. Wirbeltiere. 2:1-82.

Piana, G.P.   (1891).   Contribuzione alla conoscenza dell
     struttura e della funzione del organo di Jacobson.
     Monitore zool. ital. 1:44-47.
Pinkus, F.   (1894). Ueber einen noch nicht beschriebenen
     Hirnnerven des Protopterus annectens. Anat. Anz. 9:562-566.
Pinkus, F.   (1895).   Die Hirnnerven des Protopterus annectens.
     Morph. Arb. 4:275-346.
Read, E.A.   (1908).   A contribution to the knowledge of the
     olfactory apparatus in dog, cat, and man. Am. J. Anat.
     8:17-47.
Reese, T.S.   (1965).   Olfactory cilia in the frog. J. Cell
     Biol. 25:209-230.
Retzius, G.   (1892a).   Die Endigungsweise des Riechnerven.
     Biol. Unters., N.F. 3:25-28.
Retzius, G.   (1892b).   Zur Kentniss der Nervenendigungen in
     der Riechschleimhaut. Biol. Unters., N.F. 4:62-64.
Retzius, G.   (1894).   Die Riechzellen der Ophidier. Biol.
     Unters., N.F. 6:48-51.
Sappey, Ph. C.   (1876).   Traite d'Anatomie Descriptive. V.
     Adrien Delahaye & Cie, Libraires-Editeurs, Paris.
Scarpa, A.   (1789).   Anatomicae Disquistiones de Auditu et
     Olfacto. Ticini.
Schultze, M.   (1856).   Ueber die Endigungsweise des Geruchsner-
     ven. Monatsber. Akad. Wissen. Berlin 504-514.
Schultze, M.   (1862).   Untersuchungen ueber den Bau der
     Nasenschleimhaut. Abh. naturf. Gesell. Halle 7:1-100.
Seifert, K.   (1968).   Die Feinstruktur des Riechsaumes.
     Archiv klin. exper. Ohren-, Nasen-, und Kehlkopf-
     heilk. 192:182-213.
Seifert, K.   (1970). Die Ultrastruktur des Riechepithels
     beim Makromatiker. Normale und Pathologische Anatomie.
     W. Bargmann and W. Doerr, Georg Thieme Verlag, Stuttgart.
Seifert, K. and Ule, G.   (1965).   Elektronenoptische Unter-
     suchungen am Riechenepithel. H.N.O., Berlin 13:150-151.
Simonetta, B.   (1932).   Orgine e sviluppo del nervo terminale
     nei mammiferi; sua funzione e suoi rapporti con l'organo
     di Jacobson. Z. Anat. Entwikl. 97:425-463.
Stone, H., Carregal, E.J.A., and Williams, B.   (1966).   The
     olfactory-trigeminal response to odorants. Life Sciences
     5:195-201.
Stone, H., Williams, B., and Carregal, E.J.A.   (1968).   The role
     of the trigeminal nerve in olfaction. Exptl. Neurol.
     21:11-19.
Stone, H. and Robert, C.S.   (1970).   Observations on tri-
     geminal olfactory interactions. Brain Res. 21:138-142.
Takagi, S.F.   (1971).   Degeneration and regeneration of the
     olfactory epithelium. In: Handbook of sensory physiology
     Vol. IV (Ed. L.M. Beidler), pp. 75-94. Springer-Verlag,
     Berlin.

Thornhill, R.A. (1967). The ultrastructure of the olfactory
     epithelium of the Lamprey, Lampetra fluviatilis. J.
     Cell. Sci. 2:591-602.
Thornhill, R.A. (1970). Cell division in the olfactory epi-
     thelium of the Lamprey, Lapetra fluviatilis. Z. Zell-
     forsch. mikrosk. Anat. 109:147-57.
Trujillo-Cenoz, O. (1961). Electron microscope observa-
     tion on chemo- and mechano-receptor cells of fishes.
     Z. Zellforsch. mikrosk. Anat. 54:654-676.
Tucker, D. (1963). Olfactory, vomeronasal and trigeminal
     receptor responses to odorants. In: Proc. 1st Internat.
     Symp. Olfaction and Taste (Ed. Y. Zotterman),
     pp. 45-69. Pergamon Press, New York.
Tucker, D. (1971). Nonolfactory responses from the nasal
     cavity: Jacobson's organ and the trigeminal system.
     In: Handbook of Sensory Physiology, Vol. IV (Ed L.M.
     Beidler). Springer-Verlag, Berlin., pp. 151-181.
Van Der Stricht, O. (1909). Le Neuro-epithelium olfactif
     et sa membrane limitante interne. Mem. Acad. r. Med.
     de Belgique. 20:45-83.
Van Gehuchten, A. (1890). Contributions a l'etude de la
     Muqueuse olfactive chez les Mammiferes. La Cellule.
     6:393-407.
Vesalius, A. (1543). Epitome. Basileae
Wijha, W. van. (1894). Over de herzenzenewen der Cranioten
     bij Amphioxus. K. Akad. van Wetenschappen te Amsterdam.
     3:108-115.
Yamamoto, T., Tonosaki, A., and Kurosawa, T. (1965).
     Electron microscope studies on the olfactory epithelium
     in frogs. Acta. Anat. Nippon. 40:343-353.
Yasutake, S. (1951). The fine structure of the olfactory
     epithelium studied with the electron microscope. J.
     Kurume Med. Assoc. 22:1279-304.

MINIMUM ODORANT CONCENTRATIONS DETECTABLE BY THE DOG AND THEIR

IMPLICATIONS FOR OLFACTORY RECEPTOR SENSITIVITY[1]

David G. Moulton

Monell Chemical Senses Center & Department of Physiology,
School of Medicine, University of Pennsylvania, and
V. A. Hospital, Philadelphia, Pennsylvania 19104

INTRODUCTION

In the absence of other cues, the sooner an animal detects,
recognizes and responds to certain odorants entering its environment, the better will be its chances of survival. Faced with such
evolutionary pressures, the ability to detect odorants, in some
species at least, may well have approached or reached absolute
theoretical limits. This seems to be true of the silk moth, Bombyx
mori. The male will respond, behaviorally, to the female sex
attractant, Bombykol, when 10^4 molecules/cm^3 are present in air
currents flowing at 60cm^3/s over 2 s. Experiments with tritiated
Bombykol suggest that this corresponds to about one molecule per
sense cell (Schneider, et al., 1968; Schneider, 1969). In mammals,
however, delivery of odorant molecules to the receptors depends on
the flow of air through respiratory airways and across extensive
mucous surfaces. These and other variables compound the difficulty
of assessing the sensitivity of individual receptors or sites. It
is nevertheless valuable to attempt such an estimate not only for
its intrinsic interest but also for the light it may throw on
olfactory mechanisms, and for the framework it provides in defining
the nature and diversity of events that limit access of odorant
molecules to receptor sites. Ultimately such information may also
assist in evaluating whether, in a specific context, a given odorant is present in sufficient concentration to elicit a pattern of
behavior.

---

[1]Supported by grant No. 73-2425, U.S. Air Force, Air Force Office
of Scientific Research.

The starting point is knowledge of the minimum concentration of a given odorant that can be detected (as measured psychophysically), preferably for a species with highly developed olfactory powers. Few would deny that the dog represents such a species and while there is some conflict concerning the magnitude of the absolute threshold range, most authors agree that the dog can detect certain fatty acids in concentration at least 100 times lower than can man (Neuhaus, 1953; Moulton, et al., 1960). For the present purposes it is convenient to use absolute thresholds of German shepherds for α-ionone (Moulton & Marshall, 1976). Before applying this data, however, we shall describe briefly how it is derived.

MEASUREMENT OF MINIMUM CONCENTRATION OF α-IONONE PERCEPTIBLE BY THE DOG

The apparatus is illustrated, in simplified form, in Fig. 1. It consists, essentially, of a behavioral test chamber lined with teflon and housed inside a controlled environment room. The room, in turn, occupies part of a laboratory. Odorant from a multistage air-dilution olfactometer is delivered to one and air to the remaining two of three presentation bays at the front of the test chamber. The dog is trained to sample each bay in turn and to indicate the bay associated with the odor by inserting its snout into it for five seconds. This breaks a light beam directed across the bay to a photocell. If correct, the dog receives a water reward. If incorrect, glass doors swing down to close all bays. Concentrations are successively lowered until the dog's performance drops to chance.

Difficulties encountered with this approach include the problem of training dogs to sustain stable performances at near-threshold concentrations and the long periods required both to remove contaminants and to establish equilibrium conditions in the olfactometer and delivery lines. For example, in establishing a flow of α-ionone at a concentration of 10^{-8} of saturated vapor, we found that it was 25 hours before a dog--which had previously been trained to a stable performance on this concentration--began to detect the odor issuing from the presentation bay. A further period was required before it regained previous performance levels. Limited checks with a gas chromatograph also indicated that the delay was due to strong sorption of α-ionone on glass surfaces. Desorption required comparable periods during which the delivery lines were flushed with air until performance on repeated trials decayed to chance levels. (While gas chromatography was used to calibrate the olfactometer at higher concentrations, only the dog could detect the lowest concentrations.) Clearly, in this case, rapid switching from one concentration or compound to another would have yielded misleading results.

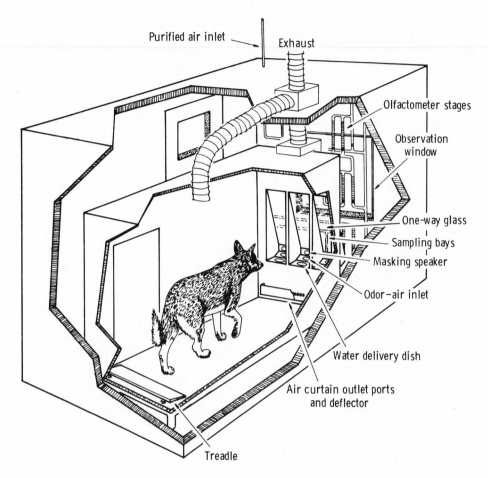

Fig. 1. Apparatus for testing dogs in an odor detection task. The
test chamber is housed in a controlled environment room occupying
part of a laboratory. For the purposes of illustration many details
have been simplified or omitted. (For example, a gas chromatograph
and water reservoir bottles are normally housed on the roof of the
chamber, and an air conditioning unit and purification stages lie on
the roof of the room.) The olfactometer is shown in semi-schematic
form. (From Moulton and Marshall, 1976)

The ability of dogs to detect and respond to non-olfactory
cues can also lead to errors. Clicks associated with solenoid
switching and other auditory cues are those most commonly encountered. On one occasion, however, an imbalance of air flows provided
the dog with the necessary cue. The dog was achieving performances
above chance on "blank" trials where air was being delivered to the
normally odorized line. Odor contamination of the line could not
have accounted for the difference. Finally an error was found in

the setting of the flow rate in one line. The dog was detecting
an imbalance when the difference between the flows, as defined by a
differential threshold fraction (Δ Flow/Flow), fell in the range:
0.12 - 0.08. When the flows were equalized, performance fell to
chance.

Using this method we derived stimulus-response curves for
α-ionone (Fig. 2). One dog could detect $4 \times 10^{4.5}$ molecules/cm^3
of α-ionone. For all dogs, however, the range was $4 \times 10^{4.5}$ -
$4 \times 10^{6.5}$ molecules/cm^3.

ESTIMATION OF SENSITIVITY OF INDIVIDUAL RECEPTORS

The range of minimum perceptible concentrations, in itself,
tells us little about the sensitivity of individual olfactory receptors; or about the number of molecules required to excite receptors
at concentrations near threshold. Such estimates have been made for
a fox terrier bitch by Neuhaus (1953). He determined that at
threshold there were 28,000 olfactory receptors available for every
molecule of butyric, valerianic, caproic or caprylic acid. But we
now have more accurate measures of receptor cell density, the course
of the nasal air stream in the dog, and the relative retention times
for different odors flowed across olfactory surfaces (see below).
In addition, the careful quantitative work of Stuiver (1958) on
human subjects has identified certain of the factors controlling
the arrival of odorous molecules at olfactory receptor sites and has
attempted to evaluate the relative magnitudes of their effects
(de Vries and Stuiver, 1960).

In the light of these developments it is appropriate to
reexamine the question of receptor sensitivity using the data we
have obtained for α-ionone. The steps outlined by de Vries and
Stuiver (1960) provide a convenient framework for calculating the
number of molecules that reach receptor sites. Having done this
we can estimate the ratio of molecules to receptors and the sensitivity of a single receptor. Finally this information allows us to
calculate the ratio of molecules to receptor sites.

To determine the number of molecules reaching receptor sites
we need to know: (1) The minimum number (N_0) of molecules of
α-ionone that the dog can detect. (We can take this to be the
amount contained in one sniff of a threshold concentration.) (2)
The fraction (f_1) of inspired air that reaches the olfactory region.
(3) The fraction (f_2) of odorous molecules left in the air that
reaches the olfactory region. (4) The fraction (f_3) of those
molecules reaching the olfactory region which actually interact
with receptor sites. (The remaining molecules escape sorption or
bind to inactive sites.)

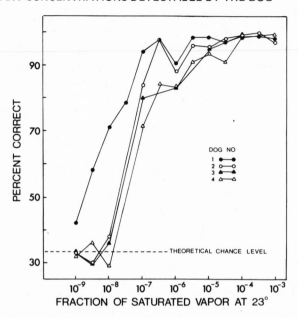

Fig. 2.   Concentration-response curves for α-ionone in four dogs.
(From Moulton and Marshall, 1976)

The number (N) of odorous molecules reaching olfactory receptor sites is then given by

$$N = N_0 f_1 f_2 f_3$$

To calculate N_0, assume the dog inspires about 60 cm^3 in one
sniff of α-ionone.   In the case of the best performing dog this
represents about 2.4 x 10$^{6.5}$ molecules of the odorant at threshold.
The magnitudes of the remaining fractions cannot be determined with
precision.   However, there are several lines of evidence on which
we can draw to derive estimates.

(a) The Fraction of Inspired Air that Reaches the Olfactory Region

This fraction (f_1) depends on the course of the nasal air
stream during active sniffing.   De Vries and Stuiver (1960) estimate it to be up to 20 per cent for the human nasal cavity on the
basis of a study of the flow of aluminum particles suspended in
water in a model of the nose.   However, there are wide variations
among mammalian species in the pattern of nasal air flow.   Thus in
the dog (as in the rabbit, cat, rat and guinea pig), there is some
evidence that a higher proportion of the total air flow courses
along the floor of the nasal chamber, away from the olfactory
region, than is the case in man (Lucas and Douglas, 1934; Becker
and King, 1957).   Assume, then, that f_1 is ten per cent.

(b) The Fraction of Odorous Molecules Left in the
Air Reaching the Olfactory Region

This fraction (f_2) is also likely to be smaller in the dog than the 50 per cent estimated by de Vries and Stuiver (1960) for man. The reason is the vastly greater sorptive surface presented to incoming air by the non-olfactory mucosa of the dog's nasal cavity. This includes the surfaces of a prominent swell body and complex maxilloturbinates--structures absent in man.

Evidence bearing on this point comes from a study of the patterns of fluorescein sodium deposition in the nasal cavity of the dog. The dogs were induced to sniff the compound in the form of an aerosol spray. When the exposed nasal passages were viewed under ultraviolet light the heaviest concentrations appeared on non-olfactory surfaces (middle and inferior meatuses, oblique sulcus and nasopharynx). On the other hand, the ethmoturbinates, which support the olfactory epithelium, also fluoresced noticeably (Becker and King, 1957). From the relevant figure in this publication, f_2 could be in the order of 30 per cent.

It might, of course, be argued that an aerosol of fluorescein would be unlikely to provide an adequate predictor of the pattern of deposition of α-ionone molecules at near-threshold concentrations. The evidence of Hornung, et al. (1975) on frogs, suggests that, in fact, f_2 would be even smaller. They drew tritiated butanol through the nasal sac, froze the frog in liquid nitrogen and sectioned the nasal area. About 84 per cent of the molecules remained close to the external nares and less than 1 per cent reached the internal nares. While such high retentivity may not be characteristic of α-ionone, we can assume, as a first approximation, that f_2 is ten per cent.

The final fraction (f_3) is the proportion of molecules reaching the olfactory region that interact with olfactory receptor sites. This is particularly difficult to estimate since it depends on several factors, none of which are well understood. Firstly, there is difficulty of knowing whether a single odorant molecule interacts with only a single site, or whether it can interact with a number of sites in succession. Hornung and Mozell (1976), in an extension of the study described above (Hornung, et al., 1973), found that a large proportion of the tritiated butanol molecules, which flowed over the olfactory sac, appeared to remain in the mucosa for at least 30 min, and 90 per cent of those were deeper than 45 μm below the surface of the mucus. On the other hand, it is a common observation that the responses of single units in the olfactory epithelium to any one of a number of odorants decay to base line within about 500 ms following cessation of the stimulus. Furthermore, the duration of a dog's sniff is probably about 100 ms. Within such time constraints the number of multiple "hits" that a single molecule can make may be limited.

A second difficulty is in knowing the extent to which binding to inactive sites occurs. Finally there is the problem of determining what proportion of molecules pass over the olfactory area but leave without coming into contact with the olfactory surface. If we assume that multiple hits compensate for failure of molecules to reach or to bind to active sites, f_3 becomes 1 and

$$N = 2.4 \times 10^{6.5} \times .1 \times .1 \times 1 = 2.4 \times 10^{4.5} \text{ molecules.}$$

(c) The Ratio of Odorant Molecules to Receptors

Having estimated the number of molecular "hits" on active receptor sites we must now consider how many receptors are available. This requires knowledge of both receptor density and field size. It is not sufficient to know the extent of the olfactory area for an unspecified breed of dog since estimates for different breeds vary by more than a factor of 16. Fortunately, the area of the olfactory epithelium has been reported for the German shepherd: 169.46 cm^2 (Lauruschkus, 1942). On the other hand, receptor density for this breed is not known. It is, however, likely to fall within the range found for cat, guinea pig, rat and an unspecified breed of dog--namely, 50,000-215,000 receptors/mm^2. (See Altner and Kolnberger, 1975). Assume, then, that it is 120,000/mm^2. This implies that the German shepherd has about 2×10^9 receptors, which is about ten times more than has been estimated for the rabbit and 100 times more than has been estimated for man.

We cannot assume, however, that receptor sites are randomly distributed according to their odor specificities. There is evidence, derived electrophysiologically from the tiger salamander, that receptors sensitive to camphor, for example, are concentrated in the medial region of the ventral olfactory surface (Kauer and Moulton, 1974). Other anatomical and electrophysiological studies also suggest that there is a non-homogeneous distribution of receptor site types (see Moulton, 1976). However, some of these latter findings may at least partly reflect a further complicating factor, namely that odorant molecules are not necessarily distributed evenly across the olfactory surface. In particular, α-ionone is likely to bind relatively strongly. Consequently, the molecules will tend to concentrate on the more anterior surfaces. Regions lying remote from the respiratory air-stream within the recesses of the turbinal folds may receive no molecules. There is no data that would allow us to estimate accurately the consequence of these features, i.e., the distribution of molecules or of sites for α-ionone molecules on the olfactory surface. As a first approximation, however, we can allow that their combined action may be comparable to restricting the number of available molecules to one tenth of the total olfactory surface.

On the basis of these calculations there are 2×10^8 receptors available to receive $2.4 \times 10^{4.5}$ molecules, or about 2,600 receptors to the molecule. Even allowing for an error of 10^3 this still implies that one receptor can probably respond to one molecule of α-ionone.

(d) The Ratio of Odorant Molecules to Receptor Sites

In the context of the odorant-receptor interaction the significant unit is the receptor site, rather than the receptor itself. It seems probable that active binding sites are concentrated on the olfactory surface--the region of the receptor membrane in contact with the surface mucus. The most prominent structures at this level are cilia of which there are an estimated 100-150 per cell in the dog (Okano, et al., 1967). Thus, if one takes the question of sensitivity to the level of the binding site, it is necessary first to know the surface area (A) of the cilia and the density of sites per unit area of cilia. From this, one can calculate the total number of sites.

The first step in estimating A is determining the total surface area of a single cilium. Fortunately, its dimensions can be calculated partly from data given by Okano, et al. (1967) for the dog: from its base, with a thickness of 0.25-0.3 μm, each ciliary shaft thins toward its distal end. Assume, then, a mean diameter of 0.2 μm. Ciliary length is uncertain but is probably comparable to that of the guinea pig--at least 50 μm (Calalano and Biondi, 1969). The surface area is, then, 31.42 μm^2 per cilium. Since there are an estimated 125 cilia per cell and about 2×10^9 receptors in the German shepherd the total ciliary surface area is in the order of 7.85 m^2, or several times the area of the dog's body surface.

The second step is to determine the density of sites per unit membrane surface. This is not known. For the present purposes, however, another membrane system specialized to detect low concentrations of a chemical may be the closest model for which pertinent information exists. This is the subsynaptic membrane of the neuromuscular junction bearing sites receptive to acetylcholine. The density of the sites is calculated to be in the order of $10^4/\mu m^2$ (see Gage, 1976). Assuming this can also be taken as an estimate of A, the total number of olfactory receptive sites becomes 8×10^{16}. If only one tenth of these sites are accessible to α-ionone there are still about 10^{11} receptor sites available for each molecule of the odorant at threshold. Low levels of site occupancy must also prevail over the entire dynamic range of the concentration-response function in odorants so far examined, since this range is generally less than three log units of concentration. The additional number of molecules inhaled in a bout of sniffing (as opposed to a single sniff) is also inconsequential in the context of a population of 10^{11} sites. However, we should bear in mind that cilia may bear no receptor sites or bear them only at certain regions (such as the basal segment). Even so, site occupancy levels could still be relatively low.

CONCLUSIONS

The conclusion that one molecule of a specified odorant is probably sufficient to excite a single olfactory receptor agrees with the estimates of both Neuhaus (1953) for the dog and de Vries and Stuiver (1960) for man. (This is despite the fact that the calculations are now based partly on much evidence unavailable to the earlier authors.) Apparently the dog has little or no advantage over man at the single receptor level. Furthermore, whatever may be the nature of the transduction process in olfaction, it would seem to require almost negligible energy changes to activate it.

Where the dog does differ from man, however, is in the receptor reserve available. Our calculations suggest that there are over a billion receptors in the olfactory epithelium of the German shepherd or somewhat more than 100 times the number given for man. This is reflected in a comparison between the ratio of receptors to odorant molecules available at threshold. In contrast to the ratio of 2,600:1 for the German shepherd it is closer to 1:1 for man (de Vries and Stuiver). This reserve may come into play in the detection of compounds having lower thresholds than α-ionone. It may also be critical in the spatial analysis of complex mixtures of odorants.

ACKNOWLEDGMENT

I thank Dr. D.A. Marshall for his comments and criticisms of the manuscript.

REFERENCES

Altner, H., and I. Kolnberger (1975). The application of transmission microscopy to the study of the olfactory epithelium of vertebrates. In Moulton, D.G., Turk, A., and Johnson, J.W., Jr. (eds.) Methods in Olfactory Research. Academic Press, London, 163-190.

Becker, R.F., and J.E. King (1957). Delineation of the nasal air streams in the living dog. A.M.A. Archiv. Otolaryngol. 65:428-436.

Calalano, G.B., and S. Biondi (1969). Aspetti ultrastrutturali del recettore olfattivo nella cavia. La Clinica Otorinolaringoiatrica 23:166-201.

De Vries, H., and M. Stuiver (1960). The absolute sensitivity of
    the human sense of smell. In Sensory Communication, W.A.
    Rosenblith (ed.) Wiley & Sons, New York. 159-167.

Gage, P.W. (1976). Generation of end-plate potentials. Physiol.
    Rev. 56:177-247.

Hornung, D.E., R.D. Lansing, and M.M. Mozell (1975). Distribution
    of butanol molecules along bull frog olfactory mucosa. Nature
    254:617-618.

Hornung, D.E., and M.M. Mozell (1976). Removal of odorants from
    the olfactory sac. Neurosci. Abstr. 1:4.

Kauer, J.S., and D.G. Moulton (1974). Responses of olfactory bulb
    neurones to odour stimulation of small nasal areas in the
    salamander. J. Physiol. 243:717-737.

Lauruschkus, G. (1942). Über Riechfeldgrösse und Riechfeldkoeffi-
    zient bei einigen Hundrassen und der Katze. Arch. Tierheilk.
    7:473-497.

Lucas, A.M., and L.C. Douglas (1934). Principles underlying ciliary
    activity in the respiratory tract II. A comparison of nasal
    clearance in man, monkey, and other mammals. Arch. Otolaryng.
    20:518-541.

Moulton, D.G. (1976). Spatial patterning of response to odors in
    the peripheral olfactory system. Physiol. Rev. 56:578-593.

Moulton, D.G., E.H. Ashton, and J.T. Eayrs (1960). Studies in
    olfactory acuity 4. Relative detectability of n-aliphatic
    acids by the dog. Anim. Behav. 8:117-128.

Moulton, D.G, and D.A. Marshall (1976). The performance of dogs
    in detecting α-ionone in the vapor phase. J. Comp. Physiol.
    110:287-306.

Neuhaus, W. (1953). Über die Riechschärfe des Hundes für Fett-
    säuren. Zeitschr. f. Vergl. Physiol. 35:527-552.

Okano, M., A.F. Weber, and S.P. Frommes (1967). Electron micro-
    scopic studies of the distal border of the canine olfactory
    epithelium. J. Ultrastructure Res. 17:487-502.

Schneider, D. (1969). Insect olfaction: deciphering system for
    chemical messages. Science 163:1031-1037.

Schneider, D., G. Kasang, and K.E. Kaissling (1968). Bestimmung
    der Riechschwelle von Bombyx mori mit Tritium-markiertem
    Bombykol. Naturwissenschaften 55:395.

PROCESSING OF OLFACTORY STIMULI AT PERIPHERAL LEVELS

Maxwell M. Mozell

Physiology Department, State University of New York

Upstate Medical Center, Syracuse, New York 13210

PUTATIVE MECHANISMS FOR OLFACTORY DISCRIMINATION

Based upon his electrophysiological work during the early 1950's, Adrian (1950, 1951, 1953, 1954) has left a legacy concerning the mechanisms which might underlie olfactory discrimination at the level of the olfactory mucosa. One such mechanism involves the selective sensitivity of individual receptor cells to particular odorants or groups of odorants. Any given cell would not be equally sensitive to all odorants, but would instead be maximally excited by some odorants and excited less, or not at all, by others. Thus, each receptor cell would signal the degree to which its particular sensitivity is matched by the molecules of incoming odorants.

A second mechanism suggested by Adrian does not depend upon the sensitivity of the receptor cells per se but instead relies upon the differential migration and distribution of molecules of different odorants along the mucosa. These in turn depend upon such molecular properties as solubility in mucus, diffusion rate, and volatility. Each receptor cell would then signal the degree to which odorant molecules reach its particular position along the olfactory mucosa, and each odorant would therefore establish a characteristic spatiotemporal pattern of response activity across the entire sheet of receptors.

Mentioned, though not emphasized by Adrian (1950), is a third possible mechanism which combines aspects of the other two. Perhaps receptors of like sensitivity are clustered together in groups, thus producing across the mucosa differential matrices of regional sensitivity. Such an arrangement would, like the molecular

465

migration of odorant molecules, also give rise to differential
spatiotemporal activity patterns across the mucosa but for quite
different reasons.

It is important to note that Adrian did not propose these
mechanisms as being in competition with each other for acceptance
as the one fundamental mechanism basic to olfactory discrimination.
Rather he saw them as being complementary to each other with their
interaction giving rise to a larger number of response permutations
for the analysis of incoming odorants than would each mechanism
alone.

These mechanisms which Adrian proposed to be operative at the
level of the olfactory mucosa were actually based upon his findings in the olfactory bulb. Since Adrian's time, however, there
has been considerable work at the level of the olfactory mucosa
itself to assess the existence of these various mechanisms. It is
a purpose of this presentation to briefly summarize the current
status of each of them using a limited number of illustrative
studies.

SELECTIVITY OF THE RECEPTOR CELLS

Gesteland and his associates (1963, 1965) were the first to
develop a technique to successfully record extracellularly from
single olfactory receptor cells. Several other laboratories have
since reported single receptor cell data (Shibuya and Shibuya,
1963; Mathews and Tucker, 1966; Mathews, 1972; Takagi and Omura,
1963; O'Connell and Mozell, 1969). The preparation is in general
the same from laboratory to laboratory. The olfactory mucosa is
exposed by removing one of the walls of the compartment housing
it. For instance, in the frog, where much of this work has been
done, the roof of the olfactory sac is most often removed and
recordings are taken from the exposed eminentia which forms the
floor of the sac. The microelectrode is placed upon the surface
of the mucosa and is driven down into it by small steps until
characteristic single unit activity is seen. Since a wall of the
olfactory compartment is removed, the odorant can be puffed
directly onto the mucosa in the vicinity of the recording elec-
trode. This allows the observation of the selective sensitivity
of receptor cells to different odorants without the possible con-
taminating effect of one of the other mechanisms suggested by
Adrian as possibly basic to olfactory discrimination, viz., the
effect of the differential migration of the molecules of different
odorants across the mucosa. That is, bringing the odorants
directly to the receptor cell at the site of the recording elec-
trode bypasses any differences there might ordinarily be among the
odorants in reaching that cell, and thus removes the possibility
that these differences might be reflected in the cell's responses

to a series of odorant stimuli.

It is likely that those investigators who first recorded single unit activity from the olfactory mucosa expected to find cells with differing sensitivities to the odorants tested. Furthermore, these investigators may have also expected the receptor cells to fall into specific categories (or types) as defined by the particular odorants to which they responded. That is, one type of receptor cell might respond to one group of odorants, another type of receptor cell to another group of odorants, etc. If this occurred, the physicochemical properties common to all those odorants defining one receptor cell type could be compared to those properties common to the odorants defining other receptor cell types. Such a cross-comparison might have been expected to give some insight into the process by which odorant molecules interact with olfactory receptor cells.

To exemplify the results of these single receptor cell studies, the data from one of them (Mathews, 1972) is given in Fig. 1. The matrix in this figure shows which units of the 19 sampled in the turtle mucosa are excited by each of the 21 odorants presented. It can be seen that each receptor cell is indeed selectively sensitive in the sense that it does not respond to all odorants but rather to a particular group of them. It can also be seen, however, that no two cells respond to the same group of odorants, thus precluding on the basis of these data their obvious categorization into receptor cell types. It must be recognized, however, that this result could be very much influenced not only by the sample sizes of both the odorants tested and the receptor cells monitored, but also by a variety of other experimental variables such as the range of the odorant intensities used and the chemical purity of those odorants (O'Connell and Mozell, 1969). Indeed, in a highly controlled study, O'Connell and Mozell (1969) did demonstrate some commonalities among units in the way they responded to a very limited number of odorants, but the design of this study was not intended to establish the existence of receptor cell types. Recognizing the limitations of the single unit data so far collected, one can still cautiously conclude that receptor cell types have not yet, at least, been demonstrated. In addition, as exemplified in Fig. 1, there does not seem to be any obvious chemical basis for the way the odorants fall together in their stimulation of given units. For instance, very similar homologs like amyl acetate and butyl acetate often do not stimulate the same receptor cells. Furthermore, in Fig. 1 there are many examples of odorants with very different chemical properties which, nevertheless, stimulate the same cells.

Although these single unit studies seem to disclose a rather asystematic basis for the analysis of odorants, the investigators point out (e.g., Gesteland, 1965) that such a system is particu-

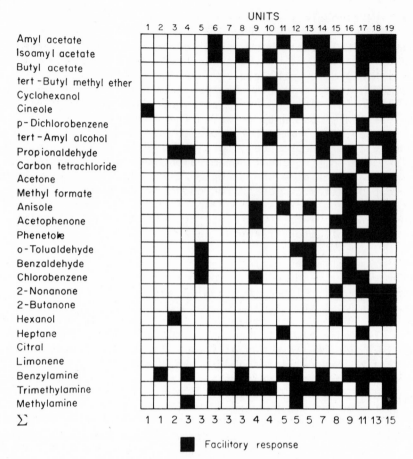

Fig. 1. Matrix of receptor cell responses by units and odorants. (Mathews, 1972, by permission of Rockefeller University Press).

larly suited to encode a large number of odorants by the rich variety of different activity profiles which can be produced by all the units taken in concert. For instance, using the data of Fig. 1, amyl acetate could be encoded by the excitation of units 6, 11, 13, 14, 17, 18, and 19, whereas the encoding of isoamyl acetate would be by the excitation of units 6, 8, 10, 14, 15, 17, 18, and 19. It should be pointed out that in the type of matrix presented in Fig. 1 the degree to which a unit is excited is not shown. If one adds to such a matrix some designation of the response level, the number of different possible profiles for the encoding of different chemicals is greatly increased.

From the electrophysiological studies there is no clear

evidence to support the concept of receptor cell types. However, does this mean that in the processing done by the receptor cells there is no evidence for the sorting of like odorants from unlike odorants? On the contrary, evidence that such sorting may indeed play a role in the analysis of odorants at the receptor cell level has been provided by a biochemical approach taken by Getchell and Gesteland (1972) which, however, focuses not on the receptor cells per se but rather upon the receptor sites they putatively bear. As a measure of olfactory response these investigators used the electro-olfactogram (EOG), a slow potential which can be recorded from the mucosal surface in response to odorant stimulation (Ottoson, 1956). It appears, at least in part, to represent the combined activity of many responding olfactory cells (Ottoson, 1971). Getchell and Gesteland observed that the application to the mucosa of a group-specific protein reagent, N-ethylmaleimide (NEM), blocked the EOG to all the odorants they tested. However, the EOG to at least some odorants could be protected against NEM blockage by subjecting the mucosa to those odorants prior to NEM application. For example, pretreatment of the mucosa with ethyl butyrate prior to NEM application was observed to protect the receptor activity (i.e., the EOG) in response to ethyl butyrate and one of its homologs, methyl butyrate, but not in response to other odorants. Later Getchell and Getchell (1974) observed the same protection for another homolog, ethyl isobutyrate. This suggested the existence of a receptor site which is selective for ethyl butyrate-like molecules.

Thus, initially these data may seem to reveal a paradox. On the one hand, the biochemical approach supports the existence of selective receptor sites for related odorants, but on the other hand, the electrophysiological approach fails to find receptor cell types. If indeed there are sites on receptor cells which show selectivity for related odorants, one might have expected to find receptor cells which are likewise selective for related odorants. This apparent contradiction can, of course, be resolved by suggesting that no one receptor cell need possess receptor sites with only one kind of selectivity. If, for instance, ten different kinds of sites are ultimately identified, any combination of these ten might be carried by any given receptor cell. In addition, even for cells which carry the same kinds of sites, the relative numbers of each kind may still differ. Therefore, in spite of the fact that selective receptor sites may play a role in the analysis of odorants, it is not inconsistent to find that no two receptor cells respond in the same way to a given set of odorants.

MUCOSAL REGIONS OF SELECTIVELY SENSITIVE RECEPTORS

Kauer and Moulton (1974) developed an odorant delivery system
specifically designed to test Adrian's suggestion that receptors
of like sensitivity might be regionally clustered along the
olfactory receptor sheet. For this test it was necessary that
the odorant delivery system stimulate discrete, punctate regions
of the mucosa. The key aspect of this system was the delivery
nozzle which consisted of two glass tubes, one inside the other.
The inner tube directed the odorant toward the olfactory epithelium whereas the outer tube provided a vacuum screen which prevented the spread of molecules beyond the prescribed region. By
successive shifts in the position of the nozzle the entire exposed
mucosal surface could be covered with a series of punctate stimulations. (Note that this technique, like the one discussed above,
also circumvents the differential migration of odorant molecules
along the mucosa.)

Using this punctate stimulator with salamanders and frogs,
Kauer and Moulton assessed the possibility of different regional
sensitivities across the mucosa by recording from single units in
the olfactory bulb. They observed each unit's discharge magnitude
as different odorants were "focused" on each punctate mucosal
region. The activity of one bulbar unit in response to punctate
stimulation by three different odorants is summarized in Fig. 2.
Each sketch represents the floor of the olfactory sac and shows
its regional responsiveness to only one of the three odorants
used. Each circle depicts a stimulated punctate region, and the
height of the cylinder arising from the circle represents the
response magnitude of the bulbar unit. It can be seen in this
figure that there is a regional variation in the responsiveness
to any given odorant. However, whether this regional variation
can serve as a basis for odorant discrimination depends upon
whether there are different regions of responsiveness for
different odorants. In this regard it would be difficult to
assert from the data presented that the regional responsiveness
for pinene differs from that for camphor. However, the regional
selectivity for both pinene and camphor does appear quite
different from that for amyl acetate. Whereas the sensitivity
to the former two odorants seems confined to a rather localized
region of the mucosa, the sensitivity to amyl acetate appears
more homogeneously distributed. Consequently, these data, at
least on the basis of three odorants, support the possibility
that there are different regions of selectivity across the mucosa
for different odorants. This might serve as a basis for olfactory
discrimination.

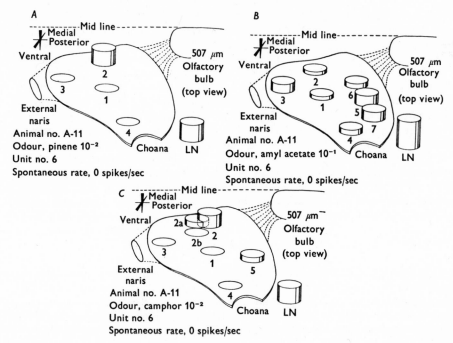

Fig. 2. The activity of one bulbar unit in response to punctate stimulation of the mucosa by three different odorants. (Kauer and Moulton, 1974, by permission of Cambridge University Press).

DIFFERENTIAL DISTRIBUTION OF ODORANT MOLECULES ACROSS THE MUCOSA

Adrian's earliest explanations for olfactory discrimination drew heavily upon the putative differential migration and resultant differential distribution of the molecules of different odorants across the mucosa. To test this concept, Mozell (1966), using frogs, compared the multiunit discharges recorded from two branches of the olfactory nerve which were previously shown (Mozell, 1964) to supply two widely separated mucosal regions on the dorsal wall of the olfactory sac. The more medial nerve branch (MB) supplied an area near the external naris and, therefore, its discharge reflected the activity of a mucosal region which would be among the earliest contacted by the incoming odorized air. The more lateral nerve branch (LB) supplied an area overhanging the internal naris. Its discharge would, therefore, reflect the activity of a mucosal region farther along the flow path near where the odorized air leaves the sac. Note that this technique allows the sampling of mucosal responses without surgically compromising the integrity of the olfactory sac. Therefore, unlike with the other techniques described above, the possi-

bility of observing differential molecular migrations along the
mucosa cannot be circumvented by the puffing of odorants directly
onto particular mucosal locations.

Mozell used electronic summators (Beidler, 1953) to quantify
the multiunit discharges recorded from the two nerve branches.
This yielded traces like those shown in Fig. 3, where the area
under each trace is proportional to the activity in the nerve
branch being monitored. The upper trace in each frame of this
figure shows the LB response to each odorant; the next lower
trace in each frame shows the MB response. The trace at the
bottom of each frame shows the onset of the artificially produced
sniff which draws the odorant into the intact olfactory sac via
the external naris. These artificially produced sniffs are controlled for volume, flow rate and concentration (Mozell, 1966).

Figure 3 (Mozell, 1966) shows but one of the 30 such arrays of
summated discharges which were recorded in response to the four
odorants listed. Each odorant was presented at several different
concentrations noted across the top of the figure in terms of
partial pressure (x 10^{-2} mm Hg). It can be seen that the summated
discharges of both nerve branches increase with concentration for
all four odorants. Of more importance, however, to the present
discussion is the apparent difference among the odorants in the
relative activity they produce at the two mucosal regions sampled.
To quantify this relative difference, Mozell calculated the ratio
of the summated discharge recorded from the lateral nerve branch to
that recorded from the medial nerve branch. This was referred to
as the LB/MB ratio and its magnitude indicated the gradient of
activity each odorant elicited across the mucosa from a region near
the external naris to one near the internal naris. As exemplified
by Fig. 3, citral and geraniol, with their smaller ratios, produced
more steeply declining gradients along the flow path than did
d-limonene and octane. In a later, more comprehensive study Mozell
(1970) confirmed that citral and geraniol produce median ratios considerably less than 1.0, again indicating greater activity near the
external naris. In contrast, for d-limonene and octane the median
ratios approached 1.0, indicating a more equal level of activity
along the mucosa. On the basis of these four chemicals it appeared
that different odorants might establish characteristic spatial
activity patterns across the mucosa. This possibility was further
supported by the LB/MB ratios of 12 additional odorants (Mozell,
1970) which, ranging from zero to slightly over unity, remained
characteristic for each odorant from animal to animal.

Mozell (1966) observed that different odorants not only
established different spatial patterns of activity across the
mucosa; they appeared to establish different temporal patterns as
well. He noted that the latency difference, i.e., the amount of
elapsed time between the onset of the MB response and the onset of

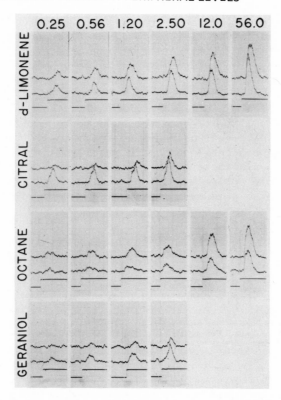

Fig. 3. Summated neural discharges recorded from the lateral
branch (upper trace) and medial branch (next lower trace) of the
olfactory nerve in response to several concentrations of four
different odorants. (Mozell, 1966, by permission of Rockefeller
University Press).

the subsequent LB response, differed for different odorants. This
occurred even though all odorants were drawn as artificially produced sniffs into the olfactory sac at the same flow rates, volumes,
and concentrations. Examples of this effect can be seen in Fig. 3
where, in general, the latency is shortest for octane and then
increases in the order of d-limonene, citral, and geraniol.

     Mozell (1966) further observed a parallel between the latency
differences of the odorants and their LB/MB ratios, viz., the
smaller an odorant's LB/MB ratio the longer its latency difference.
This relationship between these two measures gave an early indication that perhaps they were actually two views of the same underlying process. At any rate, as Adrian had predicted from his work
in the olfactory bulb, there did appear to be a spatiotemporal

differentiation of odorants across the olfactory mucosa. However, his suggestion that these spatiotemporal patterns would be based upon the differential migration of the molecules of different odorants across the mucosa was yet to be supported.

In this regard Mozell recognized that there were at least two mechanisms which could explain the various spatiotemporal patterns he observed. First, there might be a differential sorption of the molecules of different odorants by the mucosa. Thus, molecules of poorly sorbed odorants would more readily migrate across the mucosa than would those of more strongly sorbed odorants. They would therefore arrive at distant regions more rapidly and in greater numbers, producing the shorter latency differences and the larger LB/MB ratios.

The second possible mechanism, more pertinent perhaps to spatial patterns than to temporal ones, suggests that from one region to another there may be a disproportionate variation in the relative numbers of receptors selectively sensitive to each of the odorants. By this mechanism an odorant with its receptors concentrated around the external naris would generate a steeper activity gradient along the mucosa than would an odorant with its receptors more homogeneously distributed.

To choose between these two alternative mechanisms, Mozell (1964) reversed the flow direction of the odorants through the olfactory sac. Instead of having them enter the external naris and flow toward the internal naris, they were made to enter the internal naris and flow toward the external naris. This procedure resulted in reversal of the discharge magnitudes recorded from the two nerve branches. That is, the branch giving the larger response when the odorant entered the external naris became the one giving the smaller response when the odorant entered the internal naris. This reversal would seem to argue against the concept that the activity gradients are based upon regions of similarly selective receptors since regardless of the flow direction the mucosal regions of greater and lesser sensitivity would be expected to remain constant. On the other hand, this reversal would indeed be the expectation if differential sorption were the basis for the different activity gradients. The odorants most strongly sorbed by the mucosa would then be retained at either end, depending upon the naris they entered. Therefore, there was now evidence not only showing the existence of distinctive mucosal patterns for various odorants but also suggesting that these patterns might be based upon variations in the migration and distribution of the molecules of different odorants in accordance with their differential sorption by the mucosa.

However, the electrophysiological observations, upon which rests the above proposed mechanism for olfactory discrimination,

can really only give indirect evidence of molecular migration
patterns across the mucosa since these observations must be made
at a point in the olfactory process which may be several steps
beyond the molecular distribution itself. Indeed, in the chain of
events which must occur between the movement of the molecules over
the mucosa and the transmission of signals down the primary neurones, some event, other than molecular migration, may either
itself generate the LB/MB ratios or greatly alter those already
generated by the differential distribution of the molecules. It
was therefore necessary to obtain direct evidence from the behavior
of the molecules of different odorants themselves that they do
indeed migrate differentially along the mucosa and that these
differences can account for the previously reported LB/MB ratios.
To address this problem Mozell and Jagodowicz (1973, 1974) replaced
the column of a standard gas chromatograph with in vivo olfactory
sacs of anesthetized frogs. With the appropriate adaptations of
the gas chromatographic methodology to this "frog column" situation, Mozell and Jagodowicz were able to determine the relative
amount of time required (i.e., the relative retention time) for
the molecules of different odorants to cross the intact olfactory
sac. The results for 15 different odorants tested with 10
different frogs are shown in Fig. 4. A detailed explanation for
this figure can be found in previous papers (Mozell and Jagodowicz,
1973, 1974). The point to be made here is the wide range of relative retention times produced by these 15 different odorants. The
longest relative retention time (that for isovaleric acid) was over
220 times greater than the shortest relative retention time (that
for octane). (For a variety of experimental conditions including
a carrier gas flow rate of 25 cc/min, the originally measured
absolute retention times, from which the relative retention times
were calculated, were 1.2 and 274 seconds respectively.) The
retention times for the remaining odorants were rather well distributed within this range. Apparently, therefore, the molecules
of different odorants do indeed migrate at considerably different
rates along the mucosa.

     In addition, when the relative retention time of each of these
odorants was compared to its LB/MB ratio, a strong negative correlation was found (Mozell and Jagodowicz, 1973). That is, as the
relative retention time increased, the LB/MB ratio decreased. This
is exactly what would be expected if the LB/MB ratio was determined by the rate of migration of odorant molecules along the
mucosa. For example, in the time of a given sniff the molecules
of the longest retention time odorants would likely not migrate
far into the olfactory sac, most of them not getting much beyond
its entrance. This would produce small LB/MB ratios. On the
other hand, the molecules of shorter retention time odorants would
likely reach positions farther along the mucosa in greater numbers,
thus producing larger ratios.

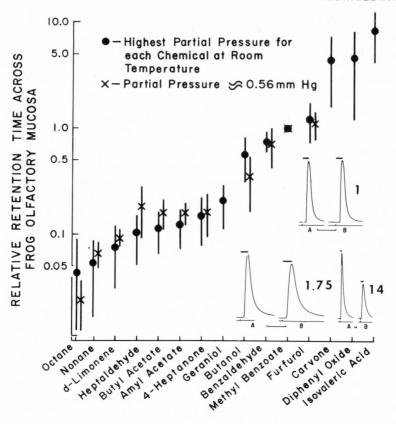

Fig. 4.   Mean relative retention times measured across the olfactory mucosa.   (Mozell and Jagodowicz, 1973, by permission of the
American Association for the Advancement of Science, Copywright
1973).

        Although, as seen above, strong inferences concerning the
probable distribution patterns of different odorants can be made
from electrophysiological and chromatographic data, conclusive
evidence for the existence of such patterns and a full appreciation of their exact topologies would require a technique which
provides a direct display of how the molecules themselves are
actually distributed.   This requirement appears to have recently
been met by using tritium-labelled odorants to quantitatively map
these putative differential patterns of molecular sorption along
the mucosa (Hornung, Lansing, and Mozell, 1975; Hornung and Mozell,
1977).

        For this work two odorants, butanol and octane, were chosen

for their markedly different LB/MB ratios. Within a rather wide
range of stimulus parameters butanol produced a median ratio of
zero, predicting that the mucosal density of its molecules,follow-
ing a sniff, would be greatest around the entrance to the olfactory
sac. On the other hand, octane's LB/MB ratio was near unity, predicting that its molecules would be more evenly distributed across
the mucosa.

To test these predictions Hornung and Mozell (1977) gave to
each frog a single, highly controlled sniff of either tritiated
butanol or tritiated octane. Immediately after stimulation each
frog was quick-frozen in liquid nitrogen to prevent further movement of the sorbed odorant molecules. With the frog still frozen
a fine jeweler's saw was used to cut the roof of the olfactory sac,
and its attached olfactory mucosa, into five sections. These sections (designated M1 through M5) ran in consecutive order from the
external naris (M1) caudally to the internal naris (M4) and beyond
(M5). Each was individually solubilized, and its radioactivity
measured with a liquid scintillation counter. The number of
odorant molecules in each section could then be estimated from
this recovered radioactivity. In order to take into account the
differences in the size of the sections, these estimates were con-
verted into number of molecules per unit surface area, i.e., the
surface area concentration.

Figure 5 summarizes the results of this approach. It shows
the relative surface area concentrations of both butanol and
octane for each of the five consecutive mucosal sections. It is
quite clear from this figure that butanol molecules drawn by an
artificially produced sniff into the olfactory sac are distributed
very unevenly along the mucosa. Indeed, the vast majority of
butanol molecules are found near the external naris in section M1,
and the concentration gradient across the mucosa (or for that
matter even to the next section, M2) is extremely steep. In sharp
contrast to butanol, the molecules of octane are quite evenly distributed across the mucosa, there being essentially a zero gradient.

These results are exactly what the electrophysiological data
had predicted for octane and butanol, and it now seems more probable than ever that the LB/MB ratios do indeed reflect mucosal
odorant distributions. Thus, the use of tritium-labelled odorants
gives a graphic impact to the earlier electrophysiological and
chromatographic studies which strongly indicated that the molecules
of different odorants become differentially distributed in their
sorption along the olfactory mucosa. Again, the existence of
another of Adrian's predicted mechanisms for olfactory discrimination seems to be supported by the current research.

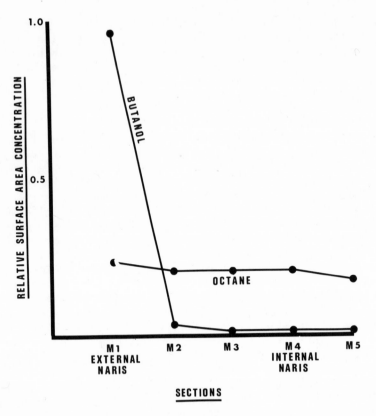

Fig. 5. The relative surface area concentrations for each mucosal section following an artificially produced sniff of either ^{3}H-butanol or ^{3}H-octane. Stimulus parameters: partial pressure, 6.78 mmHg; flow rate, 16 cc/min; volume, 0.42 cc. (Based upon data reported by Hornung and Mozell, 1977).

POSSIBLE INTERACTION OF THE VARIOUS MECHANISMS IN DISCRIMINATION

Indeed, all of Adrian's proposed mechanisms for the analysis of odorants at the level of the olfactory mucosa seem to have received support. First, selective sensitivity does appear to be present in the analyzing process in that each receptor cell responds to a particular group of odorants and, for any given cell, this particular group of odorants may be determined by the unique combination of selective receptor sites that it bears. Secondly, it appears that cells bearing similar receptor sites may indeed be clustered together imparting a regional selectivity to the mucosa.

Finally, it does appear that the molecules of different odorants are themselves separated by their differential sorption along the mucosa. If further research continues to support the existence of these various mechanisms, it will become necessary to conceptualize and investigate how they might all interact in the total process of odorant analysis. This interaction might be quite complex but out of this complexity might come the very permutability needed in the response of the peripheral olfactory system for it to analyze the vast number of chemicals (often very similar) which can apparently be discriminated.

For instance, Getchell and her coworkers demonstrated the possibility of selective receptor sites for ethyl butyrate and a few of its homologs. Although these sites might differentiate these ethyl butyrate related odorants from unrelated odorants, how might the related ones be discriminated from each other? Members of an homologous series, varying as they do in their configuration and physical properties, might very well be sorbed differentially across the mucosa. Therefore, some of the receptors tuned to a particular homologous series but positioned farther along the flow path might, for some members of the series, receive disproportionately fewer molecules than for other members. This then could establish different activity patterns across the mucosa upon which the discrimination could be based.

Likewise, the region in Fig. 2 selectively sensitive to pinene and that selectively sensitive to camphor may very well be the same region. However, if, as seems likely from some of their physicochemical properties, these two chemicals are differentially sorbed by the mucosa, the steepness of the activity gradients they each establish along this particular mucosal region may differ enough to provide a basis for discrimination. On the other hand, Fig. 3 shows that the LB/MB ratios produced by octane and d-limonene are very similar and therefore predict very similar sorption patterns across the mucosa. In this case, it may be the selectivity of receptors, perhaps regionally positioned, which will separate the odorants.

The above examples for the possible interplay of the several suggested mechanisms for olfactory discrimination are just a few that this line of thought can generate. They may be unfortunate examples in that they tend to dwell on the limiting case where the differentiation rests solely on one of the mechanisms because the others seem non-discriminatory. Although such limiting examples may make the point that several mechanisms might play a role in olfactory discrimination, they fail to show the possibly large number of cases where they may all play a role simultaneously. The differences between many chemicals are likely so subtle as to require the information available from all the mechanisms operating in concert.

From such considerations it seems futile to attempt, as has been done recently (Moulton, 1976), to argue for a particular mechanism as being the predominant one for olfactory discrimination. Perhaps, as exemplified above, the olfactory system might rely on one mechanism more than another for a particular analysis, but a different analysis might require a shift of its emphasis. In short, the olfactory system must make a vast number of very fine discriminations; it probably, without prejudice, takes advantage of all those mechanisms it has available which can separate one odorant from another.

ACKNOWLEDGMENTS

The work from the author's laboratory reported in this manuscript as well as the preparation of the manuscript itself has been supported by NIH Grant NS03904.

The author expresses his gratitude to those who read the manuscript: Drs. Robert J. O'Connell (The Rockefeller University), David E. Hornung (St. Lawrence University), and Leslie Hammack (Upstate Medical Center, SUNY). Gratitude must also be expressed to Mr. Stanley Swieck for his help in preparing the manuscript.

REFERENCES

Adrian, E.D. 1950. Sensory discrimination with some recent evidence from the olfactory organ. Brit. Med. Bull. 6:330-333.

Adrian, E.D. 1951. Olfactory discrimination. Année Psychol. 50:107-113.

Adrian, E.D. 1953. The mechanism of olfactory stimulation in the mammal. Advan. Sci. (London) 9:417-420.

Adrian, E.D. 1954. The basis of sensation - some recent studies of olfaction. Brit. Med. J. 1:287-290.

Beidler, L.M. 1953. Properties of chemoreceptors of the tongue of rat. J. Neurophysiol. 16:595-607.

Getchell, M.L. and R.C. Gesteland. 1972. The chemistry of olfactory reception: stimulus specific protection from sulfhydryl reagent inhibition. Proc. Nat. Acad. Sci. 69:1494-1498.

Getchell, T.V. and M.L. Getchell. 1974. Signal-detecting mechanisms in the olfactory epithelium: molecular discrimination. Ann. N. Y. Acad. Sci. 237:62-75.

Gesteland, R.C., J.Y. Lettvin and W.H. Pitts. 1965. Chemical transmission in the nose of the frog. J. Physiol. 181:525-559.

Gesteland, R.C., J.Y. Lettvin, W.H. Pitts and A. Rojas. 1963. Odor specificities of the frog's olfactory receptors. In Olfaction and Taste I. Y. Zotterman, Ed.: 19-34. Pergamon Press. New York, N.Y.

Hornung, D.E., R.D. Lansing and M.M. Mozell. 1975. Distribution of butanol molecules along bullfrog olfactory mucosa. Nature 254:617-618.

Hornung, D.E. and M.M. Mozell. 1977. Factors influencing the differential sorption of odorant molecules across the olfactory mucosa. J. Gen. Physiol., In press.

Kauer, J.S. and D.G. Moulton. 1974. Responses of olfactory bulb neurones to odor stimulation of small nasal areas in the salamander. J. Physiol. 243:717-737.

Mathews, D.F. 1972. Response patterns of single neurones in the tortoise olfactory epithelium and olfactory bulb. J. Gen. Physiol. 60:166-180.

Mathews, D.F. and D. Tucker. 1966. Single unit activity in the tortoise olfactory mucosa. Fed. Proc. 25:329.

Moulton, D.G. 1976. Spatial patterning of response to odors in the peripheral olfactory system. Physiol. Rev. 56:578-593.

Mozell, M.M. 1964. Evidence for sorption as a mechanism of the analysis of vapors. Nature 203:1181-1182.

Mozell, M.M. 1966. The spatiotemporal analysis of odorants at the level of the olfactory receptor sheet. J. Gen. Physiol. 50:25-41.

Mozell, M.M. 1970. Evidence for a chromatographic model of olfaction. J. Gen. Physiol. 56:46-63.

Mozell, M.M. and M. Jagodowicz. 1973. Chromatographic separation of odorants by the nose: retention times measured across in vivo olfactory mucosa. Science 181:1247-1249.

Mozell, M.M. and M. Jagodowicz. 1974. Mechanisms underlying the analysis of odorant quality at the level of the olfactory mucosa. Part I. Spatiotemporal sorption patterns. Ann. N. Y. Acad. Sci. 237:76-90.

O'Connell, R.J. and M.M. Mozell. 1969. Quantitative stimulation
of frog olfactory receptors. J. Neurophysiol. 32:51-63.

Ottoson, D. 1956. Analysis of electrical activity of the olfactory epithelium. Acta Physiol. Scand. 35(Suppl. 122):1-83.

Ottoson, D. 1971. The electro-olfactogram. In Handbook of Sensory Physiology, Olfaction. L.M. Beidler, Ed. Vol. 4, Part 1:
205-215. Springer-Verlag, New York.

Shibuya, T. and S. Shibuya. 1963. Olfactory epithelium and unitary responses in the tortoise. Science 140:495-496.

Takagi, S.F. and K. Omura. 1963. Responses of the olfactory
receptor cells to odours. Proc. Japan Acad. 39:253-255.

TASTE STIMULI AS POSSIBLE MESSENGERS

L. M. Beidler

Florida State University

Tallahassee, Florida

It is well known that insects may use highly specific pheromones to release stereotyped behavior leading to reproduction. In
this regard the insect has become the model for other animals, and
odors have become emphasized as the means of communication. Consequently, another chemosensory system, taste, is seldom considered
in the study of pheromone-released behavior by these other animals.
This neglect of taste is perhaps understandable when considering
distant communication in non-aquatic species since in such species
taste communication would most often require body contact. (In
aquatic animals several contingencies, including the arrangement of
the receptors and the water solubility of the stimuli, often convert
taste into a distance sensory system.) However, it is conceivable
that even for non-aquatic animals much information might still be
transferred during body contact using the taste system to receive
the chemical signals. Gross observations of the mating behavior of
many mammals would indicate that taste is indeed used as a cue, and
recent research of the secretion of the gerbil Harderian gland
suggests that tastes, as well as odors, may be involved in groom-
induced behavior (Thiessen, Clancy and Goodwin, 1976). If indeed
tastes can operate as conspecific messengers, several questions
become pertinent: How varied are the taste qualities? To what
kinds of chemicals and with what specificity is the gustatory system responsive? How sensitive is the system to these chemicals and
how does the gustatory system compare with the olfactory system in
its ability to act as an information channel?

QUALITIES OF TASTE SENSATION

It is often stated that there are but four primary taste qualities in humans: sweet, salty, sour, and bitter. If this is true, any complex taste could then be matched by the proper mixture of four substances, each representing one of the four primary taste qualities. This was demonstrated by von Skramlik in 1937. He proposed that the complex taste of any given salt could be matched by a mixture of the proper concentrations of quinine, sodium chloride, tartaric acid and fructose. He found, for example, that the following equation held for simulating the taste sensation produced by a 0.374 M concentration of NH_4Cl:

$$0.374 \text{ M } NH_4Cl = 0.00016 \text{ M quinine} \cdot HCl + 1.2 \text{ M NaCl}$$
$$+ 0.0039 \text{ M tartaric acid}$$

In this particular match one of the primary tastes, that represented by fructose, was not required.

In a similar manner von Skramlik matched the tastes of other inorganic salts with reasonable success. However, many substances with more complex tastes could not be so simply matched even though other taste-associated sensations such as pain, warmth or "after tastes" were considered.

This method of unscrambling components of a complex sensation is rather similar to that which occurred in the area of olfaction. The advent of gas chromatography and mass spectrometry made it possible to identify the chemical components in natural substances, and it was thought that the flavor of any given natural substance could then be matched by a mixture of its pure chemical components. This was a reasonable approach but was met with only partial success. It is even more difficult with the theory of four taste qualities to understand how the use of taste enhancers such as monosodium glutamate or certain nucleotides can enhance the flavor of meaty soups to which they are often added. Although the theory of four primary taste qualities is a useful but limited concept in the laboratory, our knowledge of primary taste qualities is still too insufficient to understand the taste sensations produced by many of the complex substances found in the real world.

KINDS OF TASTE STIMULI: QUALITATIVE AND QUANTITATIVE

The tongue of man can respond to a wide variety of chemicals. In 1914, Cohn published a book entitled Die Organischen Geschmacksstoffe which took 936 pages to describe the tastes of inorganic and organic molecules. It was customary in those days to taste any newly synthesized molecules. Since then the development of new molecules has accelerated and we can now estimate that tens of thousands of known molecules produce a taste sensation of one kind or another.

Both the kind of substance and its concentration are important parameters for a taste sensation. What magnitudes of concentrations are actually found in nature? Sodium chloride is often in foods at a concentration of 1.0 M or higher. The potassium salt in soya is about 0.43 M. The acids vary greatly: about 1.7 M acetic acid in cider vinegar, 0.4 M citric acid in lime juice, 0.19 M malic acid in gooseberries, and 0.05 M oxalic acid in rhubarb. The sugar concentrations also vary greatly: over 0.5 M in sugar cane, 0.4 M glucose in apples, 0.33 M fructose in grapes, and 0.14 M lactose in cattle milk. The above examples illustrate that some molecules, highly attractive to animals, may be present in nature at high concentrations. It is also interesting to note that salts at high concentrations are present in many body secretions. However, if communication is to be specific among animals one would expect other less ubiquitous types of molecules to be used. For instance, amino acids and fatty acids are abundant enough in certain animal secretions to enable high response specificity. Furthermore, many animals can respond to very low (10^{-10}M to 10^{-18}M) concentrations of these substances.

In recent years much information has been obtained which shows that some animals can respond to substances that are highly complex and of high molecular weight. The tick responds to ATP, ADP, DPNH, GSH, etc. (Galun & Kindler, 1968).

The discovery that one of the dipeptides, L-aspartyl-L-phenylalanine methyl ester, is very sweet to humans led to the investigation of well over fifty dipeptide structures, some being found sweet, others bitter, and still others without taste. As such substances are used more and more with subhuman species, an almost certain discovery will be that some of these complex substances can stimulate animals other than man. It is also interesting to note that amino acids stimulate the taste receptors of animals and that the proper mixture of several amino acids can become an even stronger stimulus. Furthermore, it is possible that an animal can also use the mixtures of fatty acids to detect an individual of a given species.

A most important advance in chemoreception was the discovery that certain proteins can interact with human taste buds. The acceptance of the concept that animals can detect as large a molecule as a protein did not come easily. In 1961, Dethier studied protein ingestion by blowflies and concluded that the blowfly could indeed detect proteins. He used not only natural stimuli but crystalline hemoglobin as well. In later years it was realized that other animals can detect protein. Kurihara and Beidler (1968) investigated the taste modifying properties of the berry from the

Miracle Fruit tree of Nigeria, and after lengthy chemical analysis, determined that the active ingredient was a glycoprotein with a molecular weight of about 44,000. This was the first protein known to interact with the taste receptors of man. Indeed, this protein is sweet at low pH. A number of years later, other investigators found that the sweet principles of several other fruits were also proteins. Two such sweet proteins are Monellin with a molecular weight of 11,500 and Thaumatin with a molecular weight of 21,000 (van der Wel, 1974). These discoveries revealed the enormous variations in the types of stimuli to which taste receptors can respond. In recent years a number of other proteins have been studied in other animal forms. For example, human serum albumin with a molecular weight of 60,000 can stimulate a marine snail at a threshold of about 10^{-9} molar concentration (Gurin & Carr, 1971). Since proteins are known to be excellent messengers in cell biology, it is quite possible that they could also be utilized as contact messengers among animals to initiate highly specific behaviors.

The ability to taste some of these specific proteins appears limited to man and closely related species. Such species limitations might be expected since proteins, being rather large molecules, probably involve stereospecificity in their reception, an involvement which might predict greater similarities among more closely related animal species. The taste physiologist is very familiar with such stereospecificity. Indeed, the first experiment with amino acids on biological material indicating such steric influences was performed by Piutti (1886) when he separated the D & L forms of asparagine and showed that the taste of the D form is different from that of the L.

COMPARISON OF GUSTATORY AND OLFACTORY SYSTEMS

The olfactory system can detect certain specific odors with extremely high sensitivity. For this reason it is often thought that the olfactory system is much more sensitive than the gustatory system. However, this is not always true since the judgment of which system is the more sensitive depends upon the particular molecule or substance chosen for comparison. For instance, the olfactory system of the catfish was found by Caprio (1976) to be more sensitive to certain amino acids and the gustatory to others. A better distinction between the two systems than their relative sensitivities is the method by which the molecules stimulating them are transported to their end organs, viz., direct contact between the stimulus source and the receptor sheet (gustation) or migration of the molecules over distance from the stimulus source to the receptor sheet (olfaction). As mentioned earlier, even this distinction breaks down if aquatic forms are considered.

Informational capacity can be crudely estimated by a consideration of the sensory nerves involved (Beidler, 1966). The rabbit, for example, has about 10^8 olfactory nerve fibers with a maximum firing rate of about 10/s. The same animal has about 2×10^4 -10^5 gustatory nerve fibers with a maximum firing rate of about 100/s. Thus, the olfactory system possesses a slightly greater information potential (10^9 bits/s) than does the gustatory system (10^7 bits/s).

It may be concluded that the taste system is quite sensitive to certain chemicals; that it can respond to a very wide range of different molecules ranging in molecular weight from unity of H^+ to 44,000; that animals secrete mixtures of molecules that characterize the individual; and that the molecules could be deposited in specific areas of the animal allowing another individual to "zero in" on a given part of the anatomy. One can conclude that the taste system could indeed be a good means for both distance and contact communication in aquatic animals and could complement the olfactory system with a contact dimension for communication in terrestrial animals. The frequency at which gustation is used for communication is still unknown.

REFERENCES

Beidler, L.M. 1966. Advances in Chemistry Series 56:1-28.

Cohn, G. 1914. Die Organischen Geschmacksstoffe. Franz Siemenroth SW 11, Hafenplatz 9.

Dethier, V.G. 1961. Biol. Bull. 121:456-470.

Galun, R., and S.H. Kindler. 1968. J. Insect Physiol. 14:1409-1421.

Gurin, S., and W.E. Carr. 1971. Science 174:293-295.

Kurihara, K., and L.M. Beidler. 1968. Science 161:1241-1243.

Piutti, A. 1886. Compt. Rend. Acad. Sc. 103:134.

Caprio, J.T. 1976. Doctoral Thesis: Olfactory and Gustatory Responses of Catfish to Amino Acids and Derivatives. Florida State University.

Von Skramlik, E. 1937. Handbuch der Biologischen Arbeitsmethoden 5:1727-1774.

Thiessen, D.D., A. Clancy, and M. Goodwin. 1976. J. Chem. Ecol. 2:231-238.

van der Wel, H. 1974. <u>Miracle Fruit, Katemfe, and Serendipity</u>
    <u>Berry</u>. In Symposium: Sweeteners, p. 194 (Ed. G.E. Inglett)
    Avi Publishing Co., Inc., Westport, Conn.

CENTRAL PROCESSING OF OLFACTORY SIGNALS

Gordon M. Shepherd

Yale University School of Medicine
333 Cedar Street
New Haven, Connecticut   06510

Olfactory information is transmitted from the olfactory receptors in the nose through several stages of processing in the
olfactory bulb and the olfactory regions of the brain.   Through
recent and ongoing work, we are learning a good deal about the
basic anatomy and physiology of these pathways.   I shall confine
myself to reviewing here some of the areas in which progress has
been most rapid and most promising for understanding mechanisms of
odor processing.

SYNAPTIC ORGANIZATION

The term synaptic organization has come to signify the interrelated data obtained from several disciplines - anatomy, physiology, pharmacology, biophysics - that together define the functional circuits of a given region of the nervous system.   The most
extensive analyses have been carried out on the regions that are
most accessible and have the most stereotyped structure.   In these
respects the olfactory bulb, together with the cerebellum and
retina, has been particularly attractive.   The synaptic organization of the olfactory bulb has been reviewed on several occasions
(Shepherd, 1972, 1974; Shepherd, Getchell & Kauer, 1975).   Here we
may note that because of the distinct laminar structure of the
olfactory bulb, two main stages of processing of sensory information
can be identified.   The first is in the glomerular layer, where
the input from the olfactory receptor axons to the dendrites of
mitral cells is immediately subjected to an initial processing via
synaptic circuits through the dendrites of periglomerular short-
axon cells (Pinching & Powell, 1971; White, 1973).   The second
stage is in the subjacent external plexiform layer, where the main

type of connection involves reciprocal synapses between the mitral
and granule cell dendrites (Hirata, 1964; Andres, 1965; Rall,
Shepherd, Reese & Brightman, 1966). Physiological studies suggest
that these circuits provide for inhibitory processing of the input
to the mitral (and tufted) cells (Getchell & Shepherd, 1975), and
also inhibitory control of the output from the mitral (and tufted)
cells (Phillips, Powell & Shepherd, 1963; Rall & Shepherd, 1968;
Nicoll, 1969) to the central olfactory regions.

If we turn briefly to the olfactory cortex, the arrangement
there, with the mitral and tufted cell axons entering from the
surface, is also advantageous to the experimenter. Anatomical
studies have shown that the olfactory bulb axons terminate on the
apical dendrites of pyramidal cells (Westrum, 1969; Price, 1973).
Physiological studies have shown that the pyramidal cells are
first excited by these axons, and are then subjected to inhibition
by interneuronal circuits within the cortex (Freeman, 1964;
Beidenbach & Stevens, 1969; Haberly & Shepherd, 1973). In addition,
both anatomical (Price, 1973) and physiological (Haberly & Shepherd,
1973) studies provide evidence for self-re-excitation of the
pyramidal cells through their long axon collaterals which spread
widely throughout the olfactory cortex. These excitatory and
inhibitory circuits are implicated in the generation of rhythmic
activity, a property which takes on added significance in light of
the propensity of olfactory cortical regions to undergo seizure
activity (cf. Eleftheriou, 1972).

TEMPORAL PATTERNS IN OLFACTORY PROCESSING

Analysis of signal processing mechanisms in other sensory
systems has required correlation of responses with the known time
course of step stimuli; analyses of the visual system (Kuffler,
1953; Hubel & Wiesel, 1962) and of stretch receptors (Ottoson &
Shepherd, 1971) may be cited as examples. We have recently begun
a similar study of the olfactory system. For this purpose, odors
are delivered to the olfactory mucosa through a system of concentric
pipettes which permits a rapid onset, steady plateau, and rapid
termination of stimulation. The odor is mixed with a carrier gas
containing a small amount of CO_2, so that the time course of the
step pulse can be monitored by a CO_2 analyzer. The CO_2 has little
effect by itself on unit activity in the olfactory bulb. Using
this system, we have confirmed Kauer's (1974) initial report of
response types of units in the salamander olfactory bulb. Some
responses to odor pulses take the form of prolonged excitatory
discharges, but these are found only near threshold; at higher
concentrations, the initial brief discharge is terminated by
powerful suppression (Kauer, 1974; Kauer & Shepherd, 1975b). Some
responses to a given odor consist of suppression throughout most

or all of a pulse (Kauer, 1974; Kauer & Shepherd, 1975a). By
varying the odor concentration and pulse duration, these and other
properties of a response can be correlated precisely with the
onset, plateau, and off phases of the stimulus.

The results thus far indicate that inhibitory processes may
be very powerful in shaping the response characteristics of cells
to different odors at the bulbar level. It is tempting to relate
these processes to the inhibitory circuits in the bulb that are
mentioned above. However, one must first determine the response
characteristics of the olfactory receptors which transmit the
information to the bulb. It is known that the response patterns
of olfactory receptors are complex (Duchamp, Revial, Holley &
MacLeod, 1974). Detailed comparisons between receptor and bulbar
responses will hopefully permit us to determine the specific
contributions made by olfactory bulb circuits to temporal patterns
of neuronal activity at the bulbar level.

SPATIAL PATTERNS IN OLFACTORY PROCESSING

Spatial factors have an obvious importance in other sensory
systems (e.g. vision), but their role in the olfactory system is
more subtle. Some would deny their role altogether. However,
anatomical evidence for a degree of topographical order in the
projection from olfactory receptors to olfactory glomeruli is
compelling (Land, Eager & Shepherd, 1971; Land, 1973). Other
workers have obtained evidence for spatial gradients of activity
in the olfactory epithelium and in the olfactory bulb. These
subjects have been reviewed by Mozell (1971) and Moulton (1976).

One promising approach to the analysis of spatial factors is
to spatially restrict the stimulus. Using punctate odor stimulation of the olfactory mucosa in the salamander, Kauer & Moulton
(1974) showed that single units in the olfactory bulb tend to be
excited by stimulation of relatively limited areas of the mucosa.
In contrast, suppression can be elicited from relatively wide
areas. This could reflect the widespread inhibitory circuits
through granule cells onto mitral cells described above. Further
studies are needed to assess the differences between excitatory
and inhibitory receptive fields, and establish their importance
in mammals.

In addition to analysis of spatial factors by traditional
methods of anatomy and physiology, other methods have recently
been brought to bear. Doving and Pinching (1973) have shown that
exposure of rats to continuous strong odor stimulation for periods
of a week or more leads to degeneration of mitral cells in restricted regions of the olfactory bulb, and these regions vary

with different odors. The mechanism of this 'selective degeneration' is not yet clear.

Very recently a new approach to the question of spatial
organization has been introduced by Sharp, Kauer & Shepherd
(1975), using the Sokoloff method of 2-deoxyglucose (2DG) analysis
(Sokoloff, 1975). The substance 2DG is taken up by nerve cells
and phosphorylated like glucose, but is not further metabolized.
In tracer amounts, combined with 14C labelling and autoradiography,
it can be used to map sites of activity-related glucose uptake.
Our initial results show that, in rats exposed to strong odor
stimulation for 45 minutes, there are restricted regions of activity in the olfactory bulb. In animals breathing room air, there
are occasional small dense foci of activity, located in the glomerular layer. These foci were entirely unexpected; there is no
evidence for their existence from studies using traditional methods.
It seems possible that these foci represent activity induced by
odor at background levels in the ambient air; if so, they may be
one of the most sensitive indicators of olfactory activity yet
recognized.

CENTRAL OLFACTORY PATHWAYS

The projections of centrifugal fibers to the bulb from anterior
olfactory nucleus, anterior commissure, and diagonal band were
shown by Price (1973) to terminate with specific synaptic connections onto the granule cell dendritic tree. I have already noted
the evidence that granule cells are inhibitory to mitral cells in
the bulb. It therefore appears that centrifugal inhibition of the
olfactory bulb, first demonstrated by Kerr & Hagbarth (1955), is
mediated not by inhibitory centrifugal fibers, but by centrifugal
fibers which are excitatory to the granule cells, which are then
inhibitory to the mitral cells. The key role of the granule cell
for integration within the olfactory bulb and for inhibitory
control of the mitral cell (Rall & Shepherd, 1968) is thus re-
emphasized. The contributions of noradrenergic fibers, arising in
the brain stem, to the centrifugal control of the bulb should also
be noted (Fuxe, 1965). Finally, it should be noted that some
centrifugal fibers reach the glomerular layer. There they can
exert control over periglomerular cells, but they are excluded
from the glomerular interiors wherein the initial processing of
olfactory receptor input takes place, as described above.

With regard to central olfactory connections, the study of
Powell, Cowan & Raisman (1965) defined clearly the main olfactory
tracts and regions. We now recognize five main subdivisions of
olfactory cortex: anterior olfactory nucleus, prepyriform cortex,
olfactory tubercle, corticomedial amygdala, and transitional

entorhinal cortex (Price, 1973). A great deal of effort has gone into identifying more precisely different aspects of the projections. An important step was the demonstration by Winans & Scalia (1970) that the accessory olfactory bulb has an input to the amygdala which is separate from that of the main olfactory bulb. This separation carries through to the projections from the amygdala to the hypothalamus and surrounding regions (Raisman, 1972).

The connections of the olfactory cortical regions with central limbic structures have come under increasing study, and the recent advent of new methods, employing transport of radioactive amino acids (Price, 1973) and horseradish peroxidase (Broadwell, 1975), promises to greatly enlarge our knowledge in this area. The prepyriform cortex projects to the mediodorsal thalamus (which has strong interconnections with the forebrain) and to posterior hypothalamus (Scott & Leonard, 1971). The olfactory tubercle has overlapping projection sites, though to differing extents (Heimer & de Olmos, 1975; Scott & Chafin, 1975). In addition, the olfactory tubercle receives dopamine-containing fibers from the brain stem (Ungerstedt, 1971). By virtue of this input, and a close association with the basal ganglia (Heimer & Wilson, 1975), the tubercle is in the mainstream of research on dopaminergic systems (Krieger, Kauer, Shepherd & Greengard, 1977) and their possible relation to behavioral and psychotic disorders in humans (Stevens, 1973). The amygdala has strong connections with the hypothalamus, and has been recognized for many years to be implicated in a variety of behaviors, particularly related to aggression and reproduction (Eleftheriou, 1972). Finally it may be noted that there are several studies showing reception of olfactory input in the hippocampus, and integration there with other sensory modalities (MacLean, 1972).

CONCLUSION

From this brief review it can be seen that knowledge of structure and function in the central olfactory pathways has become quite extensive. The new methods in neuroanatomy are giving us much more direct information about pathways and connections than we have had in the past. Similarly, physiological methods are providing us with increasingly precise data on the excitatory and inhibitory interactions that must be the basis for sensory processing in these pathways. It seems not unrealistic to anticipate that we are approaching the time when we will understand in main outline the mechanisms of olfactory processing and will be able to pinpoint some of the specific roles of olfactory signals in limbic system functions and behavior.

ACKNOWLEDGEMENTS

Our research has been supported in part by U.S. Public Health Service grant NS 07609.

REFERENCES

Andres, K.H. 1965. Der Feinbau des Bulbus Olfactorius der Ratte unter besonderer Berücksichtigung der Synaptischen Verbindungen. Z. Zellforsch. 65: 530–561.

Beidenbach, M.A. and C.F. Stevens. 1975. Synaptic organization of cat olfactory cortex as revealed by intracellular recording. J. Neurophysiol. 32: 204–214.

Broadwell, R.D. 1975. Olfactory relationships of the telencephalon and diencephalon in the rabbit. II. An autoradiographic and horseradish peroxidase study of the efferent connections of the anterior olfactory nucleus. J. Comp. Neurol. 164: 389–410.

Doving, K.B., and A.J. Pinching. 1973. Selective degeneration of neurones in the olfactory bulb following prolonged odour exposure. Brain Res. 52: 115–129.

Duchamp, A., M.F. Revial, A. Holley and P. MacLeod. 1974. Odor discrimination by frog olfactory receptors. Chem. Senses Flavor 1: 213–233.

Eleftheriou, B.F. (ed.) 1972. The Neurobiology of the Amygdala. New York: Plenum.

Freeman, W.J. 1964. A linear distributed feedback model for prepyriform cortex. Exptl. Neurol. 10: 525–547.

Fuxe, K. 1965. The distribution of monoamine terminals in the central nervous system. Acta physiol. scand. 64: Suppl. 247, 37–85.

Getchell, T.V. and G.M. Shepherd. 1975. Short–axon cells in the olfactory bulb: dendrodendritic synaptic interactions. J. Physiol. (London) 251: 523–548.

Haberly, L.B. and G.M. Shepherd. 1973. Current density analysis of summed evoked potentials in opossum prepyriform cortex. J. Neurophysiol. 36: 789–802.

Heimer, L. and J. de Olmos. 1975. The olfactory tubercle projections in the rat. Neurosci. Abst. p.680.

Heimer, L. and R.D. Wilson. 1975. The subcortical projections of the allocortex: similarities in the neural associations of the hippocampus, the piriform cortex, and the neocortex. in Golgi Centennial Symposium (ed. M. Santini). New York: Raven. pp. 177-193.

Hirata, Y. 1964. Some observations on the fine structure of the synapses in the olfactory bulb of the mouse, with particular reference to the atypical synaptic configuration. Arch. Histol. Japan 24: 293-302.

Hubel, D.H. and T. Wiesel. 1962. Receptive fields, binocular interaction and functional architecture in the cat's visual cortex. J. Physiol. (London) 160: 106-154.

Kauer, J.S. 1974. Response patterns of amphibian olfactory bulb neurones to odour stimulation. J. Physiol. (London) 243: 695-715.

Kauer, J.S. and D.G. Moulton. 1974. Responses of olfactory bulb neurones to odor stimulation of small nasal areas in the salamander. J. Physiol. (London) 234: 717-737.

Kauer, J.S. and G.M. Shepherd. 1975a. Olfactory stimulation with controlled and monitored step pulses of odor. Brain Res. 85: 108-113.

Kauer, J.S. and G.M. Shepherd. 1975b. Concentration-specific responses of salamander olfactory bulb units. J. Physiol. (London) 252: 49-50P.

Kerr, D.I.B. and K.-E. Hagbarth. 1955. An investigation of olfactory centrifugal fiber system. J. Neurophysiol. 18: 362-374.

Krieger, N.R., J.S. Kauer, G.M. Shepherd and P. Greengard. Dopamine-sensitive adenylate cyclase within laminae of the olfactory tubercle. Brain Res. (in press).

Kuffler, S.W. 1953. Discharge patterns and functional organization of mammalian retina. J. Neurophysiol. 16: 37.

Land, L.J. 1973. Localized projection of olfactory nerves to rabbit olfactory bulb. Brain Res. 63: 153-166.

Land, L.J., R.P. Eager, and G.M. Shepherd. 1971. Olfactory nerve projections to the olfactory bulb in rabbit: demonstration by means of a simple ammoniacal silver degeneration method. Brain Res. 23: 250-254.

MacLean, P.D. 1972. Implications of microelectrode findings on exteroceptive inputs to the limbic cortex. in Limbic System Mechanisms and Autonomic Function (ed. C.H. Hockman). Springfield: Thomas. pp. 115-136.

Moulton, D.G. 1976. Spatial patterning of responses to odors in the peripheral olfactory system. Physiol. Rev. 56: 578-593.

Mozell, M.M. 1971. Spatial and temporal patterning. in Handbook of Physiology. vol. IV. Chemical Senses 1: Olfaction (ed. by L.M. Beidler). Berlin: Springer. pp. 205-215.

Nicoll, R.A. 1969. Inhibitory mechanisms in the rabbit olfactory bulb: dendrodendritic mechanisms. Brain Res. 14: 157-172.

Ottoson, D. and G.M. Shepherd. 1971. Transducer properties and integrative mechanisms in the frog's muscle spindle. in Handbook of Sensory Physiology. vol. I. Principles of Receptor Physiology (ed. W.R. Loewenstein). New York. Springer: pp. 442-499.

Phillips, C.G., T.P.S. Powell, and G.M. Shepherd. 1963. Responses of mitral cells to stimulation of the lateral olfactory tract in the rabbit. J. Physiol. (London) 168: 65-88.

Pinching, A.J. and T.P.S. Powell. 1971. The neuropil of the glomeruli of the olfactory bulb. J. Cell Sci. 9: 347-377.

Powell, T.P.S., W.M. Cowan and G. Raisman. 1965. The central olfactory connexions. J. Anat. 99: 791-813.

Price, J.L. 1973. An autoradiographic study of complementary laminar patterns of termination of afferent fibers to the olfactory cortex. J. Comp. Neurol. 150: 87-108.

Price, J.L. and T.P.S. Powell. 1970. The synaptology of the granule cells of the olfactory bulb. J. Cell Sci. 7: 125-155.

Raisman, G. 1972. An experimental study of the projection of the amygdala to the accessory olfactory bulb and its relationship to the concept of a dual olfactory system. Exp. Brain Res. 14: 395-408.

Rall, W. and G.M. Shepherd. 1968. Theoretical reconstruction of field potentials and dendrodendritic synaptic interactions in olfactory bulb. J. Neurophysiol. 31: 884-915.

Rall, W., G.M. Shepherd, T.S. Reese and M.W. Brightman. 1965. Dendro-dendritic synaptic pathway for inhibition in the olfactory bulb. Exptl. Neurol. 14: 44-56.

Scott, J.W. and B.R. Chafin. 1975. Origin of olfactory projections to lateral hypothalamus and nuclei gemini of the rat. Brain Res. 88: 64-68.

Scott, J.W. and C.M. Leonard. 1971. The olfactory connections of the lateral hypothalamus in the rat, mouse and hamster. J. Comp. Neurol. 141: 331-344.

Sharp, F.R., J.S. Kauer and G.M. Shepherd. 1971. Local sites of activity-related glucose metabolism in rat olfactory bulb during olfactory stimulation. Brain Res. 98: 596-600.

Shepherd, G.M. 1972. Synaptic organization of the mammalian olfactory bulb. Physiol. Rev. 52: 864-917.

Shepherd, G.M. 1974. The Synaptic Organization of the Brain. New York: Oxford.

Shepherd, G.M., T.V. Getchell and J.S. Kauer. 1975. Analysis of structure-function relations in the olfactory pathway. in The Nervous System: The Basic Neurosciences (ed. R. Brady). New York: Raven. pp. 207-220.

Sokoloff, L. 1975. Influence of functional activity on local cerebral glucose utilization. in Brain Work: the Coupling of Function, Metabolism and Blood Flow in the Brain. (ed. by D.H. Ingvar and N.A. Lassen). Copenhagen: Munksgaard. pp. 385-388.

Stevens, J.R. 1973. An anatomy of schizophrenia? Arch. Gen. Psychiatr. 29: 177-189.

Ungerstedt, U. 1971. Stereotoxic mapping of the monoamine pathways in the rat brain. Acta physiol. Scand. Suppl. 367: 1-48.

Westrum, L.E. 1969. Electron microscopy of degeneration in the lateral olfactory tract and plexiform layer of the prepyriform cortex of the rat. Z. Zellforsch. 98: 157-187.

White, L.E. 1973. Synaptic organization of the mammalian olfactory glomerulus: new findings including an intraspecific variation. Brain Res. 60: 299-313.

Winans, S.S. and F. Scalia. 1970. Amygdaloid nucleus: new afferent input from the vomeronasal organ. Science 170: 330-332.

DYNAMIC ASPECTS OF CENTRAL OLFACTORY PROCESSING

Foteos Macrides

Worcester Foundation for Experimental Biology

Shrewsbury, Massachusetts   01545

I. RESPONSE MEASUREMENT IN THE VERTEBRATE OLFACTORY SYSTEM

Perhaps because the sense of smell involves the discrimination of molecules by neural tissue, most neurophysiological studies of vertebrate olfaction have used purified compounds as stimuli and have been directed at relating neural response parameters to the structural and/or physicochemical properties of molecules. In studies of odor responses by individual neurons (single units), a commonly employed paradigm has been one in which individual members from a preselected battery of "odors" (volatile molecules in otherwise highly purified air) are puffed sequentially onto the nasal mucosa and an attempt is made to derive the relevant structural and/or physicochemical parameters of odor from the relative response spectra of the units recorded. When employed in studies of peripheral olfactory neurons (cf. Gesteland, Lettvin and Pitts, 1965), this paradigm is consistent with the assumption that these neurons contain receptor sites for odoriferous molecules and in effect is aimed at a preliminary answer to the two interrelated questions of how specific are the receptor sites and do the peripheral neurons have more than one type of receptor site. This general paradigm also has been used in studies of central olfactory neurons to test hypotheses about the physical dimensions of odors. For example, Higashino, Takeuchi and Amoore (1969) attempted to verify a stereochemical theory of odor quality by using a battery of pure chemical stimuli and comparing the response spectra of single units recorded in the olfactory bulbs with generalization gradients derived from psychophysical studies.

An alternative approach which has been employed in studies of central olfactory processing in mammals (cf. Macrides and Chorover,

1972; Pfaff and Gregory, 1971; Pfaff and Pfaffmann, 1969a, 1969b; Scott and Pfaff, 1970) involves the use of stimuli which may not be defined specifically in terms of their chemical composition, but rather at the preliminary stages of research are defined in terms of their functional properties. That is, the odors are operationally defined with respect to their sources and their presumed roles in the regulation of the recording subject's normal behavior and/or endocrine function (e.g., odors of animal origin which are used in social communication and/or which influence hormone secretion). The rationale behind this approach is that the peripheral transduction and coding of odors may be directly related to the structural and/c physicochemical properties of molecules but subsequent synaptic process essing of inputs from peripheral neurons likely also involves the extraction of different classes of information or higher order features relevant for a behavioral or endocrine response by the subject. The chemical composition of operationally defined stimuli may be species-specific (e.g., the sex attractants of mice versus hamsters may differ), but the central mechanisms by which their information is extracted likely are similar across species. Thus a detailed examination of how these stimuli are processed in the central olfactory pathways of any given species may provide general insights into the functional organization of the mammalian olfactor system. An implicit expectation in this approach is that at some point(s) in the central olfactory pathways neurons will be found whose responses bear lawful relationships to the functional properties of the stimuli. Therefore, though different in its orientatio this approach is similar to that which relies exclusively on batter ies of purified chemicals as stimuli in that the experimenter hopes to "make sense" out of a matrix of neuronal response spectra. The validity of any inference drawn from such studies heavily depends o the validity of the response measures used in generating the matrix

Single unit studies of vertebrate olfaction most commonly have relied on changes in average firing rate as a measure of neuronal responsiveness to odorants. When applied to the responses of peripheral olfactory neurons, this measure is consistent with the hypo esis that the transduction process involves an interaction of odori erous molecules with receptor sites of the neuronal membrane, leadi to a change in the membrane potential which if sufficiently strong will alter the neuronal discharge frequency. Since there is no kno basis for synaptic interactions among peripheral olfactory neurons, a change in firing rate after application of an odorant to the nasa mucosa may be accepted as evidence that a recorded neuron is capabl of transducing that odorant into an electrical signal. However, th relative magnitude of change in firing rate may be expected to depe not only on a peripheral neuron's sensitivity to the odorant, but also on the ability of the odorant to penetrate the nasal mucus and exert its influence on the neuronal membrane. Normal olfactory sam pling in terrestrial vertebrates involves the drawing of air across the nasal cavity, and such sampling has been shown in some mammals

to involve stereotyped sniffing patterns associated with an atten-
tive state (cf. Komisaruk, 1970; Macrides, 1975; Welker, 1964).
Mozell and Jagodowicz (1973) have demonstrated that air-borne molecules can have markedly different retention times and consequent
migration rates in the olfactory mucosa. Thus, during normal olfactory sampling, the latency and possibly the duration of change in a
peripheral olfactory neuron's firing rate may be expected to depend
not only of factors intrinsic to the transduction process, but also
on physical interactions of odorants with the nasal mucosa, on the
relative location of the neuron in the olfactory epithelium, and on
the subject's sniffing pattern (cf. Mozell, 1966). A matrix of
response spectra for peripheral olfactory neurons, if based on
changes in average firing rate measured during arbitrarily chosen
periods after stimulus application, would fail to reflect any systematic variations in the temporal and spatial distributions of
neuronal activity in the olfactory epithelium as might normally
accompany olfactory sampling. Such a matrix might permit inferences
about molecular parameters relevant to the transduction process
(e.g., stereochemical factors associated with stimulus-receptor
site interactions), but would obscure information which might be
present (presumably "coded") in the spatiotemporal patterns of
peripheral neuron activity and which might be extracted by synaptic
mechanisms in the central nervous system.

The problems associated with response measurement, and the
need for tailoring measurements to the specific question being investigated, become more acute in the central nervous system. At
this level we must concern ourselves not only with the issue of
how the olfactory environment is represented in the activities of
neurons, but more extensively with the question of how these activities ultimately permit appropriate behavioral and/or endocrine
responses to the environment by the intact organism. That is, even
more than in the periphery we must treat the recorded neurons as
participants in neural circuits and attempt to interpret neuronal
responses with respect to their possible relevance for mechanisms
of synaptic integration. In the following section we will examine
some examples of responses to odorants by single units recorded in
the olfactory bulbs of rodents under conditions intended to simulate
aspects of normal olfactory sampling. The studies from which these
examples are taken (cf. Macrides 1970, 1971; Macrides and Chorover,
1972) employed both pure chemical and operationally defined odorants
as stimuli, and were aimed at delineating response parameters that
could be correlated with the chemical and/or functional attributes
of the stimuli. These studies have revealed inhalation-related
aspects of odor sensitivity and selectivity which are absent or
obscured in conventional frequency measures. A general review of
this and related work appears elsewhere (Macrides, 1976). In the
present discussion we will focus on the question of how the
inhalation-related response properties of bulb units may be important for processes of synaptic integration. In section III we will

attempt to relate these response properties to the synaptic organization of the olfactory bulb. We will also consider some of the
implications of the foregoing discussion regarding the nature of
olfactory stimuli and their discrimination in the central nervous
system.

II. INHALATION-RELATED RESPONSE PROPERTIES OF THE OLFACTORY BULB

Our recordings were conducted in hamsters and deermice which
were surgically prepared with a double tracheotomy procedure. One
cannula was passed caudally through a cut in the trachea to provide
for pulmonary ventilation. A second cannula was passed rostrally
to the back of the nasal cavity and artificial sniffing was produced
with negative pressure pulses under experimental control. Subjects
had a nosecone loosely fitted over the snout. Purified air or air
containing odorants flowed continuously through the nosecone and
with each inhalation (nasal sniff) this air was drawn into the nasal
cavity (cf. Macrides and Chorover, 1972 for other methodological
details).

Figure 1 compares the responses to the same odorant (amyl
acetate) for two different olfactory bulb units. The period when
the odorant was present in the inhaled air (30 seconds) is indicated
by the step in the line beneath each histogram, and by the oblique
bar to the left of each response surface. The tracing beneath the
response surfaces records the nasal air flow. The duration of each
inhalation cycle was one second. In this and in subsequent figures,
photographs are included which show many superimposed discharges of
the units upon which the analyses were performed. The height of
successive vertical bars in frequency response histograms (FRH)
represent the total number of discharges which occurred during
successive inhalation cycles, so that variations in bursting pattern
within the inhalation cycles are discounted and only "overall" or
"average" firing levels are represented. The variations in the
inhalation-related bursting pattern during successive cycles are
illustrated with the response surfaces (RS). In these photographic
displays, sweeps of an oscilloscope beam are synchronized with the
inhalation cycle and tracings for successive inhalation cycles are
incremented up and to the right. Each tracing represents the voltage across a capacitor which receives a constant quantity of charge
with each discharge of the unit, so that the voltage builds up and
decays according to the momentary firing rates of the unit (Macrides
1972). The tracings are intended to reflect the influence which the
recorded unit would exert on a postsynaptic neuron if a quantity of
neurotransmitter were released with each discharge. If the transmitter were excitatory, the moment-to-moment variations in height
within each tracing reflect the extent to which the unit would exert
a depolarizing influence (drive the postsynaptic neuron toward
threshold). Conversely, if the transmitter were hyperpolarizing,

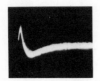
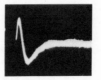

RESPONSE SURFACES

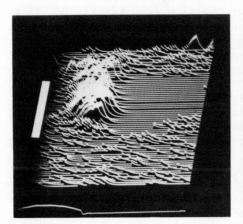
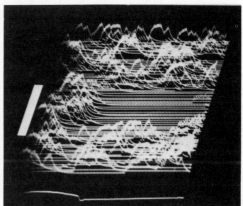

FREQUENCY RESPONSE HISTOGRAMS

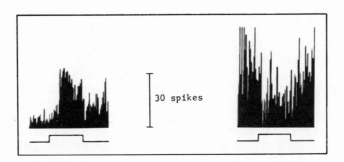

30 spikes

Fig. 1.   Responses of two bulb units to amyl acetate.

the variations in height reflect the extent to which the unit would
be preventing the postsynaptic neuron from reaching threshold.

During background conditions, when odorants were not intentionally introduced into the inhaled air, the unit illustrated on the
left in Fig. 1 had a moderate rate of firing whereas the unit on
the right had a relatively high rate. Inhalation of amyl acetate
produced a substantial increase in overall or average firing rate for
the unit on the left, and a decrease for the unit on the right (cf.
FRH's). However, both units fired predominantly during the early
portion of each inhalation cycle and their activity was clearly
suppressed during the middle and late portion (cf. RS's). To illustrate the conceptual difficulties associated with the use of changes
in average or overall firing rate as a response measure for olfactory
bulb units, let us imagine that these two units converged onto the
same postsynaptic neuron. Since the changes in average firing rate
were of opposite direction for the two units, we might expect the
changes in postsynaptic influence to cancel. However, since the
unit on the right maintained its background level of activity during
portions of each inhalation cycle when the unit on the left also
exhibited most of its discharges, there might well be an opportunity
for postsynaptic influences from both units to sum and each period
of summation would be followed by periods when neither unit could
exert a postsynaptic influence. An attempt to systematize these
responses on the basis of firing rates measured during arbitrary and
relatively long periods of time could yield a misleading impression
of their possible significance for postsynaptic neurons.

Figure 2 illustrates the responses of a third olfactory bulb
unit to amyl acetate. Upon introduction of this odorant into the
inhaled air, activity during the early portion of the inhalation
cycle diminished and in contrast to the units in Fig. 1 this unit
began to discharge predominantly during the late portion (cf. RS's).
During successive inhalations both the repetition rate for spikes
within the burst and the duration of the burst increased, so that
discharges "spilled over" into the early portion of the inhalation
cycles. The strength of the burst then subsided, but activity
continued to occur predominantly during the late portion of each
inhalation cycle. The left RS analyses activity during inhalation
of amyl acetate at a dilution of 10^{-5} of its saturated concentration
in air, and the right RS is for inhalation at a lesser dilution of
10^{-3}. Beneath each RS is an inhalation cycle histogram (ICH). The
heights of successive vertical bars in ICH's represent the number
of discharges occurring during successive portions of the inhalation
cycle, summed over many inhalation cycles. An ICH thus might be
viewed as a "probability distribution" for firing during different
portions of each cycle. The ICH's in Fig. 2 are for the period of
odor presentation. They show that inhalation of the odorant at a
higher concentration caused the unit to emit a greater overall number
of spikes, but that the distribution of spikes within the inhalation

INHALATION CYCLES

RESPONSE SURFACES

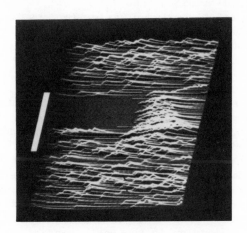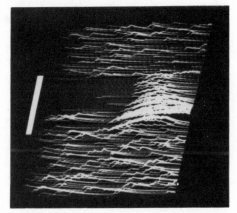

INHALATION CYCLE HISTOGRAMS

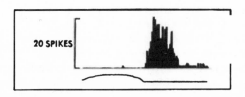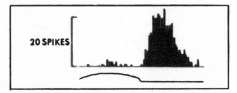

Fig. 2. Responses to two concentrations of amyl acetate.

cycles was similar to that at the lower concentration. This illustrates a common finding in our studies, that the magnitude of change
in firing rate of bulb units can be related to the concentration of
odorants. However, over a wide range of concentrations an odorant
can produce consistent alterations in the inhalation-related timing
of discharges and, presumably, in the relative timing of postsynaptic
influences by different units.

The converse also has been a common finding in our studies.
Different odorants can cause individual bulb units to discharge at
distinctly different times during the inhalation cycle. Moreover,
striking changes in the distribution of discharges within the inhalation cycle can occur when stimuli are presented at concentrations
which produce little change in average firing rate (i.e., in total
number of discharges per inhalation cycle). Figure 3 shows RS's for
a bulb unit during inhalation of cineole and amyl acetate at dilutions of 10^{-6}. This unit had an inhalation-related pattern of firing
under background conditions. Cineole accentuated the background
pattern whereas amyl acetate produced an inversion in the distribution
of discharges. It should be emphasized that though the average
firing rate did not increase during inhalation of cineole, the clus-
tering of spikes into a more restricted portion of each inhalation
cycle represents an increase in the momentary firing rate, and in
the possibility for postsynaptic influences associated with successive discharges to sum, during that portion. Similarly, odor-induced
changes in the relative timing of discharges by a group of bulb units
converging onto a common neuron conceivably could result in differential (odor-related) patterns of summation in the postsynaptic
neuron even if none of the presynaptic neurons showed marked changes
in their overall rates of firing.

We find that many olfactory bulb units show background patterns
of inhalation-related activity under the conditions of our experiments. The degree of relationship to the inhalation cycle and the
most probable timing of discharges within the cycle differ for different units, and are characteristic of each unit. That is, as
illustrated in Fig. 3, the inhalation-related patterns can be altered
by particular odorants and upon removal of these odorants from the
inhaled air the units promptly return to their characteristic background patterns. We have been able to rule out the possibility that
these background patterns might be mechanical artifacts of brain
pulsation (Macrides and Chorover, 1972), but the question of whether
they reflect some mechanical sensitivity of peripheral receptors or
autonomic responses to nasal airflow remains unanswered (cf. Walsh,
1956). Since the inhalation-related activity of bulb units can be
influenced by known odorants at extremely low concentrations, it is
possible that the background patterns are due to persistent contaminants in the purified air or to scents which normally might be
present at the subject's snout (cf. Thiessen, Clancy and Goodwin,
1976). If we intentionally present odorants to which a subject is

INHALATION CYCLES

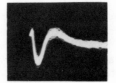

RESPONSE SURFACES

CINEOLE

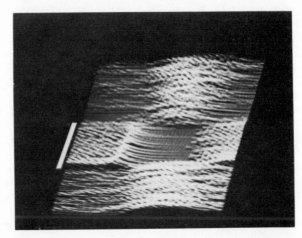

AMYL ACETATE

Fig. 3. Responses to different odorants. Adapted from Macrides and Chorover (1972); copyright 1972 by the American Association for the Advancement of Science.

chronically exposed, as by passing the purified air over the subject's cage bedding or over filter papers which were rubbed on the
subject's scent glands, then additional bulb units exhibit persistent inhalation-related patterns of firing which can be altered by
other odorants.

The point within the inhalation cycle during which a bulb unit
exhibits maximum or minimum firing is often altered when the inhalation rate is varied, but the direction and degree of these shifts
tend to be the same for the background patterns and for the stimu-
lus-specific patterns. Thus, the differential aspects of inhalation-related firing are preserved. An example is presented in
Figure 4. The RS's and ICH's were constructed with a time base of
1 second at inhalation rates of 1, 2 and 3 sniffs per second. The
RS's show the firing patterns during sniffing of amyl acetate. The
top ICH of each pair illustrates background conditions sampled over
20 second periods. The bottom ICH of each pair is for 20 second
periods when amyl acetate was being inhaled. The total number of
discharges during each of the six sample periods varied by less than
20% (amyl acetate presented at a dilution of 10^{-5}). At each sniffing
rate, however, the odorant produced an inversion in the inhalation-
related distributions of spikes. We similarly have found that when
two known odorants cause a bulb unit to discharge at different times
during sniffing with a slow repetition rate, this differential aspect
of responding tends to be preserved with more rapid rates and con-
comcomitantly shallower tidal volumes.

The tendency to adopt a more rapid inhalation rate and shallower
tidal volumes during olfactory investigation has been well documented
in rodents (Welker, 1964) and may be a general phenomenon among
macrosmatic mammals. It may also merit emphasis that the inhalation
of highly purified air is a phenomenon generally restricted to the
laboratory. Under more natural conditions, macrosmatic mammals are
known to actively impart their scents to their environment (Thiessen,
this volume) and thus must discriminate novel odorants amidst other
odors to which they are chronically exposed. If we approach the
example in Fig. 4 from the perspective that it reflects the contribution by but one of many different neural elements to the patterns
of summation in postsynaptic neurons, we see that at the more rapid
inhalation rate the change in postsynaptic influence induced by amyl
acetate could be assessed over shorter intervals and recurred more
often than at the slower inhalation rate. That is, the speed and
reliability with which a change in the olfactory environment could
be detected appears to have been enhanced.

III. THE NOTION OF PATTERN VISION IN THE NOSE

Shepherd (1970) has pointed out the striking analogy which can
be drawn between the synaptic organization of the olfactory bulb and

RESPONSE SURFACES

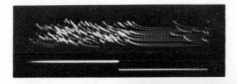

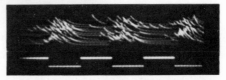

ONE SNIFF PER SEC   TWO SNIFFS PER SEC   THREE SNIFFS PER SEC

INHALATION CYCLE HISTOGRAMS

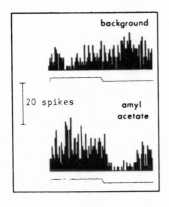

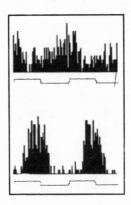

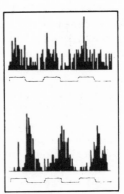

Fig. 4.   Responses to amyl acetate at different sniffing rates.

that of the vertebrate retina. The organization which he has outlined for the bulb consists of input cells (the olfactory receptors),
output neurons (the mitral and tufted cells), and two main classes
of interneurons (the glomerular and granule cells) which can exert
regulating influences on bulbar input and output processing. He
likened glomerular interneurons to the horizontal cells, and granule
cells to the amacrine cells, of the vertebrate retina. His examination of the synaptic connections among the two classes of bulbar
interneurons and the bulbar output neurons led him to suggest that
in the bulb, as in the retina, these interneurons subserve integrative functions by providing recurrent and lateral inhibition. This
type of organization in the visual system is believed to account for
functional properties such as contrast enhancement, directional
selectivity and adaptation. He speculated that a search for similar
functions in the olfactory system "may give insight into the nature
of the olfactory stimulus and suggest new experimental approaches to
the difficult problem of olfactory discrimination."

     Moulton (1976) recently reviewed anatomical and electrophysi-
ological evidence which establishes the existence of a topographical
organization of axonal projections from the olfactory epithelium to
the bulb. Like the retina, the olfactory bulb thus appears to have
access to a spatial representation of its receptor sheet. Moulton
emphasized, however, that odorant molecules are not known to be
organized in a spatial pattern as they enter the nares. If olfactory
discrimination involves some form(s) of pattern analysis, the patterns must be created within the nasal cavity. Adrian (1950, 1956)
long ago outlined two basic mechanisms by which the interactions of
odorants with the nasal organ could result in spatiotemporal patterns of activity in the receptor sheet. The first mechanism depends
on differences in the relative sensitivities of receptor cells to
odorants. Different odorants could produce different patterns of
receptor cell activity in relation to the spatial distributions of
responsive receptor cells within the sheet. The second mechanism
depends on the action of physical variables (e.g., the sorptive
properties of intranasal tissue) which could differentially influence
the distributions of inhaled molecules. These two basic mechanisms
are not mutually exclusive. As discussed in section I, both kinds
of phenomena would be expected to operate during normal olfactory
sampling in terrestrial vertebrates (cf. Mozell, this volume). In
our analogy with the retina, the differential sensitivities of peripheral olfactory receptor cells to odorants might be compared with
the differential sensitivities of visual receptor cells to the wave-
length of light. The actions of physical variables in conjunction
with sniffing might be conceptualized as being analogous to the
actions of the iris, lens and extraocular muscles in regulating the
quantity and spatiotemporal distributions of light across the retina
during inspection of a visual scene. Such analogies can be dangerous
if taken too literally, but they may be of value in helping us to
organize our thinking and to generate concrete hypotheses for future

research stemming from the more general notion that olfactory discrimination involves the analysis of firing patterns in ensembles
of peripheral receptor cells. Any attempt to draw parallels between
neural mechanisms of odor discrimination and pattern vision also must
be tempered by anatomical findings that the topographical organization of projections from the periphery to the olfactory bulb is not
well preserved in the projection patterns of the bulbar output neurons to more central olfactory structures (cf. Devor, this volume).

In the preceding section we discussed how the individual tracings in a response surface may be considered to reflect the patterns
of postsynaptic influence which the recorded neuron would exert during the course of each inhalation cycle. These tracings also may
be considered to reflect the patterns of summation within the recorded neuron in response to its peripheral and intrabulbar inputs. That
is, the moment-to-moment variations in height during the course of
each tracing reflect the manner in which excitatory and/or inhibitory
influences upon the neuron are summing at its site of spike initiation, and thus are modulating the neuron's firing rate during the
course of the inhalation cycle. When viewed with this perspective,
the observed variations in moment-to-moment firing rate are at least
consistent with the hypothesis that the neurons are integrating
sequences of excitatory influences from the periphery and inhibitory
influences within the bulb. We may further speculate that the
odorant-induced variations in timing of bursts by different bulb
units reflect (a) the spatiotemporal patterns of receptor cell activity which are elicited by the odorants as they become distributed
across the receptor sheet and are transmitted to the bulb in a topographically organized manner, and (b) lateral inhibitory interactions
which accentuate the spatiotemporal patterns of excitation in the
olfactory bulb.

If we conceive of the individual neurons in the olfactory bulb
as a spatial array of points, we can imagine that a sniff of air
containing one odorant (e.g., amyl acetate) produces one pattern of
excitation in this array, and that a comparable sniff of a different
odorant (e.g., cineole) produces a different pattern. A mixture of
the odorants could produce a conglomerate of patterns. We might
therefore be tempted to speculate that such patterns of excitation
represent an "encoding" of these odorants in the olfactory bulb.
This line of reasoning parallels that which idealizes the topographical organization of projections in the visual system into one of
"point-to-point" projections, and attempts to account for the perception of images as the production of "isomorphic representations"
of images in arrays of light-responsive neurons. In effect, we would
be assuming that each odorant or mixture of odorants can be discriminated by the kind of neural image it produces in the bulb as a consequence of the patterns of excitation it elicits in the olfactory
mucosa. The shortcoming of a scheme which treats neural coding as
simply the mirroring of the environment in arrays of neurons is that

it begs the issue of neural processing. How do the stimulus-elicited patterns of excitation in the bulb constitute a reorganization of the peripheral patterns into a more meaningful neural form?

In trying to deal with this question, let us resist the temptation to treat our various odorants as entities which are displayed in some rigid array of central neurons and instead consider them as constellations of chemical parameters or features which are embodied in the molecules and are expressed through interactions with the nasal organ. Our approach to our unit data then becomes one of trying to generate testable hypotheses about how the observed patterns of activity might serve to characterize the olfactory environment according to these features. For example, consider our observation that under a given set of sniff parameters a change in an odorant's concentration can alter the robustness of a bulb unit's bursts while little affecting the inhalation-related timing of these bursts (cf. Fig. 2). This response property might be related to the finding of Mozell and Jagodowicz (1973) that the nose separates vapors in a manner similar to a gas chromatograph. As a result, a change in the concentration of a particular odorant drawn into the nose with controlled sniffs does alter the total amounts of receptor cell activity elicited at different regions of the mucosa, but does not alter the relative amounts or relative latencies (Mozell, 1966). Might the inhalation-related response properties of bulb units permit the characterization of odorants according to features which influence their migration patterns in the mucosa? Do they serve to distinguish changes in the molecular composition of inhaled air from changes in concentration? Our observation that different odorants can cause a bulb unit to burst at distinctly different times (cf. Fig. 3) might be related to the different sequences of excitation in the periphery which result from differences in the responsiveness of receptor cells to these odorants. If so, might the inhalation-related response properties of bulb units permit the characterization of odorants according to their stereochemical features? These hypotheses are not mutually exclusive. In testing them we may well find that the responses of some bulb units are more related to the stereochemical properties of odorants, while those of other bulb units are more related to physicochemical properties. Should this indeed turn out to be the case, would we have our first glimmer of "olfactory feature detectors"?

We are, of course, merely engaging in speculation. Our intent throughout this discussion has been to illustrate a line of reasoning which assumes that olfactory neurons process "information" through dynamic patterns of change, and to introduce an experimental approach which seeks to understand the nature of olfactory stimuli through the "normal operating characteristics" of the system which characterizes them. Let us conclude by considering the broader implication of the foregoing discussion. If olfactory discrimination involves the analysis of spatiotemporal patterns of activity in ensembles of peripheral receptor cells, and if these patterns are generated through the interactions of odorants with the nasal organ as a consequence of inhalation, then through its control over inhalation a subject may be able to influence the character or quality of its olfactory information (cf. Fig. 4). In this sense, olfactory discrimination may be an active process (Macrides, 1976). The ability of the nervous system to fully characterize its olfactory environment may depend not only on the stereo- and physicochemical attributes of the stimuli themselves, but also on the manner in which these stimuli are explored and the context in which they are placed.

REFERENCES

Adrian, E. D. 1950. Sensory discrimination. Brit. Med. Bull.: 6, 330-333.

Adrian, E. D. 1956. The action of the mammalian olfactory organ. J. Laryngol. Otol.: 70, 1-14.

Gesteland, R. C., Lettvin, J. Y. & Pitts, W. H. 1965. Chemical transmission in the nose of the frog. J. Physiol. (Lond.): 181, 525-559.

Higashino, S., Takeuchi, H. & Amoore, J. E. 1969. Mechanisms of olfactory discrimination in the olfactory bulb of the bullfrog. In: Olfaction and Taste, Proc. IIIrd Int. Symp., C. Pfaffmann, Ed., 192-211. Rockefeller University Press, New York.

Komisaruk, B. R. 1970. Synchrony between limbic system theta activity and rhythmical behavior in rats. J. Comp. Physiol. Psychol.: 70, 482-492.

Macrides, F. 1972. Methods for on-line oscillographic display of spike frequency data. Physiol. Behav.: 9, 867-868.

Macrides, F. 1976. Olfactory influences on neuroendocrine function in mammals. In: Mammalian Olfaction, Reproductive Processes and Behavior, R. L. Doty, Ed., 29-65. Academic Press, New York.

Macrides, F. 1970. Single unit activity in the hamster olfactory bulb: responses to animal and pure chemical odors. Unpublished doctoral thesis, Massachusetts Institute of Technology.

Macrides, F. 1971. Single unit activity in the hamster olfactory bulb: responses to animal and pure chemical odors. Paper presented to the Eastern Psychological Association, New York.

Macrides, F. 1975. Temporal relationships between hippocampal slow waves and exploratory sniffing in hamsters. Behav. Biol.: 14, 295-308.

Macrides, F. & Chorover, S. L. 1972. Olfactory bulb units: activity correlated with inhalation cycles and odor quality. Science: 175, 84-87.

Moulton, D. G. 1976. Spatial patterning of response to odors in the peripheral olfactory system. Physiol. Rev.: 56, 578-593.

Mozell, M. M. 1966. Spatiotemporal analysis of odorants at the level of the olfactory receptor sheet. J. Gen. Physiol.: 50, 25-41.

Mozell, M. M. & Jagodowicz, M.   1973. Chromatographic separation of odorants by the nose: retention times measured across in vivo olfactory mucosa.   Science: 191, 1247-1249.

Pfaff, D. W. & Gregory, E.   1971. Olfactory coding in olfactory bulb and medial forebrain bundle of normal and castrated male rats. J. Neurophys.: 34, 208-216.

Pfaff, D. W. & Pfaffmann, C.   1969a. Behavioral and electrophysiological responses of male rats to female rat urine odors.   In: Olfaction and Taste, Proc. IIIrd Int. Symp., C. Pfaffmann, Ed., 258-267.   Rockefeller University Press, New York.

Pfaff, D. W. & Pfaffmann, C.   1969b. Olfactory and hormonal influences on the basal forebrain of the male rat.   Brain Res.: 15, 137-156.

Scott, J. W. & Pfaff, D. W.   1970. Behavioral and electrophysiological responses of female mice to male urine odors.   Physiol. Behav.: 5, 407-411.

Shepherd, G. M.   1970. The olfactory bulb as a simple cortical system: experimental analysis and functional implications. In: The Neurosciences: Second Study Program, F. O. Schmitt, Ed., 539-552.   Rockefeller University Press, New York.

Thiessen, D. D., Clancy, A. & Goodwin, M.   1976. Harderian gland pheromone in the Mongolian gerbil Meriones unguiculatus.   J. Chem. Ecol.: 2, 231-238.

Walsh, R. R.   1956. Single cell spike activity in the olfactory bulb.   Am. J. Physiol.: 186, 255-257.

Welker, W. I.   1964. Analysis of sniffing in the albino rat. Behaviour: 22, 223-244.

ON THE ANATOMICAL SUBSTRATE FOR FLAVOR

Ralph Norgren

The Rockefeller University

New York, New York 10021

Flavor contributes significantly to the control of feeding behavior and the pleasure associated with eating. It has long been recognized that flavor consists of far more than simple gustatory sensibility. Odor, texture, temperature, consistency and taste each contribute in varying degrees. The neural mechanisms underlying this amalgam of sensations, much less its influence over ingestion, have yet to be specified. Recent investigations of both the gustatory and olfactory systems, however, provide an anatomical basis for interaction between these two chemical senses. Additional evidence indicates that other, non-chemical, oral sensibilities (thermal and tactile) may converge on the same areas.

At the turn of the century, Edinger (1908) hypothesized an area for oral sensibility (olfactory and oropharyngeal) located at the base of the forebrain "in all vertebrates up to mammals." In birds the principal trigeminal nucleus projects directly to nucleus basalis in the ventral forebrain. Lesions of either nucleus in pigeons induce long lasting or permanent adipsia and aphagia (Zeigler & Karten, 1973a, b). In mammals, the principal trigeminal nucleus projects primarily to the ventrobasal complex of the dorsal thalamus (Smith, 1973, 1975). Lesions involving trigeminal afferents in mammals result in, at best, a transient aphagia, hypophagia (Zeigler & Karten, 1974), or even hyperphagia (Skultety, 1973). In one mammal, the rat, the gustatory system projects both to the dorsal thalamus and ventral forebrain. Both limbs of this bifurcated projection overlap or abut central olfactory afferents, and lead to areas implicated in the control of feeding behavior.

Since Dr. Shepard's paper deals with central pathways and processing in the olfactory system, this discussion will elaborate on the gustatory system first, then cite specific studies relevant to the central interaction of olfactory, oropharyngeal, and gustatory information.

From fish to humans, the primary central synapse for peripheral gustatory afferents is the anterior pole of the nucleus of the solitary tract (NST) in the medulla (Herrick, 1944; Oakley & Benjamin, 1966). Previously accepted descriptions of secondary gustatory fibers arising from the solitary nucleus pictured fibers crossing the midline as internal arcuates to join the medial lemniscus and ascend to the medial ventrobasal thalamus (Allen, 1923; Gerebtzoff, 1939). Gustatory responses can be recorded from the rostral pole of the solitary nucleus. Degenerating fibers traced from lesions of such recording sites neither entered the medial lemniscus, nor reached the thalamus. Instead, the fibers ascended through the ipsilateral reticular formation until they reached the caudal border of the trigeminal motor nucleus. Here they turned dorsally, ramified in the dorsomedial corner of the principal trigeminal nucleus and adjacent reticular formation, and then appeared to terminate in a small celled zone embedded in, and surrounding the brachium conjunctivum. This zone corresponds to the caudal end of the pontine parabrachial nuclei.

Electrophysiological investigation of the parabrachial area revealed neurons which responded to sapid stimuli applied on the tongue in a manner similar to gustatory afferents recorded from peripheral nerves (Norgren & Pfaffmann, 1975). Instilling Xylocaine in the middle ear to anesthetize the chorda tympani nerve blocked both the spontaneous and evoked activity of anterior tongue taste units. After lesions confined to the pontine taste area, degenerating axons ascended in the ipsilateral central tegmental tract on the ventrolateral border of the central gray until they reached the meso-diencephalic junction. After a sharp ventral flexure, some of the degenerating axons ended in a terminal zone filling the medial extension of the thalamic ventrobasal complex, an area previously identified as a relay for gustatory and lingual afferents (Norgren & Leonard, 1973).

The thalamic gustatory relay projects to a band of cortex approximately 1 mm wide by 3 mm long squeezed between the rhinal sulcus and the somatic sensory area. The thalamic lingual area projects to cortex immediately dorsal to the gustatory zone. The areas overlap, particularly in the zone rostral to the middle cerebral artery (Norgren & Wolf, 1975). Gustatory cortex in the rat sends efferents in varying degrees to each of the subcortical afferent relays, i.e., the central nucleus of the amygdala, lateral hypothalamus, ventrobasal thalamus, parabrachial nuclei, and

nucleus of the solitary tract. Therefore, although cytoarchitec-
turally distinct from somatic sensory cortex, gustatory cortex has
a projection scheme similar to other cortical sensory areas
(Norgren & Grill, 1976).

In addition to projecting to the ventrobasal thalamus,
degenerating fibers arising from lesions of the pontine taste area
could be traced ventral and anterior to the thalamic taste area
through the subthalamus and into the internal capsule. A thin
field of fine degenerating axons was visible rostral to the
internal capsule in the substantia innominata. Although control
lesions established that these degenerating axons originated in
the same area as the gustatory neurons, no assurance existed that
the fibers carried gustatory information. By implanting stimulating
electrodes in the thalamic taste area and the ventral forebrain, it
was possible to antidromically activate pontine gustatory neurons
and demonstrate not only that taste fibers reach substantia innominata but also that many were collaterals of the axons termi-
nating in the thalamus (Norgren, 1974).

Small injections of tritiated amino acid into the electrophysiologically identified gustatory zones in the pons have further
documented the bifurcated forebrain projections of the central
gustatory system (Norgren, 1976). Labeled fibers followed the same
ipsilateral pathway to the thalamus and ventral forebrain discovered
previously with degeneration techniques. In addition, this more
sensitive technique for tracing axonal pathways revealed terminal
fields in subthalamus and far-lateral hypothalamus, central nucleus
of the amygdala, and the dorsolateral quadrant of the bed nucleus
of the stria terminalis. Some of these axons convey gustatory
information, because stimulation through electrodes in the vicinity
of the central nucleus of the amygdala can antidromically drive
pontine neurons which also respond to sapid stimuli.

Secondary trigeminal axons ascend from pons and medulla in the
contralateral trigeminal lemniscus to synapse largely in the ventrobasal complex of the dorsal thalamus. Some collaterals also end in
zona incerta, but this contribution has never been considered
substantial (Smith, 1973, 1975). The ventrobasal relay projects
directly to neocortex. The gustatory system, then, is quite
distinct vis-a-vis the majority of cephalic somatosensory afferents.
It is primarily ipsilateral, not contralateral; reticular, not
neolemniscal; at least disynaptic, not monosynaptic, and has a
bifurcated projection to the forebrain.

Intraoral trigeminal afferents, or at least a subset of them,
appear to be organized more like the gustatory system than the
remainder of the trigeminal system. The course and termination of
gustatory fibers to the diencephalon follows the route of the

ipsilateral dorsal secondary ascending trigeminal tract. The very
existence of the dorsal trigeminal system, however, has been dis-
puted since the turn of the century (see Norgren & Leonard, 1973
for a more detailed discussion). The arrangement of the central
gustatory relays in conjunction with several undisputed characteristics of the trigeminal system produce a reasonable explanation
for the dorsal trigeminal controversy, as well as anatomical support for the electrophysiological evidence discussed here.

The trigeminal relays in the caudal brainstem have inverted
somatotopic arrangements with the intraoral afferents ending in
the most dorsomedial corner. At the level of the principal
nucleus, Torvik (1957) found these cells degenerated after an
ipsilateral thalamic lesion in kittens. The other neurons in the
principal trigeminal nucleus showed signs of retrograde degeneration only on the side opposite the thalamic lesion. Further
caudally, primary lingual afferents extend medially from pars
interpolaris of the descending trigeminal nucleus to synapse in
the lateral half of the rostral solitary nucleus (Astrom, 1953;
Torvik, 1955; Blomquist & Antem, 1965). Both tongue tactile and
thermal, as well as sapid stimuli drive units recorded from this
area (Makous et al., 1963). Lesions at sites which responded
exclusively to tongue thermal stimuli result in a pattern of
degeneration indistinguishable from that accompanying medullary
taste lesions. The dorsomedial nucleus of Astrom occupies roughly
the same position relative to the pontine gustatory relay as the
primary lingual termination does to the solitary nucleus. In some
species, Astrom's neurons may receive few if any primary trigeminal
afferents, but many secondary afferents from the medulla. This
would account for the failure of some experiments to demonstrate a
dorsal secondary ascending trigeminal tract when lesions were
confined to the core of the principal trigeminal nucleus. This
relationship would also redefine the dorsal secondary ascending
trigeminal tract as a tertiary lingual or oropharyngeal-gustatory
system. A similar hypothesis was first proposed by Von Economo in
1911.

Regardless of their afferentation and central course, some
pontine neurons which respond to intraoral stimuli have axons which
reach both the dorsal thalamus and ventral forebrain. Both tongue
thermal and tactile responses can be recorded from the parabrachial
area. The pure tongue thermal responses occur rostral to the
pontine gustatory neurons (Norgren & Pfaffmann, 1975) and the tongue
tactile responses lateral to the gustatory zone (Norgren, unpublished observations). In two separate studies designed to study
pontine gustatory projections, a number of other units were isolated
which could be antidromically invaded by stimulating points in the
dorsal thalamus, ventral forebrain, or both. Of 27 units which
responded to stimuli other than taste, only those activated by

intraoral stimuli (7 - tongue thermal; 2 - palatal tactile) were
antidromically driven via electrodes in the ipsilateral hypothalamus, internal capsule, substantia innominata or amygdala
(Norgren, 1974, 1976). Of those responding to other peripheral
stimuli (jaw stretch, tooth touch, vibrissae or facial hair displacement) or not activated by any stimulus, one was antidromically
invaded, and then from the thalamus.

    Such data demonstrates that some non-chemical intraoral af-
ferents have a course similar to the third order gustatory projections. The same data poses a number of other questions concerning
the organization of intraoral afferents. For instance, are there
intraoral afferents with a central course similar to the rest of
the trigeminal system -- that is, contralateral, neolemniscal, and
largely thalamo-cortical? If such a system exists, is it represented by a completely separate set of neurons (peripherally and/or
centrally), or simply collateral axons of the ipsilateral system.
For purposes of the present discussion, however, it is sufficient
to know that some gustatory and intraoral afferents have closely
related or overlapping synaptic fields at several levels in the
forebrain.

    In order to have a theoretically viable substrate for the
elaboration of flavor, we have only to demonstrate that olfactory
information has access to one or several of these oral-gustatory
zones. In fact, olfactory afferents terminate in, or immediately
adjacent to, all of the forebrain oral-gustatory zones. Contrary
to previously established doctrine, olfactory afferents do have a
thalamic relay, and therefore, a neocortical distribution as well.
Both prepiriform cortex and olfactory tubercle project to a
restricted area of the thalamic dorsomedial nucleus (Scott &
Leonard, 1971). Although separated within the thalamus, the dorsomedial nucleus and the medial (lingual-gustatory) ventrobasal
complex project to adjacent areas of neocortex. In the rat, the
olfactory core of the dorsomedial nucleus projects to the cortex
lining the dorsal side of the rhinal sulcus (Leonard, 1969).
Rostrally, thalamic gustatory neurons end 0.5 mm to 1.0 mm dorsal
to this sulcus, and caudally, as the fissure becomes less and less
distinct, the gustatory terminal area approaches even closer
(Norgren & Wolf, 1975). Based on a comparison of original material,
the caudal aspect of the neocortical olfactory area shares a common
boundary with rostral gustatory cortex (Fig. 1). A similar arrangement appears to occur in primate cortex, i.e., olfactory
neocortex just rostral to the gustatory area. Since these areas
have been studied in widely differing primate species (Macaque and
Squirrel monkey), it is not possible to assess the degree of juxta-
position (Leonard, 1972; Pribram et al., 1953; Tanabe et al., 1975;
Benjamin & Burton, 1968).

    Regardless of the intimacy of the association, no evidence

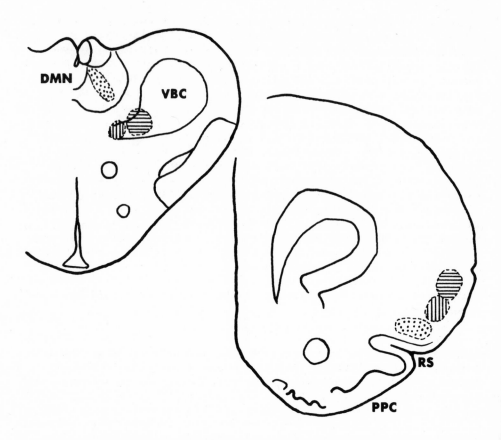

Fig. 1.   Diagrammatic comparison of three thalamic lesions and their
corresponding cortical degeneration fields.   Dots represent the
extent of the lesion centered in the olfactory core of the dorsomedial nucleus and the accompanying cortical degeneration; vertical
lines -- the thalamo-cortical gustatory projection; horizontal
lines -- the lingual representation (lesion centered in an area
inhibited when warm water was washed over the tongue).   The dorsomedial lesion extends somewhat rostral to the level depicted, and
degeneration extends rostrally in sulcal cortex almost to the
frontal pole.   The thalamic gustatory and lingual lesions are
smaller, and centered approximately 0.5 mm posterior to the plane
of the section.   The cortical degeneration extends caudally from
the section represented.   Data for the dorsomedial nucleus projection was kindly provided by Dr. Christiana M. Leonard of the
University of Florida.
Abbreviations: DMN - dorsomedial nucleus; VBC - ventrobasal
complex; RS - rhinal sulcus; PPC - prepiriform cortex.

indicates that olfactory and gustatory afferents synapse on common
cortical neurons. Even in the rat, such a relation would require
some intracortical processing. In several other areas, however,
third order olfactory and gustatory afferents do converge on a
common pool of neurons, providing at least the anatomical possibility that some neurons process both aspects of chemical
sensibility. Olfactory afferents pass into and through the lateral
hypothalamus in a distinct fascicle tucked into the angle between
the optic tract and the internal capsule (Scott & Leonard, 1971;
Broadwell, 1975b). These fibers originate from cells in prepiriform cortex and, perhaps predominantly, olfactory tubercle
(Scott & Chafin, 1975). The main terminus of this system appears
to be the nuclei gemini of Lundberg (1962) in the posterior hypothalamus. Nevertheless, neurons throughout lateral hypothalamus,
and particularly those close to the olfacto-hypothalamic tract
respond to both electrical stimulation of the olfactory bulb and
odors (Scott & Pfaffmann, 1969, 1972). Most likely not all these
olfactory responses represent direct (third order) activity, but
the high percentage of neurons responding indicates the olfactory
system has a powerful influence on the lateral hypothalamus.

Anatomical evidence indicates a direct, if somewhat sparse
projection from the pontine gustatory zone into the far lateral
hypothalamus -- the area receiving the bulk of the olfactory afferents (Norgren, 1976). In fact, a few fibers from the pons
ascend in the olfacto-hypothalamic tract (Norgren & Leonard, 1973).
Electrophysiological data seems consistent with a relatively weak
oral-gustatory input to the hypothalamus. Electrical stimulation
of the inferior alveolar nerve and strong salt solutions applied
to the tongue both drive neurons in the posterior lateral hypothalamus (Takaori et al., 1968; Wyrwicka & Chase, 1970; Kawamura
et al., 1970). When a variety of stimuli are presented to
unanesthetized preparations, however, gustatory and oral afferents
have no more influence on hypothalamic units than other stimuli
(Campbell et al., 1969; Norgren, 1970). Most units in the lateral
hypothalamus respond to more than one modality, and in that sense
resemble neurons found in the reticular formation. In the only
study known to this author in which a large sample of neurons was
tested with both olfactory and gustatory stimulus, the area
examined was confined to the medial hypothalamus, specifically the
ventromedial nucleus (Campbell et al., 1969). In this case, most of
the units which responded were specific to one sensory modality.
In the far lateral hypothalamus, where more or less direct
olfactory and oral-gustatory afferents converge, no similar experiments have been reported.

In the amygdala, little if any electrophysiological evidence
has been reported for the presence or absence of oral or gustatory
afferent influence. Despite its proximity to olfactory cortex,

very few studies have examined olfactory influences on amygdala
units. The most extensive investigation found 40% of the neurons
isolated in the amygdala could be driven by odor stimuli, but apparently none of the neurons tested were located within the
central nucleus (Cain & Bindra, 1972). Heimer and DeOlmos have
recently obtained anatomical evidence indicating a massive projection from periamygdaloid (olfactory) cortex to the central nucleus
(Fig. 2). Since pontine gustatory neurons also project to the
central nucleus, it becomes an obvious candidate for processing
information relevant to the perception of flavor. Neurons in the
amygdala, and the central nucleus in particular, have very low or
non-existent spontaneous activity, and seem to be highly susceptible
to anesthetic depression (Norgren, unpublished observations). The
electrophysiological examination of sensory influence on central
nucleus neurons, therefore, will require unanesthetized, and
probably freely moving, preparations. Such investigations are currently in progress.

     The final forebrain area which apparently receives afferent
information from two chemosensory systems is the bed nucleus of the
stria terminalis. Even more caution must be exercised here than in
the other areas, because all the inferences are based solely on
anatomical information. Every injection of tritiated amino acid
which demonstrated a projection from the pons to the amygdala, also
labeled the dorsolateral bed nucleus. Stimulation electrodes have
yet to be implanted there to test for antidromic activation of
pontine gustatory neurons. The other chemical afferents to the bed
nucleus come not from the olfactory cortex or bulb, but directly
from the accessory olfactory bulb (Broadwell, 1975a). The evidence
derives from different species (rat and rabbit, respectively), so
direct assessment of overlap in the two projections is not possible.
Since the vomeronasal projection is second order, electrophysiological confirmation should not be difficult to obtain. Should some
bed nucleus neurons prove sensitive to both gustatory and vomeronasal stimuli, it may aid in deciphering the functions of Jacobson's
organ. Perhaps more interesting, at least from an anthropocentric
viewpoint, would be an investigation of a similar population of
neurons in species without vomeronasal receptors, i.e., primates.

     The sensory modalities which contribute to the sensation of
flavor can potentially interact in at least four areas of the
brain -- sulcal cortex, lateral hypothalamus, amygdala, and the bed
nucleus of the stria terminalis. In varying degrees, each of these
areas has been shown to influence ingestive behavior in animals.
It must be noted, however, that damage to most areas of the forebrain alters intake at least temporarily. Nevertheless, afferents
from the nose and mouth monitor the moment-to-moment passage of
material through the oral cavity, providing information critical
to the decision to ingest or reject. The same information

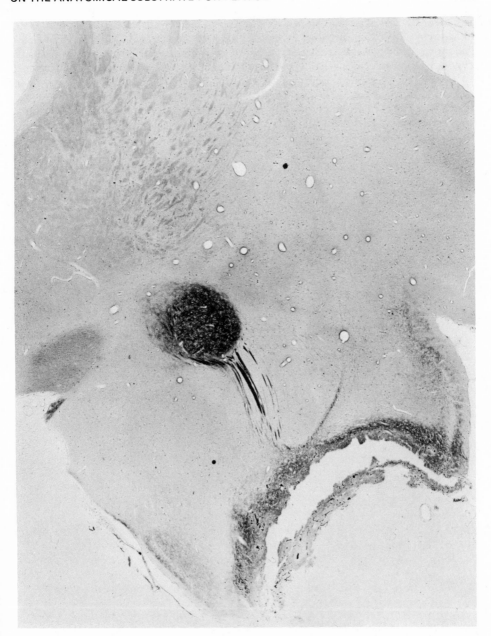

Fig. 2. Low power photomicrograph of a coronal section through the Guinea pig brain. A lesion of the periamygdaloid (olfactory) cortex -- lower right -- results in degenerating fibers and terminals filling the central nucleus of the amygdala. Cupric-silver method, three day survival. The plate was kindly provided by Dr. Lennart Heimer of the University of Virgina Medical School.*

influences feeding patterns (LeMagnen, 1967; Larue & LeMagnen, 1973), triggers adaptive metabolic reflexes (Nicolaidis, 1969; Anokhin & Sudakov, 1967; Rupe & Mayer, 1967), alters long term energy balance (Becker & Kissileff, 1973), and induces sexual behavior in hamsters (O'Connell, personal communication). In most instances there is no indication that olfactory, oral, and gustatory information must act in concert to induce these effects, but any zone in which all three converge must be considered a prime suspect in the genesis of these behavioral regulations.

Dr. Robert J. O'Connell read and improved an earlier draft, and Jean Clement typed the manuscript. They both have my thanks. Support for the author's research was provided through NIH grants NS 10150 and GM 1789.

*Figure 2 has appeared previously in a chapter entitled "Wallerian Degeneration and Anterograde Tracer Methods", by Ann M. Graybiel published in The Use of Axonal Transport for Studies of Neuronal Connectivity, W.M. Cowan and M. Cuenod, editors, Elsevier, Amsterdam, 1975. Reproduced with permission of the author and publisher.

REFERENCES

ALLEN, W. F.   1923.   Origin and destination of the secondary
    visceral fibers in the guinea-pig.   J. Comp. Neurol.   35:
    275-310.

ANOKHIN, P. K., and SUDAKOV, K. V.   1966.   Sensory mechanism of
    satiety. in Proceedings of the 7th International Congress of
    Nutr., Vol. 2.   Friedr. Vieweg and Sohn, Hamburg.

ASTROM, K. E.   1953.   On the central course of afferent fibers in
    the trigeminal, facial, glossopharyngeal, and vagal nerves
    and their nuclei in the mouse.   Acta Physiol. Scand.   29
    (suppl. 106): 209-320.

BECKER, E., and KISSILEFF, H.   1974.   Inhibitory controls of
    feeding by ventromedial hypothalamus.   Am. J. Physiol.   226:
    383-396.

BENJAMIN, R. M., and BURTON, H.   1968.   Projection of taste nerve
    afferents to anterior opercular-insular cortex in squirrel
    monkey (Saimiri sciureus).   Brain Res.   7: 221-231.

BLOMQUIST, A. J., and ANTEM, A.   1965.   Localization of the
    terminals of the tongue afferents in the nucleus of the
    solitary tract.   J. Comp. Neurol.   124: 127-130.

BROADWELL, R. D.   1975a. Olfactory relationships of the telencepha-
    lon and diencephalon in the rabbit. I. An autoradiographic
    study of the efferent connections of the main and accessory
    olfactory bulbs.   J. Comp. Neurol.   163: 329-346.

BROADWELL, R. D.   1975b. Olfactory relationships of the telencepha-
    lon and diencephalon in the rabbit. II. An autoradiographic
    and horseradish peroxidase study of the efferent connections
    of the anterior olfactory nucleus.   J. Comp. Neurol.   164:
    389-410.

CAIN, D. P., and BINDRA, D.   1972.   Responses of amygdala single
    units to odors in the rat.   Exp. Neurol.   35: 98-110.

CAMPBELL, J. F., BINDRA, D., KREBS, H., AND FERENCHAK, R. P. 1969.
    Responses of single units of the hypothalamic ventromedial
    nucleus to environmental stimuli.   Physiol. Behav.   4:
    183-187.

VON ECONOMO, C.   1911.   Über dissoziierte Empfindungslähmung bei
    Ponstumoren und über die zentralen Bahnen des sensiblen
    Trigeminus.   J. Psychiat. Neurol.   32: 107-138.

EDINGER, L. 1908. The relations of comparative anatomy to comparative psychology. J. Comp. Neurol. Psychol. 18: 437–457.

GEREBTZOFF, M. A. 1939. Les voies centrales de la sensibilité et du gout et leurs terminaisons thalamiques. Cellule 48: 91–146.

HERRICK, J. C. 1905. The central gustatory paths in the brains of bony fishes. J. Comp. Neurol. Psychol. 15: 375–456.

KAWAMURA, Y., KASAHARA, Y., and FUNAKOSHI, M. 1970. A possible brain mechanism for rejection behavior to strong salt solution. Physiol. Behav. 5: 67–74.

LARUE, C., and LE MAGNEN, J. 1973. Effets de l'interruption des voies olfacto-hypothalamiques sur la séquence alimentaire du rat. J. Physiol., Paris 66: 699–713.

LE MAGNEN, J. 1967. Habits and food intake. Pages 11–30 in C. F. Code (ed.), Handbook of Physiology. Section 6: Alimentary Canal. Vol. 1. Control of Food and Water Intake. American Physiol. Society, Washington, D. C.

LEONARD, C. M. 1969. The prefrontal cortex of the rat. I. Cortical projection of the medio-dorsal nucleus. II. Efferent connections. Brain Res. 12: 321–343.

LEONARD, C. M. 1972. The connections of the dorsomedial nuclei. Brain, Behav. Evol. 6: 524–541.

LUNDBERG, P. O. 1962. The nuclei gemini. Two hitherto undescribed nerve cell collections in the hypothalamus of the rabbit. J. Comp. Neurol. 119: 311–316.

MAKOUS, W., NORD, S., OAKLEY, B., and PFAFFMANN, C. 1963. The gustatory relay in the medulla. Pages 381–393 in Y. Zotterman (ed.), Proceedings of the First International Symposium on Olfaction and Taste. Pergamon Press, Oxford.

NICOLAIDIS, S. 1969. Early systemic responses to orogastric stimulation in the regulation of food and water balance: functional and electrophysiological data. Annals N. Y. Acad. Sci. 157: 1176–1203.

NORGREN, R. 1970. Gustatory responses in the hypothalamus. Brain Res. 21: 63–77.

NORGREN, R. 1974. Gustatory afferents to ventral forebrain. Brain Res. 81: 285-295.

NORGREN, R. 1976. Taste pathways to hypothalamus and amygdala. J. Comp. Neurol. 166: 17-30.

NORGREN, R., and GRILL, H. J. 1976. Efferent distribution from the cortical gustatory area in rats. Neuroscience Abstracts: Proceedings of the Sixth Annual Meeting of the Society for Neuroscience. Toronto, November 1976, In press.

NORGREN, R., and LEONARD, C. M. 1971. Taste pathways in rat brainstem. Science 173: 1136-1139.

NORGREN, R., and LEONARD, C. M. 1973. Ascending central gustatory pathways. J. Comp. Neurol. 150: 217-238.

NORGREN, R., and PFAFFMANN, C. 1975. The pontine taste area in the rat. Brain Res. 91: 99-117.

NORGREN, R., and WOLF, G. 1975. Projections of thalamic gustatory and lingual areas in the rat. Brain Res. 92: 123-129.

OAKLEY, B., and BENJAMIN, R. M. 1967. Neural mechanisms of taste. Physiol. Rev. 46: 173-211.

PRIBRAM, K. H., CHOW, K. L., and SEMMES, J. 1953. Limit and organization of the cortical projection from the medial thalamic nucleus in the monkey. J. Comp. Neurol. 98: 433-448.

RUPE, B. D., and MAYER, J. 1967. Endogenous glucose release by oral sucrose administration in rats. Experientia 23: 1009-1010.

SCOTT, J. W., and CHAFIN, B. R. 1975. Origin of olfactory projections to lateral hypothalamus and nuclei gemini of the rat. Brain Res. 88: 64-68.

SCOTT, J. W., and LEONARD, C. M. 1971. The olfactory connections of the lateral hypothalamus in the rat, mouse and hamster. J. Comp. Neurol. 141: 331-344.

SCOTT, J. W., and PFAFFMANN, C. 1967. Olfactory input to the hypothalamus: electrophysiological evidence. Science 158: 1592-1594.

SCOTT, J., and PFAFFMANN, C. 1972. Characteristics of responses of lateral hypothalamic neurons to stimulation of the olfactory system. Brain Res. 48: 251–264.

SKULTETY, F. M. 1973. Hyperphagia after midbrain lesions involving the medial lemniscus. Exp. Neurol. 38: 6–19.

SMITH, R. L. 1973. The ascending fiber projections from the principal sensory trigeminal nucleus in the rat. J. Comp. Neurol. 148: 423–446.

SMITH, R. L. 1975. Axonal projections and connections of the principal sensory trigeminal nucleus in the monkey. J. Comp. Neurol. 163: 347–376.

TAKAORI, S., SASA, M, and FALCUDA, N. 1968. Responses of posterior hypothalamic neurons to electrical stimulation of the inferior alveolar nerve and distention of stomach with cold and warm water. Brain Res. 11: 225–237.

TANABE, T., YARITA, H., IIMO, M., OGSHIMG, Y., and TAKAGI, S. F. 1975. An olfactory projection area in orbitofrontal cortex of monkey. J. Neurophysiol. 38: 1269–1283.

TORVIK, A. 1955. Afferent connections to the sensory trigeminal nuclei, the nucleus of the solitary tract and adjacent structures. An experimental study in the rat. J. Comp. Neurol. 106: 51–141.

TORVIK, A. 1957. The ascending fibers from the main trigeminal sensory nucleus. An experimental study in the cat. Am. J. Anat. 100: 1–15.

WYRWICKA, W., and CHASE, M. H. 1970. Projections from the buccal cavity to brain stem sites involved in feeding behavior. Exp. Neurol. 27: 512–519.

ZEIGLER, H. P., and KARTEN, H. J. 1973a. Brain mechanisms and feeding behavior in the pigeon (Columba livia): I. Quinto-frontal structures. J. Comp. Neurol. 152: 59–83.

ZEIGLER, H. P., and KARTEN, H. J. 1973b. Brain mechanisms and feeding behavior in the pigeon (Columba livia): II. Analysis of feeding behavior deficits after lesions of quinto-frontal structures. J. Comp. Neurol. 152: 83–101.

ZEIGLER, H. P., and KARTEN, H. J. 1974. Central trigeminal structures and the lateral hypothalamic syndrome in the rat. Science 186: 636–638.

CENTRAL PROCESSING OF ODOR SIGNALS: LESSONS FROM

ADULT AND NEONATAL OLFACTORY TRACT LESIONS[1]

Marshall Devor

Neurobiology Unit, Russian Compound, Life Sciences

Institute, The Hebrew University, Jerusalem, Israel

It is plain why we know so little about the neural processing of odor signals. Beginning at the receptor level there appear to operate a number of different mechanisms for stimulus separation and transduction (Beidler, 1971). But even if there were a single simple process, say an orderly array of specific receptor types or a spatiotemporal map (Davies, 1971), we would need take only one step further centrally before order was chaos once again. From receptor sheet to olfactory bulb there is only the faintest hint of a topographic mapping (Clark, 1951). Then we face the convergence of 26,000 receptor cells, on the average, into a single glomerulus and thence to second order neurons in the bulb, each participating in many glomeruli (Allison and Warwick, 1949). Finally, most every region of the bulb distributes axons to the whole of the olfactory cortex (see Fig.1; White, 1965; Lohman and Mentink, 1969; Price, 1973; Broadwell, 1975; Devor, 1976a). This looks like a structure specifically designed to scramble the message received by single receptors. Nor is there much evidence of specificity in electrophysiological recordings from single cells at receptor, bulbar or cortical levels (e.g. Gesteland, 1971; Mathews, 1972; Haberly, 1969). If we accept the premise that nature is not perverse, then there may well be a mechanism operating here that is different from that of most other sensory systems in which feature extraction and topographic mapping and remapping play such a prominent role.

---

[1]This work was performed in the Department of Psychology, M.I.T., Cambridge, Massachusetts, supported by grants from the Sloan and Grant Foundations and NIH grants #MH 07923-11 and #EY 00126. I am pleased to acknowledge the valuable contributions to this work of S.L. Chorover, M. Kaitz, M.R. Murphy and G.E. Schneider.

This in mind I chose to jump to central levels and seek order
in the cortical processes for analyzing a single defined odor. The
plan of attack, in the tradition of Allen (1943), was straightforward.
With neuroanatomy as a guide I would cut various of the candidate
pathways and examine whether or not the test odor was perceived. The
curious outcome was that signal recognition, as reflected in an
animal's behavioral response to the odor, was not an all or none
thing. Rather, the degree or probability of normal response depended
in a continuous manner upon the extent of olfactory cortex receiving
input from the olfactory bulb. Though it cannot be completely ruled
out there is once again no clear sign of a functional topography.

AN ODOR-DEPENDENT BEHAVIOR

It is both tedious and potentially misleading to examine lesion
effects on learned odor discrimination or detection tasks. The
lesion may pervert the quality of an odor without changing its
discriminability. Instead, I chose a natural behavior dependent on
a specific scent, a presumed pheromone: mating in the male Syrian
golden hamster (*Mesocricetus auratus*, Waterhouse). Sexually vigorous
male hamsters begin to mate within seconds of being paired with a
receptive female, seemingly oblivious to most outside distractions,
olfactory or otherwise. Prominant in its behavior pattern is the
hamster's tremendous interest in the strongly odorous female hamster
vaginal discharge (FHVD) (Dieterlen, 1959; Johnston, 1970). In fact,
when FHVD is placed on other male hamsters, they try to mate with
these (Murphy, 1973; Darby, Devor and Chorover, 1975). Most important,
however, Murphy and Schneider (1970) showed that male hamsters lose
all interest in the female if the olfactory bulbs have been removed
bilaterally (Fig. 2B). Thus, we have a specific odorant and an easily
observed behavioral marker for its recognition.

Can we be sure that olfactory processing is essential? Two
experiments prove the point (Devor and Murphy, 1973). First, destruction of the olfactory receptor sheet by perfusion of the nasal cavity
with a solution of $ZnSO_4$ (Alberts and Galef, 1971) eliminates mating
(Fig. 2C). This result was originally refuted by Powers and Winans
(1973) who failed to abolish mating with this technique. Indeed, they
suggested that the deficit following bilateral bulb lesions had nothing
to do with airborn odors. In a more recent study, Powers and Winans
(1975) cut the vomeronasal nerve and then in the animals that continue
to mate they perfused with $ZnSO_4$. Now mating was eliminated. Based
on these new data the authors reversed their former position and now
agree that odor perception is critical. However, they have added the
interesting twist that either main olfactory or vomeronasal channels
may suffice. Why was $ZnSO_4$ treatment alone effective in my and
Murphy's hands but not in theirs? My best guess is that our treatment
blocked main and vomeronasal function whereas theirs left vomeronasal
function intact. The difference may lie in the care we took in the

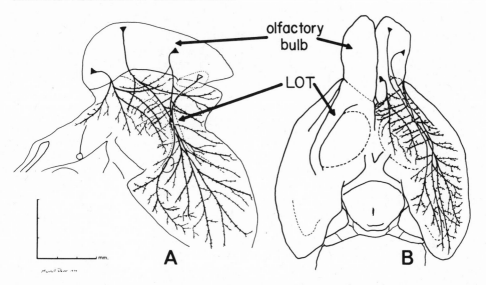

Fig.1: Axonal distribution of main and accessory olfactory bulbs
in the olfactory cortex. The axons and their terminals occupy a
thin cylindrical shell at the cortical surface (right). The shell
can be transformed into a flattened map (left).

selection of anaesthetic and preanaesthetic agents that would minimize mucosal swelling and consequent protection of patches of nasal
mucosa. Be that as it may, the exercise ought to stand as a warning
to anyone evaluating one of the now dozens of papers claiming non-
olfactory effects of olfactory system lesions based on negative
results following ZnSO₄ treatment (e.g. Rowe and Smith, 1972; Spector
and Hull, 1972). Just as with any other lesion yielding negative
results one must insist on histological verification of a complete
lesion following ZnSO₄ infusion. Even behavioral evidence of ansomia
to some test odor at some test concentration is insufficient.

Given these uncertainties, a more sure proof of the olfactory
dependence of hamster mating behavior was necessary. Thus, in a
second experiment, one olfactory bulb was removed from a series of
sexually vigorous males. This lesion in itself is without effect
(Fig. 2D). If, however, the nostril on the other side is then
occluded (Fig. 2E) mating ceases. Mating can thus be turned on or off
at will merely by alternately occluding one nostril and then the

other. The hamster will begin mating only if he can smell the female.
Curiously, if the male is rendered anosmic by nostril occlusion
while in the process of mating and is then returned rapidly to the
female, he is likely to continue mating despite anosmia. Odor perception is necessary not to sustain the behavior pattern but to turn
it on.

CENTRAL OLFACTORY CONNECTIONS

The central connections relevant to the present experiments can
be reviewed briefly (Heimer, 1968, 1972; Price, 1973; Scalia and
Winans, 1975; Broadwell, 1976; Devor, 1976a). Axons of mitral and
tufted cells of the main olfactory bulb run caudally, coalesce to
form the lateral olfactory tract (LOT) and then at intervals emit
collateral branches which innervate a large region of ventral cortex
(Fig.1B). This projection is essentially non-topographic. Axons of
accessory olfactory bulb neurons, the cells that receive their input
from the vomeronasal organ, run along with the LOT but do not emit
their terminals until they reach the medial portion of the amygdaloid
complex. Broadwell (1976) has traced some of these on into the stria
terminalis. The cortical afferents from main and accessory olfactory
bulbs terminate exclusively in the most superficial layer of cortex.
They therefore form a thin sheet bent into a half cylinder. Using a
fairly straightforward transform (Devor, 1976a) this whole projection
field can be represented on a single flattened map (Fig.1A).

I will not attempt here a full inventory of the output connections of the various subfields of the olfactory projection cortex,
efferent, association and commissural. Two classes are of special
interest, however. First, the whole of the projection cortex of the
main olfactory bulb in turn projects to the thalamic nucleus medialis
dorsalis and much of it also to a ventrolateral component of the
medial forebrain bundle (VLMFB). Some or all of the VLMFB fibers go
on to terminate in a small premammillary nucleus, the nuclei gemini
of Lundberg and probably also among lateral hypothalamic cells that
lie along their course (Scott and Leonard, 1971; Heimer, 1972).
Second, the projection cortex of the accessory olfactory bulb sends
most of its output toward medial forebrain structures along the stria
terminalis (Leonard and Scott, 1971).

TRANSECTION OF THE LATERAL OLFACTORY TRACT (LOT)

In an extensive series of hamsters the LOT was cut either unilaterally, bilaterally, or unilaterally in combination with contralateral removal of the olfactory bulb (Devor, 1973). The surgery was
performed by exposing the tract through a periorbital exposure and
severing it under direct visual control. The completeness of all cuts
was verified histologically in cresylviolet and Fink-Heimer (1967)
stained frozen serial sections.

The LOT is the olfactory bulb's sole efferent pathway. To transect it, however, is not equivalent to removing the bulb itself. No matter where the tract is cut there is always some olfactory cortex between the cut and the bulb that continues to receive input from rostral LOT collaterals (Fig.1).

All of the animals that underwent unilateral lesions continued to mate normally. The effects of bilateral lesions proved to depend on where the LOT was sectioned. When the cut was made rostrally, at the level of the anterior olfactory nuclei (Fig. 2F), all interest in the female ceased just as if both olfactory bulbs had been removed. Similarly, when the animals were deprived of food, they were unsuccessful at finding a food pellet planted in their cage bedding. If the lesion was made somewhat further caudally, at the level of the olfactory tubercle (Fig. 2G), a different syndrome emerged. Again the animals failed to mate. But unlike the former group, they showed all the signs of great sexual arousal; they became active, sniffed and licked the female, and were attracted to vaginal discharge (FHVD) both *in situ* and when presented in a glass jar. Finally, they were able to find buried food.

One possible interpretation of these results is that the processing of food odor and an arousing component of FHVD is accomplished at the level of the olfactory tubercle and that processing of a behavior triggering component of FHVD is accomplished farther caudally. FHVD may well contain more than one active compound (Singer et al., 1976). Alternatively, odor recognition may depend not upon cortical localization but upon the overall area of cortex innervated. Lesions would then have the effect of altering the quality of all odors rather than causing blindness to a few. Rostrally placed LOT cuts, for example, might render the odor quality of meaningful components of FHVD unrecognizable, while more caudal cuts would pervert their quality less severely, reducing but not eliminating their effectiveness. Certainly the animals with caudal LOT cuts were able to smell. Furthermore, it is likely that even those with rostrally placed cuts were not completely anosmic. Consider a series of experiments in which Slotnick (1972) examined the effect of lesions to the rostral olfactory cortex and LOT (level as in Fig. 2F) on a conditioned odor discrimination task. Moderately sized lesions were virtually without effect. More extensive lesions, including the complete LOT, led to failure in the odor discrimination task. On subsequent retraining, however, the rats successfully relearned the discrimination and thereafter performed at odor concentrations only slightly above their preoperative threshold. Could it be that the moderate lesions perverted odor quality but not sufficiently to render the odors indistinguishable, while the extensive lesions severely reduced discriminability so that further training was necessary to master the now relatively more difficult task?

The more caudal of the LOT cuts made in the hamster experiment eliminated mating but not arousal. It would have been useful to make cuts successively further caudally to find at which level mating is no longer impaired. Unfortunately, such surgery is difficult because just caudal to the olfactory tubercle the LOT fans out into a broad fiber sheet (Devor, 1976a). Such a preparation emerged, however, in a series of experiments in which the LOT was cut in neonatal hamsters.

NEONATAL OLFACTORY TRACT TRANSECTIONS

Using a surgical approach modified after that used on adults, I sectioned the LOT either completely or partially in a large series of male hamster pups 1 to 34 days of age (Devor, 1975). Since my intention was to examine mating behavior in these animals after they matured, early bilateral lesions would not do. If the animals showed a deficit, how could one be sure that it was related to a loss in the sense of smell rather than to abnormal development of effector or neuroendocrine mechanisms? Even some unoperated male pups mature to be indifferent or non-maters. To solve this problem, the LOT was cut on only one side. When the animals reached maturity they were given abundant sexual experience and the majority showed the normal mating pattern. The most consistent and vigorous maters were then subjected to a second lesion, removal of the olfactory bulb on the formerly unoperated side. Thus, after confirming ability to mate, the sensory competence of the neonatally operated side was tested.

Consider first the animals with complete LOT section. If the tract had been cut at 7 days of age or later, mating ceased following removal of the bulb on the intact side (Fig. 4). Some of the animals however, continued to show interest in the female and some succeeded in finding buried food. By contrast, most of the animals initially operated within the first 3 or 4 days after birth continued to mate and find buried food. Even these, however, showed a deficit. Prior to their second (bulb) lesion they all mated vigorously on almost every pairing. That is, they began to sniff and lick immediately upon being placed with a receptive female, mounted usually within 30 sec but always within 120 sec, and proceeded with the rigidly timed sequence of mounts, intromission and grooming so characteristic of this species (Reed and Reed, 1946). Afterwards, however, mating became fitful. Latency to first mount, for example, was often prolonged by several minutes. And even once the male began to mount, it might stop and then display no apparent interest in the female for a short interval or even for the duration of the test. Sometimes the sequence of mating was disorderly with cycles of grooming or intromission omitted. And on some test days a male might not mate at all. Inconsistent as it was, however, mating was always abolished completely if the appropriate nostril was occluded or if, in a final

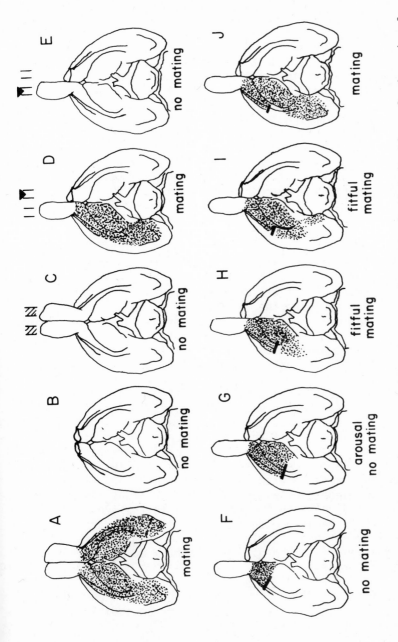

Fig. 2: Mating behavior in male hamsters depends upon the central distribution of odor signals. (A) Intact hamster brain; (B) bilateral bulbectomy; (C) olfactory mucosa destroyed; (D) unilateral bulbectomy with ipsilateral nostril occlusion; (E) unilateral bulbectomy with contralateral nostril occlusion; (F-J) unilateral LOT section combined with contralateral bulbectomy. The LOT was cut: (F) in adulthood, rostrally; (G) in adulthood, farther caudally; (H) complete cut neonatally; (I) partial cut neonatally; (J) partial cut in adulthood. Stippling indicates the area (not necessarily the density) of projection of olfactory bulb with access to odor information.

operation, the remaining olfactory bulb was removed (Fig. 4).
Thus, even this fitful form of mating depended on the recognition
of an olfactory signal.

Now consider the animals that suffered only partial LOT lesions.
If the operation was performed in adulthood or adolescence, mating
continued unchanged after the contralateral bulbectomy, no matter
whether the medial, middle or lateral part of the tract was cut. If
the lesion was performed during the first 10 days of life, however,
mating performance following the second operation became inconsistent
as after complete LOT cuts at 1 to 4 days of age. In brief, if the
lesion was complete the outcome was more favorable if done early;
if the lesion was incomplete, the outcome was more favorable if done
in adulthood. Can these results be explained in terms of surviving
connections of the previously cut olfactory tract?

MORPHOLOGICAL PLASTICITY

The projections of the olfactory bulb on the side of the
sectioned LOT were examined without knowledge of the animals'
behavioral performance. This was done by removing the remaining bulb
surgically and then applying the Fink-Heimer procedure to stain the
resulting degeneration. Since only freshly produced terminal debris
is stained and not that generated by the tract cut itself, the
resulting maps represent the bulbar distribution that was present
at the time of behavioral testing (Devor, 1976b).

If the LOT had been severed completely in adulthood or adolescence, the bulb's distribution was simply abridged. Areas distal to
the cut received no bulb input, areas proximal received their normal
projection (Fig. 2F,G, 3A). If the cut had been made during the
second or third week of life, dramatic changes occurred. Regions
distal to the cut again received no input. Proximally, however,
collaterals of cut axons sent sprouts beyond their normal boundaries
both in depth and tangentially, up the medial wall of the hemisphere
and laterally around the rhinal sulcus and into the frontal (sulcal)
neocortex (Fig. 3B). Recalling that LOT section at 7 days of age
does not lead to functional sparing, it would seem that these
anomalous projections into non-olfactory cortex do not serve an
adaptive olfactory function.

Similar, if less prominent proximal sprouting occurred in cases
of complete LOT section at 1 to 4 days of age. In addition, however,
fibers now entered the olfactory cortex distally either by growing
over the cut or by following an anomalous route laterally from the
olfactory tubercle (Fig. 2H). The extended bulbar distribution
covered only part of the denervated cortex, never more than 1/2 to
2/3. Furthermore, there was often no surviving projection of
accessory olfactory bulb to amygdala. Nonetheless, the moderate

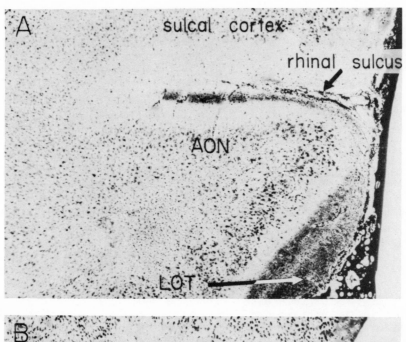

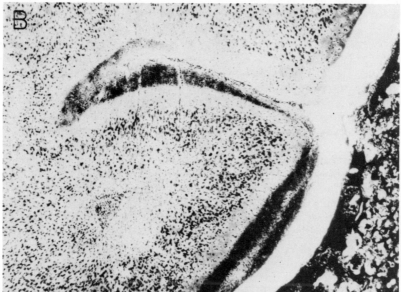

Fig. 3: Proximal sprouting of LOT fibers into sulcal neocortex following neonatal transection. (A) Frontal section through the anterior olfactory nuclei (AON) rostral (proximal) to the level at which the LOT had been sectioned at 43 days of age. (B) Same but after LOT section at 7 days of age. Fink-Heimer stain 4 days after ipsilateral bulbectomy.

additional bulbar distribution in these animals almost certainly
accounts for their relative behavioral sparing. This was demonstrated
by an additional experiment performed on 5 hamsters that showed
clear functional sparing. These animals underwent a third operation
in which the olfactory cortex was re-exposed and a second cut made,
this aimed at severing the sprouted fibers (Fig. 4B). Now mating
ceased. Since the new cut introduced Fink-Heimer stainable debris
of its own, final bulbar distribution was determined autoradiographically following an intrabulbar injection of S^{35}-labeled
methionine. Indeed, no label could be found distal to the second cut.
Finally, a group of hamsters was prepared in which one bulb was
removed at 3 days of age and the contralateral LOT severed in
adulthood. None of these animals continued to mate.

The anatomical consequences of partial LOT section are still
more surprising. Partial section in mature hamsters, a lesion that
leaves mating unaffected, denervated completely only a small patch
of cortex immediately distal to the cut. The remainder of the
bulb's distribution was normal in extent, if somewhat sparse
(Fig. 2J). This is to be expected as each of the uncut fibers
continues to innervate essentially the whole cortex (Fig.1). Partial
lesions in the neonate had a different effect. There was a moderate
degree of proximal sprouting, as described above. Secondly, the
zone just distal to the cut was filled in. And finally, the most
caudal extent of the cortex, though far removed from the lesion
itself, failed to receive bulb fibers (Fig. 2I). The longest branches
of the uncut LOT axons either failed to grow this far, or were
secondarily retracted. In any event, the final outcome is a bulbar
distribution foreshortened by 1/3 to 1/2 of its normal extent, an
effect which apparently accounts for the poor performance of these
animals relative to ones that underwent the same lesion later in life.

The neuromorphological implications of these data have been
discussed in detail elsewhere (Devor and Schneider, 1975; Devor,
1976b). Briefly, I believe that they support an idea originally
proposed by Schneider (1973) that developing neurons tend to conserve
their total amount of axonal arborization. According to this principl
if the distal branches of axons are cut, as in early complete LOT
lesions, the proximal branches tend to emit extra sprouts in compen-
sation. Conversely, if uncut axons emit extra sprouts in one location
such as in the region vacated by a prior partial LOT cut, they do
so only at the cost of an equal number of branches lost elsewhere.
Whatever the mechanism we have two conditions, early complete and
partial LOT section, in which bulbar distribution is intermediate
between full innervation and the innervation surviving complete
caudal LOT section in adulthood. Correspondingly the behavioral
outcome can be viewed as intermediate between vigorous mating and
arousal without mating.

PROBLEMS FOR A THEORY OF CENTRAL ODOR PROCESSING

There is little reason to suppose that the axonal rearrangements that follow early LOT lesions leave terminal connections of bulb efferents precisely as they would have been had no lesion occurred. It is therefore all the more remarkable to see continued the pattern that first emerged from the series of rostral and caudal adult LOT cuts. The greater the area of olfactory bulb distribution, the more complete is the hamster's mating performance (Fig. 2F-J). With only a rostral sliver of cortex innervated the female is of no interest. With slightly more she is attractive, but the stereotyped mating sequence is not triggered. Still more and mating occurs but inconsistently. And finally, with only a small patch of cortex denervated, mating is normal and vigorous. Fully aware of the logical leap involved, I would like to propose that behind this relationship lies a more fundamental one; the greater the area of olfactory cortex innervated the more qualitatively normal are odors perceived. With only a small region of olfactory cortex innervated the odor of FHVD is severely distorted. With a larger area innervated, there is less perversion of quality and a greater likelihood of the odor's functional action on any given test. The possibility of an odor topography across the cortex has not been ruled out. The gradual transition from no interest in the female to vigorous mating, with increasing area of bulbar distribution is not suggestive, however, of functional localization.

If olfactory processing of the presumptive pheromone that kindles mating behavior is not accomplished in a restricted portion of cortex, where then? With each sniff of odor-laden air a massive and simultaneous burst of impulses is generated by the receptors, transformed in the bulb, and then sent crashing onto the beach of the olfactory cortex This process is repeated on each sniff cycle. And for each cycle a two-dimensional matrix of synaptic currents and neural discharge sweeps over the cortex and is then damped out (Howland, Lettvin, McCulloch, Pitts and Wall, 1955; Freeman, 1972). Looked at from above, each volley in effect would generate a complex pattern of spatial and temporal interaction, a fingerprint of neural activity. It is not unreasonable to suppose that odors set up characteristic patterns, discriminable at some level of resolution from those of other odors. The code for individual odor quality would then take the form of rapidly fading fingerprints stamped anew across the face of the cortex with each successive sniff. This image in mind, it is immediately clear that the fingerprint of any particular odor, say vaginal discharge, would be read with less and less certainty as the area of innervated cortex were whittled away.

This framework was proposed by Adrian (1942) long ago but seems to have had relatively little influence. I refer to it here because it accommodates many observations about olfactory anatomy, physiology and psychophysics that are perplexing if one thinks only in terms of

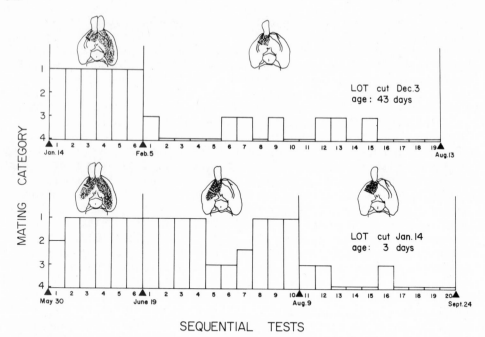

SEQUENTIAL   TESTS

Fig. 4:   Summary data of two representative hamsters. Top: LOT cut
at 43 days of age. Bottom: LOT cut at 3 days of age. Mating ceased
after a second cut. Insets illustrate bulbar distribution. Mating
categories are: (1) vigorous mating, (2) fitful mating, (3) interest,
no mating, (4) no interest. For criteria see Devor (1975).

functional localization. For example, there have long been and
there remain a number of candidate theories of odor separation and
transduction at the receptor level, each incomplete, and yet each
too strongly supported to be waived entirely (Beidler, 1971). This
system could tolerate the simultaneous operation of several overlapping stimulus separation mechanisms because there is no need to
organize channels by odor type. As long as receptor cells do not
all respond identically to different scents, an odor-specific
fingerprint will be set up. Indeed, there would be little advantage
in segregating cells by location or specificity, or in having a
limited number of classes of odor response spectra. By the same
token, since the code would be represented in the activity of cell
ensembles and never by single cells, our frustrating failure to find
cells well tuned to single odors would find a new explanation. So
too would the baffling structure of olfactory connections with its
tremendous convergence and lack of topography. Finally, we would
have an explanation as to why clinical manifestations of severe
uncinate damage stress global perversions of odor sense (parosmia,
kakosmia) rather than islands of selective anosmia. Perhaps in cases

of mild or gradually progressive damage, subtle changes in odor quality would not be noticed at all. Indeed, given the unique power of rapid learning in the olfactory mode (Engen, Kuisma and Eimas, 1973) one might effectively recode familiar objects in terms of their newly altered cortical fingerprints on an ongoing basis.

CENTRAL CONVERGENCE

If odor quality is coded as a two dimensional array of cortical cell activity, then there is no cell whose firing pattern can convey the identity of a single odor. How then can the nervous system "see" the fingerprints? Here again we are provided with an answer by a feature of central anatomy unique among sensory systems. As discussed above, there are two convergent axonal systems that can be thought of as reading off large areas of the olfactory cortex. These project to loci that must be nodal points for olfactory processing: the thalamic nucleus medialis dorsalis and the VLMFB tract with special importance directed at the small premammillary nuclei gemini. I do not suspect that these are sites bearing highly tuned single units. They could, however, be points where the broad cortical patterns are focused for ultimate use in higher central nervous system processes. I know of no studies in which both of these nodal points have been subjected to experimental lesions. There may be a clinical example of this, however, in Korsakoff's syndrome. In this disease there is a variable pattern of necrosis throughout the brain. One involvement common to most cases, however, is bilateral destruction of the mediodorsal thalamus and the mammillary bodies. Finally, it has recently been shown that these patients, to a far greater extent than a matched sample of non-Korsakoff alcoholics, are unable to perceive odors (Jones, Moskowitz and Butters, 1975). Few other conditions, short of frank sheering of the olfactory nerves, result in such severe dysosmia.

Are different scents represented centrally as topographically separated patches or as sets of superimposed, widely distributed spatiotemporal fingerprints? At present each of the lines of evidence favoring the latter possibility is circumstantial at best, and there is strong precedence from other senses in favor of the former. And yet, few of us believe that even in the most topographic of systems that orderly mapping and remapping continues indefinitely or that the ultimate in functional localization, the proverbial "grandmother cell" actually exists. Rather, higher order information in all systems is probably represented in the activity of cell ensembles (Hebb, 1949). The olfactory system may simply skip over the mapping stages and code odor signals early in the form of cortical spatio-temporal patterns. If correct, this possibility is optimistic. For unlike other sensory systems, the ensemble here is both geometrically simple (two-dimensional) and directly accessible to experimental investigation. One may never find single cells at any level

specifically tuned to the odor of FHVD, or make a restricted lesion
that renders an animal specifically anosmic to this pheromone. By
recording from a matrix of electrodes positioned on the cortical
surface, however, one might find a pattern of neural discharge
distinctive of this scent.

SUMMARY

Mating behavior in male hamsters depends on the recognition of
a specific scent produced by the female. By transecting the lateral
olfactory tract at various levels and in mature and young hamsters,
selected areas of olfactory cortex were isolated from olfactory
bulbar input. Mating performance appeared to depend on the total
area of cortex innervated rather than on any specific region. As a
tentative explanation it is proposed that different scents are
represented on the cortex as widely overlapping spatiotemporal
fingerprints. The less cortex is innervated, the poorer is odor
resolution and recognition.

REFERENCES

Adrian, E.D. (1942) Olfactory reactions in the brain of the hedgehog.
J. Physiol. 100:459-473.

Alberts, J.R. and Galef, B.G. Jr. (1971) Acute anosmia in the rat:
A behavioral test of a peripherally induced olfactory deficit.
Physiol. Behav. 6:619-621.

Allen, W.F. (1941) Effects of ablating the pyriform-amygdaloid areas
and hippocampi on positive and negative conditioned reflexes and on
conditioned olfactory differentiation. Am. J. Physiol. 132:81-92.

Allison, A.C. and Warwick, R.T. (1949) Quantitative observations of
the olfactory system of the rabbit. Brain 72:186-197.

Beidler, L.M. (1971) ed. Handbook of Sensory Physiology, Vol.IV:
Chemical Senses, 1. Olfaction. Springer-Verlag, New York.

Broadwell, R.D. (1975) Olfactory relationships of the telencephalon
and diencephalon in the rabbit. I. An autoradiographic study of the
efferent connections of the main and accessory olfactory bulbs.
J. Comp. Neur. 163:329-345.

Clark, W.E. LeGros (1951) The projection of the olfactory epithelium
on the olfactory bulb in the rabbit. J. Neurol. Neurosurg. Psychiat.
14:1-10.

Darby, E.M., Devor, M., and Chorover, S.L. (1975) A presumptive sex pheromone in the hamster: Some behavioral effects. J. Comp. Physiol. Psychol. 88:496-502.

Davies, J.T. (1971) Olfactory Theories. In: Handbook of Sensory Physiology, Vol.IV: Chemical Senses 1. Olfaction. Biedler, L.M., ed. Springer-Verlag, New York.

Devor, M. (1973) Components of mating dissociated by lateral olfactory tract transection in male hamsters. Brain Res. 64:437-441.

Devor, M. (1975) Neuroplasticity in the sparing or deterioration of function after early olfactory tract lesions. Science 190:998-1000.

Devor, M. (1976a) Fiber trajectories of olfactory bulb efferents in the hamster. J. Comp. Neur. 166:31-48.

Devor, M. (1976b) Neuroplasticity in the rearrangement of olfactory tract fibers after neonatal transection in hamsters. J. Comp. Neur. 166:49-72.

Devor, M. and Murphy, M.R. (1973) The effect of peripheral olfactory blockade on the social behavior of the male golden hamster. Behav. Biol. 9:31-42.

Devor, M. and Schneider, G.E. (1975) Neuroanatomical plasticity: The principle of conservation of total axonal arborization. In: Aspects of Neural Plasticity, Vital-Durand, F. and Jeannerod, M., eds. INSERM 43:191-200.

Dieterlen, F. (1959) Das Verhalten des Syrischen Goldhamsters (Mesocricetus auratus, Waterhouse). Z. Tierpsychol. 16:47-103.

Engen, T., Kuisma, J.E. and Eimas, P.D. (1973) Short term memory of odors. J. Exp. Psychol. 99:222-225.

Fink, R.P. and Heimer, L. (1967) Two methods for selective silver impregnation of degenerating axons and their synaptic endings in the central nervous system. Brain Res. 4:369-374.

Freeman, W.J. (1972) Waves, pulses and the theory of neural masses. Prog. Theoret. Biol., 2:88-165.

Gesteland, R.C. (1971) Neural coding in olfactory receptor cells. In: Handbook of Sensory Physiology, Vol.IV: Chemical Senses 1. Olfaction. Biedler, L.M., ed. Springer-Verlag, New York.

Haberly, L.B. (1969) Single unit responses to odor in the prepiriform cortex of the rat. Brain Res. 12:481-484.

Hebb, D.O. (1949) The Organization of Behavior. Wiley, New York.

Heimer, L. (1968) Synaptic distribution of centripedal and centrifugal nerve fibers in the olfactory system of the rat. An experimental
anatomical study. J. Anat. 103:413-432.

Heimer, L. (1972) The olfactory connections of the diencephalon in
the rat. Brain, Behav. Evol., 6:484-523.

Howland, R., Lettvin, J.Y., McCulloch, W.S., Pitts, W. and Wall, P.D.
(1955) Reflex inhibition by dorsal root interaction. J. Neurophysiol.
18:1-17.

Johnston, R.E. (1970) Scent marking, olfactory communication and
social behavior in the golden hamster, Mesocricetus auratus. Ph.D.
dissertation, Rockefeller University, New York.

Jones, B.P., Moskowitz, H.R. and Butters, N. (1975) Olfactory
discrimination in alcoholic Korsakoff patients. Neuropsychologia
13:173-179.

Leonard, C.M. and Scott, J.W. (1971) Origin and distribution of
amygdalofugal pathways in the rat: An experimental neuroanatomical
study. J. Comp. Neur. 141:313-330.

Lohman, A.H.M. and Mentink, G.M.(1969) The lateral olfactory tract,
the anterior commissure, and the cells of the olfactory bulb. Brain
Res., 12-396-413.

Mathews, D.F. (1972) Response patterns of single units in the olfactory bulb of the rat to odor. Brain Res. 47:389-400.

Murphy, M.R. (1973) Effects of female hamster vaginal discharge on
the behavior of male hamsters. Behav. Biol. 9:367-375.

Murphy, M.R. and Schneider, G.E. (1970) Olfactory bulb removal eliminates mating behavior in male hamsters. Science 167:302-303.

Powers, J.B. and Winans, S.S. (1973) Sexual behavior in peripherally
anosmic male hamsters. Physiol. Behav. 10:361-368.

Powers, J.B. and Winans, S.S. (1975) Vomeronasal organ: Critical
role in mediating sexual behavior of the male hamster. Science
187:961-963.

Price, J.L. (1973) An autoradiographic study of complementary laminar
patterns of termination of afferent fibers to the olfactory cortex.
J. Comp. Neur. 150:87-108.

Reed, A.C. and Reed, R. (1946) The copulatory behavior of the golden hamster. J. Comp. Psychol. 39:7-12.

Rowe, F.A. and Smith, W.E. (1972) Effects of peripherally induced anosmia on mating behavior of male mice. Psychon. Sci. 27:33-34.

Scalia, F. and Winans, S.S. (1975) The differential projections of the olfactory bulb and the accessory olfactory bulb in mammals. J. Comp. Neur., 161:31-56.

Schneider, G.E. (1973) Early lesions of superior colliculus: factors affecting the formation of abnormal retinal projections. Brain, Behav. Evol., 8:73-109.

Scott, J.W. and Leonard, C.M. (1971) The olfactory connections of the lateral hypothalamus in the rat, mouse and hamster. J. Comp. Neur. 141:331-344.

Singer, A.G., Agosta, W.C., O'Connell, R.J., Pfaffman, C., Bowen, D.V. and Field, F.H. (1976) Dimethyl disulfide: An attractant pheromone in hamster vaginal secretion. Science 191:948-950.

Slotnick, B.M. (1972) Odor detection and discrimination in rats following section of the lateral olfactory tract. Presented before the Society for Neuroscience, Houston, Texas.

Spector, S.A. and Hull, E.M. (1972) Anosmia and mouse killing by rats: A non-olfactory role for the olfactory bulbs. J. Comp. Physiol. Psychol. 80:354-356.

White, L.E. Jr. (1965) Olfactory bulb projections of the rat. Z. Zellforsch. 98:157-187.

CENTRAL CONTROL OF SCENT MARKING

Pauline Yahr

Department of Psychobiology

University of California, Irvine

Despite the growing interest in olfactory communication, few studies have examined neural mechanisms controlling scent marking. Compared to efforts to understand how odor cues affect recipient animals, little attention has focused on any aspect of signal sending. Interest in the signaling animal is often limited to identifying the chemical structure of the odor involved and to determining how and where it is produced. Yet to understand olfactory communication processes, we must also study the mechanisms that determine when and where animals actively signal by releasing or depositing odors.

Most data on how the brain controls scent marking were collected in an attempt to learn how gonadal steroid hormones, particularly androgens, modify marking behavior in Mongolian gerbils (Meriones unguiculatus). Gerbils scent mark by rubbing a ventral sebaceous gland over objects on the ground. Gerbils are well suited to laboratory studies because it is easy to score their marking behavior by observing them skim pegs on the floor of an open field. Because individual variation is large, we pretest gerbils for scent marking frequency so that reliable scent markers can be selected for study. During a 5-min test, a gerbil may scent mark anywhere from zero to sixty times.

The scent gland and marking behavior of gerbils are sexually dimorphic and sensitive to gonadal steroid hormones. Males have notably larger scent glands and obtain higher marking scores than females, unless the female is pregnant or lactating (Thiessen, Blum and Lindzey, 1970; Wallace, Owen and Thiessen, 1973). The scent gland and marking behavior of males develop at puberty (Lindzey,

Thiessen and Tucker, 1968) and are maintained in adulthood by
testicular androgens. Castrating adult males reduces their marking
scores, often to zero, and causes the scent gland to atrophy.
Androgen replacement therapy partially or completely reverses both
effects depending on the dose of hormone administered (Blum and
Thiessen, 1971; Thiessen, Friend and Lindzey, 1968; Yahr and
Thiessen, 1972). Estrogens can also fully reinstate scent marking
and scent gland function in castrated males (Nyby and Thiessen,
1971; Yahr and Thiessen, 1972).

The low marking scores obtained by most females are seemingly
not affected by gonadal steroid hormones (Thiessen and Lindzey,
1970; Whitsett and Thiessen, 1972; Yahr, 1976a), but ovarian steroids, particularly estrogens, maintain scent marking in females that
do mark frequently (Owen and Thiessen, 1973, 1974; Wallace et al.,
1973; Yahr and Thiessen, 1975). The ovaries also support lactation-
induced scent marking in female gerbils (Wallace et al., 1973).

Although scent gland function and marking behavior are coordinated by hormone stimulation, they are controlled independently.
Gonadally intact males scent mark normally even after surgical
removal of their scent glands (Blum and Thiessen, 1970). Also, some
hormonal treatments promote scent gland growth in castrated males
without stimulating marking behavior (Yahr, 1976b; Yahr and
Thiessen, 1972).

NEURAL LOCI INVOLVED IN SCENT MARKING

Androgens modulate marking behavior in male gerbils by acting
directly on the brain, particularly in the preoptic area (Thiessen
and Yahr, 1970; Thiessen, Yahr and Owen, 1973; Yahr and Thiessen,
1972). Small amounts of testosterone propionate (TP) implanted
directly into the preoptic area (using 21 gauge cannulae) reinstate
scent marking in castrated males; however, similar implants into the
amygdala, hippocampus, reticular formation, septum, caudate nucleus
or cortex do not affect marking behavior. These data are illustrated in Figure 1. Cholesterol implanted into the preoptic area
does not induce scent marking, so the behavioral effects of the
hormone implants can not be attributed to lesions produced during
implantation. They can not be attributed to seepage of hormone into
the general circulation either, because behaviorally effective and
ineffective hormone implants produce similar small effects on
peripheral target tissues such as the scent gland and seminal
vesicles.

Estrogen implants into the preoptic area also stimulate scent
marking in castrated males (Yahr and Thiessen, 1972), but implants

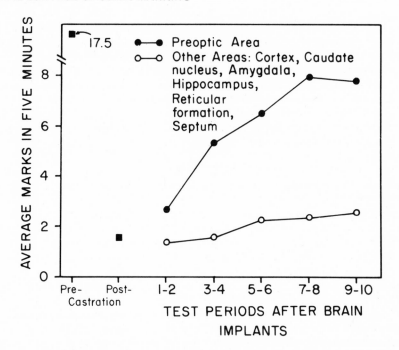

Fig. 1. The effects of implanting testosterone propionate into various brain sites on the scent marking scores of castrated male gerbils. The males were tested every other day for 20 days beginning the day after cannula implantation. (After Thiessen et al., 1973).

of 5α-dihydrotestosterone (DHT) do not (Thiessen et al., 1973). Thus distinct differences exist between hormonal control of scent marking and hormonal control of the scent gland. DHT and estrogen are both potent stimulants of scent gland secretion (see Ebling's chapter in this volume).

Data gathered recently in my laboratory suggest that the anterior aspect of the medial preoptic area is the most sensitive neural locus for androgen stimulation of scent marking in male gerbils (Yahr and Martin, unpublished). By repeatedly applying testosterone to this area of the brain via small cannulae (26 gauge), we elicited marking behavior in castrated males ($\overline{X}$ marks per 5-min test = 8.1). Implants into the middle portion of the medial

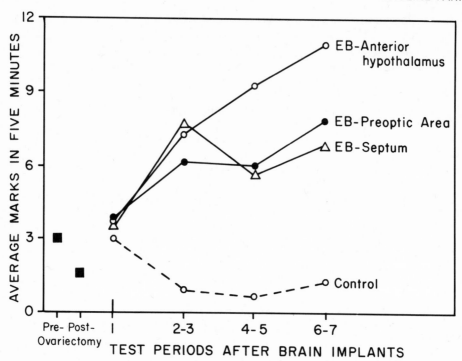

Fig. 2. The effects of implanting pellets of estradiol benzoate
(EB) or the paraffin vehicle (controls) into various brain sites on
the scent marking scores of ovariectomized female gerbils. The
females were tested every 3 days for 15 days beginning the day
after implantation. Tests 6 and 7 were done 1 and 2 weeks later.
(After Owen et al., 1974).

preoptic area ($\overline{X}$ = 1.6), into the lateral preoptic area ($\overline{X}$ = 3.1),
or near the anterior hypothalamus ($\overline{X}$ = 3.6) were much less effective, as were cholesterol implants ($\overline{X}$ = 1.6).

    Only one study has explored central control of scent marking in
female gerbils. By implanting pellets of estradiol benzoate (EB)
into the brains of ovariectomized females, Owen, Wallace and
Thiessen (1974) found that the most responsive site for eliciting
marking behavior was the anterior hypothalamus, though implants into
the preoptic area and septum were also effective (see Figure 2).
Females did not scent mark when EB was implanted into the hippocampus, amygdala, thalamus, or olfactory nucleus. In this study,
the preoptic area implants were placed into the lateral preoptic
area and this may explain why they were not as effective as implants
into the anterior hypothalamus.

Thiessen and Dawber (personal communication) detected other neural loci involved in the control of scent marking by electrically stimulating the brains of gonadally intact male gerbils. They triggered scent marking within minutes after stimulating (50 Hz, 100 μ amp, 75 msec) the olfactory bulb, the dorsal anterior olfactory nucleus, the dorsal anterior and medial aspects of the medial forebrain bundle, and the basal and anterior portions of the amygdala. Electrical stimulation of the lateral preoptic area or the reticular formation did not provoke marking behavior.

In various other species, including rats, cats, rabbits, ring doves and domestic fowl, the anterior hypothalamus and/or preoptic area mediate hormonal control of courtship and sexual behavior (Barfield, 1965, 1967; Davidson, 1966; Harris and Michael, 1964; Hutchison, 1967; Lisk, 1962; Palka and Sawyer, 1966). We have just begun studying hormonal control of sexual behavior in gerbils and do not know yet what neural loci are involved. Still, it is interesting to speculate that hormonal control of reproductive behavior and scent marking may be integrated in this area of the brain.

Hart (1974a) has already presented evidence that this is true in dogs. He tested male dogs for mating behavior and urine marking before and after lesions were made in the anterior hypothalamic-medial preoptic area of their brains. Nine dogs with large lesions in this area stopped mating completely and mating was partially impaired in four others. Twelve of these 13 dogs were also tested for urine marking and all of them marked less after surgery than before. In fact, the ranges of the preoperative and postoperative scores did not overlap for any dog. Only one male adopted the female urination stance after surgery. The rest used the normal male leg-lift stance, but urinated in one place until, presumably, the bladder was empty. Seven more dogs with smaller or more posterior lesions displayed a small impairment in sexual behavior and no reduction in their urine marking scores.

Hart (personal communication) has also studied the effects of anterior hypothalamic-medial preoptic area lesions on urine spraying in male and female cats. The owners of these cats complained that they sprayed urine indoors even though they were gonadectomized. The cats were given medroxyprogesterone, a long-acting progestin, to suppress the behavior. Nonetheless, the problem persisted in five males and three females, so lesions were made. Although the cats were not tested systematically, the owners reported that four of the male cats stopped spraying. In contrast, the two females that sprayed preoperatively (the third was merely suspected of spraying) continued to do so.

Thus in three species the anterior hypothalamic-preoptic area of the brain and related tissues, such as the medial forebrain

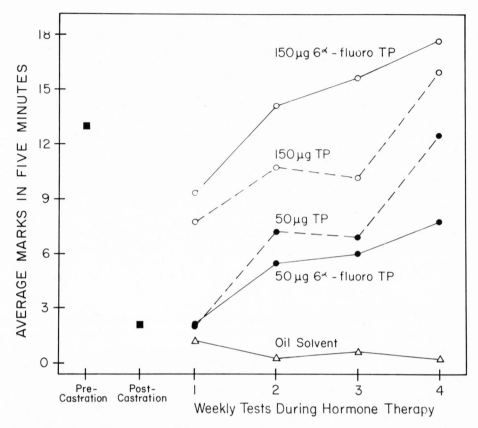

Fig. 3. The effects on the scent marking scores of castrated male gerbils of subcutaneously injecting, twice weekly, 50 or 150 μg of testosterone propionate (TP) or 6α–fluoro TP. The effect of the oil vehicle is also shown. (After Yahr, 1976b).

bundle, have been implicated to some degree in the neural control of scent marking. This consistency across species may reflect similarities in the sexual dimorphism of these scent marking behavior and in their sensitivity to androgens. Like male gerbils, male cats and dogs spray and urine mark more than females do. Castration a lso reduces spraying in most cats (Hart and Barrett, 1975), though the role of androgens in maintaining urine marking in dogs is less clear. According to pet owners, castration reduces urine marking by dogs in the house (Hopkins, Schubert and Hart, 1976), but it does not reduce marking frequency in an outdoor test arena (Hart, 1974b).

MECHANISMS OF ANDROGEN ACTION IN THE BRAIN

The remaining research on neural control of scent marking concerns how androgens act on brain cells to modulate behavior. Unfortunately, this is also a research area with few firm answers. Although considerable progress has been made toward understanding how estrogens act on the female reproductive organs (O'Malley and Means, 1974) and on the brain (Eisenfeld, 1969; Whalen and Massicci, 1975), much less is known about androgen action. With respect to androgen control of scent marking, data are available on the role of androgen metabolism to estrogen (aromatization), on the possible role of RNA and protein synthesis, and on the neurotransmitters that may mediate the behavior. All of these data pertain to gerbils.

My interest in exploring the role of androgen aromatization in the control of gerbil scent marking stemmed from an hypothesis put forward by McDonald and coworkers (1970) to account for androgen's effects on the sexual behavior of male rats. Noting that either estrogen or testosterone (which can be aromatized to form estrogen) would enhance mating in male rats whereas DHT (which can not be aromatized) would not, they suggested that aromatization might be essential for androgen stimulation of behavior. Since the data on hormonal control of gerbil scent marking were consistent with this idea, I tested the hypothesis using 6α-fluorotestosterone, another nonaromatizable androgen. As shown in Figure 3, TP and 6α-fluoro TP are equally potent at two dose levels for triggering marking behavior in male gerbils; hence, aromatization is not essential for androgen to induce this behavior (Yahr, 1976b). More recently, Baum and Vreeberg (1973) suggested that estrogen and DHT, formed by the metabolism of testosterone in the brain, synergize to induce mating behavior in male rats. While this may be true, such synergism is not needed to elicit scent marking in gerbils. Besides being non-aromatizable, 6α-fluoro TP promotes only half as much growth of the seminal vesicles as TP does (Yahr, 1976b), suggesting that it is not reduced to DHT very easily. Thus neither aromatization nor reduction are closely linked to androgen induction of scent marking in male gerbils.

Another series of studies on the mechanism of androgen action was prompted by observations that androgen action on the male reproductive tract involves changes in RNA and protein synthesis (Ahmed and Ishida, 1971; Davies and Griffiths, 1973; Mainwaring and Jones, 1975). Working in Del Thiessen's lab, Keith Owen and I attempted to determine whether androgen action on the brain involves RNA transcription and translation by using drugs to disrupt these processes (Thiessen et al., 1973). We used puromycin hydrochloride to block protein synthesis and actinomycin D to prevent transcription of RNA. In each case, we implanted hormone or hormone plus the

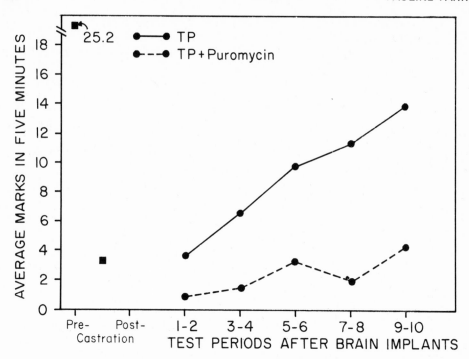

Fig. 4. The effects of implanting testosterone propionate (TP) or
TP plus puromycin hydrochloride into the preoptic area of the brain
on the scent marking scores of castrated male gerbils. The males
were tested every other day for 20 days beginning the day after
cannula implantation. (After Thiessen et al., 1973).

drug into the preoptic area of castrated male gerbils and tested
them for scent marking during the next 20 days. As seen in Figures
4 and 5, both actinomycin D and puromycin blocked androgen stimulation of marking behavior. Hence our data supported the idea that
androgens alter marking behavior by modulating RNA and protein
synthesis in preoptic area cells. Besides replicating these observations several times, we also found that ribonuclease, an enzyme
that degrades RNA, would attenuate hormonal stimulation of scent
marking if implanted with testosterone into the preoptic area. I
should point out that only animals that remained healthy throughout
these studies and whose activity levels were normal were included in
the analyses. Nonetheless, it is possible that each of these drugs
inhibited marking behavior by generally disrupting cellular activity
or by other nonspecific processes. Our observation that magnesium
pemoline, an RNA stimulant, also antagonized androgen induction of
scent marking raises this question, though this drug could have

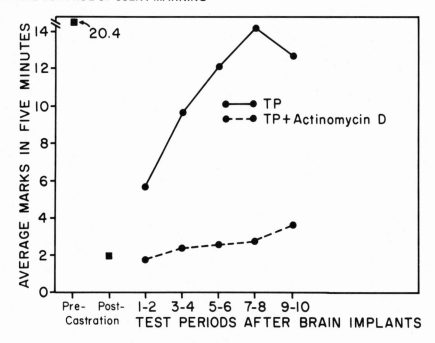

Fig. 5. The effects of implanting testosterone or testosterone plus actinomycin D into the preoptic area of the brain on the scent marking scores of castrated male gerbils. The males were tested every other day for 20 days beginning the day after cannula implantation. (After Thiessen et al., 1973).

competed with the androgen for receptor sites in the cells. Perhaps studies using drugs that inhibit RNA or protein synthesis via different mechanisms will clarify these issues.

Because the pharmacological data were at least consistent with the notion that androgens induce scent marking by inducing protein synthesis, I attempted a different test of this hypothesis (Yahr, 1972). The rationale was that if androgens alter protein synthesis in preoptic area cells, then antigenic differences should exist between preoptic area tissue from castrated males and the same tissue taken from castrates receiving hormone therapy. To test this I developed antibodies against preoptic area tissue from castrated males and against preoptic area tissue from castrates treated with TP. The tissues were separately homogenized and centrifuged and then the saline-soluble fractions were injected into rabbits.

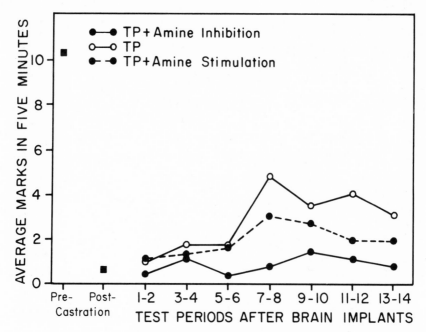

Fig. 6. The effects of implanting testosterone propionate (TP) or
TP plus drugs that affect catecholamine or serotonin action (see
text for details) into the preoptic area of the brain on the scent
marking scores of castrated male gerbils. The males were tested
every other day for 20 days beginning the day after cannula implantation. (After Yahr, 1972).

Later, antisera were collected from the rabbits and tested for their
interactions with each type of tissue homogenate by immunodiffusion
and immunoelectrophoresis. Initially it appeared that males not
exposed to hormone had an additional antigenic component in their
preoptic area cells, but a more recent replication suggests that the
difference may be a small quantitative one, if it exists at all.
Since failure to detect antigenic differences does not rule out the
possibility that the tissues differ in proteins, the hypothesis has
still not been adequately tested.

Since hormonal stimulation of scent marking must ultimately
involve changes in synaptic processes, I attempted to determine which
neurotransmitters within the brain might mediate hormonal control of
this behavior in gerbils (Yahr, 1972). As a first approach to this
problem, I administered drugs which would counteract or enhance the

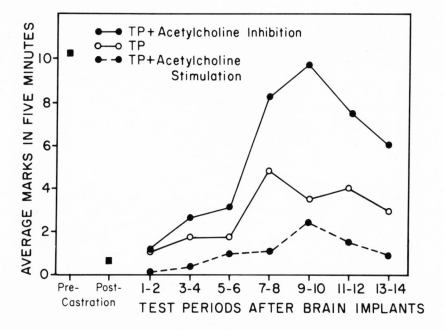

Fig. 7. The effects of implanting testosterone propionate (TP), TP plus atropine sulfate (acetylcholine inhibition), or TP plus pilocarpine nitrate (acetylcholine stimulation) into the preoptic area of the brain on the scent marking scores of castrated male gerbils. The males were tested every other day for 20 days beginning the day after cannula implantation. (After Yahr, 1972).

effectiveness of various transmittter substances. As discussed above for the studies using puromycin or actinomycin D, the drugs were mixed with hormone and implanted into the preoptic area of castrated males. One group received TP plus atropine sulfate, to inhibit acetylcholine action. Another group received TP plus pilocarpine nitrate, to enhance or mimic acetylcholine action. Other groups received TP mixed with phenoxybenzamine hydrochloride or with catechol, to modulate catecholamine action. Still others received TP mixed with parachlorophenylalanine or with 5-hydroxytryptophan plus Iproniazid phosphate, to affect serotonin action. A seventh group received TP plus both parachlorophenylalanine and catechol to simultaneously inhibit formation of serotonin and enhance catecholamine activity. The control group received TP only. Figure 6 shows that all drugs, inhibitors and stimulants, affecting amines inhibited androgen stimulation of gerbil scent marking. In contrast, drugs with opposite effects on acetylcholine action

produced opposite effects on androgen induction of marking behavior.
Atropine sulfate, an acetylcholine inhibitor, pushed scent marking
scores above control TP levels whereas pilocarpine, which mimics acetylcholine's action, inhibited the marking response. (See Figure 7).
These data suggest that androgen induction of scent marking in male
gerbils may involve changes in acetylcholine synthesis or turnover
rate; however, careful dose response studies must be done on these
and related drugs before any firm conclusions can be drawn.

SUMMARY

Central control of scent marking is a challenging, and open,
research field for anyone interested in bridging psychobiology and
chemical ecology. It will benefit from neuroanatomical, neurophysiological, neurochemical and neuroendocrinological analyses
provided they are coupled with awareness, as best possible, of the
adaptive function of the behavior. Various species besides gerbils
(e.g., rats, mice, degus and hamsters) can be studied in the laboratory, but many psychobiological techniques can also be used
outside the laboratory, particularly if animals can be confined in
pens. Rabbits, cats or dogs could be studied this way.

It is tempting to conjecture from the available data that
neural systems underlying sexual behavior control olfactory signaling whenever the chemical signals convey information pertinent
to reproductive success. This is plausible, economical in terms of
neural processing and follows from the fact that gonadal hormones
often mediate both behavior patterns. Yet more species must be
studied before any generalizations will emerge about the neural
processes that control scent marking and before interesting exceptions can be recognized as testing the rules.

We also need, of course, data on the central control of
marking behaviors that are less sexually dimorphic and less influenced by gonadal hormones. It would be surprising if the central
control of scent marking were the same when the behavior signals
distress or lays a trail, as when it denotes territorial boundaries
or attracts mates. Also, studies on the central control of other
forms of chemosignaling are required. It will be interesting to
know, for example, how the nervous system controls scent glands
that can be actively opened or closed, such as in black-tailed deer
(Odocoileus hemionous columbianus) and pronghorns (Antilocapra
americana).

Although it disappoints me to conclude that more research is
needed, it seems I must. In essence, we can not compare neural
mechanisms controlling olfactory communication with mechanisms
underlying forms of communication that serve similar adaptive

functions (e.g., bird song, ultrasounding or visual displays) until olfactory signaling processes are studied more thoroughly.

REFERENCES

Ahmed, L. and Ishida, H. 1971. Effect of testosterone on nuclear phosphoproteins of rat ventral prostate. Molec. Pharmacol. 7:323–327.

Barfield, R.J. 1965. Induction of courtship and aggressive behavior by intracranial implants of androgen in capons. Amer. Zool. 5:203.

Barfield, R.J. 1967. Activation of sexual and aggressive behavior by androgen implants in the brain of the male ring dove. Amer. Zool. 7:800.

Baum, M.J. and Vreeberg, J.T.M. 1973. Copulation in castrated male rats following combined treatment with estradiol and dihydro-testosterone. Science 182:283–285.

Blum, S.L. and Thiessen, D.D. 1970. Effect of ventral gland excision on scent marking in the male Mongolian gerbil. J. Comp. Physiol. Psychol. 73:461–464.

Blum, S.L. and Thiessen, D.D. 1971. The effect of different amounts of androgen on scent marking in the male Mongolian gerbil. Horm. Behav. 2:93–105.

Davidson, J.M. 1966. Activation of the male's sexual behavior by intracranial implantation of androgen. Endocrinol. 79:783–794.

Davies, P. and Griffiths, K. 1973. Stimulation of ribonucleic acid polymerase activity in vitro by prostatic steroid-protein receptor complexes. Biochem. J. 136:611–622.

Eisenfeld, A.J. 1969. Hypothalamic estradiol binding macromolecules. Nature 224:1202–1203.

Harris, G.G. and Michael, R.P. 1964. The activation of sexual behavior by hypothalamic implants of oestrogen. J. Physiol. 171:275–301.

Hart, B.L. 1974a. Medial preoptic-anterior hypothalamic area and sociosexual behavior of male dogs: A comparative neuropsychological analysis. J. Comp. Physiol. Psychol. 86:328–349.

Hart, B.L. 1974b. Environmental and hormonal influences on urine marking behavior in the adult male dog. Behav. Biol. 11:167-176.

Hart, B.L. and Barrett, R.E. 1973. Effects of castration on fighting, roaming, and urine spraying in adult male cats. J. Amer. Vet. Med. Assoc. 163:290-292.

Hopkins, S.G., Schubert, T.A., and Hart, B.L. 1976. Castration of adult male dogs: Effects on roaming, aggression, urine marking, and mounting. J. Amer. Vet. Med. Assoc., in press.

Hutchison, J.B. 1967. Initiation of courtship by hypothalamic implants of testosterone propionate in castrated doves. Nature 216:591-592.

Lindzey, G., Thiessen, D.D., and Tucker, A. 1968. Development and hormonal control of territorial marking in the male Mongolian gerbil (Meriones unguiculatus). Develop. Psychobiol. 1:97-99.

Lisk, R.D. 1962. Diencephalic placement of estradiol and sexual receptivity in the female rat. Amer. J. Physiol. 203:403-496.

Mainwaring, W.I.P. and Jones, D.M. 1975. Influence of receptor complexes on the properties of prostate chromatin including its transcription by RNA polymerase. J. Steroid Biochem. 6:475-481.

McDonald, C., Beyer, C., Newton, F., Brien, B., Baker, R., Tan, H.S., Sampson, P., Kitching, P., Greenhill, R. and Pritchard, D. 1970. Failure of 5α-dihydrotestosterone to initiate sexual behavior in the castrated male rat. Nature 227:964-965.

Nyby, J. and Thiessen, D.D. 1971. Singular and interactive effects of testosterone and estrogen on territorial marking in castrated male Mongolian gerbils (Meriones unguiculatus). Horm. Behav. 2:279-285.

O'Malley, B.W. and Means, A.R. 1974. Female steroid hormones and target cell nuclei. Science 183:610-620.

Owen, K. and Thiessen, D.D. 1973. Regulation of scent marking in the female Mongolian gerbil Meriones unguiculatus. Physiol. Behav. 11:441-445.

Owen, K. and Thiessen, D.D. 1974. Estrogen and progesterone interaction in the regulation of scent marking in the female Mongolian gerbil (Meriones unguiculatus). Physiol. Behav. 12:351-355.

Owen, K., Wallace, P., and Thiessen, D.D. 1974. Effects of intra
    cerebral implants of steroid hormones on scent marking in the
    ovariectomized female gerbil (Meriones unguiculatus). Physiol.
    Behav. 12:755-760.

Palka, Y.S. and Sawyer, C.H. 1966. The effects of hypothalamic
    implants of ovarian steroids on estrous behavior in rabbits. J.
    Physiol. 180:251-269.

Thiessen, D.D., Blum, S.L., and Lindzey, G. 1970. A scent marking
    response associated with the ventral sebaceous gland of the
    Mongolian gerbil (Meriones unguiculatus). Anim. Behav. 18: 26-
    30.

Thiessen, D.D., Friend, H.C., and Lindzey, G. 1968. Androgen
    control of territorial marking in the Mongolian gerbil.
    Science 160: 432-434.

Thiessen, D.D. and Lindzey, G. 1970. Territorial marking in the
    female Mongolian gerbil: Short-term reactions to hormones.
    Horm. Behav. 1:157-160.

Thiessen, D.D. and Yahr, P. 1970. Central control of territorial
    marking in the Mongolian gerbil. Physiol. Behav. 5:275-278.

Thiessen, D.D., Yahr, P., and Owen, K. 1973. Regulatory mechanisms
    of territorial marking in the Mongolian gerbil. J. Comp.
    Physiol. Psychol. 82:382-393.

Wallace, P., Owen, K., and Thiessen, D.D. 1973. The control and
    function of maternal scent marking in the Mongolian gerbil.
    Physiol. Behav. 10:463-466.

Whalen, R.E. and Massicci, J. 1975. Subcellular analysis of the ac-
    cumulation of estrogen by the brain of male and female rats.
    Brain Res. 89:255-264.

Whitsett, J.M. and Thiessen, D.D. 1972. Sex difference in the
    control of scent-marking behavior in the Mongolian gerbil
    (Meriones unguiculatus). J. Comp. Physiol. Psychol. 78:381-385.

Yahr, P. 1972. Molecular regulation of territorial marking in the
    Mongolian gerbil (Meriones unguiculatus). Unpublished Ph.D.
    dissertation, University of Texas at Austin.

Yahr, P. 1976a. Effects of hormones and lactation on gerbils that
    seldom scent mark spontaneously. Physiol. Behav. 16:395-399.

Yahr, P. 1976b. The role of aromization in androgen stimulation of gerbil scent marking. Horm. Behav. 7:259-265.

Yahr, P. and Thiessen, D.D. 1972. Steroid regulation of territorial scent marking in the Mongolian gerbil (Meriones unguiculatus). Horm. Behav. 3:359-368.

Yahr, P. and Thiessen, D.D. 1975. Estrogen control of scent marking in female Mongolian gerbils (Meriones unguiculatus). Behav. Biol. 13:95-101.

Hornung, D.E., 460, 464, 476, 477, 478, 481
Horrall, R., 149, 161, 164, 167
Horridge, G.A., 446, 449
Horvath, C., 129, 133
Howe, N.R., 312, 316
Howland, R., 539, 544
Hrbaček, J., 6, 13, 312, 316
Hubel, D.H., 490, 495
Huber, G.C., 436, 451
Hucklebridge, F.H., 192, 202
Hudgens, G.A., 173, 179, 181
Huggins, C., 22, 32
Huggins, G.R., 39, 42, 90, 104, 105, 108, 113, 269, 270, 281, 282, 284, 285
Hughes, P.R., 66, 70, 83, 92
Hughes, R.L., 325, 331
Hull, E.M., 171, 175, 181, 531, 545
Hummel, H., 413, 430
Hummel, R., 81, 88
Hunt, D.F., 82, 88
Hurley, H.J., 35, 42, 275, 277, 285
Hurley, R., 36, 41
Hutchison, J.B., 551, 560
Hüttel, R., 305, 316
Iimo, M., 519, 528
Ikan, R., 62, 68
Ilg, L., 6, 13
Immler, K., 237, 247
Ingram, W.R., 198, 202
Ionescu, M.D., 136, 165
Isaacson, R.L., 197, 202
Isacks, N., 61, 69, 73, 91, 97, 114, 117, 133, 379, 389, 396, 412, 415, 432
Ishay, J., 62, 68
Ishida, H., 553, 559
Ivanov, V.P., 437, 449
Jackson, G., Jr., 143, 157, 163
Jackson, T., 444, 449
Jacobson, M., 62, 68, 436, 451
Jagodowicz, M., 501, 512, 514, 575, 576
Jagodowski, K.P., 437, 451
Jäkel, A., 11, 13
James, J.D., 67, 68

Jameson, D.L., 148, 164
Jarret, A., 21, 32
Jaynes, J., 333, 355
Jeffrey, R., 175, 181
Jensen, D.D., 172, 184
Jensen, E.V., 22, 32
Jerhoff, B., 192, 200, 202
Johnessee, J.S., 75, 87
Johns, B.E., 73, 90, 100, 103, 114, 360, 368, 373, 374
Johnson, J.A., 136, 162
Johnson, R.P., 72, 88, 169, 181, 321, 330, 378, 387
Johnson, V.E., 105, 113
Johnson, W., 239, 249
Johnston, J.B., 436, 451
Johnston, J.W., Jr., 392, 406, 411
Johnston, R.E., 172, 176, 178, 181, 196, 202, 225, 226, 227, 228, 229, 231, 233, 235, 236, 237, 238, 239, 240, 242, 243, 245, 246, 247, 249, 264, 368, 369, 373, 376, 417, 431, 530, 544
Jolly, A., 406, 411
Jones, B.P., 541, 544
Jones, D.M., 553, 560
Jones, R.B., 2, 13, 172, 176, 181, 188, 190, 191, 193, 194, 202, 203, 352, 354, 357, 373
Jonsson, P., 217, 222
Jugovich, J., 383, 387
Juvet, R.S., 81, 86, 126, 133
Kahn, M.W., 192, 193, 203
Kaissling, K.E., 455, 464
Kalkowski, W., 170, 181
Kallius, E., 437, 451
Källquist, L., 414, 429
Kamon, K., 437, 451
Kannowski, P.B., 383, 387
Kargar, B.L., 129, 133
Karli, P., 175, 179
Karlsen, J., 80, 88
Karlson, P., 1, 13, 71, 88, 186, 203
Kart, R., 25, 30, 170, 172, 180
Karten, H.J., 515, 528
Kasahara, Y., 521, 526
Kasang, G., 555, 564